T0317537

Wind Effects on Structures

Wind Effects on Structures

Modern Structural Design for Wind

Emil Simiu, P.E., Ph.D.
NIST Fellow
National Institute of Standards and Technology, USA

DongHun Yeo, P.E., Ph.D.
Research Structural Engineer
National Institute of Standards and Technology, USA

Fourth Edition

Edition History
John Wiley & Sons (1e, 1978), John Wiley & Sons (2e, 1986), John Wiley & Sons (3e, 1996)

Registered Offices
John Wiley & Sons, Inc., 111 River Street, Hoboken, NJ 07030, USA
John Wiley & Sons Ltd, The Atrium, Southern Gate, Chichester, West Sussex, PO19 8SQ, UK

Editorial Office
9600 Garsington Road, Oxford, OX4 2DQ, UK

For details of our global editorial offices, customer services, and more information about Wiley products visit us at www.wiley.com.

Wiley also publishes its books in a variety of electronic formats and by print-on-demand. Some content that appears in standard print versions of this book may not be available in other formats.

Library of Congress Cataloging-in-Publication Data

Names: Simiu, Emil, author. | Yeo, DongHun, author.
Title: Wind effects on structures : modern structural design for wind / Emil
  Simiu, P.E., Ph.D., NIST Fellow, National Institute of Standards and
  Technology, DongHun Yeo, P.E., Ph.D., Research Structural Engineer, National
  Institute of Standards and Technology.
Description: Fourth edition. | Hoboken, NJ : John Wiley & Sons, 2019. |
  Includes bibliographical references and index. |
Identifiers: LCCN 2018038948 (print) | LCCN 2018040522 (ebook) | ISBN
  9781119375906 (Adobe PDF) | ISBN 9781119375937 (ePub) | ISBN 9781119375883
  (hardcover)
Subjects: LCSH: Wind-pressure. | Buildings–Aerodynamics. | Wind resistant
  design.
Classification: LCC TA654.5 (ebook) | LCC TA654.5 .S55 2019 (print) | DDC
  624.1/75 – dc23
LC record available at https://lccn.loc.gov/2018038948

Cover Design: Wiley
Cover Image: © Jackal Pan / Getty Images

Set in 10/12pt WarnockPro by SPi Global, Chennai, India

10 9 8 7 6 5 4 3 2 1

For Devra, SueYeun,
     Zohar, Nitzan, Abigail, and Arin

Contents

Preface to the Fourth Edition

The quarter of a century that elapsed since the publication of the third edition of *Wind Effects on Structures* has seen a number of significant developments in micrometeorology, extreme wind climatology, aerodynamic pressure measurement technology, uncertainty quantification, the optimal integration of wind and structural engineering tasks, and the use of "big data" for determining and combining effectively multiple directionality-dependent time series of wind effects of interest. Also, following a 2004 landmark report by Skidmore Owings and Merrill LLP on large differences between independent estimates of wind effects on the World Trade Center towers, it has increasingly been recognized that transparency and traceability are essential to the credibility of structural designs for wind. A main objective of the fourth edition of *Wind Effects on Structures* is to reflect these developments and their consequences from a design viewpoint. Progress in the developing Computational Wind Engineering field is also reflected in the book.

Modern pressure measurements by scanners, and the recording and use of aerodynamic pressure time series, have brought about a significant shift in the division of tasks between wind and structural engineers. In particular, the practice of splitting the dynamic analysis task between wind and structural engineers has become obsolete; performing dynamic analyses is henceforth a task assigned exclusively to the structural engineering analyst, as has long been the case in seismic design. This eliminates the unwieldy, time-consuming back-and-forth between wind and structural engineers, which typically discourages the beneficial practice of iterative design. The book provides the full details of the wind and structural engineers' tasks in the design process, and up-to-date, user-friendly software developed for practical use in structural design offices. In addition, new material in the book concerns the determination of wind load factors, or of design mean recurrence intervals of wind effects, determined by accounting for wind directionality.

The first author contributed Chapters 1–3; portions of Chapter 4; Chapters 5, 7, and 8; Sections 9.1 and 9.3; Chapters 10–12 and 15; portions of Chapter 17 and Part III; Part IV; and Appendices A, B, D, and E. The second author contributed Chapter 6; Section 9.2; and Section 23.5. The authors jointly contributed Chapters 13, 14, 16, and 18. They reviewed and are responsible for the entire book. Professor Robert H. Scanlan contributed parts of Chapter 4 and of Part III. Appendix F, authored by Skidmore Owings and Merrill LLP, is part of the National Institute of Standards and Technology World Trade Center investigation. Chapter 17 is based on a doctoral thesis by Dr. F. Habte supervised by the first author and Professor A. Gan Chowdhury.

Dr. Sejun Park made major contributions to Chapters 14 and 18 and developed the attendant software. Appendix C is based on a paper by A. L. Pintar, D. Duthinh, and E. Simiu.

We wish to pay a warm tribute to the memory of Professor Robert H. Scanlan (1914–2001) and Dr. Richard D. Marshall (1934–2001), whose contributions to aeroelasticity and building aerodynamics have profoundly influenced these fields. The authors have learned much over the years from Dr. Nicholas Isyumov's work, an example of competence and integrity. We are grateful to Professor B. Blocken of the Eindhoven University of Technology and KU Leuven, Dr. A. Ricci of the Eindhoven University of Technology, and Dr. T. Nandi of the National Institute of Standards and Technology for their thorough and most helpful reviews of Chapter 6. We thank Professor D. Zuo of Texas Tech University for useful comments on cable-stayed-bridge cable vibrations. We are indebted to many other colleagues and institutions for their permission to reproduce materials included in the book.

The references to the authors' National Institute of Standards and Technology affiliation are for purposes of identification only. The book is not a U.S. Government publication, and the views expressed therein do not necessarily represent those of the U.S. Government or any of its agencies.

Rockville, Maryland

Emil Simiu
DongHun Yeo

Introduction

The design of buildings and structures for wind depends upon the wind environment, the aerodynamic effects induced by the wind environment in the structural system, the response of the structural system to those effects, and safety requirements based on uncertainty analyses and expressed in terms of wind load factors or design mean recurrence intervals of the response. For certain types of flexible structure (slender structures, suspended-span bridges) aeroelastic effects must be considered in design.

I.1   The Wind Environment and Its Aerodynamic Effects

For structural design purposes the wind environment must be described: (i) in meteorological terms, by specifying the type or types of storm in the region of interest (e.g., large-scale extratropical storms, hurricanes, thunderstorms, tornadoes); (ii) in micrometeorological terms (i.e., dependence of wind speeds upon averaging time, dependence of wind speeds and turbulent flow fluctuations on surface roughness and height above the surface); and in extreme wind climatological terms (directional extreme wind speed data at the structure's site, probabilistic modeling based on such data). Such descriptions are provided in Chapters 1–3, respectively.

The description of the wind flows' micrometeorological features is needed for three main reasons. First, those features directly affect the structure's aerodynamic and dynamic response. For example, the fact that wind speeds increase with height above the surface means that wind loads are larger at higher elevations than near the ground. Second, turbulent flow fluctuations strongly influence aerodynamic pressures, and produce in flexible structures fluctuating motions that may be amplified by resonance effects. Third, micrometeorological considerations are required to transform measured or simulated wind speed data at meteorological stations or other reference sites into wind speed data at the site of interest.

Micrometeorological features are explicitly considered by the structural designer if wind pressures or forces acting on the structure are determined by formulas specified in code provisions. However, for designs based on wind-tunnel testing this is no longer the case. Rather, the structural designer makes use of records of non-dimensional aerodynamic pressure data and of measured or simulated directional extreme wind speeds at the site of interest, in the development of which micrometeorological features were taken into account by the wind engineer and are implicit in those records. However, the integrity of the design process requires that the relevant micrometeorological features on which those records are based be fully documented and accounted for.

To perform a design based on aerodynamic data obtained in wind-tunnel tests (or in numerical simulations) the structural engineer needs the following three products:

1) Time series of pressures at large numbers of taps, non-dimensionalized with respect to the wind tunnel (or numerical simulation) mean wind speed at the reference height (commonly the elevation of the building roof) (Chapters 4–6).
2) Matrices of directional mean wind speeds at the site of interest, at the prototype reference height.
3) Estimates of uncertainties in items (1) and (2) (Chapter 7).

These products, and the supporting documentation consistent with Building Information Modeling (BIM) requirements to allow effective scrutiny, must be delivered by the wind engineering laboratory to the structural engineer in charge of the design. The wind engineer's involvement in the structural design process ends once those products are delivered. The design is then fully controlled by the structural engineer. In particular, as was noted in the Preface, dynamic analyses need no longer be performed partly by the structural engineer and partly by the wind engineer, but are performed solely, and more effectively, by the structural engineer. This eliminates unwieldy, time-consuming back-and-forth between the wind engineering laboratory and the structural design office, which typically discourages the beneficial practice of iterative design. Chapters 1–7 constitute Part I of the book.

I.2   Structural Response to Aerodynamic Excitation

The structural designer uses software that transforms the wind engineering data into applied aerodynamic loads. This transformation entails simple weighted summations performed automatically by using a software subroutine. Given a preliminary design, the structural engineer performs the requisite dynamic analyses to obtain the inertial forces produced by the applied aerodynamic loads. The effective wind loads (i.e., the sums of applied aerodynamic and inertial loads) are then used to calculate demand-to-capacity indexes (DCIs), inter-story drift, and building accelerations with specified mean recurrence intervals. This is achieved by accounting rigorously and transparently for (i) directionality effects, (ii) combinations of gravity effects and wind effects along the principal axes of the structure and in torsion, and (iii) combinations of weighted bending moments and axial forces inherent in DCI expressions. Typically, to yield a satisfactory design (e.g., one in which the DCIs are not significantly different from unity), successive iterations are required. All iterations use the same applied aerodynamic loads but different structural members sizes. Part II of the book presents details on of the operations just described, software for performing them, and examples of its use supported by a detailed user's manual and a tutorial. Also included in Part II is a critique of the high-frequency force balance technique, commonly used in wind engineering laboratories before the development of multi-channel pressure scanners, material on wind-induced discomfort in and around buildings, tuned mass dampers, and requisite wind load factors and design mean recurrence intervals of wind effects.

Part III presents fundamentals and applications related to aeroelastic phenomena: vortex-induced vibrations, galloping, torsional divergence, flutter, and aeroelastic response of slender towers, chimneys and suspended-span bridges. Part IV contains material on trussed frameworks and plate girders, offshore structures, tensile membrane structures, tornado wind and atmospheric pressure change effects, and tornado- and hurricane-borne missile speeds.

Appendices A–E present elements of probability and statistics, elements of the theory of random processes, the description of a modern peaks-over-threshold procedure that yields estimates of stationary time series peaks and confidence bounds for those estimates, elements of structural dynamics based on a frequency-domain approach still used in suspended-span bridge applications, and elements of structural reliability that provide an engineering perspective on the extent to which the theory is, or is not, useful in practice. The final Appendix F is a highly instructive Skidmore Owings and Merrill report on the estimation of the World Trade Center towers response to wind loads.

Part I

Atmospheric Flows, Extreme Wind Speeds, Bluff Body Aerodynamics

1

Atmospheric Circulations

Wind, or the motion of air with respect to the surface of the Earth, is fundamentally caused by variable solar heating of the Earth's atmosphere. It is initiated, in a more immediate sense, by differences of pressure between points of equal elevation. Such differences may be brought about by thermodynamic and mechanical phenomena that occur in the atmosphere both in time and space.

The energy required for the occurrence of these phenomena is provided by the sun in the form of radiated heat. While the sun is the original source, the source of energy most directly influential upon the atmosphere is the surface of the Earth. Indeed, the atmosphere is to a large extent transparent to the solar radiation incident upon the Earth, much in the same way as the glass roof of a greenhouse. That portion of the solar radiation that is not reflected or scattered back into space may therefore be assumed to be absorbed entirely by the Earth. The Earth, upon being heated, will emit energy in the form of terrestrial radiation, the characteristic wavelengths of which are long (in the order of $10\,\mu$) compared to those of heat radiated by the sun. The atmosphere, which is largely transparent to solar but not to terrestrial radiation, absorbs the heat radiated by the Earth and re-emits some of it toward the ground.

1.1 Atmospheric Thermodynamics

1.1.1 Temperature of the Atmosphere

To illustrate the role of the temperature distribution in the atmosphere in the production of winds, a simplified version of model circulation will be presented. In this model the vertical variation of air temperature, of the humidity of the air, of the rotation of the Earth, and of friction are ignored, and the surface of the Earth is assumed to be uniform and smooth.

The axis of rotation of the Earth is inclined at approximately 66° 30′ to the plane of its orbit around the sun. Therefore, the average annual intensity of solar radiation and, consequently, the intensity of terrestrial radiation, is higher in the equatorial than in the polar regions. To explain the circulation pattern as a result of this temperature difference, Humphreys [1] proposed the following ideal experiment (Figure 1.1).

Assume that the tanks A and B are filled with fluid of uniform temperature up to level *a*, and that tubes 1 and 2 are closed. If the temperature of the fluid in A is raised while the temperature in B is maintained constant, the fluid in A will expand and reach the

Wind Effects on Structures: Modern Structural Design for Wind, Fourth Edition. Emil Simiu and DongHun Yeo.
© 2019 John Wiley & Sons Ltd. Published 2019 by John Wiley & Sons Ltd.

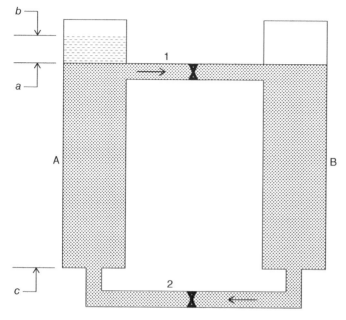

Figure 1.1 Circulation pattern due to temperature difference between two columns of fluid. Source: From Ref. [1]. Copyright 1929, 1940 by W. J. Humphreys.

level *b*. The expansion entails no change in the total weight of the fluid contained in A. The pressure at *c* therefore remains unchanged, and if tube 2 were opened, there would be no flow between A and B. If tube 1 is opened, however, fluid will flow from A to B, on account of the difference of head $(b - a)$. Consequently, at level *c* the pressure in A will decrease, while the pressure in B will increase. Upon opening tube 2, fluid will now flow through it from B to A. The circulation thus developed will continue as long as the temperature difference between A and B is maintained.

If tanks A and B are replaced conceptually by the column of air above the equator and above the pole, in the absence of other effects an atmospheric circulation will develop that could be represented as in Figure 1.2. In reality, the circulation of the atmosphere is vastly complicated by the factors neglected in this model. The effect of these factors will be discussed later in this chapter.

The temperature of the atmosphere is determined by the following processes:

- Solar and terrestrial radiation, as discussed previously
- Radiation in the atmosphere
- Compression or expansion of the air
- Molecular and eddy conduction
- Evaporation and condensation of water vapor.

1.1.2 Radiation in the Atmosphere

As a conceptual aid, consider the action of the following model. The heat radiated by the surface of the Earth is absorbed by the layer of air immediately above the ground (or the

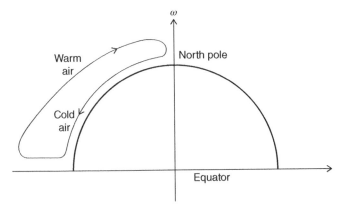

Figure 1.2 Simplified model of atmospheric circulation.

Figure 1.3 Transport of heat through radiation in the atmosphere.

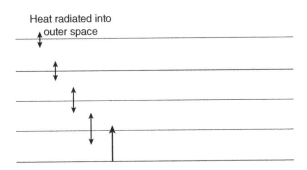

surface of the ocean) and reradiated by this layer in two parts, one going downward and one going upward. The latter is absorbed by the next higher layer of air and again reradiated downward and upward. The transport of heat through radiation in the atmosphere, according to this conceptual model, is represented in Figure 1.3.

1.1.3 Compression and Expansion. Atmospheric Stratification

Atmospheric pressure is produced by the weight of the overlying air. A small mass (or particle) of dry air moving vertically thus experiences a change of pressure to which there corresponds a change of temperature in accordance with the Poisson dry adiabatic equation

$$\frac{T}{T_0} = \left(\frac{p}{p_0}\right)^{0.288} \tag{1.1}$$

A familiar example of the effect of pressure on the temperature is the heating of compressed air in tire pump.

If, in the atmosphere, the vertical motion of an air particle is sufficiently rapid, the heat exchange of that parcel with its environment may be considered to be negligible, that is, the process being considered is *adiabatic*. It then follows from Poisson's equation that since ascending air experiences a pressure decrease, its temperature will also decrease.

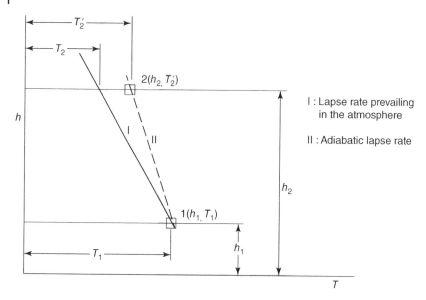

Figure 1.4 Lapse rates.

The temperature drop of adiabatically ascending dry air is known as the dry *adiabatic lapse rate* and is approximately 1°C/100 m in the Earth's atmosphere.

Consider a small mass of dry air at position 1 (Figure 1.4). Its elevation and temperature are denoted by h_1 and T_1, respectively. If the particle moves vertically upward sufficiently rapidly, its temperature change will effectively be adiabatic, regardless of the lapse rate (temperature variation with height above ground) prevailing in the atmosphere. At position 2, while the temperature of the ambient air is T_2, the temperature of the element of air mass is $T_2' = T_1 - (h_2 - h_1)\, \gamma_a$, where γ_a is the adiabatic lapse rate. Since the pressure of the element and of the ambient air will be the same, it follows from the equation of state that to the difference $T_2' - T_2$ there corresponds a difference of density between the element of air and the ambient air. This generates a buoyancy force that, if $T_2 < T_2'$, acts upwards and thus moves the element farther away from its initial position (superadiabatic lapse rate, as in Figure 1.4), or, if $T_2 > T_2'$, acts downwards, thus tending to return the particle to its initial position. The stratification of the atmosphere is said to be *unstable* in the first case and *stable* in the second. If $T_2 = T_2'$, that is, if the lapse rate prevailing in the atmosphere is adiabatic, the stratification is said to be *neutral*. A simple example of the stable stratification of fluids is provided by a layer of water underlying a layer of oil, while the opposite (unstable) case would have the water above the oil.

1.1.4 Molecular and Eddy Conduction

Molecular conduction is a diffusion process that effects a transfer of heat. It is achieved through the motion of individual molecules and is negligible in atmospheric processes. Eddy heat conduction involves the transfer of heat by actual movement of air in which heat is stored.

1.1.5 Condensation of Water Vapor

In the case of unsaturated moist air, as an element of air ascends and its temperature decreases, at an elevation where the temperature is sufficiently low condensation will occur and heat of condensation will be released. This is equal to the heat originally required to change the phase of water from liquid to vapor, that is, the latent heat of vaporization stored in the vapor. The temperature drop in the saturated adiabatically ascending element is therefore slower than for dry air or moist unsaturated air.

1.2 Atmospheric Hydrodynamics

The motion of an elementary air mass is determined by forces that include a vertical *buoyancy force*. Depending upon the temperature difference between the air mass and the ambient air, the buoyancy force acts upwards (causing an updraft), downwards, or is zero. These three cases correspond to unstable, stable, or neutral atmospheric stratification, respectively. It is shown in Section 2.3.3 that, depending upon the absence or a presence of a stably stratified air layer above the top of the atmospheric boundary layer, called capping inversion, neutrally stratified flows can be classified into truly and conventionally neutral flows.

The horizontal motion of air is determined by the following forces:

1) The *horizontal pressure gradient force* per unit of mass, which is due to the spatial variation of the horizontal pressures. This force is normal to the lines of constant pressure, called *isobars*, that is, it is directed from high-pressure to low-pressure regions (Figure 1.5). Let the unit vector normal to the isobars be denoted by **n**, and consider an elemental volume of air with dimensions dn, dy, dz, where the coordinates n, y, z are mutually orthogonal. The net force per unit mass exerted by the horizontal pressure gradient along the direction of the vector **n** is

$$dy\,dz\frac{\left[p - \left(p - \frac{\partial p}{\partial n}dn\right)\right]}{(dn\,dy\,dz\,\rho)} = \frac{1}{\rho}\frac{\partial p}{\partial n} \tag{1.2}$$

where p denotes the pressure, and ρ is the air density.

2) The *deviating force due to the Earth's rotation*. If defined with respect to an absolute frame of reference, the motion of a particle not subjected to the action of an external force will follow a straight line. To an observer on the rotating Earth, however, the path described by the particle will appear curved. The deviation of the particle with

Figure 1.5 Direction of pressure gradient force.

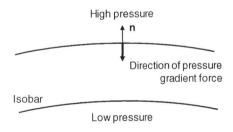

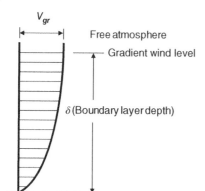

V_{gr}

Free atmosphere

Gradient wind level

δ (Boundary layer depth)

Figure 1.6 The atmospheric boundary layer.

respect to a straight line fixed with respect to the rotating Earth may be attributed to an apparent force, the *Coriolis force*

$$F_c = mf\,v \tag{1.3}$$

where m is the mass of the particle, $f = 2\omega \sin \varphi$ is the Coriolis parameter, $\omega = 0.7292 \times 10^{-4}$ s^{-1} is the angular velocity vector of the Earth, φ is the angle of latitude, and v is the velocity vector of the particle referenced to a coordinate system fixed with respect to the Earth. The force F_c is normal to the direction of the particle's motion, and is directed according to the vector multiplication rule.

3) The *friction force*. The surface of the Earth exerts upon the moving air a horizontal drag force that retards the flow. This force decreases with height and becomes negligible above a height δ known as *gradient height*. The atmospheric layer between the Earth's surface and the gradient height is called *the atmospheric boundary layer* (see Chapter 2). The wind velocity speed at height δ is called the *gradient velocity*,[1] and the atmosphere above this height is called the *free atmosphere* (Figure 1.6).

In the free atmosphere an elementary mass of air will initially move in the direction of the pressure gradient force – the driving force for the air motion – in a direction normal to the isobar. The Coriolis force will be normal to that incipient motion, that is, it will be tangent to the isobar. The resultant of these two forces, and the consequent motion of the particle, will no longer be normal to the isobar, so the Coriolis force, which is perpendicular to the particle motion, will change direction, and will therefore no longer be directed along the isobar. The change in the direction of motion will continue until the particle will move steadily *along* the isobar, at which point the Coriolis force will be in equilibrium with the pressure gradient force, as shown in Figure 1.7.

Within the atmospheric boundary layer the direction of the friction force, denoted by S, coincides with the direction of motion of the particle. During the particle's steady motion the resultant of the mutually orthogonal Coriolis and friction forces will balance the pressure gradient force, that is, will be normal to the isobars, meaning that the friction force – and therefore the motion of the particle – will cross the isobars (Figure 1.8). Since the friction force, which retards the wind flow and vanishes at the gradient height, decreases as the height above the surface increases, the velocity increases

---

1 For "straight winds" (i.e., winds whose isobars are approximately straight), the term "geostrophic" is substituted in the meteorological literature for "gradient."

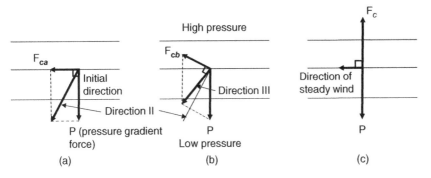

Figure 1.7 Frictionless wind balance in geostrophic flow.

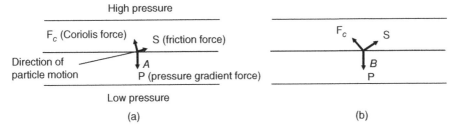

Figure 1.8 Balance of forces in the atmospheric boundary layer.

Figure 1.9 Wind velocity spiral in the atmospheric boundary layer.

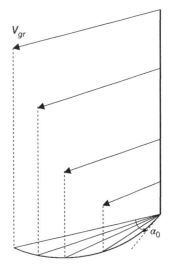

with height (Figure 1.6). The Coriolis force, which is proportional to the velocity, also increases with height. The combined effect of the Coriolis and friction forces causes the angle between the isobars and the direction of motion within the ABL, shown as α_0 in Figures 1.8 and 1.9, to increase from zero at the gradient height to its largest value at the Earth's surface. The wind velocity in the boundary layer can therefore be represented by a spiral, as in Figure 1.9. Under certain simplifying assumption regarding the effective flow viscosity the spiral is called the Ekman spiral (see Section 2.3.1).

If the isobars are curved, the horizontal pressure gradient force as well as the centrifugal force associated with the motion on a curved path will act on the elementary mass of air in the direction normal to the isobars, and the resultant steady wind will again flow along the isobars. Its velocity results from the relations

$$V_{gr}f \pm \frac{V_{gr}^2}{r} = \frac{dp/dn}{\rho} \tag{1.4}$$

where r is the radius of curvature of the air trajectory. If the mass of air is in the Northern Hemisphere, the positive or the negative sign is used according as the circulation is cyclonic (around a center of low pressure) or anticyclonic (around a center of high pressure).

1.3   Windstorms

1.3.1   Large-Scale Storms

Large-scale wind flow fields of interest in structural engineering may be divided into two main types of storm: extratropical (synoptic) storms, and tropical cyclones. *Synoptic storms* occur at and above mid-latitudes. Because their vortex structure is less well defined than in tropical storms, their winds are loosely called "straight winds."

Tropical cyclones, known as *typhoons* in the Far East, and *cyclones* in Australia and the Indian Ocean, generally originate between 5° and 20° latitudes. *Hurricanes* are defined as tropical cyclones with sustained surface wind speeds of 74 mph or larger. Tropical cyclones are translating vortices with diameters of hundreds of miles and counterclockwise (clockwise) rotation in the Northern (Southern) hemisphere. Their *translation speeds* vary from about 3–30 mph. As in a stirred coffee cup, the column of fluid is lower at the center than at the edges. The difference between edge and center atmospheric pressures is called *pressure defect.* Rotational speeds increase as the pressure defect increases, and as the *radius of maximum wind speeds,* which varies from 5 to 60 miles, decreases.

The structure and flow pattern of a typical tropical cyclone is shown in Figure 1.10. The *eye of the storm* (Region I) is a roughly circular, relatively dry core of calm or light winds surrounded by the *eye wall.* Region II contains the storm's most powerful winds. Far enough from the eye, winds in Region V, which decrease in intensity as the distance from the center increases, are parallel to the surface. Where Regions V and II intersect the wind speed has a strong updraft component that alters the mean wind speed profile and is currently not accounted for in structural engineering practice. The source of energy that drives the storm winds is the warm water at the ocean surface. As the storm makes landfall and continues its path over land, its energy is depleted and its wind speeds gradually decrease. Figure 1.11 shows a satellite image of Hurricane Irma. In the United States hurricanes are classified in accordance with the Saffir–Simpson scale (Table 1.1).[2]

1.3.2   Local Storms

Foehn winds (called chinook winds in the Rocky Mountains area) develop downwind of mountain ridges. Cooling of air as it is pushed upwards on the windward side of a

---

2 See Commentary, ASCE 7-16 Standard [2].

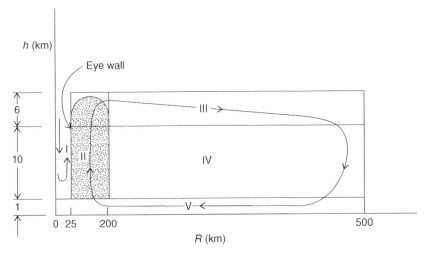

Figure 1.10 Structure of a hurricane.

Figure 1.11 Satellite view of hurricane Irma. Source: National Oceanic and Atmospheric Administration photo.

mountain ridge causes condensation and precipitation. The dry air flowing past the crest warms as it is forced to descend, and is highly turbulent (Figure 1.12). A similar type of wind is the *bora*, which occurs downwind of a plateau separated by a steep slope from a warm plain.

Jet effect winds are produced by features such as gorges.

Table 1.1 Saffir-Simpson scale and corresponding wind speeds[a].

Category	Damage potential	1-min speed at 10 m over open water (mph)	3-s gust speed at 10 m over open terrain exposure (mph)	N. Atlantic examples
1	Minimal	74–95	81–105	Agnes 1972
2	Moderate	96–110	106–121	Cleo 1974
3	Extensive	111–129	122–142	Betsy 1965
4	Extreme	130–156	143–172	David 1979
5	Catastrophic	≥157	≥173	Andrew 1992

a) For the definition of 1-minute and 3-second wind speeds see Section 2.1. Official speeds are in mph.

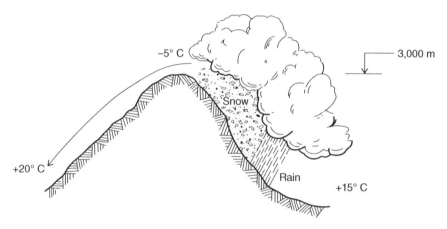

Figure 1.12 Foehn wind.

Thunderstorms occur as heavy rain drops, due to condensation of water vapor contained in ascending warm, moist air, drag down the air through which they fall, causing a *downdraft* that spreads on the earth's surface (Figure 1.13). The edge of the spreading cool air is the *gust front*. If the wind behind the gust front is strong, it is called a *downburst*. Notable features of downbursts are the typical difference between the profiles of their peak gusts near the ground and those of large-scale storms, and the differences among the time histories of various thunderstorms [3] (Figure 1.14). According to [5], the maximum winds (i.e., design level winds) rarely occur at the locations where profiles differ markedly from the logarithmic law.

Microbursts were defined by Fujita [4] as slow-rotating small-diameter columns of descending air which, upon reaching the ground, burst out violently (Figure 1.15). A number of fatal aircraft accidents have been caused by microbursts. According to [5], "because of the higher frequency and large individual area of a microburst, probabilities of structural damage by microbursts with 50–100 mph wind speeds could be much higher than those of tornadoes."

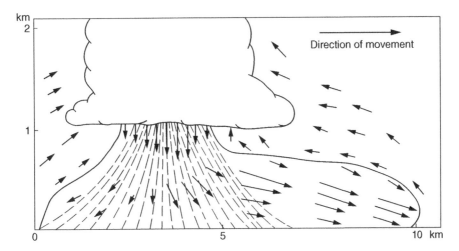

Figure 1.13 Section through a thunderstorm in the mature stage.

Tornadoes are small vortex-like storms, and can contain winds in excess of $100 \, \text{m s}^{-1}$ (Figure 1.16) [6, 7].

For unvented or partially unvented structures, the difference between atmospheric pressure at the tornado periphery and the tornado center (i.e., the pressure defect) typical of cyclostrophic storms is a significant design factor. For such structures, the difference between the larger atmospheric pressure that persists inside the structure and the lower atmospheric pressure acting on the structure during the tornado passage results in large, potentially destructive net pressures that must be accounted for in design (see Chapter 27).

The National Weather Service and the U.S. Nuclear Regulatory Commission are currently classifying tornado intensities in accordance with the Enhanced Fujita Scale (EF-scale), agreed upon in a forum organized by Texas Tech University in 2001. The EF-scale, shown in Table 1.2, replaced the original Fujita scale following a consensus opinion that the latter overestimated tornado wind speeds (see, e.g., [8]). The EF scale is based on the highest 3-second wind speed estimated to have occurred during the tornado's life, and is shown in Table 1.2.

As noted in [9], "no tornado has been assigned an intensity of EF6 or greater, and there is some question whether an EF6 or greater tornado would be identified if it did occur." For tornadoes that occur in areas containing no objects capable of resisting events with intensity EF0 (e.g., in a corn field), no intensity estimate is possible. An additional difficulty is that intensity estimates depend upon quality of construction. Since there are no measurements of tornado speeds at heights above ground comparable to typical building heights, it is necessary to rely on largely subjective estimates, based primarily on observations of damage.

For additional material on tornadoes, see Sections 3.4 and 5.3, and Chapters 27 and 28.

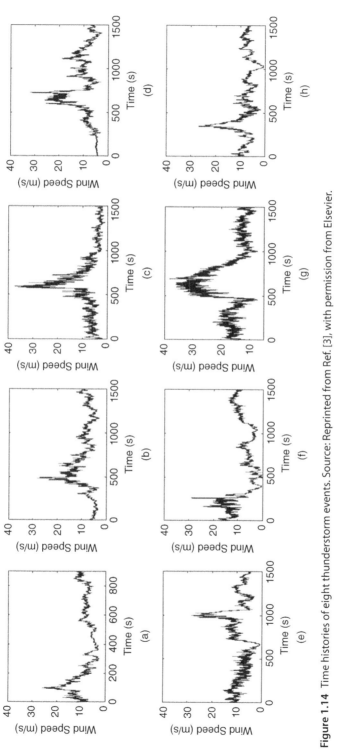

Figure 1.14 Time histories of eight thunderstorm events. Source: Reprinted from Ref. [3], with permission from Elsevier.

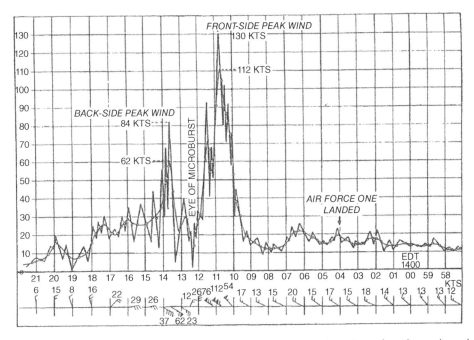

Figure 1.15 Andrews Air Force Base microburst on 1 August 1 1983. Its 149.7 mph peak speed was the highest recorded in a microburst in the U.S [4].

Figure 1.16 Tornado funnel (Source: National Oceanic and Atmospheric Administration photo).

Table 1.2 Tornado enhanced Fujita Scale.

Intensity	Description	Enhanced Fujita Scale 3-s peak gust speed (mph)
EF0	Light damage	65–85
EF1	Moderate damage	86–110
EF2	Considerable damage	111–135
EF3	Severe damage	136–165
EF4	Devastating damage	166–200
EF5	Incredible damage	>200

References

1 Humphreys, W.J. (1940). *Physics of the Air*. New York: McGraw-Hill.
2 ASCE, "Minimum design loads for buildings and other structures (ASCE/SEI 7-16)," in ASCE Standard ASCE/SEI 7-16, Reston, VA: American Society of Civil Engineers, 2016.
3 Lombardo, F.T., Smith, D.A., Schroeder, J.L., and Mehta, K.C. (2014). *Journal of Wind Engineering and Industrial Aerodynamics* 125: 121–132. http://dx.doi.org/10.1016/j.jweia.2013.12.004.
4 Fujita, T.T. (1990). Downbursts: meteorological features and wind field characteristics. *Journal of Wind Engineering and Industrial Aerodynamics* 36: 75–86.
5 Schroeder, J. L., Personal communication, Nov. 21, 2016.
6 Lewellen, D.C., Lewellen, W.S., and Xia, J. (2000). The influence of a local swirl ratio on tornado intensification near the surface. *Journal of the Atmospheric Sciences* 57: 527–544.
7 Hashemi Tari, P., Gurka, R., and Hangan, H. (2010). Experimental investigation of tornado-like vortex dynamics with swirl ratio: the mean and turbulent flow fields. *Journal of Wind Engineering and Industrial Aerodynamics* 98: 936–944.
8 Phan, L. T. and Simiu, E., "Tornado aftermath: Questioning the tools," *Civil Engineering*, December 1998, 0885-7024-/98-0012-002A. https://www.nist.gov/wind.
9 Ramsdell, J. V. Jr., and Rishel, J. P., *Tornado Climatology of the Contiguous United States*, A. J. Buslik, Project Manager, NUREG/CR-4461, Rev. 2, PNNL-15112, Rev. 1, Pacific Northwest National Laboratory, 2007.

2

The Atmospheric Boundary Layer

As indicated in Chapter 1, the Earth's surface exerts on the moving air a horizontal drag force whose effect is to retard the flow. This effect is diffused by turbulent mixing throughout a region called the atmospheric boundary layer (ABL). In strong winds the depth of the ABL ranges from a few hundred meters to a few kilometers, depending upon wind speed, roughness of terrain, angle of latitude, and the degree to which the stratification of the free flow (i.e., the flow above the ABL) is stable. Within the ABL the mean wind speed varies as a function of elevation.

This chapter is devoted to studying aspects of ABL flow of interest from a structural engineering viewpoint. Section 2.1 is concerned with the dependence of the wind speed on averaging time. Section 2.2 presents the equations of mean motion in the ABL. Sections 2.3 and 2.4 pertain to horizontally homogeneous flows over flat uniform surfaces, and contain, respectively, theoretical as well as empirical results on the dependence of wind speeds on height above the Earth's surface, and the structure of atmospheric turbulence. Section 2.5 concerns horizontally non-homogeneous flows (i.e., flows affected by changes of surface roughness or by topographic features, and flows in tropical storms and thunderstorms). Since the structural engineer is concerned primarily with the effect of strong winds, it will be assumed that the ABL flow is neutrally stratified. Indeed, in strong winds turbulent transport dominates the heat convection by far, so that thorough turbulent mixing tends to produce neutral stratification, just as in a shallow layer of incompressible fluid mixing tends to produce an isothermal state. In flows of interest in structural engineering, a layer of stably stratified flow, called the capping inversion, is present above the ABL and significantly affects the ABL's height.

2.1 Wind Speeds and Averaging Times

If the flow were laminar wind speeds would be the same for all averaging times. However, owing to turbulent fluctuations, such as those recorded in Figure 2.1, the definition of wind speeds depends on averaging time.

The peak 3-second gust speed is the peak of a storm's speeds averaged over 3 seconds. In 1995 it was adopted in the ASCE Standard as a measure of wind speeds. Similarly, the peak 5-second gust speed is the largest speed averaged over 5 seconds. The 5-second speed is reported by the National Weather Service ASOS (Automated Service Observing

Wind Effects on Structures: Modern Structural Design for Wind, Fourth Edition. Emil Simiu and DongHun Yeo.
© 2019 John Wiley & Sons Ltd. Published 2019 by John Wiley & Sons Ltd.

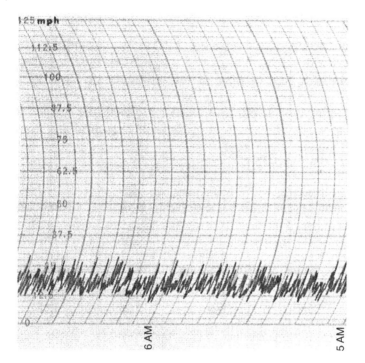

Figure 2.1 Wind speed record.

System), and is about 2% less than the 3-second speed. The 28-mph peak of Figure 2.1 is, approximately, a 3-second speed.

The hourly wind speed is the speed averaged over 1 hour. It is commonly used as a reference wind speed in wind tunnel simulations. Hence the need to estimate the hourly speed corresponding to a 3-second (or a 1-minute, or a 10-minute) speed specified for design purposes or recorded at weather stations. In Figure 2.1 the statistical features of the record do not vary significantly (i.e., the record may be viewed as *statistically stationary*, see Appendix B) over an interval of almost two hours; the hourly wind speed is about 18.5 mph, or about 1/1.52 times the peak 3-second gust.

Sustained wind speeds, defined as wind speeds averaged over intervals in the order of 1 min, are used in both engineering and meteorological practice. The *fastest 1-minute wind speed* or, for short, the 1-minute speed, is the storm's largest 1-minute average wind speed. The *fastest-mile wind speed* U_f is the storm's largest speed in mph averaged over a time interval $t_f = 3600/U_f$. For example, a 60 mph fastest-mile wind speed is averaged over a 60-second time interval.

Ten-minute wind speeds are wind speeds averaged over 10 min, and are used in World Meteorological Organization (WMO) practice as well as in some standards and codes.

The ratio between the peak gust speed and the mean wind speed is called the *gust factor*. Expressions for the relation between wind speeds with different averaging times are provided in Section 2.3.7 as functions of parameters defined subsequently in this chapter.

2.2   Equations of Mean Motion in the ABL

The motion of the atmosphere is governed by the fundamental equations of continuum mechanics, which include the equation of continuity – a consequence of the principle of mass conservation, – and the equations of balance of momenta, that is, the Navier–Stokes equations (see also Chapters 4 and 6). These equations must be supplemented by phenomenological relations, that is, empirical relations that describe the specific response to external effects of the medium being considered. (For example, in the case of a linearly elastic material the phenomenological relations consist of the so-called Hooke's law.)

If the equations of continuity and the equations of balance of momenta are averaged with respect to time, and if terms that can be shown to be negligible are dropped, the following equations describing the mean motion in the boundary layer of the atmosphere are obtained:

$$U\frac{\partial U}{\partial x} + V\frac{\partial U}{\partial y} + W\frac{\partial U}{\partial z} + \frac{1}{\rho}\frac{\partial p}{\partial x} - f\,V - \frac{1}{\rho}\frac{\partial \tau_u}{\partial z} = 0 \tag{2.1}$$

$$U\frac{\partial V}{\partial x} + V\frac{\partial V}{\partial y} + W\frac{\partial V}{\partial z} + \frac{1}{\rho}\frac{\partial p}{\partial y} + f\,U - \frac{1}{\rho}\frac{\partial \tau_v}{\partial z} = 0 \tag{2.2}$$

$$\frac{1}{\rho}\frac{\partial p}{\partial z} + g = 0 \tag{2.3}$$

$$\frac{\partial U}{\partial x} + \frac{\partial V}{\partial y} + \frac{\partial W}{\partial z} = 0 \tag{2.4}$$

where U, V, and W are the mean velocity components along the axes x, y, and z of a Cartesian system of coordinates whose z-axis is vertical; p, ρ, f, and g are the mean pressure, the air density, the Coriolis parameter, and the acceleration of gravity, respectively; and τ_u, τ_v are shear stresses in the x and y directions, respectively. The x-axis is selected, for convenience, to coincide with the direction of the shear stress at the surface, denoted by τ_0 (Figure 2.2).

It can be seen, by differentiating Eq. (2.3) with respect to x or y, that the vertical variation of the horizontal pressure gradient depends upon the horizontal density gradient. For the purposes of this text it will be sufficient to consider only flows in which the horizontal density gradient is negligible. The horizontal pressure gradient is then invariant

Figure 2.2 Coordinate axes.

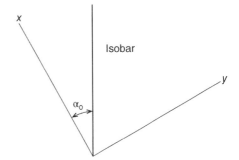

with height and thus has, throughout the boundary layer, the same magnitude as at the boundary layer's top:

$$\frac{\partial p}{\partial n} = \rho \left(fV_{gr} \pm \frac{V_{gr}^2}{r} \right)$$ (2.5)

where V_{gr} is the gradient velocity, r is the radius of curvature of the isobars, and n is the direction of the gradient wind (see Eq. [1.4]).

The geostrophic approximation corresponds to the case where the curvature of the isobars can be neglected. The gradient velocity is then called the geostrophic velocity and is denoted by G. Eq. (2.5) then becomes

$$\frac{1}{\rho}\frac{\partial p}{\partial x} = fV_g$$
$$\frac{1}{\rho}\frac{\partial p}{\partial y} = -fU_g$$ (2.6a,b)

where U_g and V_g are the components of the geostrophic velocity G along the x- and y-axes.

The boundary conditions for Eqs. (2.1)–(2.4) may be stated as follows: at the ground surface the velocity vanishes, while at the top of the ABL the shear stresses vanish and the wind flows with the gradient velocity V_{gr}. In addition, an interaction between the ABL and the capping inversion occurs (see Section 2.3.3).

2.3 Wind Speed Profiles in Horizontally Homogeneous Flow Over Flat Surfaces

It may be assumed that in large-scale non-tropical storms, within a flat site of uniform surface roughness with sufficiently long fetch, a region exists over which the flow is horizontally homogeneous. The existence of horizontally homogeneous atmospheric flows is supported by observations and distinguishes ABLs from two-dimensional boundary layers such as occur along flat plates. In the latter case the flow in the boundary layer is decelerated by the horizontal stresses, so that the boundary-layer thickness grows as shown in Figure 2.3 [1]. In atmospheric boundary layers. In atmospheric boundary layers, however, the horizontal pressure gradient – which below the free atmosphere is only

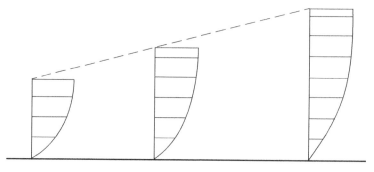

Figure 2.3 Growth of a two-dimensional boundary layer along a flat plate.

partly balanced by the Coriolis force (Figure 1.8) – re-energizes the flow and counteracts the boundary-layer growth. Horizontal homogeneity of the flow is thus maintained [2].

Under equilibrium conditions, in horizontally homogeneous flow Eqs. (2.1) and (2.2), in which Eq. (2.6a,b) are used, become

$$V_g - V = \frac{1}{\rho f} \frac{\partial \tau_u}{\partial z}$$

$$U_g - U = \frac{1}{\rho f} \frac{\partial \tau_v}{\partial z} \tag{2.7a,b}$$

The Ekman spiral was the first attempt to describe the ABL in mathematical terms, and is presented in Section 2.3.1 for the sake of its historical interest. In the 1960s and 1970s a major advance was achieved in the field of boundary-layer meteorology, based on an asymptotic approach. As shown in Section 2.3.2, the asymptotic approach yields the unphysical result that the mean speed component V vanishes throughout the boundary layer's depth, except at its top, where it has the value V_g. In addition, the 1960s and 1970s work did not consider the important effect of the capping inversion on the ABL height. Section 2.3.3 introduces the contemporary classification of neutrally stratified ABLs as functions of the Brunt-Väisälä frequency. The latter characterizes the interaction between the ABL and the capping inversion, and provides expressions for the height of the ABL that account for that interaction. Section 2.3.4 presents the logarithmic description of the mean wind speed within the lower layer of the ABL, called the surface layer, as well as estimates of the surface layer's depth. Section 2.3.5 presents the power law representation of the wind speed profile which, though obsolete, is still being used in some codes and standards, including the ASCE 7-16 Standard [3]. Section 2.3.6 discusses the relation between characteristics of the ABL flows in different surface roughness regimes. Section 2.3.7 provides details on the relation between wind speeds with different averaging times.

2.3.1 The Ekman Spiral

The Ekman spiral model is obtained if it is assumed in Eq. (2.7a,b) that the shear stresses are proportional to a fictitious constant K, called eddy viscosity, such that

$$\tau_u = \rho K \frac{\partial U}{\partial z}$$

$$\tau_v = \rho K \frac{\partial V}{\partial z} \tag{2.8a,b}$$

Equations (2.7) and (2.8) then become a system of differential equations with constant coefficients. With the boundary conditions $U = V = 0$ for height above the surface $z = 0$, and $U = U_g$, $V = V_g$ for $z = \infty$, the solution of the system is

$$U = \frac{1}{\sqrt{2}} G[1 - e^{-az}(\cos az - \sin az)]$$

$$V = \frac{1}{\sqrt{2}} G[1 - e^{-az}(\cos az + \sin az)] \tag{2.9a,b}$$

where $a = [f/(2K)]^{1/2}$. Equations (2.9a,b), which describe the Ekman spiral, are represented schematically in Figure 1.9. Observations are in sharp disagreement with these equations. For example, while according to Eq. (2.9a,b) the angle α_0 between the surface stress τ_0 and the geostrophic wind direction is 45°, observations indicate that this angle may range approximately between approximately 5° and 30° (see Section 2.3.3). The cause of the discrepancies is the assumption, mathematically convenient but physically incorrect, that the eddy viscosity is independent of height.

2.3.2 Neutrally Stratified ABL: Asymptotic Approach

A vast literature is available on the numerical solution of the equations of motion of the fluid. A different type of approach, based on similarity and asymptotic considerations, was developed in [2]. The starting point of the asymptotic approach is the division of neutral boundary layers into two regions, a surface layer and an outer layer. In the surface layer the shear stress τ_0 induced by the boundary-layer flow at the Earth's surface must depend upon the flow velocity at a distance z from the surface, the roughness length z_0 that characterizes the surface roughness, and the density ρ of the air, that is,

$$\tau_0 \mathbf{i} = F\left(U\mathbf{i} + V\mathbf{j},\ z,\ z_0,\ \rho\right) \tag{2.10}$$

where U and V are the components of the mean wind speed along the x and y directions and $\mathbf{i}, \mathbf{j}$ are unit vectors. Eq. (2.10) can be written in the non-dimensional form:

$$\frac{U\mathbf{i} + V\mathbf{j}}{u_*} = \psi_{1x}\left(\frac{z}{z_0}\right)\mathbf{i} + \psi_{1y}\left(\frac{z}{z_0}\right)\mathbf{j} \tag{2.11}$$

where

$$u_* = \left(\frac{\tau_0}{\rho}\right)^{1/2} \tag{2.12}$$

is the friction velocity and $\boldsymbol{\Psi}_1 = \psi_{1x}\mathbf{i} + \psi_{1y}\mathbf{j}$ is a vector function to be determined. Eq. (2.11), known as the *law of the wall*, is applicable in the surface layer, and can be written in the form:

$$\frac{U\mathbf{i} + V\mathbf{j}}{u_*} = \psi_{1x}\left(\frac{z}{H}\frac{H}{z_0}\right)\mathbf{i} + \psi_{1y}\left(\frac{z}{H}\frac{H}{z_0}\right)\mathbf{j} \tag{2.13}$$

where

$$H = cu_*/f. \tag{2.14}$$

H denotes the boundary-layer depth (i.e., the height to which the effect of the surface shear stress has diffused in the flow), f is the Coriolis parameter, and on the basis of data available in the 1960s it was assumed in [2] $c \approx 0.25$. As indicated earlier, the mean velocity components $U(H)$ and $V(H)$ are denoted by U_g and V_g, respectively, and their resultant, denoted by G, is the geostrophic velocity.

In the outer layer it can be asserted that, at height z, the velocity reduction with respect to G must depend upon the surface shear stress τ_0 and the air density ρ. The non-dimensional expression for this dependence is the *velocity defect law*:

$$\frac{U\mathbf{i} + V\mathbf{j}}{u_*} = \frac{U_g\mathbf{i} + V_g\mathbf{j}}{u_*} + \psi_{2x}\left(\frac{z}{H}\right)\mathbf{i} + \psi_{2y}\left(\frac{z}{z_0}\right)\mathbf{j} \tag{2.15}$$

where $\boldsymbol{\Psi}_2$ is a vector function to be determined.

Consider, in Eqs. (2.13) and (2.15), the x components

$$\frac{U\mathbf{i}}{u_*} = \psi_{1x}\left(\frac{z}{H}\frac{H}{z_0}\right)\mathbf{i} \tag{2.16}$$

$$\frac{U\mathbf{i}}{u_*} = \frac{U_g\mathbf{i}}{u_*} + \psi_{2x}\left(\frac{z}{H}\right)\mathbf{i} \tag{2.17}$$

From the observation that a multiplying factor inside the function ψ_{1x} must be equivalent to an additive function outside the function ψ_{2x}, the following equations are obtained:

$$\frac{U}{u_*} = \frac{1}{k}\left(\ln\frac{z}{H} + \ln\frac{H}{z_0}\right) \tag{2.18}$$

$$\frac{U}{u_*} = \frac{U_g}{u_*} + \frac{1}{k}\left(\ln\frac{z}{H}\right) \tag{2.19}$$

for the surface and the outer layer, respectively. In Eqs. (2.18) and (2.19), $k \approx 0.40$ is the von Kármán constant, and the height z is measured from the elevation z_0 above the surface.

From Eq. (2.18) it follows immediately

$$\frac{U}{u_*} = \frac{1}{k}\left(\ln\frac{z}{z_0}\right) \tag{2.20}$$

By equating Eqs. (2.18) and (2.19) in the overlap region, there results

$$\frac{U_g}{u_*} = \frac{1}{k}\left(\ln\frac{H}{z_0}\right) \tag{2.21}$$

The logarithmic law is seen to apply to the U component of the wind velocity throughout the depth of the boundary layer.

Consider now the components

$$\frac{V\mathbf{j}}{u_*} = \psi_{1y}\left(\frac{z}{H}\frac{H}{z_0}\right)\mathbf{j} \tag{2.22}$$

$$\frac{V\mathbf{j}}{u_*} = \frac{V_g\mathbf{j}}{u_*} + \psi_{2y}\left(\frac{z}{H}\right)\mathbf{j} \tag{2.23}$$

It was assumed in [2, 4–6] that $\psi_{1y} \equiv 0$. Then, Eqs. (2.22) and (2.23) yield in the overlap region

$$\frac{V_g\mathbf{j}}{u_*} + \psi_{2y}\left(\frac{z}{H}\right)\mathbf{j} = 0 \tag{2.24}$$

that is,

$$\psi_{2y}\left(\frac{z}{H}\right) = -\frac{V_g}{u_*}$$

$$\psi_{2y}\left(\frac{z}{H}\right) = \frac{B}{k} \tag{2.25a,b}$$

where, based on measurements available in the 1960s, it was assumed $B/k \approx 4.8$ (e.g., [6]). It follows from Eqs. (2.23) and (2.25a,b) that

$$V(z) = 0 \quad (z < H). \tag{2.26}$$

Since, for $z = H$, $V(H) = V_g$, Eq. (2.23) yields

$$\Psi_{2y}(H/H) = 0 \tag{2.27}$$

and, by virtue of Eq. (2.26),

$$V(z) = V_g \delta(H), \tag{2.28}$$

where δ denotes the Dirac delta function. This physically unrealistic result is an artifact of the asymptotic approach, which transforms the actual profile $V(z)$ into the non-physical profile represented by Eq. (2.28).

2.3.3 Brunt-Väisäla Frequency. Types of Neutrally Stratified ABLs

Brunt-Väisäla Frequency. In much of the theoretical work on ABL flow performed until the 1990s or so, ABL flows for which the buoyancy flux at the surface, denoted by μ, is $\mu = 0$ and $\mu < 0$ were defined as neutral and stable, respectively. This classification did not consider the interaction between the ABL and the free flow (i.e., the flow above the ABL) that, when stably stratified, can have a significant effect on the height of the ABL [7–9].

The interaction between the ABL and the stably stratified free flow above the ABL is characterized by the non-dimensional parameter $\mu_N = N/|f|$, where N is the Brunt-Väisäla frequency. Consider an air particle with density $\rho(z)$ at elevation z in a stably stratified flow. If the particle is displaced by a small amount z', it will be subjected to an incremental pressure $g[\rho(z+z') - \rho(z)]$. The motion of the particle will be governed by the equation

$$\rho(z)\frac{\partial^2 z'}{\partial t^2} = g[\rho(z + z') - \rho(z)] \tag{2.29}$$

$$\frac{\partial^2 z'}{\partial t^2} = \frac{g}{\rho(z)}\frac{\partial \rho(z)}{\partial z}z' \tag{2.30}$$

Let

$$-\frac{g}{\rho(z)}\frac{\partial \rho(z)}{\partial z} = N^2 \tag{2.31}$$

It follows from Eqs. (2.30) and (2.31) that, for positive values of $\partial \rho(z)/\partial z$ (i.e., for a stable stratification of the free flow), z' is a harmonic function with frequency N, which drives the interaction between the stably stratified free flow and the ABL. (See also [10, p. 136].)

Truly Neutral and Conventionally Neutral ABL Flows. Based on the dependence of the ABL flow upon both μ and the non-dimensional parameter $\mu_N = N/|f|$, neutrally stratified ABL flows are classified into two categories [7–9]:

1) *Truly neutral flows* ($\mu \approx 0$, $N = 0$), "observed during comparatively short transition periods after sunset on a background of residual layers of convective origin," "often treated as irrelevant because of their transitional nature, and usually excluded from data analysis."
2) *Conventionally neutral flows* ($\mu \approx 0$, $N > 0$) (i.e., neutrally stratified *and* interacting with the stably stratified layer above the ABL), are characterized by negligible buoyancy and a number $\mu_N \neq 0$; typically $50 < \mu_N < 300$. Recall that, in strong winds, the buoyancy in the ABL may be assumed to be negligible owing to strong mechanical, as opposed to thermal, turbulent mixing.

Of these two categories it is the conventionally neutral flows that are of interest in structural engineering applications.

Models of the ABL flow used in structural engineering applications have been based on the assumption that the flow stratification is truly neutral. The failure of the asymptotic similarity approach to consider the effect of the capping inversion results in the incorrect prediction of the ABL height, as is shown subsequently.

Integral Measures of the Conventionally Neutral ABL. The integral measures of the ABL are the geostrophic drag coefficient, the cross-isobaric angle, and the ABL height.

For μ_N values typical of conventionally neutral flows (i.e., $50 < \mu_N < 300$), the dependence of the geostrophic drag coefficient

$$C_g = \frac{u_*}{G} \tag{2.32}$$

and of the cross-isobaric angle α_0 upon the Rossby number

$$Ro = \frac{G}{|f|z_0} \tag{2.33}$$

can be represented by the following expressions, based on measurements by Lettau [11]:

$$C_g = \frac{0.205}{\log_{10}(Ro) - 0.556} \tag{2.34}$$

$$\alpha_0 = \frac{173.58}{\log_{10}(Ro)} - 3.03 \tag{2.35}$$

[12, 13, p. 338]. Also, for conventionally neutral ABLs,

$$\frac{1}{H^2} = \left[\frac{f^2}{C_R^2} + \frac{N|f|}{C_{CN}^2}\right]\frac{1}{u_*^2} \tag{2.36}$$

where $C_R \approx 0.6$ and $C_{CN} \approx 1.36$ [7–9]. Therefore the ABL height is

$$H = \frac{C_h(\mu_N)u_*}{f} \tag{2.37}$$

where $C_h(\mu_N) = (1/C_R^2 + \mu_N/C_{CN}^2)^{-1/2}$. Note the difference with the expression for H in Eq. (2.14) For any given friction velocity u_*, Coriolis parameter f and surface roughness length z_0, the quantities G, α_0 and H are obtained by using Eqs. (2.32)–(2.36).

Example 2.1 *ABL integral measures. Mean wind speed and veering angle profiles.* Consider the following parameters: $f = 10^{-4}\,\text{s}^{-1}$, $N = 0.018\,\text{s}^{-1}$, so $\mu_N = 180$, and $z_0 = 0.3\,\text{m}$ (suburban terrain exposure), $u_* = 1.5$ m s^{-1}. It can be verified by using Eq. (2.36) that $C_h \approx 0.10$, so $H = 0.10 \times 1.5/10^{-4} = 1500\,\text{m}$. (According to Eq. (2.14), $H = 3750\,\text{m}$.) The trial value $G = 41$ m s^{-1} yields $\log_{10}(Ro) = 6.14$, $u_*/G \approx 0.037$, to which there corresponds $G = 41$ m s^{-1} and $\alpha_0 \approx 25°$. For $z = 300$ m, $z/H = 0.20$; for $z = 800$ m, $z/H = 0.53$. Figures 2.4 and 2.5 show the dependence on height z of the speeds $U(z)$ and $V(z)$, their resultant, and the angle $\alpha_0(z)$, as obtained in [14] by Computational Fluid Dynamics techniques. Note that the component $V(800\,\text{m})$ and, *a fortiori*, the component $V(300\,\text{m})$, have negligible contributions to the resultant mean wind speed, and that the veering angles $\alpha_0(300\,\text{m})$ and $\alpha_0(800\,\text{m})$ are approximately 2 and 6°, respectively. Results for $C_h = 0.19$, based on [15, figure 7], are also included in Figures 2.4 and 2.5.

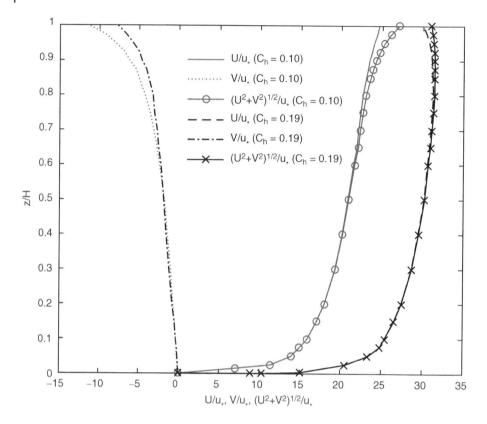

Figure 2.4 Dependence of U/u_*, V/u_*, and $\sqrt{U^2 + V^2}$ on z/H.

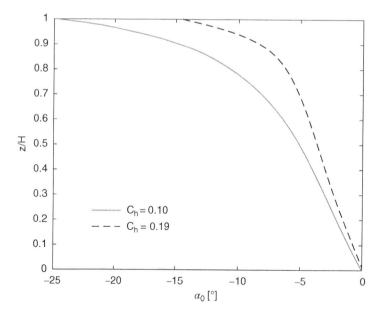

Figure 2.5 Dependence of veering angle on z/H.

No mathematical expression that uses the parameters z_0 and u_* is available for the description of the wind profile throughout the depth of the ABL. However, Section 2.3.4 presents the relation between the friction velocity u_* and the mean wind speed $U(z)$ in the lower portion of the ABL, and information on surface roughness lengths z_0 for various types of surface.

2.3.4 The Logarithmic Mean Wind Profile

The Logarithmic Law. Within the lower layer of the ABL whose height is denoted by z_s, the component $V(z)$ of the mean wind velocity is at least one order of magnitude smaller than the component $U(z)$ and is therefore negligible in practice (see Figure 2.4). The logarithmic law (Eq. [2.20]), renumbered here as Eq. (2.38),

$$\frac{U(z)}{u_*} = \frac{1}{k} \ln \frac{z}{z_0} \tag{2.38}$$

is valid for all heights z above the Earth's surface within the region $z_0 < z < z_s$. By virtue of Eq. (2.38)

$$u_* = \frac{U(z)}{2.5 \ln(z/z_0)} \tag{2.39}$$

where $z < z_s$.

According to a belief predating modern ABL research but still persisting among some wind engineers [16], $z_s \approx 100$ m. Also, according to the ASCE 7-16 Standard [3], the ABL depth is independent of wind speed. In fact, the depth H of the ABL is proportional to u_* (see Eq. [2.37]). The relation

$$z_s \approx 0.02 \frac{u_*}{f} \tag{2.40}$$

where f is the Coriolis parameter (see Section 1.2) [2, 4–6], is a lower bound for the height z_s. Eq. (2.40) follows from the assumption that, in the region $z_0 < z < z_s$, the shear stress τ_u differs little from the surface stress τ_0, and the component V of the velocity is small. Integration of Eq. (2.7a,b) over the height z_s yields

$$\tau_u = \tau_0 + \rho f \int_{z_0}^{z_s} (V_g - V)dz \approx \tau_0 + \rho f V_g z_s \tag{2.41}$$

or

$$|\rho f V_g z_s| \approx \eta \tau_0 \tag{2.42}$$

where η is a small number. Since $\tau_0 = \rho u_*^2$ (Eq. [2.12]) and $V_g/u_* = -B/k \approx 4.8$ (Eq. 2.25a,b),

$$z_s = \frac{\eta u_*^2}{f V_g} = \frac{\eta k}{f B} u_* = b \frac{u_*}{f} \tag{2.43}$$

According to [6] the logarithmic law holds, for practical purposes, even beyond heights at which η is in the order of 30%, meaning that $b > 0.02$.

Equations (2.39) and (2.40) show that the height z_s over which the logarithmic law is valid is approximately proportional to the wind speed $U(z)$ ($z_0 < z < z_s$).

Example 2.2 *Estimation of friction velocity u_*.* Assume $z = 10$ m, $U(z) = 30$ m s^{-1} and $z_0 = 0.03$ m (open exposure). Eq. (2.39) yields $u_* = 2.07$ m s^{-1}.

Example 2.3 *Estimation of surface layer depth z_s.* Assume $u_* = 2.07$ m s^{-1} and $f = 10^{-4}$ s^{-1}. According to Eq. (2.40) $z_s = 414$ m.

Surface Roughness Lengths z_0 and Surface Drag Coefficients. Tables 2.1–2.3 list surface roughness lengths z_0 based, respectively, on measurements included in the Commentary to the ASCE 7-16 Standard [3], and specified in the Eurocode [21].

Table 2.1 Values of surface roughness length z_0 and surface drag coefficients κ for various types of terrain.

Type of Surface	z_0 (cm)	$10^{-3}\,\kappa$
Sand[a]	0.01–0.1	1–2
Snow surface	0.1–0.6	2–3
Mown grass ($\approx$0.01 m)	0.1–1	2–3
Low grass, steppe	1–4	3–5
Fallow field	2–3	4–5
High grass	4–10	5–8
Palmetto	10–30	8–13
Pine forest (mean height of trees: 15 m; one tree per 10 m^2; $z_d = 12$ m[b])	90–100	28–30
Sparsely built-up suburbs[c]	20–40	11–15
Densely built-up suburbs, towns[c]	80–120	25–36
Centers of large cities[c]	200–300	62–110

a) [17].
b) [18].
c) Values of z_0 to be used in conjunction with the assumption $z_d = 0$ [19].

Table 2.2 Surface roughness lengths z_0 as listed in ASCE 7-16 Commentary [3].

Type of surface	z_0, ft. (m)
Water[a]	0.016–0.033 (0.005–0.01)
Open terrain[b]	0.033–0.5 (0.01–0.15)
Urban and suburban terrain, wooded areas[c]	0.5–2.3 (0.15–0.7)

a) The larger values apply over shallow waters (e.g., near shore lines). Approximate typical value corresponding to ASCE 7-16 Exposure D: 0.016 ft. (0.005 m) (ASCE Commentary). According to [20], for strong hurricanes $z_0 \approx 0.001$–0.003 m.
b) Approximate typical value corresponding to ASCE 7-16 Exposure C: 0.066 ft. (0.02 m) (ASCE Commentary).
c) Value corresponding approximately to ASCE 7-16 Exposure B: 0.5 ft. (0.15 m); this value is smaller than the typical value for ASCE 7-16 Exposure B: 1 ft. (0.3 m) (ASCE Commentary).

Table 2.3 Roughness lengths z_0 as specified in Eurocode [21].

Type of surface	z_0 (m)
Sea or coastal areas exposed to the open sea	0.003
Lakes or flat and horizontal area with negligible vegetation and no obstacles	0.01
Areas with low vegetation and isolated obstacles like trees or buildings with separations of maximum 20 obstacle heights (e.g., villages, suburban terrain, permanent forest)	0.05
Areas with regular cover of vegetation or buildings or with isolated obstacles with separations of maximum 20 obstacle heights (villages, suburban terrain, forests)	0.30
Areas in which at least 15% of the surface is covered with buildings whose average height exceeds 15 m	1.0

The surface drag coefficient is defined as

$$\kappa = \left[\frac{k}{\ln(10/z_0)} \right] \tag{2.44}$$

where $k = 0.4$ is the von Kármán constant, and z_0 is expressed in meters. Values of κ corresponding to various values of z_0 are given in Table 2.1.

The surface drag coefficient κ for wind flow *over water surfaces* depends upon wind speed. On the basis of a large number of measurements, the following empirical relations were proposed for the range $4 < U(10) < 20 \text{ m s}^{-1}$ [22]:

$$\kappa = 5.1 \times 10^{-4} [U(10)]^{0.46}$$
$$\kappa = 10^{-4} [7.5 + 0.67 U(10)] \tag{2.45a,b}$$

where $U(10)$ is the mean wind speed in m s^{-1} at 10 m above the mean water level [23]. According to [24], for wind speeds $U(10) < 40 \text{ m s}^{-1}$,

$$\kappa = 0.0015 \left[1 + \exp \left(-\frac{U(10) - 12.5}{1.56} \right) \right]^{-1} + 0.00104 \tag{2.46}$$

For additional information on the wind flow over the ocean, see [20, 25–27].

The following relation proposed by Lettau [28] may be used to estimate z_0 for built-up terrain:

$$z_0 = 0.5 H_{ob} \frac{S_{ob}}{A_{ob}} \tag{2.47}$$

where H_{ob} is the average height of the roughness elements in the upwind terrain, S_{ob} is the average vertical frontal area presented by the obstacle to the wind, and A_{ob} is the average area of ground occupied by each obstruction, including the open area surrounding it.

Example 2.4 *Application of the Lettau formula.* Check the Eurocode value $z_0 = 1 \text{ m}$ indicated in Table 2.3 against Eq. (2.47), assuming the average building height is $H_{ob} = 15 \text{ m}$, the average dimensions in plan of the buildings are $16 \times 16 \text{ m}$, and $A_{ob} = 1600 \text{ m}^2$. We have $S_{ob} = 15 \times 16 = 240 \text{ m}^2$, so the average area occupied by buildings is $16 \times 16/1600 = 16\%$. Eq. (2.47) yields $z_0 = 1.125 \text{ m}$.

The surface roughness length z_0 is a conceptual rather than a physical entity, and cannot therefore be measured directly. It can in principle be determined by measuring the mean wind speeds $U(z_1)$ and $U(z_2)$ at the elevations z_1 and z_2, respectively. However, small errors in the measurement of the speeds can lead to large errors in the estimation of the roughness length.

Example 2.5 *Errors in roughness length estimates based on mean wind speed measurements.* Assume measurements of mean wind speeds $U(z_1)$ and $U(z_2)$ are available at elevations z_1 and z_2 above ground. Eq. (2.38) yields $U(z_2)/U(z_1) \equiv r_{21} = \ln(z_2/z_0)/\ln(z_1/z_0)$. After some algebra it follows that

$$z_0 = \exp\left(\frac{r_{21} \ln z_1 - \ln z_2}{r_{21} - 1}\right) \tag{2.48}$$

Let $z_1 = 10$, $z_2 = 25$ and $z_0 = 0.026$ m. Eq. (2.38) yields $U(z_2)/U(z_1) = 1.154$. It follows then from Eq. (2.48) that, indeed, $z_0 = 0.026$ m. However, if measurement errors resulted in a 5% error in r_{21}, that is, if in Eq. (2.46) the ratio $r_{21} = 1.05 \times 1.154$ is used, the result obtained is $z_0 = 0.13$ m, rather than 0.026 m.

For a more effective approach to estimating roughness length, based on measurements of turbulence intensity, see [29] and Example 2.14.

Zero-plane Displacement. On account of the finite height of the roughness elements, the following empirical modification of Eq. (2.38) is required. The quantity z, rather than denoting height above ground, is defined as

$$z = z_{gr} - z_d \tag{2.49}$$

where z_{gr} is the height above ground and z_d is a length known as the zero-plane displacement. The quantity z is called the effective height. It is suggested in [30] that reasonable values of the zero plane displacement in cities may be obtained using the formula

$$z_d = \bar{h} - \frac{z_0}{k} \tag{2.50}$$

where $\bar{h}$ is the general roof-top level.

2.3.5 Power Law Description of ABL Wind Speed Profiles

The logarithmic law has long superseded the power law in meteorological practice. Unlike the logarithmic law, the power law is strictly empirical. It was first proposed about a century ago for open terrain in [31] and for built-up terrain in [32]. It is still used in the United States [3], Canada [33], and Japan [34], primarily owing to the earlier belief that the logarithmic law is only valid up to 50–100 m, even in strong winds.

The variation of wind speed with height can be expressed approximately as

$$U(z) = U(z_{ref})\left(\frac{z}{z_{ref}}\right)^{1/\alpha} \tag{2.51}$$

where z_{ref} is a reference height, for example 10 m above ground in open terrain. In Eq. (2.51) the exponent $1/\alpha$ depends upon surface roughness and upon averaging time, the profiles being flatter as the averaging time decreases. The power law applied to

3-second peak gust wind profiles has the same form as Eq. (2.51), however, in the ASCE 7 Standard its exponent is denoted by $\hat{a}$ rather than by α. Five-second peak gusts may in practice be assumed to differ negligibly from 3-second gusts. Eq. (2.51) is assumed in the ASCE 7 Standard and the National Building Code (NBC) of Canada to be valid up to a height z_g purported to represent the geostrophic height and referred to therein as the gradient speed. Table 2.4 lists power law exponents and gradient heights z_g specified in the ASCE Standard and the NBC specified for four surface exposure categories: A (centers of large cities), B (suburban terrain), C (open terrain), and D (open water). Category A was excluded from later versions of the ASCE 7 Standard, on account of the poor agreement of the power law with actual wind speeds over centers of large cities. It is shown in [62] that the values of z_g assumed in the power law model can result in strongly unconservative estimates of wind effects on super-tall buildings designed in accordance with ASCE 7-16 provisions.

Example 2.6 *Application of the power law.* Let $z_{ref} = 32.8$ ft. (10 m), $U_{3s}(z_{ref}) = 55$ mph, $\hat{a} = 1/9.5$ (open terrain). From Eq. (2.51), at 100 ft. above ground $U_{3s}(100\text{ ft}) = 55(100/32.8)^{1/9.5} = 62$ mph.

As noted by Panofsky and Dutton [35, p. 131], the power law can be fitted reasonably well to the log law *only over small height ranges*.

2.3.6 ABL Flows in Different Surface Roughness Regimes

Wind speed maps are developed for structural engineering purposes for open terrain exposure. Since most structures are not built in open terrain, it is necessary to determine wind speeds corresponding to the speeds specified in wind maps for exposures other than open. This is done by using the fact that, in any given large-scale storm, the geostrophic speed is independent of surface friction and therefore of terrain roughness (Eq. [1.4]). We first consider the case in which wind profiles are described by the logarithmic law. Next we consider the power law case.

Table 2.4 Power law exponents and gradient heights specified in the 1993–2016 versions of ASCE 7 Standard, and in the National Building Code of Canada (NBCC) [33].

	Exposure	A[a]	B[b]	C[c]	D[d]
ASCE 7-93[e]	$1/\alpha$	1/3	1/4.5	1/7	1/10
	z_g ft. (m)	1500 (457)	1200 (366)	900 (274)	700 (213)
NBC[f]	$1/\alpha$	0.4	0.28	0.16	–
	z_g ft. (m)	1700 (520)	1300 (400)	900 (274)	–
ASCE 7[g]	$1/\hat{a}$	–	1/7	1/9.5	1/11.5
(1995–2016)	z_g ft. (m)	–	1200 (366)	900 (274)	700 (213)

a) Centers of large cities.
b) Suburban terrain, towns.
c) Open terrain (e.g., airports).
d) Water surfaces.
e) Sustained speeds.
f) Mean hourly speeds.
g) Peak 3-second gust speeds.

Wind speeds described by the logarithmic law. Examples 2.7 and 2.8 consider, respectively, the cases of suburban and ocean versus open exposure.

Example 2.7   It can be verified that, for $f = 10^{-4}\,\mathrm{s}^{-1}$, given a surface with open exposure ($z_0 = 0.03\,\mathrm{m}$), to a storm that produces a friction velocity $u_* = 2.5\,\mathrm{m\,s}^{-1}$ there corresponds a geostrophic speed $G \approx 83\,\mathrm{m\,s}^{-1}$. In accordance with the definition of Ro, for suburban terrain exposure ($z_{01} = 0.3\,\mathrm{m}$), to $G = 83\,\mathrm{m\,s}^{-1}$ there corresponds $\mathrm{Ro}_1 = \log$ $[83/(10^{-4} \times 0.3)] = 6.44$. From Eq. (2.34), $C_{g1} = 0.035$, so $u_{*1} = 83 \times 0.035 \approx 2.9\,\mathrm{m\,s}^{-1}$ (Eq. [2.32]), and the cross-isobaric angle is $\alpha_{01} \approx 24\,°$ (Eq. [2.35]). From Eqs. (2.36) and (2.37) there follows, for $N = 0.01\,\mathrm{s}^{-1}$, $C_{h1} = 0.13$ and $H_1 = 2.9 \times 0.13/10^{-4} \approx 3800\,\mathrm{m}$ (i.e., about half the asymptotic estimate $H_1 = 7250\,\mathrm{m}$ (Eq. [2.14]).

Example 2.8   For ocean surfaces, assuming $G = 83\,\mathrm{m\,s}^{-1}$ and $z_{01} = 0.003\,\mathrm{m}$, $\log_{10}\mathrm{Ro}_1 = \log\,(83/[10^{-4} \times 0.003]) = 8.44$, and $C_{g1} \approx 0.026$, so $u_{*1} = 83 \times 0.026 = 2.15\,\mathrm{m\,s}^{-1}$, and $\alpha_{01} \approx 18°$. Eq. (2.36) yields $H_1 = 2800\,\mathrm{m}$ (vs. the asymptotic estimate $H_1 = 5400\,\mathrm{m}$) and $C_{h1} = 0.13$.

Results close to those obtained by the relatively elaborate procedure used in Examples 2.7 and 2.8 can be obtained by Biétry's equation, adopted with a minor modification in the Eurocode [21]:

$$\frac{u_{*1}}{u_*} = \left(\frac{z_{01}}{z_0}\right)^{0.0706} \tag{2.52}$$

Example 2.9   *Application of Eq.* (2.52). Let $z_0 = 0.03\,\mathrm{m}$. If $z_{01} = 0.3\,\mathrm{m}$, $u_{*1}/u_* = 1.18$, versus $2.9/2.5 = 1.16$ as shown in Example 2.7; if $z_{01} = 0.003\,\mathrm{m}$, then $u_{*1}/u_* = 0.86$, versus $2.15/2.5 = 0.86$ as shown in Example 2.8.

Wind speeds described by the power law. For strong winds, given the mean hourly speed $U(z_{open})$ at the reference height z_{open} above open terrain with power law exponent $1/\alpha_{open}$, the mean hourly wind speed at height z above built-up terrain with power law exponent $1/\alpha$ is

$$U(z) = U(z_{open})\left(\frac{z_{g,open}}{z_{open}}\right)^{1/\alpha_{open}}\left(\frac{z}{z_g}\right)^{1/\alpha} \tag{2.53}$$

where the product of the first two terms in the right-hand side is the gradient speed above open terrain, $U(z_{g,open})$. Since gradient speeds are not affected by surface roughness, the gradient speed over built-up terrain, $U(z_g)$, is equal to $U\,(z_{g,open})$. The last factor in Eq. (2.53) transforms $U(z_g)$ into $U(z)$ at height z above built-up terrain. A relation similar to Eq. (2.53) is also used (with the appropriate values of the parameters z_g and $\hat{\alpha}$ from Table 2.4) for 3-second peak gust speeds, denoted here by U_{3s} (and in the ASCE 7 Standard by V), and for sustained wind speeds such as fastest-mile speeds or 1-minute speeds. In the ASCE 7 Standard, $U_{3s}(z_{open} = 10\,\mathrm{m})$ is the 3-second basic wind speed, and the product of the last two terms in Eq. (2.53) is denoted in the Standard by $\sqrt{K_z}$.

Example 2.10   *Relation between wind speeds in different roughness regimes, power law description.* Denote the 3-second peak gust speed by U_{3s}. Let $U_{3s}(32.8\,\mathrm{ft}) = 86\,\mathrm{mph}$

above open terrain ($\hat{a} = 9.5$, $z_g = 274$ m, Table 2.4). Eq. (2.53) yields $U_{3s}(45\,\text{m}) = 45$ m s^{-1} (*open terrain*). Using Table 2.4 and Eq. (2.53), above *suburban terrain* ($\hat{a} = 7.0$ and $z_g = 366$ m), $U_{3s}(10\,\text{m}) = 33$ and $U_{3s}(45\,\text{m}) = 40$ m s^{-1}.

2.3.7 Relation Between Wind Speeds with Different Averaging Times

The *mean ratio* $r(t, z_0, z)$ between the largest average t-second speed during a storm with a 1-hour duration and that storm's mean hourly (3600 s) speed is a function of the averaging time t, the terrain roughness length z_0, and the height above ground z (Table 2.5). As noted in Section 2.1, the ratio U_{3s}/U is called the gust factor.

Terrain with open exposure. For the particular case of open terrain exposure ($z_0 \approx 0.03$–0.05 m) and a height above ground $z = 10$ m, the approximate ratio r is listed for selected values of t as follows [36]:

These values are applicable to large-scale, non-tropical storms, over open terrain with open exposure, and at the standard (10 m) height above ground. These values are applicable only at the standard reference height over terrain with open exposure.

Example 2.11 *Conversion of fastest-mile wind speed to mean hourly speed and to peak 3-second gust for open terrain.* For a fastest-mile wind speed at 10 m over open terrain of 90 mph, the averaging time is 3600/90 = 40 s, and the corresponding hourly speed and peak 3-second gust are 90/1.29 = 69.8 and 69.8 × 1.52 = 106 mph, respectively.

Example 2.12 *Conversion of peak 3-second gust speed to mean hourly speed for open terrain.* Let the peak 3-second gust speed at 10 m above ground in open terrain be 30 m s^{-1}. For wind tunnel testing and structural purposes, winds characterized by that gust speed are modeled by winds with a 30/1.52 = 20 m s^{-1} mean hourly speed at 10 m above ground in open terrain.

Terrain with Exposure Other than Open. The following approximate relation may be used:

$$U_t(z) = U(z) + c(t)[\overline{u^2(z, z_0)}]^{1/2}$$

$$U_t(z) = U(z)\left[1 + \frac{\sqrt{\beta(z, z_0)}c(t)}{2.5\ln(z/z_0)}\right] \tag{2.54a,b}$$

where $U_t(z)$ is the peak speed averaged over t s within a record of approximately one hour, $U(z)$ is the mean wind speed for that record over terrain with surface roughness z_0, $\beta(z_0)$, $c(t)$ are given in Tables 2.6 and 2.7. Following [10, Eq. (18.25b)],

$$\beta(z, z_0) = \beta(z_0)\exp\left[-1.5\frac{z}{H}\right] \tag{2.55}$$

where H is the ABL depth and z, z_0, and H are in meters.

Table 2.5 Ratios r between t-s and mean hourly speeds at 10 m above open terrain.

t (s)	3	5	40	60	600	3600
	1.52	1.49	1.29	1.25	1.1	1.0

Table 2.6 Factor $\beta(z_0)$.

z_0 (m)	0.005	0.03	0.30	1.00
$\beta(z_0)$	6.5	6.0	5.25	4.9

Table 2.7 Factor $c(t)$.

t (s)	1	10	20	30	50	100	200	300	600	1000	3600
$c(t)$	3.00	2.32	2.00	1.73	1.35	1.02	0.70	0.54	0.36	0.16	0

Note: coefficient $c(t)$ is an approximate empirical peak factor which increases as t decreases.

Example 2.13 *Conversion of Saffir–Simpson scale 1-minute speeds at 10 meters over water to peak wind speeds at 10 m above open terrain, Category 4 hurricane.* From Table 1.1, the 1-minute speeds at 10 m above open water that define the weakest and strongest Category 4 hurricanes are 130 and 156 mph, respectively. The conversion depends on the assumed values of the surface roughness lengths z_0 for open water and open terrain. Relative large values of z_0 are applicable to wind flow over water near shorelines where the water is shallow, as opposed to flow over open water. Assuming that for hurricane winds over open water $z_0 = 0.003$ m, Eq. (2.54a,b) yields, with $\beta(z_0 \approx 0.003 \text{ m}) \approx 6.5$, and c (60 s) ≈ 1.29 (Tables 2.6 and 2.7):

$$U_{60\,s}^w(10 \text{ m}) = U^w(10 \text{ m}) \left[1 + \frac{2.55 \times 1.29}{2.5\ln(10/0.003)}\right]$$

$$U^w(10 \text{ m}) = 0.86 U_{60\,s}^w(10 \text{ m})$$

where the superscript w signifies "over open water."

Assuming that over open terrain $z_0 = 0.04$ m, Eqs. (2.38) and (2.52) yield

$$U^w(10 \text{ m}) = U(10 \text{ m})\left[\frac{0.003}{0.04}\right]^{0.0706} \frac{\ln(10/0.003)}{\ln(10/0.04)}$$

$$U(10 \text{ m}) = 0.816 U^w(10 \text{ m})$$

where $U(10 \text{ m})$ is the mean hourly wind speed over open terrain. It follows that

$$U(10 \text{ m}) = 0.86 \times 0.816 U_{60\,s}^w(10 \text{ m})$$
$$= 0.7 U_{60\,s}^w(10 \text{ m})$$

Therefore, the peak 3-second gust over open terrain is (Table 2.7):

$$U_{3\,s}(10 \text{ m}) = 1.52 \times 0.7 \times U_{60\,s}^w(10 \text{ m})$$
$$= 1.06 U_{60\,s}^w(10 \text{ m}).$$

To the speed $U_{60\,s}^w(10 \text{ m}) = 155$ mph there then corresponds a calculated peak 3-second gust at 10 m over open terrain $U_{3\,s}(10 \text{ m}) \approx 164$ mph. In the preceding calculations it was assumed that relations that apply to horizontally homogeneous wind flow (i.e., flow in synoptic storms) are also applicable to hurricanes, in which the isobars are curved, rather than straight, and the flow is therefore horizontally inhomogeneous.

2.4 ABL Turbulence in Horizontally Homogeneous Flow Over Smooth Flat Surfaces

Except for winds with relatively low speeds under special temperature conditions, the wind flow is not laminar (smooth). Rather, it is *turbulent* – it fluctuates in time and space; that is, at any one point in space, the wind speed is a random function of time (Figure 2.1), and at any one moment in time the wind speed is a random function of position in space.

Atmospheric flow turbulence characterization is of interest in structural engineering applications for the following reasons. First, turbulence affects the definition of the wind speed specified in engineering calculations, as shown in Sections 2.1 and 2.3.7. Second, by transporting particles from flow regions with high momentum into low-speed regions, turbulence can influence significantly the wind flow around a structure and, therefore, the aerodynamic pressures acting on the structure (Chapters 4 and 5). Therefore, to simulate correctly full-scale aerodynamic effects in the laboratory, it is necessary to achieve laboratory flows that simulate the features of atmospheric turbulence (Chapter 5). Third, turbulence produces resonant dynamic effects in flexible structures that must be accounted for in structural design (Chapter 11).

Descriptors of the turbulence used in applications include the turbulence intensity (Section 2.4.1), integral scales of turbulence (Section 2.4.2); and the spectra and the cross-spectra of the turbulent velocity fluctuations (Sections 2.4.3 and 2.4.4).

2.4.1 Turbulence Intensities

The longitudinal turbulence intensity at a point with elevation z is defined as

$$I_u(z) = \frac{\overline{u^2(z,z_0)}^{1/2}}{U(z)} \tag{2.56a}$$

that is, as the ratio of the r.m.s. of the longitudinal wind speed fluctuations $u(z, t)$ to the mean speed $U(z)$, $u(z, t)$ being parallel to $U(z)$. Since

$$\overline{u^2(z,z_0)}^{1/2} = \sqrt{\beta(z,z_0)}\, u_* \tag{2.56b}$$

where approximate values of $\beta(z, z_0)$ are given by Eq. (2.55) and Table 2.6, and by virtue of the log law,

$$I_u(z) \approx \frac{\sqrt{\beta(z,z_0)}}{2.5\ln(z/z_0)} \tag{2.56c}$$

Example 2.14 *Calculation of longitudinal turbulence intensity.* For $z_0 = 0.03$ m, $z = 20$ m, Eq. (2.56c) and Table 2.6 yield $I_u(z) \approx 0.15$.

Equation (2.56c) allows an approximate estimate of the roughness z_0 based on the measurement of $I_u(z)$. Note that if the calculated roughness length z_0 were significantly different from 0.03 m, then a corresponding value of $\beta \neq 6.0$ would be assumed on the basis of Table 2.6, and z_0 would be obtained by successive approximations.

In the surface layer the decrease of $\overline{u^2(z,z_0)}^{1/2}$ with height is relatively slow (see, e.g., [35, p. 185]) and is, conservatively, typically neglected in structural engineering calculations. The averaging time in Eq. (2.56) should be equal to the duration of strong winds in a storm. Typical durations being considered are 1 hour, and 10 minutes. The turbulence intensity decreases as the height above the surface increases, and vanishes near the top of the ABL. Definitions similar to Eq. (2.56) are applicable to the lateral and vertical turbulence intensities $I_v(z)$ and $I_w(z)$. In both these definitions the denominator is $U(z)$.

Measurements suggest that the turbulence intensity is typically higher by roughly 10% in tropical cyclone than in extratropical storms [37, 38], see Section 2.5.3.

2.4.2 Integral Turbulence Scales

The velocity fluctuations in a flow passing a point are associated with an overall flow disturbance consisting of a superposition of conceptual eddies transported by the mean wind. Each eddy is viewed as causing at that point a periodic fluctuation with circular frequency $\omega = 2\pi n$. The integral turbulence scales are measures of the spatial extent of the overall flow disturbance.

In particular, the integral turbulence scale L_u^x is a measure of the size of the longitudinal velocity components of the turbulent eddies. In a structural engineering context, L_u^x is a measure of the extent to which the overall fluctuating disturbance associated with the longitudinal wind speed fluctuation u will engulf a structure in the along-wind direction, and will thus affect at the same time both its windward and leeward sides. If L_u^x is large in relation to the along-wind dimension of the structure, the gust will engulf both sides. The scales L_u^y and L_u^z are measures of the transverse and vertical spatial extent of the fluctuating longitudinal component u of the wind speed. The scale L_w^x is a measure of the longitudinal spatial extent of the vertical fluctuating component w. If the mean wind is normal to a bridge span and L_w^x is large in relation to the deck width, the vertical wind speed fluctuation w will act at any given time on the whole width of the deck. If we now consider a panel normal to the mean wind direction, small values of L_u^y and L_u^z compared with the dimensions of the panel indicate that the effect of the longitudinal velocity fluctuations upon the overall wind loading is small. However, if L_u^y and L_u^z are large, the eddy will envelop the entire panel, and that effect will be significant.

Mathematically, the integral turbulence scale L_u^x (also called integral turbulence length) is defined as follows:

$$L_u^x = \int_0^\infty \frac{1}{\overline{u^2}} R_{u_1 u_2}(\xi) d\xi \tag{2.57}$$

where the overbar denotes mean value. The function $R_{u_1 u_2}(\xi)$ is defined as the autocorrelation function of the longitudinal velocity components $u(x_1, y_1, z_1, t)$ and $u(x_1 + \xi, y_1, z_1, t)$ Eq. (2.57) may be interpreted as follows. At any given time t, the fluctuation $u(x + \xi, y, z)$ differs from $u(x, y, z)$. The difference increases as the distance ξ increases. If $\xi = 0$, the autocorrelation function is unity; if ξ is small the two fluctuations are nearly the same, so in Eq. (2.57) the autocorrelation function is close to unity and its product by the elemental length $d\xi$ is therefore close to $d\xi$. On the other hand, if ξ is large, the fluctuations $u(x, y, z)$ and $u(x + \xi, y, z)$ differ randomly from each other, and their products are positive for some values of ξ and negative for others, so that their mean values tend to be vanishingly small and contribute negligibly to L_u^x. This interpretation is equivalent to stating that L_u^x is a measure of the size of the largest turbulent eddies of the flow, that is, of the eddies characterized by large autocorrelation functions.

Taylor Hypothesis. Frequency Space and Wavenumber Space. According to the Taylor hypothesis it may be assumed, approximately, that the flow disturbance is "frozen" as it travels with the mean velocity $U(z)$, that is,

$$u(x_1, \tau + t) \approx u(x_1 - x/U, \tau) \tag{2.58}$$

where $x = Ut$, $\tau = $ time, and t is a finite time increment. This assumption implies that every frequency component of the disturbance also travels essentially unchanged with the mean velocity U. During a period T, an eddy whose harmonic motion at fixed x has circular frequency $\omega = 2\pi/T = 2\pi n$, where $n = 1/T$ denotes the frequency, travels with velocity U a distance $UT = \lambda$, where $\lambda = U/n$ is the wavelength. The wavenumber is defined as $\kappa = 2\pi/\lambda = 2\pi n/U = \omega/U$. The motion is defined by a cosine function with argument $\omega t - \kappa x$ or, equivalently, $\kappa(Ut - x)$, meaning that for fixed t it is a harmonic wave in the wavenumber space, and for fixed x it is harmonic function in the frequency space.

By virtue of Taylor's hypothesis, the integral turbulence length L_u^x, defined in Eq. (2.57) by following a particle's path (i.e., in Lagrangian terms) can alternatively be defined at a fixed point (i.e., in Eulerian terms) as

$$L_u^x = U \int_0^\infty \frac{1}{u^2} R_u(\tau) d\tau \tag{2.59}$$

where the autocorrelation function is defined by Eq. (B.21).

Measurements of L_u^x. Measurements show that L_u^x increases with height above ground and as the terrain roughness decreases. The following strictly empirical expression was proposed in [39] for L_u^x:

$$L_u^x \approx Cz^m \tag{2.60}$$

where the constants C and m are obtained from Figure 2.6.

Table 2.8 lists measured values of L_u^x and estimates based on Eq. (2.60).

The uncertainties in the value of L_u^x are seen to be significant.

On the basis of recent measurements at elevations z of up to about 95 m in open sea exposure at mean speeds $U(z) = 10$ to $25\,\mathrm{m\,s^{-1}}$, it was suggested in [40], on a strictly

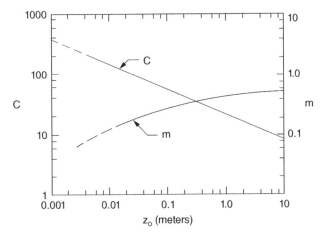

Figure 2.6 Values of C and m as functions of z_0. Source: Reprinted from [39], with permission from Elsevier.

Table 2.8 Measurements of integral turbulence scales L_u^x (m).

Exposure	z	z_0	Range	Avg.	Eq. (2.60)
Open[a]	31	0.03	60–460	200	180
Open[a]	51	0.03	130–450	200	200
Open[a]	81	0.03	60–650	300	230
Open[a]	110	0.03	110–690	350	240
Open[a]	151	0.03	120–630	400	250
Open[b]	15	0.01		82	220
Open[b]	17	0.04–0.10		55	120–160
Sub.[b]	16	1.00		36	70

a) Measurements reported in [61].
b) Measurements reported in [39].

empirical basis, that $L_u^x(z) \approx 3.3\, U(z) + 30$, where L_u^x is in meters and U is in m s^{-1}, and it was noted that L_u^x increased in the intervals 5–10, 10–20, 20–40, 40–60, and 60–80 m elevation by approximately 7, 10, 10, 8 and 5%, respectively. The dependence of the integral length scale on the velocity at all elevations is not supported by theory, however.

According to [39], it may be assumed

$$L_u^y \approx 0.33\, L_u^x; \qquad L_u^z \approx 0.5\, L_u^x; \quad L_w^y \approx 0.4z \qquad\qquad (2.61\text{a,b,c})$$

Section 2.4.3 presents the derivation of the integral turbulence length L_u^x from an expression for the spectrum of the longitudinal velocity fluctuations, based on theory and validated by measurements reported in [41] (see Eq. [2.77]).

2.4.3 Spectra of Turbulent Wind Speed Fluctuations

As indicated in Section 2.4.2, integral turbulence scales are measures of the average size of the largest turbulent eddies of the flow. In some applications a more detailed description of the turbulent fluctuations is needed. For example, the resonant response of a flexible structure is induced by velocity fluctuation components with frequencies equal or close to the structure's natural frequencies of vibration. To calculate that response, measures are needed of the size of the turbulent eddies as a function of frequency, and of the degree to which the turbulent fluctuations differ from each other as functions of their relative position in space. These measures are provided by the spectral density and the cross-spectral density functions.

The Energy Cascade. Turbulent velocity fluctuations in a flow with mean velocity U may be viewed as a result of a superposition of eddies, each characterized by a periodic motion with circular frequency $\omega = 2\pi n$ (or of wavenumbers κ). From the equations of balance of momenta for the mean motion, the following equation may be derived:

$$\left[U\frac{\partial}{\partial x}\left(\frac{\overline{q^2}}{2}\right) + V\frac{\partial}{\partial y}\left(\frac{\overline{q^2}}{2}\right) + W\frac{\partial}{\partial z}\left(\frac{\overline{q^2}}{2}\right) \right] - \left[\frac{\tau_u}{\rho}\frac{\partial U}{\partial z} + \frac{\tau_v}{\rho}\frac{\partial V}{\partial z} \right]$$

$$+ \frac{\partial}{\partial z}\left[\overline{w\left(\frac{p'}{\rho} + \frac{q^2}{2}\right)} \right] + \varepsilon = 0 \qquad\qquad (2.62)$$

where the bars indicate averaging with respect to time,

$$q = (\overline{u^2} + \overline{v^2} + \overline{w^2})^{1/2} \tag{2.63}$$

is the resultant fluctuating velocity, u, v, and w are turbulent velocity fluctuations in the x, y, and z directions, respectively, p' is the fluctuating pressure, and ε is the rate of energy dissipation per unit mass. Eq. (2.62) is the *turbulent kinetic energy equation*, and expresses the balance of turbulent energy advection (the terms in the first bracket), production (the terms in the second bracket), diffusion (the terms in the third bracket), and dissipation.

It can be shown that the inertial terms in these equations are associated with transfer of energy from larger eddies to smaller ones, while the viscous terms account for energy dissipation [42]. The latter is effected mostly by the smallest eddies in which the shear deformations, and therefore the viscous stresses, are large. In the absence of sources of energy, the kinetic energy of the turbulent motion will decrease, that is, the turbulence will decay. If the viscosity effects are small, the decay time is long if compared with the periods of the eddies in the high wavenumber range. The energy of these eddies may therefore be considered to be approximately steady. This can only be the case if the energy fed into them through inertial transfer from the larger eddies is balanced by the energy dissipated through viscous effects. The small eddy motion is then determined by the rate of energy transfer (or, equivalently, by the rate of energy dissipation, denoted by ε), and by the viscosity. The assumption that this is the case is known as *Kolmogorov's first hypothesis*. It follows from this assumption that, since small eddy motion depends only upon the internal parameters of the flow, it is independent of external conditions such as boundaries and that, therefore, *local isotropy* – the absence of preferred directions of small eddy motion – obtains.

It may further be assumed that the energy dissipation is produced almost in its entirety by the smallest eddies of the flow. Thus, at the lower end of the wavenumber subrange to which Kolmogorov's first hypothesis applies, the influence of the viscosity is small. In this subrange, known as the *inertial subrange,* the eddy motion may be assumed to be independent of viscosity, and thus determined solely by the rate of energy transfer, ε, which is equal to the rate of energy dissipation. This assumption is known as the *Kolmogorov second hypothesis.*

The total kinetic energy of the turbulent motion may, correspondingly, be regarded as a sum of contributions by each of the eddies of the flow. The function $E(\kappa)$ representing the dependence upon wavenumber κ of these energy contributions is defined as the energy spectrum of the turbulent motion.

It follows that, for sufficiently high κ

$$F[E(\kappa), \ \kappa, \ \varepsilon] = 0 \tag{2.64}$$

The dimensions of the quantities within brackets in Eq. (2.64) are $[L^3 T^{-2}]$, $[L^{-1}]$, and $[L^2 T^{-3}]$, respectively. From dimensional considerations it follows that

$$E(\kappa) = a_1 \varepsilon^{2/3} \kappa^{-5/3} \tag{2.65}$$

in which a_1 is a universal constant. On account of the isotropy, the expression for the spectral density of the longitudinal velocity fluctuations,[1] denoted by $E_u(\kappa)$, is to within

---

1 A mathematical definition of spectra is presented in Appendix B.

a constant similar to the constant in Eq. (2.65). Thus,

$$E_u(\kappa) = a\varepsilon^{2/3}\kappa^{-5/3} \tag{2.66}$$

in which it has been established by measurements that $a \approx 0.5$.

If expressed in terms of the frequency n, the spectral density is denoted by $S_u(n)$. Its expression is determined by noting that

$$\int_0^\infty E_u(\kappa)d\kappa = \int_0^\infty S_u(n)dn = \overline{u^2} \tag{2.67}$$

(see Eq. [B.15]), and $\kappa = 2\pi n/U$. Therefore,

$$S_u(n)dn = E_u(\kappa)d\kappa \tag{2.68}$$

Mathematically, the ordinates of a spectral density function are counterparts of the squares of the amplitudes of a Fourier series. In a Fourier series the frequencies are discrete, and the contribution of each harmonic component to the signal's variance is finite. In a spectral density plot the frequencies are continuous, and given a signal $g(t)$, each component $S_g(n)$ has an infinitesimal contribution to the variance of $g(t)$. Spectral density plots thus have to plots of squares of Fourier series harmonic components a relation similar to the relation of a probability density function to a discrete probability plot.

Spectra in the Inertial Subrange. Measurements performed in the surface layer of the atmosphere confirm the assumption that in horizontally homogeneous, neutrally stratified flow the energy production is approximately balanced by the energy dissipation. It then follows from Eq. (2.62) that the expression for this balance is, approximately,

$$\varepsilon = \frac{\tau_0}{\rho}\frac{dU(z)}{dz} \tag{2.69}$$

where

$$\frac{U(z)}{u_*} = \frac{1}{k}\ln\left(\frac{z}{z_0}\right) \tag{2.38}$$

If Eqs. (2.12), (2.67), and (2.38) are used,

$$\varepsilon = \frac{u_*^3}{kz} \tag{2.70}$$

For the inertial subrange, we substitute Eq. (2.70) in Eq. (2.66). Since $\kappa = 2\pi n/U(z)$, there results

$$\frac{nS_u(n)}{u_*^2} \approx 0.26\,f^{-2/3} \tag{2.71}$$

The left-hand side of Eq. (2.71) and the variable

$$f = \frac{nz}{U(z)} \tag{2.72}$$

are called, respectively, the *reduced spectrum* of the longitudinal velocity fluctuations and, in honor of Kolmogorov's student who developed Eq. (2.72), the Monin similarity coordinate. Equation (2.71) was validated by extensive measurements, for example, [43]. Its dependence on height above ground is significant for structural engineering purposes since spectral ordinates within the inertial subrange typically cause the resonant response of tall structures to wind loads. As is the case for the logarithmic law, for mean wind speeds at 10 m above ground greater than, say, 15 m s^{-1}, it is reasonable to apply Eq. (2.71) throughout the height range of interest to the structural engineer.

Spectra in the Lower-Frequency Range. The lower-frequency range, also called the energy containing range, is defined between $n \approx 0$ and the lower end of the inertial sub-range, n_s. Velocities in this range contribute the bulk of the quasi-static along-wind fluctuating loading on structures. According to theoretical and numerical results reported in [44] and [45], and to measurements reported in [41], for $0 \leq n \leq n_l$, where n_l is small (i.e., in the order of 0.02 Hz or less), the spectral density may be assumed to be constant. In particular, it follows from Eq. (2.59) and (B.25) that

$$S_u(z, 0) = \frac{4\beta u_*^2 L_u^x(z)}{U(z)}. \tag{2.73}$$

For frequencies $n_l \leq n \leq n_s$, $S_u(z, n) = a(z)/n$, where $a(z)$ is determined from the condition that, for $n = n_s$, $S_u(z, n)$ is continuous, that is, satisfies Eq. (2.71).

Expressions for the Spectrum Proposed in the 1960s and 1970s. Kaimal's spectrum has the form [46]:

$$\frac{nS_u(z, n)}{u_*^2} = \frac{105f}{(1 + 33f)^{2/3}} \tag{2.74}$$

where f is the Monin coordinate (Eq. [2.72]).

For open terrain, Eq. (2.74) does not satisfy the widely accepted requirement that the area under the spectral curve should be approximately $6u_*^2$. To satisfy this requirement the coefficients 105 and 33 are replaced in Eq. (2.74) by the coefficients 200 and 50, respectively:

$$\frac{nS_u(z, n)}{u_*^2} = \frac{200f}{(1 + 50f)^{2/3}} \tag{2.75}$$

An expression for the spectrum proposed by Davenport [47] is no longer in use because (i) it does not account for the dependence of the spectrum on height, and (ii) it implies $= 0$. The ASCE 49-12 Standard has adopted the following expression, referred to as the von Kármán spectrum [16, 48, 49]:

$$\frac{nS_u(z, n)}{u_*^2} = \frac{4\beta(nL_u^x(z)/U(z))}{[1 + 70.8(nL_u^x(z)/U(z))^2]^{5/6}} \tag{2.76}$$

Equation (2.76) was developed for aeronautical applications in conjunction with a value $L_u^x = 760$ m [48] at mid to high altitudes. It yields the correct expression for the spectrum at $n = 0$, and reflects correctly the decay of the spectrum as a function of n in the inertial subrange. However, it is universally accepted in the boundary-layer meteorological community that spectral ordinates in that subrange are well represented by Eq. (2.71). For Eq. (2.76) to be consistent with Eq. (2.71) it would be necessary that

$$L_u^x = 0.3\beta^{3/2}z \tag{2.77}$$

According to Eq. (2.77), for open terrain at 10 m above ground ($\beta = 6.0$, see Table 2.6), $L_u^x = 44$ m, whereas according to ASCE 49-12 [16] $L_u^x = 110$ m.

Reference [35, p. 176] states: "We recommend that integral scales be avoided in applications to atmospheric data. Many investigators have computed integral scales from atmospheric data, but the results are badly scattered and cannot be organized."

For this reason it has been proposed to base the estimation of the integral scale L_u^x on the frequency n_{max} for which the curve $nS_u(n)$ is a maximum. "Unfortunately, the curves $nS_u(n)$ tend to be quite flat and sufficiently variable that n_{max} is not well defined" [35]. Reference [50] also warns against the use of this approach, and notes that it likely underestimates L_u^x by a factor of 2 or 3.

Spectral Density $S_u(z, n)$ and Integral Scale [63]. A model of the spectrum $S_u(n)$ was recently developed on the basis of theoretical studies (e.g., [44, 45]) and measurements reported in [41]. Based on [41, figures 6 and 7], the spectral density of the longitudinal velocity fluctuations can be written as

$$S_u(z, z_0, n) = \begin{cases} \dfrac{a(z, z_0)}{n_l} & 0 < n < n_l \\[2ex] \dfrac{a(z, z_0)}{n} & n_l \le n < n_s \\[2ex] 0.26u_*^2\left(\dfrac{z}{U(z, z_0)}\right)^{-2/3} n^{-5/3} & n_s \le n \end{cases} \qquad (2.78a,b,c)$$

Equation (2.78c) was obtained from Eqs. (2.72) and (2.73). Using the notation

$$\frac{n_s z}{U(z)} = f_s \qquad (2.79)$$

where, according to the measurements of [41, figure 8], $f_s \approx 0.125$, For $n = n_s$, Eq. (2.78c) becomes

$$S_u(z, z_0, n_s) = 0.26u_*^2\left(\frac{z}{U(z, z_0)}\right)^{-2/3} n_s^{-5/3} \qquad (2.80)$$

The condition that the functions defined by Eqs. (2.78b) and (2.78c) be continuous at $n = n_s$ then yields

$$a(z) = 0.26u_*^2 f_s^{-2/3} \qquad (2.81)$$

The areas under the spectral curve in the intervals $0 \le n \le n_l$ is $[a(z)/n_l]n_l$. The areas under the spectral curve in the intervals $n_l \le n \le n_s$, and $n \ge n_s$ are, respectively,

$$\int_{n_l}^{n_s} 0.26u_*^2 f_s^{-2/3} \frac{dn}{n} = 0.26u_*^2 f_s^{-2/3} \ln\frac{n_s}{n_l} \qquad (2.82)$$

$$\int_{n_s}^{n_d} 0.26u_*^2\left[\frac{z}{U(z)}\right]^{-2/3} n^{-5/3} dn \approx 0.39 u_*^2 f_s^{-2/3} \qquad (2.83)$$

where n_d is the very large frequency corresponding to the onset of dissipation by molecular friction.

The total area under the spectral curve is $\beta(z_0)u_*^2$. Therefore

$$\beta(z_0)u_*^2 = 0.26u_*^2 f_s^{-2/3} + 0.26u_*^2 f_s^{-2/3} \ln\frac{n_s}{n_l} + 0.39u_*^2 f_s^{-2/3} \qquad (2.84)$$

Equation (2.84) yields

$$\begin{aligned} n_l &= n_s \exp\left[-\frac{\beta(z_0) - 0.26f_s^{-2/3} - 0.39f_s^{-2/3}}{0.26f_s^{-2/3}}\right] \\[2ex] &= f_s \frac{U(z)}{z} \exp\left[-\frac{\beta(z_0) - 0.65f_s^{-2/3}}{0.26f_s^{-2/3}}\right] \end{aligned} \qquad (2.85)$$

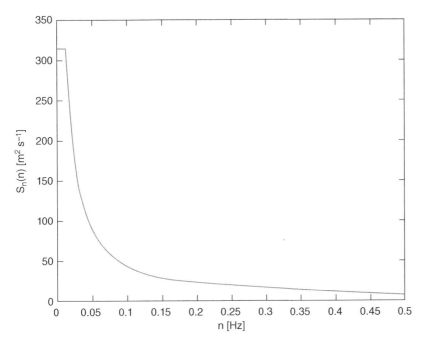

Figure 2.7 Spectral density plot, $z = 10$ m, $U(10\,\text{m}) = 30$ m s^{-1}, $z_0 = 0.03$ m, $f_s = 0.125$.

Equations (2.78a,b,c) are plotted in Figure 2.7 for $z = 10$ m, $U(10\,\text{m}) = 30$ m s^{-1}, $z_0 = 0.03$ m, $f_s = 0.125$.

The integral turbulence scale is

$$L_u^x(z) = \frac{S_u(0, z)U(z)}{4\beta(z_0)u_*^2} \tag{2.86}$$

$$L_u^x(z) = \frac{0.26 f_s^{-5/3}}{4\beta(z_0)} \exp\left[\frac{\beta(z_0) - 0.65 f_s^{-\frac{2}{3}}}{0.26 f_s^{-\frac{2}{3}}}\right] z \tag{2.87}$$

The expression for the integral scale is based on values of β that are well established and on validated models of the spectrum for both the inertial subrange and the low frequency ranges. For sufficiently low values of z, β is assumed to be constant and $L_u^x(z)$ is independent of wind speed.

Example 2.15 Let $z = 10$ m, $U(10\,\text{m}) = 5.39$ m s^{-1}, $z_0 = 0.04$ m. Therefore $\beta \approx 6.0$ and $u_* = 0.39$ m s^{-1}. According to measurements reported in [41], $f_s = 0.125$. Then $n_s = 0.0674$ Hz (Eq. [2.79]), $n_l = 2.56 \times 10^{-3}$ Hz (Eq. [2.85]), $a(10\,\text{m}) = 0.158$ (Eq. [2.81]), $S_u(n_l, 10\,\text{m}) = 61.7$ m^2 s^{-1} $= S_u(n = 0, 10\,\text{m})$ (Eqs. [2.77ab] and [2.77b]). The calculated integral length is $9.11z = 91.1$ m (Eq. [2.86]). The value provided in the ASCE 49-12 Standard is 110 m.

The measurements of [41] have consistently yielded the value $f_s = 0.125$ at all six elevations for which data were obtained. Note, however, that the calculated length is sensitive to the value of the frequency f_s. In Example 2.15, assuming $f_s = 0.1, 0.125$, and, as suggested in [44], $f_s = 0.16$, for $z_0 = 0.04$ m and $\beta = 6$, Eq. (2.87) yields (10 m) = 56.5, 91, and 171 m, respectively. This suggests that the recommendation by Panofsky and Dutton [35] quoted earlier is indeed warranted. In addition, the finding that the curve $nS_u(n)$ is flat in the range $n_l < n < n_s$ confirms the statement in [35] and [50] that the frequency for which that curve attains a maximum yields no useful information on the integral length.

Dependence of L_u^x on wind speed at higher elevations, z. It was noted that, throughout the sublayer within which the parameter β is approximately constant, the integral turbulence length is independent of wind speed. However, this appears to be no longer the case for higher elevations z.

Let the height of the ABL be denoted by H. Since H is proportional to the friction velocity u_* (Eq. [2.37]), for given z the ratio z/H is lower for higher winds, meaning that $\beta(z, z_0)$ decreases with height (Eq. [2.55]). Consider for example, the case $z_0 = 0.04$ m, $z = 55$ m, and $U(z = 55\ \text{m}) = 6.78\ \text{m s}^{-1}$ (as in [41]). The logarithmic law yields $u_* = 0.38\ \text{m s}^{-1}$. If the order of magnitude of the boundary layer depth is $H \approx 0.1 u_*/f$, where f is the Coriolis parameter (see, e.g., Examples 2.7 and 2.8), to $f = 10^{-4}\ \text{s}^{-1}$ there corresponds $H \approx 380$ m. Assuming the validity of Eq. (2.55), $\beta(z, z_0) = 6.0\ \exp(-1.5 \times 55/380) = 4.8$. On the other hand, if $U(z = 55\ \text{m}) = 68\ \text{m s}^{-1}$, $u_* = 3.8\ \text{m s}^{-1}$, $H = 3800$ m, and $\beta(z, z_0) = 6.0\ \exp(-1.5 \times 55/3800) = 5.9$. It follows from Eq. (2.87) with $f_s = 0.125$ that the calculated value of the integral scale is 230 m if $U(z = 55\ \text{m}) = 6.78\ \text{m s}^{-1}$, and 463 m if $U(z = 55\ \text{m}) = 68\ \text{m s}^{-1}$.

This example suggests that estimates of the integral scale at higher elevations depend upon the wind speed at which the measured data were obtained, and that the measurement reports should therefore include that speed.

Spectra of Vertical and Lateral Velocity Fluctuations. According to [51], up to an elevation of about 50 m, the expression for the *vertical* velocity fluctuations, which may be required for the design of some types of bridges, is

$$\frac{nS_w(z,\ n)}{u_*^2} = \frac{33.6\ f}{1 + 10\ f^{5/3}} \tag{2.88}$$

Equation (2.88) can be used for suspended-span bridge design. The expression for the spectrum of the *lateral* turbulent fluctuations proposed in [46] is

$$\frac{nS_v(z,\ n)}{u_*^2} = \frac{15\ f}{(1 + 10\ f)^{5/3}} \tag{2.89}$$

In Eqs. (2.88) and (2.89) the variable f is defined as in Eq. (2.72).

2.4.4 Cross-spectral Density Functions

The *cross-spectral density function* of turbulent fluctuations occurring at two different points in space indicates the extent to which harmonic fluctuation components with frequencies n at those points are in tune with each other or evolve at cross-purposes (i.e., are or are not mutually *coherent*). For components with high frequencies, the distance in space over which wind speed fluctuations are mutually coherent is small. For low-frequency components that distance is relatively large – in the order of integral turbulence scales. An eddy corresponding to a component with frequency n is said to envelop a structure if the distance over which the fluctuations with frequency n are relatively coherent is comparable to the relevant dimension of the structure.

The expression for the cross-spectral density of two signals u_1 and u_2 is

$$S_{u_1 u_2}^{cr}(r, n) = S_{u_1 u_2}^{C}(r, n) + i S_{u_1 u_2}^{Q}(r, n) \tag{2.90}$$

in which $i = \sqrt{-1}$, r is the distance between the points M_1 and M_2 at which the signals occur, and the subscripts C and Q identify the co-spectrum and the quadrature spectrum of the two signals, respectively. The coherence function is defined as

$$Coh(r, n) = c^2_{u_1 u_2}(r, n) + q^2_{u_1 u_2}(r, n) \tag{2.91}$$

where

$$c^2_{u_1 u_2}(r, n) = \frac{[S^C_{u_1 u_2}(r, n)]^2}{S(z_1, n)S(z_2, n)}, \qquad q^2_{u_1 u_2}(r, n) = \frac{[S^Q_{u_1 u_2}(r, n)]^2}{S(z_1, n)S(z_2, n)} \tag{2.92a,b}$$

In Eqs. (2.91) and (2.92a,b), $S(z_1, n)$, and $S(z_2, n)$ are the spectra of the signals at points M_1 and M_2. To larger integral turbulence scales there correspond increased values of the coherence.

For ABL applications it is typically assumed that the quadrature spectrum is negligible. The following expression for the cospectrum is used in applications:

$$S^C_{u_1 u_2}(r, n) = S^{1/2}(z_1, n)S^{1/2}(z_2, n) \exp(-\hat{f}) \tag{2.93}$$

where

$$\hat{f} = \frac{n[C^2_z(z^2_1 - z^2_2) + C^2_y(z^2_1 - z^2_2)]^2}{\frac{1}{2}[U(z_1) + U(z_2)]} \tag{2.94}$$

y_i, z_i are the coordinates of point M_i ($i = 1, 2$), and according to wind tunnel measurements the values of the exponential decay coefficients may be assumed to be, very approximately, $C_z \approx 10$, $C_y \approx 16$ [52]. Eqs. (2.93) and (2.94) reflect the intuitively obvious fact that the cross-spectrum decreases as (i) the frequency n increases (since, for given distance between the points M_1 and M_2, the mutual coherence is lower for small eddies than for larger eddies), and/or (ii) the distance between the points increases. For *lateral* fluctuations the expression for the cospectrum is similar, except that values $C_z \approx 7$, $C_y \approx 11$ have been proposed [53]. For two points with the same elevation, the expression for the co-spectrum of the *vertical* fluctuations is also assumed to be similar, with $C_y \approx 8$ [53]. The exponential decay coefficients are in fact dependent upon surface roughness and upon wind speed; these dependences are typically not accounted for in practice.

2.5 Horizontally Non-Homogeneous Flows

Horizontal non-homogeneities of atmospheric flows are due either to conditions at the Earth's surface (e.g., changes in surface roughness, topographic features) or to the meteorological nature of the flow (as in the case of tropical cyclones, thunderstorms or downbursts). While the structure of horizontally homogeneous flows is basically well understood, the modeling of horizontally non-homogeneous flows is to a large extent still incomplete or tentative. Computational Fluid Dynamics methods are increasingly being used for a variety of surface roughness and topographic configurations. This section contains information of interest for structural engineering purposes.

2.5.1 Flow Near a Change in Surface Roughness. Fetch and Terrain Exposure

Sites with uniform surface roughness are limited in size; the flows near their boundaries are therefore affected by the surface roughness of adjoining sites. Therefore, the surface roughness is not the sole factor that determines the wind profile at a site. The profile also depends upon the distance (the *fetch*) over which that surface roughness prevails upwind of the site. The terminology used in the ASCE 7 Standard therefore distinguishes between *surface roughness* and *exposure*. For example, a site is defined as having Exposure B if it has surface roughness B *and* surface roughness B prevails over a sufficiently long fetch; for design purposes the wind profile at a site with Exposure B may be described by the power law with parameters corresponding to surface roughness B. Sections 2.3 and 2.4 consider only the case of long fetch. The ASCE 7 Standard provides criteria on the fetch required to assume a given exposure.

Useful information on the flow in transition zones can be obtained by considering the simple case of an abrupt roughness change along a line perpendicular to the direction of the mean flow. Upwind of the discontinuity the flow is horizontally homogeneous and, near the ground, is governed by the parameter z_{01}. Downwind of the discontinuity the flow will be affected by the surface roughness z_{02} over a height $h(x)$, where x is the downwind distance from the discontinuity. This height, known as the height of the *internal boundary layer*, increases with x until the entire flow adjusts to the roughness length z_{02}. A well-accepted model of the internal boundary layer, which holds for both smooth-to-rough and rough-to smooth transitions, is

$$h(x) \approx 0.28 \, z_{0r} \left(\frac{x}{z_{0r}} \right)^{0.8} \tag{2.95}$$

[53], where z_{0r} is the largest of the roughness lengths z_{01} and z_{02}. The validity of Eq. (2.95) is limited to $h < 0.2 H$, where H is the ABL height for very large x. Within the internal boundary layer the flow adjusts to the new surface roughness as shown in Figure 2.8.

Example 2.16 Consider a zone with roughness length $z_{02} = 0.30$ m downwind of a zone with roughness length $z_{01} = 0.03$ m. The estimated height of the internal boundary layer at a distance $x = 10\,000$ m downwind of the line of separation between the two zones is $h(10\,000\,\text{m}) \approx 350$ m. The same result is valid if the zone with roughness length $z_{01} = 0.03$ m is downwind of the zone with $z_{02} = 0.30$ m.

2.5.2 Wind Profiles over Escarpments

Topographic features alter the local wind environment and create wind speed increases (*speed-up* effects), since more air has to flow through an area decreased, with respect to the case of flat land, by the presence of the topographical feature. The procedure that follows is specified in the ASCE 7-16 Standard [3] for the calculation of speed-up effects on 2- or 3-D (two- or three-dimensional) isolated hills and 2-D ridges and escarpments.

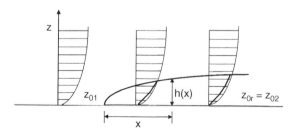

Figure 2.8 Internal boundary layer $h(x)$. Mean wind speed profile within the internal boundary layer is adjusted to the terrain roughness $z_{02} > z_{01}$.

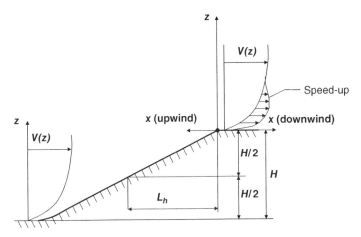

Figure 2.9 Two-dimensional escarpment.

The increase in the wind speeds due to the topography is reflected in the exposure-dependent factor K_{zt}. The Standard provides speed-up models applicable to 2-D ridges, 3-D isolated hills, and 2-D escarpments, provided that all the following conditions are satisfied (see Figure 2.9 for notations):

1) No topographic features of comparable height exist for a horizontal distance of 100 times the height of the hill H or 3.2 km, whichever is less, from the point at which the height H is determined.
2) The topographic feature protrudes above the height of upwind terrain features within a 3.2 km radius by a factor of two or more in any quadrant.
3) The structure is located in the upper half of a hill or ridge or near the crest of an escarpment.
4) $H/L_h \geq 0.2$.
5) The height of the hill H exceeds 5.25 m for Exposures C and D, and 21 m for Exposure B.

If any of the conditions 1–5 above is not satisfied, $K_{zt} = 1$.

The topographic factor is defined as $K_{zt} = [V(z,x)/V(z)]^2$, where $V(z) = 3$-second peak gust speed at height z above ground in horizontal terrain with no topographic feature. The expression for K_{zt} is:

$$K_{zt} = (1 + K_1 K_2 K_3)^2 \tag{2.96}$$

where the factor K_1 accounts for the shape of the topographic feature, K_2 accounts for the variation of the speed-up as a function of distance from the crest, and K_3 accounts for the variation of the speed-up as a function of height above the surface of the topographic feature. Values of and expressions for K_1, K_2, K_3 are given in ASCE 7-16. For example, for $H/L_h \leq 0.5$,

$$K_1 = \frac{aH}{L_h}, \qquad K_2 = 1 - \frac{|x|}{\mu L_h}, \qquad K_3 = \exp\left(-\frac{\gamma z}{L_h}\right) \tag{2.97a,b,c}$$

where for 2-D escarpments, $\gamma = 2.5$; $\mu = 1.5$ (upwind of crest), $\mu = 4.0$ (downwind of crest); $a = 0.75$ (Exposure B), $a = 0.85$ (Exposure C), and $a = 0.95$ (Exposure D).

Example 2.17 *Topographic factor for a 2-D escarpment.* The escarpment is assumed to have Exposure B and dimensions $H = 30.5$ m, $L_h = 122$ m. The topography upwind of the escarpment is assumed to satisfy conditions 1 and 2. The building is located at the top of the escarpment, and the downwind distance (see Figure 2.9) between the crest and the building's windward face is $x = 12.2$ m. (In Figure 2.9 the building would be located to the right of the crest.) We seek the quantity K_{zt} for elevation $z = 7.6$ m above ground at $x = 12.2$ m.

Condition 4 is satisfied, since $H/L_h = 30.5/122 = 0.25 > 0.2$, as is condition 5, since $H = 30.5$ m > 21 m.

Since $H/L_h < 0.5$, Eqs. (2.97a,b,c) yield:

$$K_1 = 0.75 \times 30.5/122 = 0.1875,$$
$$K_2 = 1 - 12.2/(4.0 \times 122) = 0.975,$$
$$K_3 = \exp.(-2.5 \times 7.6/122 = 0.855.$$

The topographic factor is

$$K_{zt} = (1 + 0.1875 \times 0.975 \times 0.855)^2 = 1.16^2 = 1.35.$$

This result implies that at $x = 12.2$ m downwind of the crest and $z = 7.6$ m above ground, the increased peak 3-second gust is 1.16 times larger than the peak 3-second gust at 7.6 m above ground upwind of the escarpment, and the corresponding pressures are $(1.16)^2 = 1.35$ times larger than upwind of the escarpment.

2.5.3 Hurricane and Thunderstorm Winds

In current structural engineering practice it is assumed that flow models used for synoptic storms are acceptable for hurricanes and thunderstorms as well. Although they are not yet sufficient for codification purposes, a number of research results on these two types of storm have been obtained in recent years, of which the most significant are briefly summarized or cited herein.

Hurricanes. Geophysical Positioning System (GPS) dropwindsonde (or dropsonde) measurements of hurricane wind speed profiles yielded the following results: (i) On average, in the storm's outer vortex, wind speeds increase monotonically up to an elevation of about 1 km, where they attain about 1.4 times their strength at 10 m; they then decrease monotonically between 1 and 3 km, where they attain about 1.3 times their strength at 10 m. (ii) On average, in the storm's eyewall, wind speeds increase monotonically up to an elevation of about 400 m, where they attain about 1.3 times their value at 10 m, after which they decrease monotonically between 400 and 3 km, where they attain about 1.1 times their value at 10 m [54].

The turbulence intensity in hurricane winds was found to be larger by about 10% in hurricanes than in synoptic storms [37, 38, 55]. Values of the longitudinal integral turbulence scale L_u^x measured at 10 m elevation in hurricane Bonnie varied from 40 to 370 m [37]. Table 2.9 [37] lists measured values of L_u^x based on 10- and 60-min long records at 5 and 10 m above ground, as well as values specified in the ASCE 49-12 Standard [16].

As expected, L_u^x decreases as the roughness length increases; it increases, in most cases modestly, as the height z increases from 5 to 10 m. It is seen in Table 2.9 that the

Table 2.9 Longitudinal integral length scales at 5 and 10 m elevations (m).

Hurricane	z_{0min}	z_{0max}	z	Record (length) [54]		Eq. (2.60)	ASCE 49-12 [16]	Eq. (2.87)[a]
				10 min.	60 min.			
Isidore	0.0011	0.0060	5	98	310	210–400		
			10	140	450	220–420	190	150
Gordon	0.0002	0.0014	10	176	365	370–450	190	165
Ivan	0.0080	0.0551	5	126	197	120–180		
			10	154	240	140–190	110	100
Ivan	0.0116	0.0497	5	105	314	120–130		
			10	123	366	130–140	110	100
Lili	0.0082	0.0589	5	82	189	90–180		
			10	94	226	110–190	110	100

a) Values obtained by using Eq. (2.87) were multiplied by 1.1 to account for the fact that fluctuations are stronger in hurricanes than in extratropical storms.

ASCE 49-12 Standard [16] values are considerably smaller than the reported 60-min measurements. It may be assumed that measurements of integral length scales are affected by significant uncertainties, as was noted also in Section 2.4.2.

A hurricane wind speed record that clearly reflects the passage of the eye is shown in Figure 2.10. The record was obtained at 15 m above ground by an ultrasonic anemometer unit with a wind speed range of $0–65$ m s^{-1} with a resolution of 0.01 m s^{-1}, capable of measuring instantaneous u, v, and w wind velocity components with a maximum sampling rate of 32 Hz. The traces shown are 10-minute and 3-second moving averages of data with a 10 Hz sampling rate. Note its non-stationary character, which contrasts with the stationarity of Figure 2.1.

Thunderstorms. The cold air downdraft that, in a thunderstorm, spreads horizontally over the ground, can be compared to a wall jet. Just as in a wall jet, the surface friction retards the spreading flow.

Of particular interest is the *first gust* (or *gust front*) (Figures 1.14 and 2.11), that is, the thunderstorm wind that can exhibit a considerable and relatively rapid change of speed and direction. The wind speed increase and the time interval during which it occurs have been called by some authors the *gust size* ΔV and the *gust length* Δt, respectively [55]. Depending upon thunderstorm intensity, the gust size may vary approximately from 3 to 30 m s^{-1}, while the gust length may range from approximately less than $1–10$ min.

According to numerical and laboratory simulations [56–58], as well as full-scale measurements [59], near the ground the wind speed profiles along a thunderstorm gust front can be quite different from a log-law profile. However, in current design practice it is assumed that thunderstorm characteristics may for practical purposes be assumed to be the same as those of large-scale storms. This assumption may be warranted, given that, according to [60], the maximum winds (i.e., design level winds) within the thunderstorm are rarely due to storms in which significant deviations from the log law occur. Definitive statements on the micrometeorological and statistical characterization of thunderstorms appear to be unwarranted at this time owing to the lack of sufficient full-scale high-speed data.

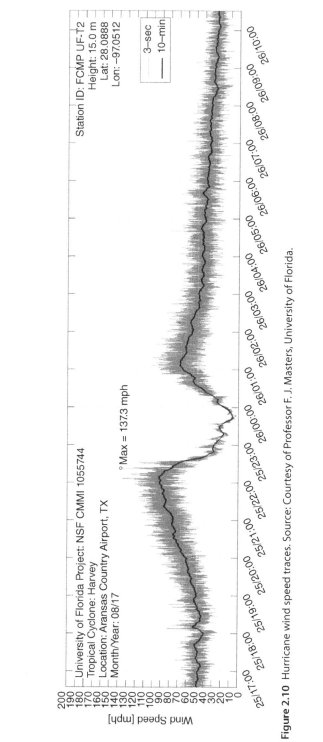

Figure 2.10 Hurricane wind speed traces. Source: Courtesy of Professor F. J. Masters, University of Florida.

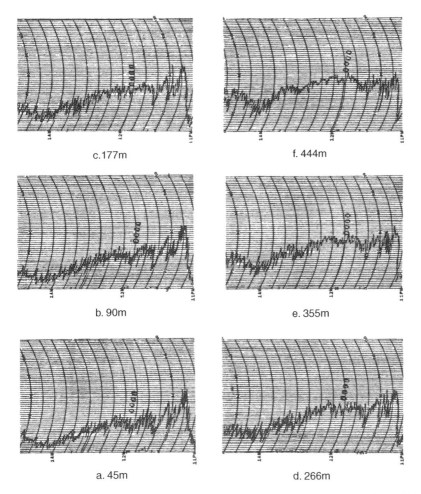

Figure 2.11 Thunderstorm wind speed records at six elevations above ground near Oklahoma City.
Source: Courtesy of National Severe Storms Laboratory, National Oceanic and Atmospheric Laboratory.

References

1 Schlichting, H. (1987). *Boundary Layer Theory*, 7th ed. New York: McGraw-Hill.

2 Csanady, G.T. (1967). On the 'resistance law' of a turbulent Ekman layer. *Journal of the Atmospheric Sciences* 24: 467–471.

3 ASCE, "Minimum design loads for buildings and other structures (ASCE/SEI 7-16)," in *ASCE Standard ASCE/SEI 7-16*, Reston, VA: American Society of Civil Engineers, 2016.

4 Tennekes, H. and Lumley, J.L. (1972). *A First Course in Turbulence*. Cambridge: MIT Press.

5 Blackadar, A.K. and Tennekes, H. (1968). Asymptotic similarity in neutral Barotropic planetary boundary layers. *Journal of the Atmospheric Sciences* 25: 1015–1020.

6 Tennekes, H. (1973). The logarithmic wind profile. *Journal of the Atmospheric Sciences* 30: 234–238.

7 Zilitinkevich, S.S. and Esau, I.N. (2002). On integral measures of the neutral barotropic planetary boundary layer. *Boundary-Layer Meteorology* 104: 371–379.

8 Zilitinkevich, S., Esau, I., and Baklanov, A. (2007). Further comments on the equilibrium height of neutral and stable planetary boundary layers. *Quarterly Journal of the Royal Meteorological Society* 133: 265–271.

9 Zilitinkevich, S. (2012). The height of the atmospheric planetary boundary layer: state of the art and new development. In: *National Security and Human Health Implications of Climate Change* (ed. H.J.S. Fernando, Z. Klaić and J.L. McCulley), 147–161. Netherlands: Springer.

10 Stull, R., *Practical meteorology: an algebra-based survey of atmospheric science*. University of British Columbia, 2015.

11 Lettau, H. (1962). Theoretical wind spirals in the boundary layer of a barotropic atmosphere. *Beitraege zur Physik der Atmosphaere* 35: 195–212.

12 Kung, E.C. (1966). Large-scale balance of kinetic energy in the atmosphere. *Monthly Weather Review* 94: 627–640.

13 Hess, G.D. and Garratt, J.R. (2002). Evaluating models of the neutral, barotropic planetary boundary layer using integral measures: part I. overview. *Boundary-Layer Meteorology* 104: 333–358.

14 Hess, G.D. (2004). The neutral, barotropic planetary boundary layer, capped by a low-level inversion. *Boundary-Layer Meteorology* 110: 319–355.

15 Simiu, E., Shi, L., and Yeo, D. (2016). Planetary boundary-layer modelling and tall building design. *Boundary-Layer Meteorology* 159: 173–181. https://www.nist.gov/wind.

16 ASCE, "Wind tunnel testing for buildings and other structures (ASCE/SEI 49-12)," in *ASCE Standard ASCE/SEI 49-12*, Reston, VA: American Society of Civil Engineers, 2012.

17 Chamberlain, A.C. (1983). Roughness length of sea, sand, and snow. *Boundary-Layer Meteorology* 25: 405–409.

18 Oliver, H.R. (1971). Wind profiles in and above a forest canopy. *Quarterly Journal of the Royal Meteorological Society* 97: 548–553.

19 Biétry, J., Sacré, C., and Simiu, E. (1968). Mean wind profiles and changes of terrain roughness. *Journal of the Structural Division, ASCE* 104: 1585–1593.

20 Powell, M.D., Vickery, P.J., and Reinhold, T.A. (2003). Reduced drag coefficient for high wind speeds in tropical cyclones. *Nature* 422: 279–283.

21 CEN, "Eurocode 1: Actions on structures – Parts 1–4: General actions – Wind actions," in *EN 1991-1-4*: European Committee for Standardization (CEN), 2005.

22 Smith, S.D. and Banke, E.G. (1975). Variation of the sea surface drag coefficient with wind speed. *Quarterly Journal of the Royal Meteorological Society* 101: 665–673.

23 Wu, J. (1969). Wind stress and surface roughness at air-sea interface. *Journal of Geophysical Research* 74: 444–455.

24 Amorocho, J. and DeVries, J.J. (1980). A new evaluation of the wind stress coefficient over water surfaces. *Journal of Geophysical Research: Oceans* 85: 433–442.

25 Garratt, J.R. (1977). Review of drag coefficients over oceans and continents. *Monthly Weather Review* 105: 915–929.

26 Smith, S.D. (1980). Wind stress and heat flux over the ocean in gale force winds. *Journal of Physical Oceanography* 10: 709–726.

27 Krügermeyer, L., Grünewald, M., and Dunckel, M. (1978). The influence of sea waves on the wind profile. *Boundary-Layer Meteorology* 14: 403–414.

28 Lettau, H. (1969). Note on aerodynamic roughness-parameter estimation on the basis of roughness-element description. *Journal of Applied Meteorology* 8: 828–832.

29 Masters, F.J., Vickery, P.J., Bacon, P., and Rappaport, E.N. (2010). Toward objective, standardized intensity estimates from surface wind speed observations. *Bulletin of the American Meteorological Society* 91: 1665–1681.

30 Helliwell, N. C., "Wind over London," in *Third International Conference on Wind Effects on Buildings and Structures*, Tokyo, 1971, pp. 23–32.

31 Hellmann, G., *Über die Bewegung der Luft in den untersten Schichten der Atmosphäre*: Königlich Preussischen Akademie der Wissenschaften, 1917.

32 Pagon, W.W. (1935). Wind velocity in relation to height above ground. *Engineering News-Record* 114: 742–745.

33 NRCC, "National Building Code of Canada," Ottawa, Ontario, Canada: Institute for Research in Construction, National Research Council of Canada, 2010.

34 AIJ, "AIJ Recommendations for loads on buildings (Chapter 6: Wind loads)," Architectural Institute of Japan, 2004.

35 Panofsky, H.A. and Dutton, J.A. (1984). *Atmospheric Turbulence: Models and Methods for Engineering Applications*, 1e. New York: Wiley-Interscience.

36 Durst, C.S. (1960). Wind speed over short periods of time. *Meteorological Magazine* 89: 181–187.

37 Schroeder, J.L. and Smith, D.A. (2003). Hurricane Bonnie wind flow characteristics as determined from WEMITE. *Journal of Wind Engineering and Industrial Aerodynamics* 91: 767–789.

38 Krayer, W.R. and Marshall, R.D. (1992). Gust factors applied to hurricane winds. *Bulletin of the American Meteorological Society* 73: 613–618.

39 Counihan, J. (1975). Adiabatic atmospheric boundary layers: a review and analysis of data from the period 1880–1972. *Atmospheric Environment* 9: 871–905.

40 Oh, S. and Ishihara, T., "A modified von Karman model for the spectra and the spatial correlations of the offshore wind field," presented at *Offshore 2015, Copenhagen, Denmark*, 2015.

41 Drobinski, P., Carlotti, P., Newsom, R.K. et al. (2004). The structure of the near-neutral atmospheric surface layer. *Journal of the Atmospheric Sciences* 61: 699–714.

42 Hinze, J.O. (1975). *Turbulence*, 2nd ed. New York: McGraw-Hill.

43 Fichtl, G.H. and McVehil, G.E. (1970). Longitudinal and lateral spectra of turbulence in the atmospheric boundary layer at the Kennedy Space Center. *Journal of Applied Meteorology* 9: 51–63.

44 Hunt, J.C.R. and Carlotti, P. (2001). Statistical structure at the wall of the high Reynolds number turbulent boundary layer. *Flow, Turbulence and Combustion* 66: 453–475.

45 Carlotti, P. (2002). Two-point properties of atmospheric turbulence very close to the ground: comparison of a high resolution les with theoretical models. *Boundary-Layer Meteorology* 104: 381–410.

46 Kaimal, J.C., Wyngaard, J.C., Izumi, Y., and Coté, O.R. (1972). Spectral characteristics of surface-layer turbulence. *Quarterly Journal of the Royal Meteorological Society* 98: 563–589.

47 Davenport, A.G. (1961). The spectrum of horizontal gustiness near the ground in high winds. *Quarterly Journal of the Royal Meteorological Society* 87: 194–211.

48 DOD, "Department of Defense Interface Standard: Flying Qualities of Piloted Aircraft (MIL-STD-1797A)," Department of Defense, 2004.

49 Lumley, J.L. and Panofsky, H.A. (1964). *The Structure of Atmospheric Turbulence*. New York: Interscience Publishers.

50 Pasquill, F. and Butler, H.E. (1964). A note on determining the scale of turbulence. *Quarterly Journal of the Royal Meteorological Society* 90: 79–84.

51 Vickery, B. J., "On the reliability of gust loading factors," in *Technical Meeting Concerning Wind Loads on Buildings and Structures*, Washington, DC, 1970.

52 Kristensen, L. and Jensen, N.O. (1979). Lateral coherence in isotropic turbulence and in the natural wind. *Boundary-Layer Meteorology* 17: 353–373.

53 Wood, D.H. (1982). Internal boundary layer growth following a step change in surface roughness. *Boundary-Layer Meteorology* 22: 241–244.

54 Yu, B., Chowdhury, A.G., and Masters, F. (2008). Hurricane wind power spectra, cospectra, and integral length scales. *Boundary-Layer Meteorology* 129: 411–430.

55 Sinclair, R. W., Anthes, R. A., and Panofsky, H. A., "Variation of the low level winds during the passage of a thunderstorm gust front," NASA-CR-2289 NASA, Washington, DC, 1973.

56 Chay, M.T. and Letchford, C.W. (2002). Pressure distributions on a cube in a simulated thunderstorm downburst – Part A: stationary downburst observations. *Journal of Wind Engineering and Industrial Aerodynamics* 90: 711–732.

57 Letchford, C.W. and Chay, M.T. (2002). Pressure distributions on a cube in a simulated thunderstorm downburst—Part B: moving downburst observations. *Journal of Wind Engineering and Industrial Aerodynamics* 90: 733–753.

58 Jubayer, C., Elatar, H., and Hangan, H., "Pressure distributions on a low-rise building in a laboratory simulated downburst," presented at the 8th International Colloquium on Bluff Body Aerodynamics and Applications, Boston, 2016.

59 Lombardo, F.T., Smith, D.A., Schroeder, J.L., and Mehta, K.C. (2014). *Journal of Wind Engineering and Industrial Aerodynamics* 125: 121–132. doi: 10.1016/j.jweia.2013.12.004.

60 Schroeder, J. L. Personal communication, November, 2016.

61 Shiotani, M., *Structure of Gusts in High Winds, Parts 1–4*. Namashino, Funabashi, Chiba, Japan: Physical Science Laboratory, Nihon University, 1967–1971.

62 Simiu, E., Heckert, N.A., and Yeo, D. (2017). "Planetary Boundary Layer Modeling and Standard Provisions for Supertall Building Design," *Journal of Structural Engineering*, 143, 06017002. https://www.nist.gov/wind.

63 Simiu, E., Potra, F.A. and Nandi, T.N. (2018) "Determining longitudinal integral turbulence scales in the near-neutral atmospheric surface layer," *Boundary-layer Meteorology*. doi: 10.1007/s10546-018-0400-4. https://www.nist.gov/wind.

3

Extreme Wind Speeds

Structures are designed to be safe and serviceable, meaning that their probabilities of exceeding specified strength and serviceability limit states must be acceptably small. These probabilities are functions of the wind speeds to which the structures are exposed. The present chapter is concerned with the probabilistic estimation of extreme wind speeds. Uncertainties in such estimates are discussed in Chapter 7. Materials that complement this chapter are provided in Appendices A and C.

Section 3.1 provides simple, intuitive definitions of exceedance probabilities and mean recurrence intervals (MRIs), and extends those definitions to wind speeds in mixed wind climates (e.g., climates with both hurricane and non-hurricane winds, or with large-scale extratropical storm and thunderstorm winds). Section 3.2 defines non-directional and directional wind speed data in non-hurricane and hurricane-prone regions, and reviews main sources of such data for the conterminous United States. Section 3.3 describes and illustrates methods for estimating extreme wind speeds with specified MRIs. Section 3.4 is devoted to tornado climatology.

3.1 Cumulative Distributions, Exceedance Probabilities, Mean Recurrence Intervals

Section 3.1.1 introduces these topics intuitively by using the example of a fair die, and shows its relevance to the probabilistic characterization of extreme wind speeds. Section 3.1.2 considers the case of mixed wind climates, in regions with, for example, hurricane winds and significant non-hurricane winds, or large-scale extratropical storm and thunderstorm winds.

3.1.1 Probability of Exceedance and Mean Recurrence Intervals

3.1.1.1 A Case Study: The Fair Die

We denote the outcome of throwing a fair die once by O. The probability, denoted by $P(O \leq n)$ ($n = 1, 2, \ldots, 6$), that the outcome (i.e., the event) O is less than or equal to n is called the *cumulative distribution function* (CDF) of the event O. The CDF of the outcome $O \leq n$ is $P(O \leq n) = n/6$. The *probability of exceedance* of the outcome n is $P(O > n) = 1 - P(O \leq n) = 1 - n/6$. The MRI of the event $O > n$ is defined as the inverse of the probability of exceedance of that event, and is the average number of trials (throws) required for $O > n$. Therefore MRI $(O > n) = 1/(1 - n/6)$. The MRI is also called the *mean return period* (see also Section A.5.1).

Example 3.1 *Mean recurrence interval of the outcome of throwing a die*. For a fair die the probability of exceedance $P(O > 5) = 1 - P(O \le 5) = 1 - 5/6 = 1/6$. The MRI of the event $O > 5$ is $1/(1/6) = 6$ trials, that is, the outcome "six" occurs, on average, once in six trials.

The probability of exceedance of an outcome n increases as the number of trials increases. If the probability of non-exceedance of the outcome n in one trial is $P(O \le n)$, owing to the independence of the outcomes (Section A.2.5), the probability of non-exceedance of the outcome n in m trials is $[P(O \le n)]^m$. The probability of exceedance of the outcome n in m trials is $1 - [P(O \le n)]^m$. For example, the probability of non-exceedance of the outcome "five" in two throws of a die is $(5/6)^2 = 25/36$, and the probability of exceedance of that outcome is $1 - 25/36 = 11/36$.

3.1.1.2 Extension to Extreme Wind Speeds

Conceptually, the difference between the statement "the outcome of throwing a die once exceeds n" and the statement "the largest wind speed V occurring in any one year exceeds v," is that the CDF of the largest speed in a year, $P(V \le v)$, is continuous, whereas $P(O \le n)$ is discrete. For any given n, $P(O \le n)$ is the same for any one trial (throw of a die), and is independent of the outcomes of other trials. Similarly, except for, say, possible global warming effects, $P(V \le v)$ is the same for any one trial (any one year), and is independent of speeds occurring in other years.

The speed v with an $\overline{N}$-year MRI is called the $\overline{N}$-*year speed*. The MRI, in years, is

$$\overline{N}(v) = \frac{1}{1 - P(V \le v)} \tag{3.1}$$

Example 3.2 *Probability of exceedance of the largest wind speed in a given data sample.* Consider the sample of size nine of the largest measured yearly wind speeds 20, 18, 21, **25**, 17, 24, 22, 20, 15 (in m s^{-1}; the largest speed in the sample is shown in bold type). There are $n = 9$ outcomes for which $V \le 25$ mph, out of $n + 1 = 10$ possible outcomes (the 10th outcome being $V > 25$ m s^{-1}). Hence the estimated probability $P(V \le 25 \text{ m s}^{-1}) = 9/10 = 0.9$. The probability of exceedance of a 25 m s^{-1} largest yearly speed is $1 - 0.9 = 0.1$. The MRI of the event that the 25 m s^{-1} wind speed is exceeded in any one year is $1/0.1 = 10$ years. The probability of the event $V \le 25$ m s^{-1} in 30 years is equal to the probability that $V \le 25$ m s^{-1} in the first year, *and* in the second year, *and* in the 30th year, that is $0.9^{30} = 0.04$. The probability that $V > 25$ m s^{-1} in 30 years is then $1 - 0.04 = 0.96$.

3.1.2 Mixed Wind Climates

We now consider wind speeds in regions exposed to both non-hurricane and hurricane winds. We are interested in the probability that, in any one year, wind speeds regardless of their meteorological nature are less than or equal to a specified speed, v.

Let the random variables V_H and V_{NH} denote, respectively, the largest hurricane wind speed and the largest non-hurricane wind speed in any one year. Further, let the probability that $V_H \leq v$ and the probability that $V_{NH} \leq v$ be denoted, respectively, by $P(V_H \leq v)$ and $P(V_{NH} \leq v)$. The random variable of interest is the maximum yearly speed regardless of whether it is a hurricane or a non-hurricane wind speed, and is denoted by $\max(V_H, V_{NH})$. The statement "$\max(V_H, V_{NH}) \leq v$" and the statement "$V_H \leq v$ and $V_{NH} \leq v$" are equivalent. Therefore, $P[\max(V_H, V_{NH}) \leq v] = P(V_H \leq v$ and $V_{NH} \leq v)$. If it assumed that V_H and V_{NH} are independent random variables, it follows (see Section A.2.5) that

$$P[\max(V_H, V_{NH}) \leq v] = P(V_H \leq v)P(V_{NH} \leq v) \tag{3.2}$$

The probability distributions $P(V_{NH} \leq v)$ and $P(V_H \leq v)$ can be obtained as shown in Section 3.1.1. With an appropriate change of notation, Eq. (3.2) is also applicable to non-thunderstorm and thunderstorm wind speeds.

The probability of occurrence of the event $V_H > v$ or $V_{NH} > v$ is (Section A.2.1):

$$P(V_H > v \text{ or } V_{NH} > v) = P(V_H > v) + P(V_{NH} > v)$$
$$= 1 - P(V_H \leq v)P(V_{NH} \leq v) \tag{3.3a,b}$$

Example 3.3 *Mean recurrence interval of the event $V_H > v$ and $V_{NH} > v$.* Assume that the MRI of the event that non-hurricane wind speeds exceed 45 m s^{-1} is $\overline{N}_{NH} = 120$ years, and that the MRI of the event that hurricane wind speeds exceed 45 m s^{-1} is $\overline{N}_H = 50$ years. The respective CDFs are $P(V_{NH} \leq 45 \text{ m s}^{-1}) = 1 - 1/\overline{N}_{NH} = 0.99167$, and $P(V_H \leq 45 \text{ m s}^{-1}) = 1 - 1/\overline{N}_H = 0.98$. By Eq. (3.2) the CDF of the 45 m s^{-1} wind speed due to non-hurricane and hurricane winds is $P(V_H \leq 45$ and $V_{NH} \leq 45 \text{ m s}^{-1}) = P(V_H \leq 45 \text{ m s}^{-1}) P(V_{NH} \leq 45 \text{ m s}^{-1}) = 0.99167 \times 0.98 = 0.972$. By Eq. (3.1) the MRI of the 45 m s^{-1} wind speed at the site is $1/(1 - 0.972) = 35.7$ years.

Example 3.4 *Probability of occurrence of the event $V_H > v$ or $V_{NH} > v$.* Assuming again $\overline{N}_{NH} = 120$ years, $\overline{N}_H = 50$ years, Eq. (3.3a,b) yields $P(V_H > v \text{ or } V_{NH} > v) = P(V_H > v) + P(V_{NH} > v) = (1 - 0.98) + (1 - 0.99167) = 0.028/$year (Eq. A.1).

3.2 Wind Speed Data

3.2.1 Meteorological and Micrometeorological Homogeneity of the Data

Extreme wind speed distributions differ depending upon the meteorological nature of the storms being considered. For this reason, hurricane, synoptic storm, and thunderstorm data should be analyzed separately. In addition, wind speed data within a data sample must be micrometeorologically homogeneous, meaning that all the data in a set must correspond to the same (i) height above the surface, (ii) surface exposure (e.g., open terrain), and (iii) averaging time (e.g., 3 s for peak wind gust speeds, 1 min, 10 min, or 1 h). Wind speeds at 10 m above terrain with open exposure, and with the specified averaging time (typically 3 seconds in the United States) are referred to as *standardized wind speeds*. If data do not satisfy the micrometeorological homogeneity requirement, they have to be transformed so that the requirement is satisfied (see Sections 2.3.4–2.3.7,

Section 2.4.1, and Ref. [1], which show that as far as the surface exposure is concerned, this task can be far from trivial).

3.2.2 Directional and Non-Directional Wind Speeds

Standard provisions for wind loads are based primarily on the use of *non-directional* extreme wind speeds, that is, largest wind speeds in any one year or storm event, regardless of their direction. *Directional* extreme wind speeds, that is, largest wind speeds in any one year or storm event for each of the directional sectors being considered, are used to estimate wind effects on special structures at sites for which aerodynamic data are available for a sufficient number of wind directions.

Denote the directional wind speeds by U_{ij} (e.g., $i = 1, 2; j = 1, 2, ..., 8$), where the subscript i indicates the year or the storm event, and the subscript j indicates the wind direction. For fixed i the corresponding non-directional wind speed is $U_i = \max_j(U_{ij})$.

Example 3.5 *Directional and non-directional wind speeds.* To illustrate the definitions of directional and non-directional wind speeds we consider the following largest peak 3-second gusts in m s^{-1} recorded in two consecutive 1-year periods:[1]

			Directional speed U_{ij}						Non-directional speed $\max_j(U_{ij})$
j	1 (NE)	2 (E)	3 (SE)	4 (S)	5 (SW)	6 (W)	7 (NW)	8 (N)	
$i = 1$	45	**50**	41	48	43	44	47	39	50
$i = 2$	39	47	43	**54**	40	42	36	38	54

The non-directional speeds are also shown (in bold type) in the list of directional speeds.

3.2.3 Wind Speed Data Sets

3.2.3.1 Data in the Public Domain

Peak Directional Gust Speeds at 10 m Above Open Terrain (Standardized Wind Speeds). Standardized peak gust speeds averaged over five seconds extracted from Automated Surface Observing Systems (ASOS) records and transformed to correspond to a 10 m elevation over terrain with open exposure are listed on the site https://www.nist.gov/wind. The difference between 5-second peak gusts and the 3-second peak gusts specified in the ASCE 7-16 Standard [2] is, in practice, negligibly small. The standardized data are separated into thunderstorm and large-scale extratropical wind speeds. This was accomplished using a procedure described in [3] and software available on https://www.nist.gov/wind.

Simulated (Synthetic) Directional Tropical Storm/Hurricane Wind Speeds. Directional wind speeds are available for 55 coastline locations ("milestones") along the

---

1 In the statistical literature a fixed time period is called an *epoch*.

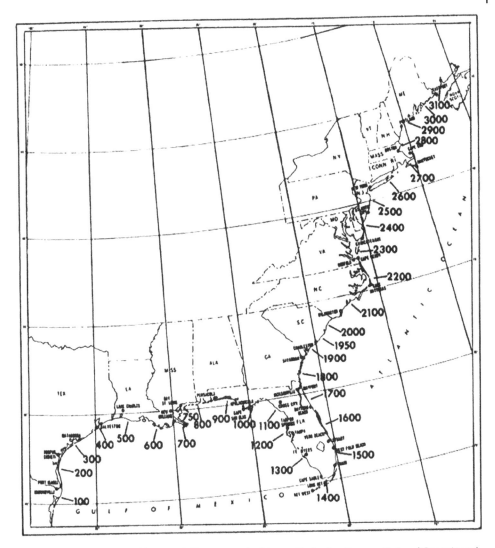

Figure 3.1 Locator map with coastal distance marked, in nautical miles. Source: National Oceanic and Atmospheric Administration.

Gulf and Atlantic coasts, shown in Figure 3.1 (see [4, 5]). The speeds were obtained by Monte Carlo simulation (see Section A.8) from approximately 100-year records of hurricane climatological data (pressure defects, radii of maximum wind speeds, and translation speeds and directions; see Section 1.3.1). Probabilistic descriptions of those data were developed and used in conjunction with the physical model described by Eq. (1.4) to obtain probabilistic models of the gradient speeds and directions. These models were then transformed via empirical expressions into probabilistic models of surface wind speeds and directions, and used for the Monte Carlo simulation of directional speed data at each of the milestones. The simulated data based on [4] are listed on https://www.nist.gov/wind. They consist of (i) estimated hurricane mean arrival rates, and (ii) sets of 999 1-min coastline wind speeds in knots at 10 m above open terrain for 16 directions at 22.5° intervals (1 knot ≈ 1.15 mph; 1 mph = 0.447 m s^{-1}; nominal ratios between 3-second speeds and 1-minute speeds and between 1-minute speeds and 1-hour speeds are 1.22 and 1.25, respectively, see Table 2.5). At any given site, as many of 20–40% of the total number of simulated hurricane wind speeds are negligibly small. Such small or vanishing wind speeds occur, for example, where the hurricane translation velocity counteracts the rotational velocity. For each of the 55 milestones shown in Figure 3.1, the respective 999 simulated data can be used to obtain, by Monte Carlo simulation, datasets of any desired size, see Section 3.3.7.

Non-directional hurricane wind speeds based on more recent simulations than those described in [5] can be obtained, both for the coastline and for regions adjacent to the coastline, from wind maps in ASCE 7-16 [2] for MRIs of up to 3000 years, and from wind maps in [6] for MRIs of up to 10^7 years.

3.2.3.2 Data Available Commercially

Peak Directional Gust Speeds for Each of 36 Directions at 10° Intervals, recorded at ASOS stations for periods of about 20 years or less (www.ncdc.noaa.gov/oa/ncdc.html).

Simulated Hurricane Directional Wind Speed Data. The methodology for obtaining directional hurricane wind speeds described in [7] is similar to the methodology used in [4], except that the various climatological and probabilistic models used therein have been refined and are based on a larger number of data. Unlike the data based on [4], the data based on [7] cover both coastlines and regions adjacent thereto.

Figure 3.2 shows approximate estimates of 2000-year (or 1700-year) mean hourly hurricane wind speeds at 10 m above open terrain as estimated in [4], the ASCE 7-10

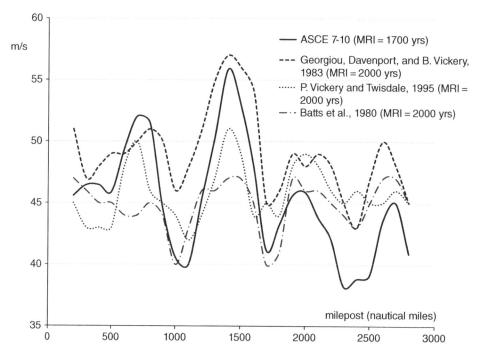

Figure 3.2 Approximate estimates of mean hourly hurricane wind speeds at 10 m above ground over terrain with open exposure. Source: After Refs. [4, 8–10].

Standard [8] and Refs. [9, 10]. Note that there are no major differences among the various estimates, except for: milestones 1100 and 2600, where speeds are likely overestimated in [10]; milestones 700 and 1400, where speeds are likely underestimated in [4]; and milestones 2300–2600, where wind speeds are likely underestimated in the ASCE 7-10 Standard.

3.3 *N̄*-year Speed Estimation from Measured Wind Speeds

Estimates of extreme wind speeds based on sets of measured wind speeds can be performed by using two types of datasets. In the traditional *epochal* approach the dataset being analyzed consists of the largest wind speeds recorded at the site of interest in each of a number of consecutive fixed epochs. To avoid seasonality effects, the epoch most commonly chosen is 1 year. The dataset then consists of the largest yearly wind speed for each year of the period of record. In the more modern *peaks-over-threshold* (POT) approach, the dataset considered in the analysis consists of wind speeds that exceed an optimal threshold.

Section 3.3.1 explains the advantages of the peaks-over-threshold (POT) over the epochal approach. Sections 3.3.2 discusses the probability distributions of the largest values and their use in structural engineering. Section 3.3.3 presents methods for estimating extreme speeds with any specified MRI $\overline{N}$, based on the epochal approach. Section 3.3.4 provides information on sampling errors in the estimation of extreme wind speeds modeled by the Type I Extreme Value distribution. Section 3.3.5 concerns the POT approach. Section 3.3.6 briefly discusses the spatial smoothing of extreme wind speed estimates performed at multiple stations within meteorologically homogeneous areas. Section 3.3.7 concerns the development of large extreme wind speed databases from relatively short records. Non-parametric estimation methods applicable to extreme wind speeds are presented in Section A.9.

3.3.1 Epochal Versus Peaks-Over-Threshold Approach to Estimation of Extremes

One advantage of the POT approach is that it allows the use of larger data samples than the epochal approach, since speeds other than the largest annual speeds can also be included in the data sample. This is illustrated in the following example.

Example 3.6 *Sample sizes in epochal and POT approaches.* Assume that in Year 1 the largest speed is 36 m s^{-1} and the second largest speed is 34 m s^{-1}, and that in Year 2 the largest speed is 43 m s^{-1} and the second and third largest speeds are 35 m s^{-1} and 31 m s^{-1}, respectively. If a threshold of 32 m s^{-1} is chosen, the speeds during Years 1 and 2 included in the sample are 43 m s^{-1}, 36 m s^{-1}, 35 m s^{-1}, and 34 m s^{-1} (four speeds). In the epochal approach only two speeds are included in the sample: 36 m s^{-1} (Year 1) and 43 m s^{-1} (Year 2). If the threshold is very high, the advantage of a larger sample size is lost. For example, if the threshold were 40 m s^{-1}, only one speed – 43 m s^{-1} – would be included in the two-year sample. If the threshold were very low, the sample would include non-extreme wind speeds; this would result in incorrect – biased – estimates of the extreme wind speeds.

An additional advantage of the POT approach is that it allows an optimal selection of the dataset being analyzed, by (i) excluding from the analysis data lower than an optimal threshold that would result in biased estimates of the extremes, and (ii) ensuring that the size of the dataset is sufficiently large to minimize sampling errors.

3.3.2 Extreme Value Distributions and Their Use in Wind Climatology

As indicated in Section A.6, a theoretical and empirical basis exists for the assumption that probability distributions of the largest values are adequate for describing extreme wind speeds probabilistically. It has been proven mathematically that three types of such distributions exist, characterized by the length of the distribution tail: the *Gumbel* distribution (also known as the Fisher-Tippett Extreme Value Type I or EV I distribution), the *Fréchet* (Fisher-Tippett EV II) distribution, and the *reverse Weibull* distribution (Fisher-Tippett EV III distribution of the largest values).

The EV I and the EV II distributions have infinitely long distribution tails. This means that their use can lead to estimates of large extremes, whose probabilities of being exceeded depend upon the thickness of the distribution upper tails. The EV I distributions tails are less thick than the tails of the EV II distributions, and entail negligibly small probabilities of exceedance of very large extremes. However, for EV II distributions, the distribution tails are thicker, and may result in estimates of unrealistically high extreme wind speeds. The EV III distribution has finite tails, meaning that, for wind speeds larger than the finite value of the distribution tail, the probabilities of exceedance are zero.

Uncertainties inherent in the estimation process can result in extreme wind speed data samples being spuriously best fitted by an EV II distribution when in fact an EV I distribution would be appropriate. For this reason, the assumption that extreme wind speeds are best fitted by an EV II distribution, used in the 1970s for the development of the extreme wind speed maps of the American National Standard A58.1, was abandoned by consensus of the ASCE 7 Standard Committee on Loads in favor of the EV I distribution.

Statistical estimates suggested that the EV III distribution may fit extreme wind speed data samples better than the EV I distribution; on the basis of such estimates the Australian/New Zealand Standard [11, Commentary C3.2] adopted the assumption that the EV III distribution is representative of the behavior of extreme wind speeds. However, estimates of the tail length of the EV III distribution are in practice prone to large errors, and to avoid the underestimation of extreme wind speeds due to spurious best fits, the ASCE 7 Standards Committee on Loads also decided against the use of the EV III distribution. Unless otherwise indicated, it will be assumed in this chapter that the EV I distribution is an appropriate probabilistic model of the extreme wind speeds.

The CDF of the EV I distribution is

$$F_I(x) = \exp\left[-\exp\left(-\frac{x-\mu}{\sigma}\right)\right] \quad (-\infty < x < \infty; \ -\infty < \mu < \infty; \ 0 < \sigma < \infty)$$

$$(3.4)$$

where μ and σ, called the location and scale parameter, respectively, are related to the mean value $E(X)$ and standard deviation $SD(X)$ of X by the expressions

$$E(X) = \mu + 0.5772\sigma \tag{3.5a}$$

$$SD(X) = \frac{\pi}{\sqrt{6}}\sigma \qquad (3.5b)$$

Inversion of Eq. (3.4) yields

$$x(F_I) = \mu - \sigma \ln(-\ln F_I) \qquad (3.6)$$

or, by virtue of Eq. (3.1),

$$x(\overline{N}) = \mu - \sigma \ln \left[-\ln \left(1 - \frac{1}{\overline{N}} \right) \right]$$
$$\approx \mu + \sigma \ln \overline{N} \qquad (3.7a,b)$$

for large $\overline{N}$.

3.3.3 Wind Speed Estimation by the Epochal Approach

This section presents two of the methods for estimating $\overline{N}$-year wind speeds under the assumption that the EV I distribution is appropriate: the method of moments and Lieblein's BLUE (Best Linear Unbiased Estimator) method.

3.3.3.1 Method of Moments

This method relies on calculated sample means $E(V)$ and standard deviations $SD(V)$ of the sample of n wind speeds. The wind speed corresponding to an MRI $\overline{N}$ is obtained from Eqs. (3.7) in which the parameters μ and σ are obtained from Eqs. (3.5).

Example 3.7 *EV I Extreme Wind Estimation, Epochal Approach*
Method of Moments. Assume that in a $n = 14$-year record at a site, the non-directional largest yearly peak 3-second gust speeds from any direction (in m s^{-1}) are: 36, 34, 35, 37, 33, 36, 40, 39, 41, 43, 33, 31, 28, 34. The epochal approach makes use of the mean $E(V) = 35.71$ m s^{-1} and standard deviation $SD(V) = 4.07$ m s^{-1} of the n largest annual speeds. From Eqs. (3.5) we obtain $\sigma = 3.17$ and $\mu = 33.90$ (in m s^{-1}). Equations (3.7a,b) yield $v(\overline{N} = 50$ years$) = 46.27$ m s^{-1} and 46.30 m s^{-1}, $v(\overline{N} = 3000$ years$) = 59.28$ m s^{-1} and 59.28 m s^{-1}, respectively.

BLUE Method. In the BLUE method the data are arranged in ascending order, that is,

$$v_1 \le v_2 \le \cdots \le v_n$$

The estimated parameters of the EV I distribution are then given by the expressions

$$\mu = \sum_{i=1}^{n} a_i v_i, \qquad \sigma = \sum_{i=1}^{n} b_i v_i \qquad (3.8)$$

where the vectors a_i, b_i are listed for $n \le 16$ in [12, p. 20], and for $n \le 100$ in the MATLAB implementation of the BLUE method, which includes a user's manual and an example https://www.nist.gov/wind.

Example 3.8 EV I Extreme Wind Estimation, Epochal Approach, BLUE Method.
Consider the dataset of Example 3.7. The rank-ordered data are
28, 31, 33, 33, 34, 34, 35, 36, 36, 37, 39, 40, 41, 43.

For the sake of clarity we follow in this example the BLUE method as presented in [12]. Using the coefficients a_i ($i = 1, 2, \ldots, 14$) [12, p. 20]:

$$\mu = 28 \times 0.163309 + 31 \times 0.125966 + 33 \times 0.108230 + 33 \times 0.095233$$
$$+ 34 \times 0.084619 + 34 \times 0.075484 + 35 \times 0.067331 + 36 \times 0.059866$$
$$+ 36 \times 0.052891 + 37 \times 0.046260 + 39 \times 0.039847 + 40 \times 0.033526$$
$$+ 41 \times 0.027131 + 43 \times 0.020317 = 33.64$$
$$\sigma = 28 \times (-0.285316) + 31 \times (-0.098775) + 33 \times (-0.045120)$$
$$+ 33 \times 0.013039 + 34 \times 0.008690 + 34 \times 0.024282 + 35 \times 0.035768$$
$$+ 36 \times 0.044262 + 36 \times 0.050418 + 37 \times 0.054624 + 39 \times 0.057083$$
$$+ 40 \times 0.057829 + 41 \times 0.056652 + 43 \times 0.052642 = 3.96$$

Equation (3.7a) then yields

$$v(\overline{N} = 50 \text{ years}) = 49.09 \text{ m s}^{-1}, \qquad v(\overline{N} = 3000 \text{ years}) = 65.33 \text{ m s}^{-1}$$

The reader can verify that the same result is obtained by using the MATLAB software referenced in this section. The method of moments, which is less efficient than the BLUE method, produces in this case estimates of the 50 and 3000-year wind speeds lower than the BLUE estimates by approximately 6 and 9%, respectively.

3.3.4   Sampling Errors in the Estimation of Extreme Speeds

The standard deviation of the errors in the estimation of extreme wind speeds with a MRI $\overline{N}$ may be obtained from the following expression [13]:

$$SD(v_{\widetilde{N}}) \approx 0.78[1.64 + 1.46(\ln \overline{N} - 0.577) + 1.1(\ln \overline{N} - 0.577)^2]^{1/2} \frac{s}{\sqrt{n}} \qquad (3.9)$$

where s is the sample standard deviation of the largest yearly wind speeds for the period of record, and n is the sample size.

Example 3.9   At Great Falls, Montana, the largest yearly sustained fastest-mile wind speeds in the period 1944–1977 (sample size $n = 34$) were
57, 65, 62, 58, 64, 65, 59, 65, 59, 60, 64, 65, 73, 60, 67, 50, 74,
60, 66, 55, 51, 60, 55, 60, 51, 51, 62, 51, 54, 52, 59, 56, 52, 49 (mph).
The sample mean and the standard deviation of for these data are $\overline{V} = 59.1$ mph and $SD(V) = 6.41$ mph. From Eqs. (3.5), (3.7) and (3.9) it follows that for $\overline{N} = 50$ years and $\overline{N} = 1000$ years,

$$v_{50} \approx 75.8 \text{ mph} \quad SD(v_{50}) \approx 3.71 \text{ mph}$$
$$v_{1000} \approx 90.8 \text{ mph} \quad SD(v_{1000}) \approx 6.36 \text{ mph}.$$

The probabilities that $v_{\widetilde{N}}$ is contained in the intervals $v_{\widetilde{N}} \pm SD(v_{\widetilde{N}})$ and $v_{\widetilde{N}} \pm 2\,SD(v_{\widetilde{N}})$ are approximately 68 and 95%, respectively. These intervals are called the 68 and 95% confidence intervals for $v_{\widetilde{N}}$ (see Section A.7.1).

3.3.5 Wind Speed Estimation by the Peaks-Over-Threshold Approach

Among the methods available for estimating extreme wind speeds by the POT approach we mention the method of moments and the de Haan method, both of which are described in Section A.7.2, and the POT Poisson-processes methods used in [14], which provide information on the uncertainty in the estimates. The plots of Figure 3.3 show estimates by the de Haan method of 100, 1000, and 100,000-year fastest-mile wind speeds at 6.1 m above ground in terrain with open exposure at Green Bay, Wisconsin. The estimates are functions of threshold speeds (in mph). The data consisted of the maximum wind speed for each of the successive 8-day intervals within a 15-year record, and included no wind speed separated by less than 5 days. For thresholds between about 38 and 32 mph (sample sizes of about 35–127), the estimated 100-year speeds are stable around 60 mph. The reliability of the estimate is poorer as the MRI increases (this is clearly seen for the 100,000-year estimates). For thresholds higher than 38 mph, the estimates are less stable for all three MRIs; that is, they vary fairly strongly as a function of threshold. For thresholds lower than about 32 mph, the estimates of the 100-yr speed are increasingly biased with respect to the 60 mph estimate, owing to the presence in the data sample of low speeds unrepresentative of the extremes. *Including low speeds in a sample used for inferences on extreme speeds can result in biased estimates,* as would be the case if the heights of children were included in a sample used to estimate the height of adults. For example, estimates of extreme wind speeds based on wind speed data recorded every hour, the vast majority

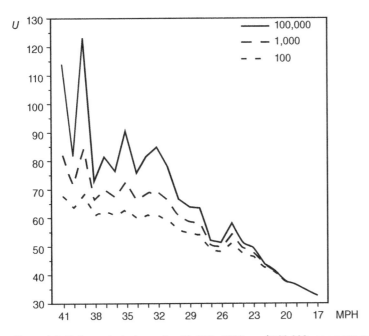

Figure 3.3 Estimated wind speeds with 100-, 1000-, and 100,000-year mean recurrence interval at Green Bay, Wisconsin, as functions of threshold (mph).

of which are low and meteorologically unrelated to the extreme wind speeds, would be unrealistic. Modern extreme value statistics recognizes that to obtain dependable estimates of extreme values it is necessary to "let the tails speak for themselves," instead of allowing estimates to be biased by data with small values, as is the case in Figure 3.3 for wind speeds below about 32 mph.

3.3.6 Spatial Smoothing

In developing wind maps results it is appropriate to apply spatial smoothing techniques to reduce discrepancies among results obtained for stations contained in a meteorologically homogeneous area of appropriate size. Such a technique was applied to the development of wind speed maps specified in the ASCE 7-16 Standard, see Section 3.2 of [14].

A technique used for the development of U.S. maps specified in the ASCE 7-10 maps consisted of considering groups of stations called "superstations," and including identical subgroups of stations in more than one "superstation." The application of this technique led to the demonstrably incorrect result that extreme wind speeds are uniform throughout most of the contiguous United States.

3.3.7 Development of Large Wind Speed Datasets

A number of structural engineering applications require the use of large wind speed datasets for use in non-parametric estimates of wind effects with long MRIs. A detailed procedure for generating such data, including directional data, is presented in [15]. For material on Monte Carlo methods used for the development of large wind speed databases, see Section A.8.

3.4 Tornado Characterization and Climatology

Tornado climatology studies and design criteria on tornado action on structures require the characterization of tornadoes from the point of view of their flow modeling and their intensities. Section 3.4.1 discusses tornado flow modeling based on atmospheric science considerations, laboratory testing, numerical methods, and observations of tornadoes. Section 3.4.2 is devoted to the use of tornado models, observations of tornadoes and their effects, and statistical methods, for the estimation of wind speeds and associated atmospheric pressure defects. Section 3.4.3 summarizes simplified, conservative models of tornado structure that the U.S. Nuclear Regulatory Commission Office of Nuclear Regulatory Research considers acceptable for the design of nuclear power plants.

3.4.1 Tornado Flow Modeling

Tornadoes are translating cyclostrophic flows that develop within severe thunderstorms. Because their horizontal dimensions are relatively small (typically in the order of 300 m), the probability that their maximum speeds at heights above ground in the order of a few tens of meters or less will be measured by a sufficiently strong instrument with fixed location, or any other instrument, is small. For his reason reliable measurements of such wind speeds are not available to date. Laboratory measurements (see Chapters 5 and 27) have shed useful light on tornado flow structure, but are only the beginning of efforts to improve current knowledge in this area of research.

A highly readable generic guide on tornado climatology is available in [16]. An analysis of information on more than 46 000 tornado segments (i.e., portions of or entire tornadoes) reported in the contiguous United States from January 1950 through August 2003 was performed in [17] with a view to determining tornado strike probabilities and maximum wind speeds for use in the development of design criteria for nuclear power plants.

Section 3.4.2 briefly summarizes salient features of [17]. Section 3.4.3 summarizes U.S. Nuclear Regulatory Commission (NRC) requirements on atmospheric pressure defects and tornado wind speeds based on the recommendations of [17].

3.4.2   Summary of NUREG/CR-4461, Rev. 2 Report [17]

Of the 46 000 segments, more than 39 600 had sufficient information on location, intensity, length and width to be used in the analysis. Estimates of and confidence intervals for expected values are based in [17] on the assumption, first suggested in 1963 [18], that lognormal distributions are appropriate. As in [16], it is noted in [17] that, even though the number of reported tornadoes has been increasing since 1950 owing to improved tornado observation techniques (Figure 3.4), the increase was limited to the least intense tornadoes; however, the missing information on weaker tornadoes appears not to affect significantly estimates of strike probabilities or maximum wind speeds.

Comprehensive estimates of tornado characteristics are presented in [17] for the entire contiguous United States, for regions thereof, and for 1°, 2°, and 4° latitude and

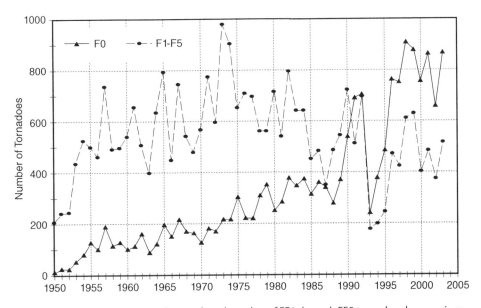

Figure 3.4 Number of EF0 tornadoes and total number of EF1 through EF5 tornadoes by year since 1950 [19].

longitude boxes. The effect of the variation of the wind speed along and across the tornado footprint was modeled by using results of studies summarized in [19].

Methods for estimating (i) tornado strike probabilities and (ii) conditional probabilities that the maximum wind speed will exceed a specified value given that a tornado strike has occurred, differ for point and finite-sized structures. For point structures only estimates of tornado impact areas are necessary. For finite-size structures, in addition to estimates of tornado impact areas, estimates of lengths of tornado paths associated with various wind speeds are needed. These were based on [20]. For example, while for EF0 tornadoes 100% of the length of the tornado path has EF0 speeds, for an EF5 tornado it was estimated that on average 0.135, 0.100, 0.190, 0.240, 0.185, and 0.150 of the total path length have EF0, EF1, EF2, EF3, EF4, and EF5 wind speeds, respectively.

For point structures, the annual probability of exceedance of the speed u_o at a point is defined as the probability that a tornado will strike that point times the annual probability that the speed u will exceed the speed u_o given that a tornado strike has occurred, that is,

$$P_p(u \geq u_o) = P_{s,p} \times P_p(u \geq u_o \mid s) \tag{3.10}$$

The annual strike probability is

$$P_{s,p} = \frac{A_t}{NA_r} \tag{3.11}$$

A_t is the total area in square miles impacted in N years by tornadoes in the region A_r of interest, that is, the product of the expected area of a tornado in the region A_r by the total number of tornado events that occurred in that region in N years, and N is the number of years of record.

The probability of exceeding a speed u_o given that a tornado has occurred is

$$P_p(u \geq u_o \mid s) = \frac{A_{u \geq u_o}}{A_t} \tag{3.12}$$

where $A_{u \geq u_o}$ is the total area impacted by wind speeds greater than u_o; see also [18]. It is assumed in [17] that $P_p(u \geq u_o \mid s)$ is described by a Weibull distribution.

For the probability of exceedance of a speed u_o within a finite-size structure, see [17].

Uncertainties in the estimation of the tornado strike probabilities and conditional probabilities of tornado wind speeds are due to errors in the tornado footprint modeling as a rectangle and in the estimation of the length, width and area of the tornado footprint, the assumption that the structure's characteristic dimension is 200 ft, and the assignment of an incorrect EF (enhanced Fujita) scale to tornadoes in the database being used. Adjustments for those errors are discussed in [19].

Recommendations in [17] of tornado design wind speeds with 10^5-, 10^6-, and 10^7-year MRIs for the three regions defined in Figure 3.5 are based on the spatially averaged estimated speeds for 2° longitude/latitude boxes and are shown in Table 3.1.

The American Nuclear Society ANSI/ANS-2.3-2011; R2016 Standard's regionalization of tornado wind speeds [22] differs somewhat from the regionalization of Figure 3.5.

3.4.3 Design-Basis Tornado for Nuclear Power Plants

The NRC Regulatory Guide 1.76 (Revision 1 March 2007) [21] provides guidance on design-basis tornado and design-basis tornado-generated missiles for nuclear power

Figure 3.5 Recommended design wind speeds with 10^7 years mean recurrence intervals [17].

Table 3.1 Recommended tornado design wind speeds.

Mean Recurrence Interval (years)	Wind Speed (mph)		
	Region I	Region II	Region III
10^5	160	140	100
10^6	200	170	130
10^7	230	200	160

Table 3.2 Design-basis tornado wind field characteristics [21].

Region	Maximum Wind Speed m s^{-1} (mph)	Translational Speed m s^{-1} (mph)	Maximum Rotational Speed m s^{-1} (mph)	Radius of Maximum Rotational Speed m (ft)	Pressure Drop mb (psi)	Pressure Drop Rate mb s^{-1} (psi s^{-1})
I	103 (230)	21 (46)	82 (184)	45.7 (150)	83 (1.2)	37 (0.5)
II	89 (200)	18 (40)	72 (160)	45.7 (150)	63 (0.9)	25 (0.4)
III	72 (160)	14 (32)	57 (128)	45.7 (150)	40 (0.6)	13 (0.2)

plants in the contiguous United States. For the regions shown in Figure 3.5, Table 3.2 reproduces the characteristics of the design-basis tornadoes provided in [21] and based on the Rankine model combined with a translational velocity (Chapter 27). Design-basis tornado-generated missiles are considered in Chapter 28.

For tornado vertical wind speeds, see Chapter 27. Wind field characterization of tornadoes in the ANSI/ANS-2.3-2011; R2016 Standard [22] differs in some respects to that of [21].

References

1 Masters, F.J., Vickery, P.J., Bacon, P., and Rappaport, E.N. (2010). Toward objective, standardized intensity estimates from surface wind speed observations. *Bulletin of the American Meteorological Society* 91: 1665–1681.

2 ASCE, "Minimum design loads for buildings and other structures (ASCE/SEI 7–16)," in *ASCE Standard ASCE/SEI 7–16*, Reston, VA: American Society of Civil Engineers, 2016.

3 Lombardo, F.T., Main, J.A., and Simiu, E. (2009). Automated extraction and classification of thunderstorm and non-thunderstorm wind data for extreme-value analysis. *Journal of Wind Engineering and Industrial Aerodynamics* 97: 120–131.

4 Batts, M. E., Russell, L. R., Cordes, M. R., Shaver, J. R., and Simiu, E., Hurricane wind speeds in the United States, Building Science Series 124, National Bureau of Standards, Washington, DC, 1980. https://www.nist.gov/wind.

5 Batts, M.E., Russell, L.R., and Simiu, E. (1980). Hurricane wind speeds in the United States. *Journal of the Structural Division-ASCE* 106: 2001–2016. https://www.nist.gov/wind.

6 Vickery, P.J., Wadhera, D., and Twisdale, L.A., "Technical basis for regulatory guidance on design-basis hurricane wind speeds for nuclear power plants," NUREG/CR-7005, U.S. Nuclear Regulatory Commission, Washington, DC, 2011.

7 Vickery, P.J., Wadhera, D., Twisdale, L.A. Jr., and Lavelle, F.M. (2009). U.S. hurricane wind speed risk and uncertainty. *Journal of Structural Engineering* 135: 301–320.

8 ASCE, "Minimum design loads for buildings and other structures (ASCE/SEI 7–10)," in *ASCE Standard ASCE/SEI 7–10*, Reston, VA: American Society of Civil Engineers, 2010.

9 Vickery, P. and Twisdale, L. (1995). Prediction of hurricane wind speeds in the United States. *Journal of Structural Engineering* 121: 1691–1699.

10 Georgiou, P.N., Davenport, A.G., and Vickery, B.J. (1983). Design wind speeds in regions dominated by tropical cyclones. *Journal of Wind Engineering and Industrial Aerodynamics* 13: 139–152.

11 AS/NZS, *Structural design actions: wind actions: commentary (supplement to AS/NZS 1170.2:2002)*, Sydney, Wellington: Standards Australia International, Standards New Zealand, 2002.

12 Lieblein, J., "Efficient Methods of Extreme-Value Methodology," NBSIR 74–602, National Bureau of Standards, Washington, DC, 1974. https://www.nist.gov/wind.

13 Gumbel, E.J. (1958). *Statistics of Extremes*. New York: Columbia University Press.

14 Pintar, A.L., Simiu, E., Lombardo, F. T., and Levitan, M. L., "Maps of Non-Hurricane Non-Tornadic Wind Speeds with Specified Mean Recurrence Intervals for the Contiguous United States Using a Two-Dimensional Poisson Process Extreme Value Model and Local Regression," NIST Special Publication 500-301, National Institute of Standards and Technology, Gaithersburg, 2015. https://www.nist.gov/wind.

15 Yeo, D. (2014). Generation of large directional wind speed data sets for estimation of wind effects with long return periods. *Journal of Structural Engineering* 140: 04014073. https://www.nist.gov/wind.

16 *U.S. Tornado Climatology*, National Climatic Data Center, Asheville, NC, 2008,

17 Ramsdell, J.V., Jr., and Rishel, J.P., *Tornado Climatology of the Contiguous United States*, A.J. Buslik, Project Manager, NUREG/CR-4461, Rev. 2, PNNL-15112, Rev. 1, Pacific Northwest National Laboratory, 2007.

18 Thom, H.C.S. (1963). Tornado probabilities. *Monthly Weather Review* 91: 730–736.

19 Reinhold, T.A. and Ellingwood, B.R., *Tornado Risk Assessment*, NUREG/CR-2944, U.S. Nuclear Regulatory Commission, Washington, DC, 1982.

20 Twisdale, L.A. and Dunn, W.L., *Tornado Missile Simulation and Design Methodology, Vols. 1 and 2. EPRI NP-2005*, Electric Power Research Institute, Palo Alto, California.

21 U.S. Nuclear Regulatory Commission, *Regulatory Guide 1.76, Design-Basis Tornado and Tornado Missiles for Nuclear Power Plants*, Revision 1, 2007.

22 American Nuclear Society, ANSI/ANS-2.3-2011. *Estimating tornado, hurricane, and extreme straight wind characteristics at nuclear facility sites*. La Grange Park, Illinois, reaffirmed Jun 29, 2016.

4

Bluff Body Aerodynamics

Aerodynamics is the study of air flows that interact with solid bodies. Streamlined bodies have shapes that help to reduce drag forces. Bodies that are not streamlined are called bluff.

Bluff body aerodynamics of interest in structural engineering applications is associated with atmospheric flows, which are incompressible owing to their relatively low speeds. With rare exceptions associated with stably stratified flows (see Section 1.1.3), atmospheric flows of interest in structural design are turbulent. In addition to the turbulence present in atmospheric flows, "signature turbulence" is generated by the presence of the body in the flow. Turbulence significantly complicates the study of bluff body aerodynamics.

Certain types of engineering structures can be subjected to aerodynamic forces generated by structural motions. These motions, called self-excited, are in turn affected by the aerodynamic forces they generate. The structural behavior associated with self-excited motions is termed aeroelastic, and is considered in Part III of the book.

As pointed out by Roshko, "the problem of bluff-body flow remains almost entirely in the empirical, descriptive realm of knowledge" [1]. Although much progress is being made in Computational Fluid Dynamics (CFD) and its application to wind engineering (Computational Wind Engineering, or CWE), its application in structural engineering practice remains limited [2]. Indeed, the simulation of flows over bluff bodies in turbulent shear flows is a formidable problem, and the approximations required in modeling the flow numerically can produce results that differ significantly and unpredictably from each other depending upon those approximations. To follow Schuster [3], conservative CFD applications are based on the paradigm "Develop, Validate, Apply," wherein end-users apply validated software to problems that fall within or at least not too far from its range of validation. As pointed out in [3], a modified paradigm "Develop, Apply, Validate" may be required under certain circumstances. This paradigm entails large uncertainties that must be accounted for; how CFD methods may be applied and ultimately developed and validated under those circumstances is discussed in [3] in the context of NASA applications. In a civil engineering context, an informal "Develop, Apply, Validate" approach has been implicit in low-risk CFD applications wherein the effect of relatively large uncertainties is tolerable: for example, the prediction of wind flows that cause easily remediable pedestrian discomfort around buildings (see Chapter 15).

Section 4.1 reviews fundamental fluid dynamics equations. Section 4.2 considers flows in a curved path and vortex flows. Section 4.3 discusses boundary layers and flow

Wind Effects on Structures: Modern Structural Design for Wind, Fourth Edition. Emil Simiu and DongHun Yeo.
© 2019 John Wiley & Sons Ltd. Published 2019 by John Wiley & Sons Ltd.

separation. Section 4.4 is devoted to wake and vortex formations in two-dimensional (2-D) flow. Section 4.5 concerns pressure, lift, drag and moment effects on 2-D bodies. Section 4.6 presents information on flow effects in three dimensions.

4.1 Governing Equations

4.1.1 Equations of Motion and Continuity

Consider a fixed elemental volume dV in a fluid. The velocity vector is expressed as

$$\mathbf{u} = u_1\mathbf{i}_1 + u_2\mathbf{i}_2 + u_3\mathbf{i}_3 \tag{4.1}$$

where $\mathbf{i}_1$, $\mathbf{i}_2$, $\mathbf{i}_3$ are unit vectors along the usual three fixed orthogonal axes.

The force acting on the fluid contained in the volume dV consists of two parts. The first part is the body force caused by gravity, and is denoted by $\mathbf{F}\rho dV$, where ρ is the fluid density. The second part is due to the net action on the fluid of the internal stresses σ_{ij} $(i, j = 1, 2, 3)$. For example, the contribution to this action of the normal stress σ_{11} (see Figure 4.1) is

$$-\sigma_{11}dx_2dx_3 + \left(\sigma_{11} + \frac{\partial\sigma_{11}}{\partial x_1}dx_1\right)dx_2dx_3 = \frac{\partial\sigma_{11}}{\partial x_1}dx_1dx_2dx_3$$

$$= \frac{\partial\sigma_{11}}{\partial x_1}dV \tag{4.2}$$

It can be similarly shown that the net force component in the i direction due to the action of all stresses σ_{ij} is

$$\sum_{j=1}^{3}\frac{\partial\sigma_{ij}}{\partial x_j}dV \tag{4.3}$$

Denoting the components of $\mathbf{F}$ by F_i $(i = 1, 2, 3)$, the force balance equations, given by Newton's second law, are

$$\frac{Du_i}{Dt}\rho\,dV = F_i\,\rho\,dV + \sum_{j=1}^{3}\frac{\partial\sigma_{ij}}{\partial x_j}dV \quad (i = 1, 2, 3) \tag{4.4}$$

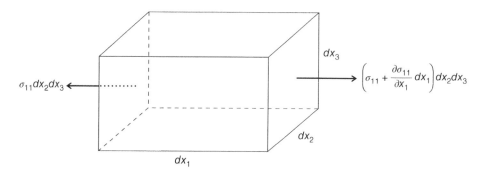

Figure 4.1 Forces along the *i* direction on an elementary volume of fluid.

where the operator D/Dt, known as the *substantial* or *material* derivative, is defined as follows:

$$\frac{D}{Dt} = \frac{\partial}{\partial t} + \sum_{i=1}^{3} \frac{\partial}{\partial x_i} \frac{dx_i}{dt}$$

$$= \frac{\partial}{\partial t} + \sum_{i=1}^{3} u_i \frac{\partial}{\partial x_i} \tag{4.5}$$

Since Eq. (4.4) is true for all volume elements, it may be divided by the factor dV, and the equations of motion of a fluid particle can be written in component form as

$$\rho \frac{Du_i}{Dt} = \rho F_i + \sum_{j=1}^{3} \frac{\partial \sigma_{ij}}{\partial x_j} \tag{4.6}$$

We now consider the principle of mass conservation, which states that the rate at which mass enters a system is equal to the rate at which mass leaves the system. If ρ is constant, mass conservation can be shown to imply

$$\sum_{i=1}^{3} \frac{\partial u_i}{\partial x_i} = 0 \tag{4.7}$$

Equation (4.7) is called the equation of continuity.

4.1.2 The Navier–Stokes Equation

Unlike a solid, under static conditions a fluid cannot support any stresses other than normal pressures. However, in dynamic situations, it may support shear in a time-dependent manner. In most fluid-mechanical applications it is adequate to assume that the stresses involved are normal pressures or ascribable to *viscosity*. Fluids with internal shear stress proportional to the rate of change of velocity with distance normal to that velocity are termed *viscous* or *Newtonian*. For example, the shear stress σ_{12} in a simple 2-D flow is expressed as

$$\sigma_{12} = \mu \frac{\partial u_1}{\partial x_2} \tag{4.8}$$

where the proportionality factor is defined as the fluid viscosity.
The units of viscosity are

$$\mu = \frac{\text{force}}{\text{area}} \times \frac{\text{length}}{\text{velocity}} = \frac{\text{force} \times \text{time}}{\text{length}^2} = \frac{\text{mass}}{\text{length} \times \text{time}}$$

Typical values of μ for air and water at $15°C$ are

$$\mu_{\text{air}} = 1.783 \times 10^{-5} \text{ kg m}^{-1} \text{s}^{-1}, \mu_{\text{water}} = 1.138 \times 10^{-3} \text{ kg m}^{-1} \text{s}^{-1}$$

By distinguishing in the stress tensor σ_{ij} at a fluid point the normal stress p (i.e., *pressure*) and the *deviatoric* stress, defined as

$$d_{ij} = 2\mu \left(e_{ij} - \frac{1}{3} \delta_{ij} \sum_{k=1}^{3} e_{kk} \right) \quad (i, j = 1, 2, 3) \tag{4.9}$$

where

$$e_{ij} = \frac{1}{2}\left(\frac{\partial u_i}{\partial x_j} + \frac{\partial u_j}{\partial x_i}\right) \tag{4.10}$$

and

$$\delta_{ij} = \begin{cases} 1, & i = j \\ 0, & i \neq j \end{cases} \tag{4.11}$$

The following expression for the stress σ_{ij} can be obtained:

$$\sigma_{ij} = -p\,\delta_{ij} + 2\mu\left(e_{ij} - \frac{1}{3}\delta_{ij}\sum_{k=1}^{3} e_{kk}\right) \tag{4.12}$$

Using the expressions for stress in a Newtonian fluid results in the equations of motion, known as Navier–Stokes equations:

$$\rho\frac{Du_i}{Dt} = \rho F_i - \frac{\partial p}{\partial x_i} + \mu\left(\sum_{j=1}^{3}\frac{\partial^2 u_i}{\partial x_j^2} + \frac{1}{3}\frac{\partial \sum_{k=1}^{3}\frac{\partial u_k}{\partial x_k}}{\partial x_i}\right) \tag{4.13}$$

For an incompressible fluid (Eq. [4.7]), Eq. (4.13) can be written as

$$\frac{\partial u_i}{\partial t} + u_j\frac{\partial u_i}{\partial x_j} = -\frac{1}{\rho}\frac{\partial p}{\partial x_i} + F_i + v\sum_{j=1}^{3}\frac{\partial^2 u_i}{\partial x_j^2} \tag{4.14}$$

where $v = \mu/\rho$ is called the *kinematic viscosity*.
For air and water, at 15°C

$$v_{\text{air}} = 1.455 \times 10^{-5}\ \text{m}^2\ \text{s}^{-1}, \quad v_{\text{water}} = 1.139 \times 10^{-3}\ \text{m}^2\ \text{s}^{-1} \tag{4.15}$$

4.1.3 Bernoulli's Equation

Consider an incompressible, inviscid flow experiencing negligible body forces. If the flow is steady, the fluid element of Figure 4.2 is subjected in the direction of the streamline (i.e., along the tangent at any instant to the flow velocity) to the force $p\,dy\,dz$, the force $-(p+dp)\,dy\,dz$, and the inertial force

$$\rho\,dx\,dy\,dz\,\frac{dU}{dt} = \rho\frac{dx}{dt}\,dy\,dz\,dU$$
$$= \rho\,dy\,dz\,U\,dU \tag{4.16}$$

where $dx/dt = U$. The equation of equilibrium among those three forces yields $-dp = \rho\,U\,dU$ and, upon integration,

$$\frac{1}{2}\rho U^2 + p = \text{const} \tag{4.17}$$

Equation (4.17) is known as Bernoulli's equation. The quantity $1/2\rho U^2$ has the dimensions of pressure and is called *dynamic pressure*. The quantity dp/dx is called the *pressure gradient* in the x direction.

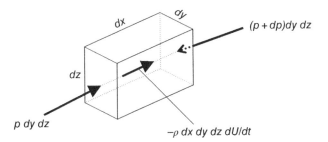

Figure 4.2 Flow-induced pressures and inertial force on an elemental volume of a fluid in motion.

Consider the streamline between two points, one of which is the *stagnation point* on the windward face of a body immersed in the flow where $U = 0$, while the other is located in the undisturbed flow far upstream of the body where the static pressure is p_0 and the velocity is U_0. The pressure at the stagnation point (i.e., the *stagnation pressure*) is

$$p_{st} = p_0 + \frac{1}{2}\rho U_0^2 \tag{4.18}$$

Bernoulli's equation is widely used to interpret the relation between pressure and velocity in atmospheric and wind tunnel flows. Detailed comments on Bernoulli's equation and its applicability, including to viscous flows, are provided in section 3.5 of [4].

4.2 Flow in a Curved Path: Vortex Flow

Consider a 2-D flow between two locally concentric streamlines with radii of curvature r and $r + dr$ (Figure 4.3). For the flow to maintain its curved path with tangential velocity U at radius r, it must experience an acceleration U^2/r toward the center of curvature. Let the pressure acting on the fluid element under consideration be denoted by p. The pressure differential between the streamlines at radii r and $r + dr$, which is responsible for this acceleration, is dp. The equation of motion for a fluid element shown in Figure 4.3 is then

$$dp\,dA = \rho\,dr\,dA\frac{U^2}{r} \tag{4.19}$$

where dA is the area of the element in a plan normal to the plan of Figure 4.3. Therefore

$$dp = \rho\,U^2\frac{dr}{r} \tag{4.20}$$

Bernoulli's equation allows the calculation of the pressure along a curved path of the flow. In particular, one may consider the case wherein the flow is circular and the value of p in Eq. (4.17) is the same on all streamlines. This is the case of *vortex flow*. Differentiation of Eq. (4.17) yields

$$\rho U\frac{dU}{dr} + \frac{dp}{dr} = 0 \tag{4.21}$$

From Eqs. (4.20) and (4.21) there follows

$$\frac{dU}{U} = -\frac{dr}{r} \tag{4.22}$$

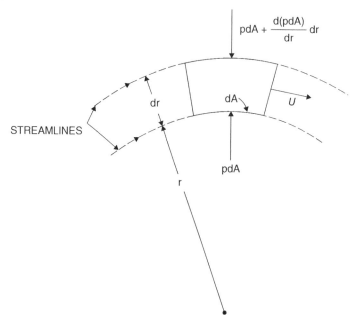

Figure 4.3 Flow in a curved path.

Integration of Eq. (4.22) yields

$$Ur = \text{const.} \tag{4.23}$$

This law states, for an incompressible and inviscid fluid, the theoretical hyperbolic relation between radius r and tangential velocity U in a free vortex. In an actual free vortex,
however, the effects of viscosity are present as well. Viscosity "locks" together a portion
of the fluid near the center and causes it to rotate as a rigid body, instead of as an inviscid
fluid described by Eq. (4.23). Thus, at the center of a free vortex the velocity increases
with radius, whereas according to Eq. (4.23) it decreases with increasing r. This decrease
actually occurs outward from a transition region in which U attains its maximum value.
The value of U in this region depends upon the fluid viscosity and the total angular
momentum of the vortex. Figure 4.4 illustrates qualitatively the pressure and velocity
dependence on radius in a free vortex occurring in a real fluid.

The free vortex is of interest in many flows that occur in engineering applications.
For example, atmospheric flows along curved isobars are described by generalizations
of Eq. (4.20). These have been described in Chapter 1, where additional Coriolis forces
have been included.

4.3 Boundary Layers and Separation

The viscosity of air at normal atmospheric pressures and temperatures has a relatively
small value. Nonetheless, in some circumstances this small viscosity plays an important
role. In particular, a consequential effect of the viscosity is the formation of boundary
layers.

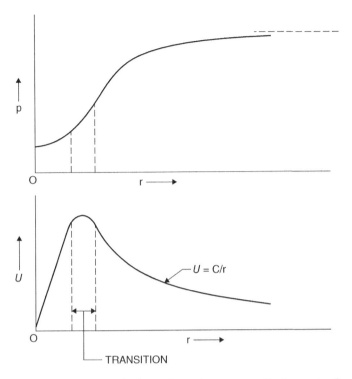

Figure 4.4 Pressure and velocity dependence upon radius in a vortex flow.

Figure 4.5 Typical boundary-layer velocity profile.

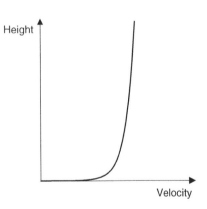

Consider an air flow over and along a stationary smooth surface. It is an experimental fact that the air in contact with the surface adheres to it. This "no slip" condition causes a retardation of the air motion in a layer near the surface called the boundary layer. Within the boundary layer the velocity of the air increases from zero at the surface to its value in the outer flow (as opposed to the boundary-layer flow). A boundary-layer velocity profile is shown in Figure 4.5.

Since air has mass, its motion exhibits inertial effects, in accordance with Newton's second law and its application to fluids, the Navier–Stokes equations. Viscous flows are therefore subjected to both inertial and viscous effects. The relation between these two effects is an index of the type of flow phenomena that may be expected to occur. The non-dimensional parameter Re called the Reynolds number is a measure of the ratio of inertial to viscous forces. For example, consider a volume of fluid with a typical dimension L. By Bernoulli's theorem, the net pressure $p - p_0$ caused by the fluid velocity U is in the order of $1/2\rho U^2$, and creates inertial forces on the fluid element enclosed by that volume in the order of $\rho U^2 L^2$. The viscous stresses on the element are in the order of $\mu U/L$, so viscosity-related forces are in the order of $(\mu U/L)L^2 = \mu U L$. The ratio of inertial to viscous forces is then in the order of

$$Re = \frac{\rho U^2 L^2}{\mu U L} = \frac{\rho U L}{\mu} = \frac{U L}{\nu} \tag{4.24}$$

A useful approximate value of the Reynolds number in air at about $20°C$ and $760\,mm$ atmospheric pressure is $67{,}000\,UL$. If Re is large, inertial effects are predominant; if Re is small, viscous effects predominate. L is a representative dimension of the body being considered.

Boundary-layer separation occurs if the kinetic energy of the fluid particles in the lower region of the boundary layer are no longer sufficient to overcome the pressures that increase in the direction of the flow and thus produce adverse pressure gradients. The flow in that region then becomes reversed, that is, separation is taking place (Figure 4.6). Shear layers generate discrete vortices that are shed into the wake flow behind the bluff body (Figure 4.7). Such vortices can cause high suctions near separation points such as corners or eaves. A flow around a building with sharp edges is shown schematically in Figure 4.8. The injection by turbulent fluctuations of high-momentum particles from the outer layer into the zone of separated flow can produce flow reattachment. Figure 4.9 shows an age-old streamlining measure aimed at reducing flow separation and strong local roof suctions near the ridge under winds normal to the end wall.

A visualization of flow separation for a bluff shape, and of the turbulent flow in the separation zone, is shown in Figure 4.10a, in which the separation zone starts close to windward edge. If the shape of the deck is streamlined, as opposed to being bluff, the separation zone is narrower, and the turbulent flow about the upper face of the deck almost disappears (Figure 4.10b).

Figure 4.11a shows the visualization of flow around a counterclockwise spinning baseball moving from left to right. Figure 4.11b is a schematic of the forces acting on the baseball with velocity U and angular velocity ω. The relative velocity of the flow with respect to the ball is directed from right to left. Entrainment of fluid due to friction at the surface of the spinning body increases the relative flow velocities with respect to the

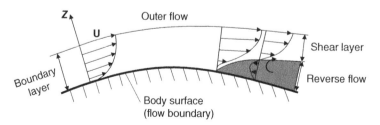

Figure 4.6 Velocity profile in the boundary layer and in the separation zone of a flow near a curved body surface. Source: After [5].

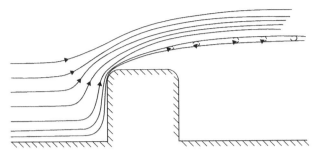

Figure 4.7 Flow separation at corner of obstacle.

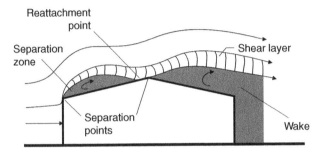

Figure 4.8 Flow about a building with sharp edges. Source: After [5].

Figure 4.9 *Three thatched cottages by a road*, Rembrandt van Rijn (1606–1669), photo Nationalmuseum, Sweden. Source: Count Kessin collection.

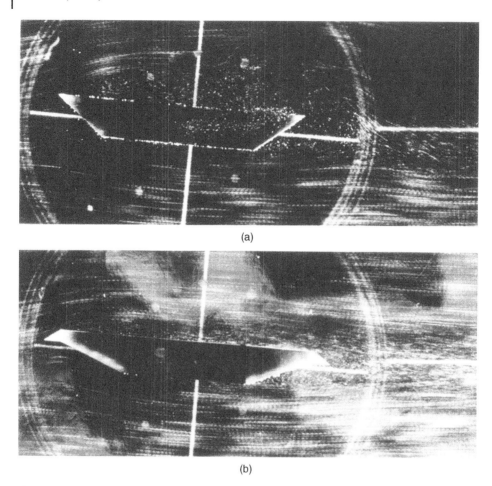

(a)

(b)

Figure 4.10 Visualization of water flow over (a) a model bridge deck section and (b) a partially streamlined model bridge deck section. Flow velocity is oriented from left to right. Source: Courtesy of the National Aeronautical Establishment, National Research Council of Canada.

body near its top and decreases them near the bottom. By virtue of Bernoulli's equation, the static pressures are therefore lower near the top and higher near the bottom. The flow asymmetry induced by spinning therefore results in a net lift force denoted by F_M in Figure 4.11b, called the Magnus force. In different aerodynamic contexts, flow asymmetries due to body motions can under certain conditions be the cause of galloping and other aeroelastic motions.

4.4 Wake and Vortex Formations in Two-Dimensional Flow

In the following discussion, the flow is assumed to be smooth (laminar) and 2-D, that is, independent of the coordinate normal to the cross section of the body.

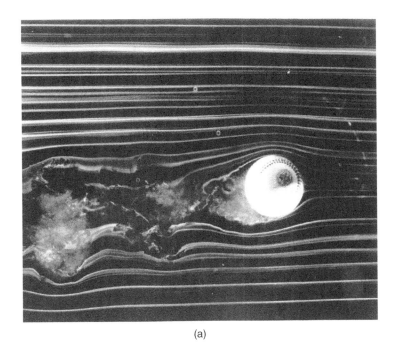

(a)

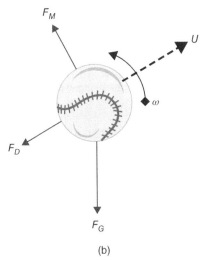

(b)

Figure 4.11 (a) Flow around a spinning baseball. Source: Courtesy of the National Institute of Standards and Technology. (b) Schematic showing forces acting on baseball with velocity $\vec{U}$ and angular velocity ω. Source: Reproduced from [6], with the permission of the American Association of Physics Teachers.

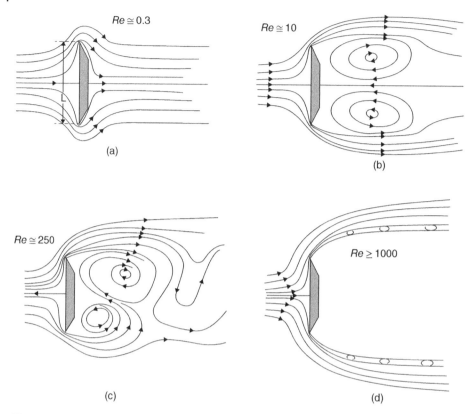

Figure 4.12 (a) Flow past a sharp-edged plate, $Re \cong 0.3$. (b) Flow past a sharp-edged plate, $Re \cong 10$. (c) Flow past a sharp-edged plate, $Re \cong 250$. (d) Flow past a sharp-edged plate, $Re \geq 1000$.

Consider a sharp-edged flat plate shown in Figure 4.12a. At a very low Reynolds number (e.g., $Re \cong 0.3$, based on the characteristic length L shown in Figure 4.12a), the flow turns the sharp corner and follows both front and rear contours of the plate. At $Re \cong 10$, obtained by increasing the flow velocity over the same plate, the flow separates at the corners and creates two large, symmetric vortices that remain attached to the back of the plate (Figure 4.12b). At $Re \cong 50$, the symmetrical vortices are broken, and replaced by cyclically alternating vortices that form by turns at the top and at the bottom of the plate and are swept downstream (Figure 4.12c). A full cycle of this phenomenon is defined as the activity between the occurrence of some instantaneous flow configuration about the body and the next identical configuration. At $Re \geq 1000$ (Figure 4.12d), the inertia forces predominate; large distinct vortices have little possibility of forming and, instead, a generally turbulent wake is formed behind the plate, its two outer edges forming each a shear layer consisting of a long series of smaller vortices that accommodate the wake region to the adjacent smooth flow regions. These results dramatically illustrate the changes in the flow with Reynolds number, proceeding from predominantly viscous effects to predominantly inertial effects.

Next, the renowned case of 2-D flow about a circular cylinder (Figure 4.13) is briefly examined. At extremely low Reynolds number based on the diameter of the cylinder

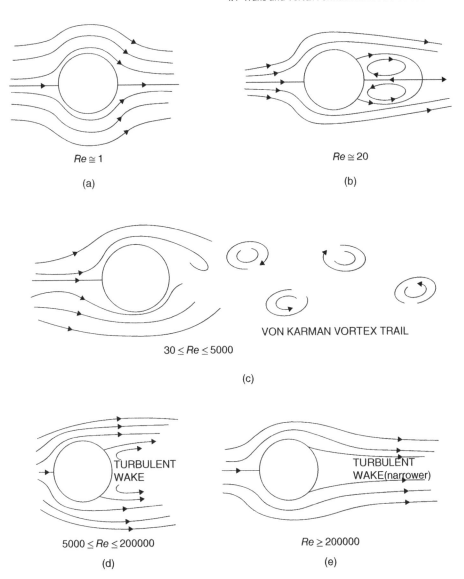

$Re \cong 1$

(a)

$Re \cong 20$

(b)

VON KARMAN VORTEX TRAIL

$30 \leq Re \leq 5000$

(c)

TURBULENT
WAKE

$5000 \leq Re \leq 200000$

(d)

TURBULENT
WAKE(narrower)

$Re \geq 200000$

(e)

Figure 4.13 Flow past a circular cylinder. (a) $Re \cong 1$; (b) $Re \cong 20$; (c) $30 \leq Re \leq 5000$; (d) $5000 \leq Re \leq 200\,000$; (e) $Re > 200\,000$. Source: From [6], by permission of the author and the American Journal of Physics.

$(Re \cong 1)$ the flow, assumed laminar as it approaches, remains attached to the cylinder throughout its complete periphery, as shown in Figure 4.13a. At $Re \cong 20$, the flow form remains symmetrical but flow separation occurs and large wake eddies are formed that reside near the downstream surface of the cylinder, as suggested in Figure 4.13b. For $30 \leq Re \leq 5000$, alternating vortices are shed from the cylinder and form a clear "vortex street" downstream. This phenomenon was first reported by Bénard in 1908 [7]; in the English-speaking world its discovery is attributed to von Kármán, who reported it in 1911 [8] – the alternating vortices are universally referred to as a von Kármán street, although some facetious aerodynamicists use the term boulevard Bénard. The finer details of this striking occurrence are still not fully understood, and have been the object of both experimental and theoretical studies (e.g., [9]). For $30 \leq Re < 5000$. say, there is established behind the cylinder a staggered, stable arrangement of vortices that moves off downstream at a velocity somewhat less than that of the surrounding fluid.

As the Reynolds number increases into the range $5000 \leq Re \leq 200\,000$, the attached flow upstream of the separation flow is laminar. In the separated flow, 3-D patterns are observed, and transition to turbulent flow occurs in the wake – farther downstream from the cylinder for the lower Reynolds numbers, and nearer the cylinder surface as the Reynolds numbers increase. For the larger Reynolds numbers in this range, the cylinder wake undergoes transition immediately after separation, and a turbulent wake is produced between the separated shear layers (Figure 4.13d). Beyond $Re = 200\,000$ (Figure 4.13e) the wake narrows appreciably, resulting in less drag.

Other bluff bodies, notably prisms with triangular, square, rectangular, and other cross sections, give rise to analogous vortex-shedding phenomena (Figure 4.14).

The pronounced regularity of such wake effects was first reported by Strouhal [11], who pointed out that the vortex shedding phenomenon can be described in terms of a non-dimensional number, the Strouhal number:

$$St = \frac{N_s D}{U} \tag{4.25}$$

where N_s is the frequency of full cycles of vortex shedding, D is a characteristic dimension projected on a plane, typically, normal to the wind velocity, and U is the

Figure 4.14 Flow around a rectangular cylinder ($Re = 200$). Source: Reprinted from [10], with permission from Elsevier.

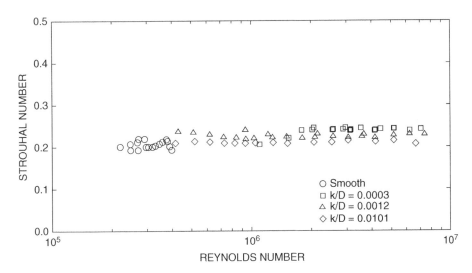

Figure 4.15 Relation between the Strouhal number and the Reynolds number for circular cylinder. Source: Reprinted from [12], with permission from Elsevier.

velocity of the oncoming flow, assumed laminar. The Strouhal number depends upon the cross-sectional shape of the cylindrical body enveloped by the flow.

Figure 4.15 shows the relation of St to Re for a circular cylinder in the range $10^5 \leq Re \leq 10^7$. Coherent vortex shedding was noted to disappear at Reynolds numbers beyond 4×10^5, and contrary to results reported by some observers and summarized in [13], there was no significant increase of the Strouhal number.

Table 4.1 lists values of St for different cross-sectional shapes for Reynolds numbers in the clear vortex-shedding range, the approach flow being laminar.

Figure 4.16 shows a vortex trail made visible by clouds over Jan Mayen Island (Arctic Ocean). For additional material on vortex trails over oceans, see also [15].

As pointed out in [16], the establishment of a vortex trail can be inhibited by a splitter plate, as shown in Figure 4.17. The action of the plate is to prevent flow cross-over between the two rows of vortices aft of the cylinder and thus to quiet the entire wake flow. Qualitatively, the presence of the plate has the same type of effect as lengthening the body in the stream direction and causing it to approach the form of a symmetric airfoil. Following this type of approach, it can be seen that elongated bodies, oriented with their long dimension parallel to the main flow, tend to elicit relatively narrow wakes.

If flows about square and rectangular prisms at high Reynolds numbers are compared (Figure 4.18), the square is seen to produce flow separation followed by a wide, turbulent wake, whereas the more elongated shapes may exhibit separation at leading corners followed downstream by flow reattachment and, finally, once more, by flow separation at the trailing edge. In contrast to the case of Figure 4.18b, if the rectangle is placed with its long dimension normal to the flow, the wake exhibits a strong vortex-shedding characteristic, followed at higher Re by a turbulent wake similar to that produced by the sharp-edged plate (Figures 4.12c and d).

Table 4.1 Strouhal number for a variety of shapes.

Wind	Profile dimensions, in mm	Value of St	Wind	Profile dimensions, in mm	Value of St
	$t = 2.0$	0.120		$t = 1.0$	0.147
		0.137			
	$t = 0.5$	0.120		$t = 1.0$	0.150
	$t = 1.0$	0.144		$t = 1.0$	0.145
					0.142
					0.147
	$t = 1.5$	0.145		$t = 1.0$	0.131
					0.134
					0.137
	$t = 1.0$	0.140		$t = 1.0$	0.121
		0.153			0.143
	$t = 1.0$	0.145		$t = 1.0$	0.135
		0.168			
	$t = 1.5$	0.156		$t = 1.0$	0.160
		0.145			
	Cylinder 11800 < Re < 19100	0.200		$t = 1.0$	0.114
					0.145

Source: From [14], ASCE.

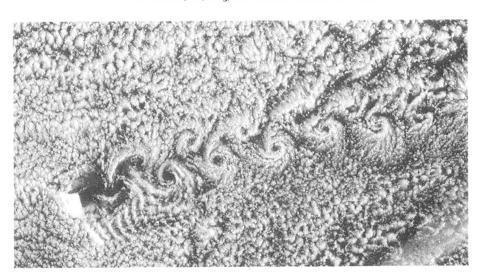

Figure 4.16 Satellite photo of Jan Mayen Island (Arctic Ocean). Source: Credits: NASA/CSFC/LaRC/JPL, MISR Team.

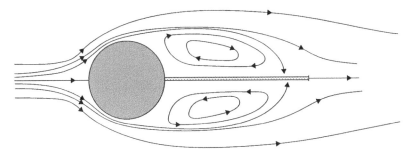

Figure 4.17 Effect of splinter plate on flow behind a circular cylinder. Source: After [16].

4.5 Pressure, Lift, Drag, and Moment Effects on Two-Dimensional Structural Forms

Figure 4.19 shows a section of a bluff body immersed in a flow of velocity U. The flow will develop local pressures p over the body in accordance with the Bernoulli equation:

$$\frac{1}{2}\rho U^2 + p = \text{const} \tag{4.26}$$

where the constant holds along a streamline and U is the velocity on the streamline immediately outside the boundary layer that forms on the body's surface. The integration of the pressures over the body results in a net force and moment. The components of the force in the along-wind and across-wind directions are called *drag* and *lift*, respectively. The drag, lift, and moment are affected by the shape of the body, the Reynolds number, and the incoming flow turbulence.

The body may be designed with the purpose of minimizing drag and maximizing lift, resulting in an airfoil-like shape. In many civil engineering applications the shape of the body is typically fixed by other design objectives than purely aerodynamic ones.

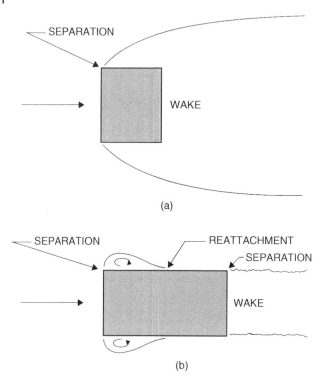

(a)

(b)

Figure 4.18 Flow separation and wake regions, square and rectangular cylinders.

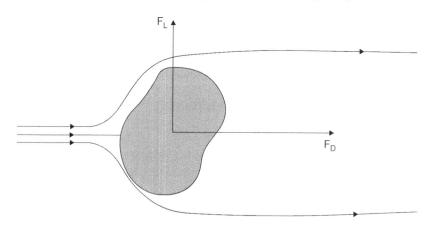

Figure 4.19 Lift and drag on an arbitrary bluff body.

Nevertheless, the lift, drag, and moment induced by the fluid flows will remain of strong interest because these are effects that must be designed against.

It is usual to refer to all pressures measured on a structural surface to the mean dynamic pressure $\frac{1}{2}\rho U^2$ of the far upstream wind or the free stream wind at a distance from the structure. Thus, non-dimensional *pressure coefficients*, C_p, are defined as

$$C_p = \frac{p - p_0}{(1/2)\rho U^2} \tag{4.27}$$

where U is the mean value of the reference wind speed and $p - p_0$ is the pressure difference between local and far upstream pressure p_0. Such non-dimensional forms enable the transfer of model experimental results to full scale, and the establishment of reference values for cataloging the aerodynamic properties of given geometric forms.

Similarly, the net aerodynamic lift and drag forces per unit span F_L and F_D in the across-wind and along-wind direction, respectively, can be rendered dimensionless and expressed in terms of *lift* and *drag coefficients*, C_L and C_D:

$$C_L = \frac{F_L}{(1/2)\rho U^2 B} \tag{4.28}$$

$$C_D = \frac{F_D}{(1/2)\rho U^2 B} \tag{4.29}$$

where B is some typical reference dimension of the structure. For the net flow-induced moment M about the elastic center the corresponding coefficient is

$$C_M = \frac{M}{(1/2)\rho U^2 B^2} \tag{4.30}$$

Figure 4.20 shows the dependence of the mean drag coefficient C_D of circular cylinders immersed in smooth flow. C_D drops sharply in the range $2 \times 10^5 \leq Re \leq 5 \times 10^5$. This is called the *critical region* and corresponds to the transition from laminar to turbulent flow in the boundary layer that forms on the surface of the cylinder. The turbulent mixing that takes place in the boundary layer helps transport fluid with higher momentum toward the surface of the cylinder. Separation then occurs much farther back and the wake consequently narrows, producing a time averaged C_D that is only about one third of its highest value. As Re increases into the supercritical and then the transcritical range $(Re > 4 \times 10^6)$, C_D increases once more but remains much lower than its subcritical values. According to [12], drag coefficients in the transcritical range are about 25% lower than those indicated in Figure 4.20.

Figure 4.21 depicts a typical distribution of the mean pressure coefficient about the circular cylinder in smooth flow as a function of angular position. The pressures corresponding to $\theta = 0°$ and $\theta = 180°$ are referred as the stagnation point and the base pressure, respectively.

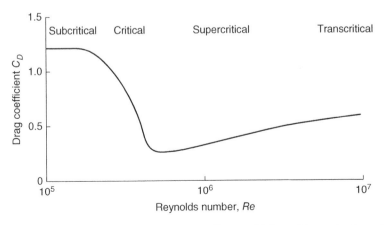

Figure 4.20 Evolution of the mean drag coefficient with Reynolds number for a circular cylinder. Source: After [13]. Courtesy of National Physical Laboratory, UK.

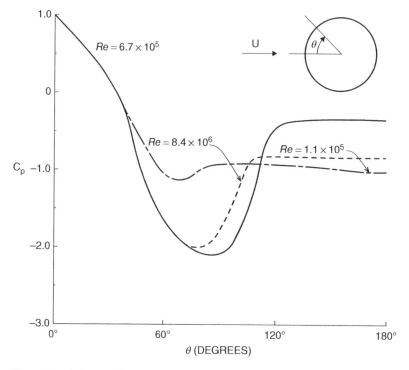

Figure 4.21 Influence of Reynolds number on pressure distribution over a circular cylinder. Source: After [17].

Figure 4.22 illustrates the evolution with Reynolds number of the mean drag coefficient of a square cylinder in smooth flow during successive modifications of its corners. Only the sharp-cornered square exhibits practically unchanging drag with change of Reynolds number. This is accounted for by the early separation of the flow at the upstream corners and the shortness of the afterbody that practically prevents flow reattachment. Squares with rounded corners tend to possess the same kind of critical region as the circular cylinder. Note also, for the circular cylinder, the dependence of the drag upon the roughness of the cylinder surface – see [19].

Because of such effects, certain features of the flow in tests of wind tunnel models can be assumed to be independent of the Reynolds number. This will be the case in some situations in which the flow breaks cleanly away at some identifiable flows past a curved body (e.g., a circular cylinder), this assumption is not warranted.

Table 4.2 shows mean values of C_D and C_L obtained in smooth flow for sectional shapes used in construction. Experiments have shown that for the shapes of Table 4.2 the effects of turbulence are small.

The r.m.s. value of the fluctuating normal force coefficient C_{Nrms} on a square cylinder with side B is shown in Figure 4.23 as a function of angle of attack α with respect to the mean wind direction. Here, the turbulence (with longitudinal integral scale $1.4B$, lateral integral scale $0.4B$, and 10% turbulence intensity) lowers the highest normal force below, and raises the lowest normal force slightly above, the respective values in laminar flow. For the effects of turbulence on the aerodynamics of a square prism, see also [21]. For a study of unsteady forces acting on rigid circular cylinders, see [22].

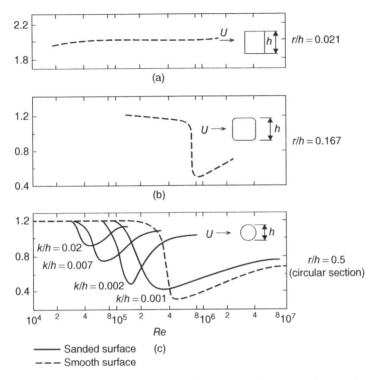

Figure 4.22 Influence of Reynolds number, corner radius, and surface roughness on drag coefficient, square to circular cylinders (r is the corner radius; k is the grain size of sand). Source: After [18].

For members with rectangular cross section the drag force depends upon (i) the ratio b/h between the sides of the cross section and (ii) the turbulence in the oncoming flow. If the ratio b/h is small, no flow reattachment occurs. Depending upon its intensity, the turbulence can enhance the flow entrainment in the wake and, therefore, cause stronger suctions and larger drag (Figure 4.24a). If the ratio b/h is sufficiently large, the turbulence can cause flow reattachment that could not have occurred in smooth flow, and thus results in lower drag (Figure 4.24b).

The dependence of the drag coefficient upon along-wind turbulence intensity in flow with homogeneous turbulence is shown for two ratios b/h in Figure 4.25.

The effect of turbulence in the case of bodies with rounded shapes is, essentially, to reduce the Reynolds number at which the critical region sets in. (The roughness of the body surface (Figure 4.22) has a similar effect, since it promotes turbulence in the boundary layer that forms on the body surface.) Fluid particle moments with higher momentum are thus transported into the lower regions of the boundary layer and help to overcome the adverse pressure gradient responsible for flow separation.

4.6 Representative Flow Effects in Three Dimensions

Most flows have a 3-D character. For example, if a hypothetical laminar flow consisting of an air mass displaced uniformly as a single unit encounters an object, it will be

Table 4.2 Two-dimensional drag and lift coefficients.

Profile and wind direction	C_D	C_L
(square section, with C_L and C_D axes indicated)	2.03	0
(vertical flat plate)	1.96 – 2.01	0
(I-beam section)	2.04	0
(H-section)	1.81	0
(L-angle, upright)	2.0	0.3
(angle section)	1.83	2.07
(L-angle)	1.99	−0.09
(angle section)	1.62	−0.48
(double-flange section)	2.01	0
(T-section)	1.99	−1.19
(channel/box section)	2.19	0

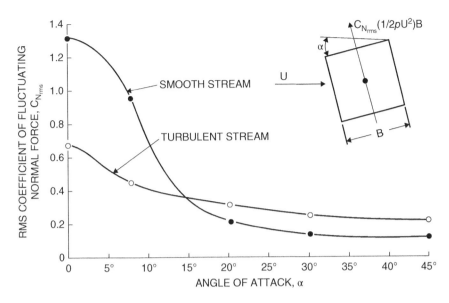

Figure 4.23 Variation of the coefficient of fluctuating normal force, C_{Nrms}, with angle of attack for a square prism. Source: From [20], reproduced with permission.

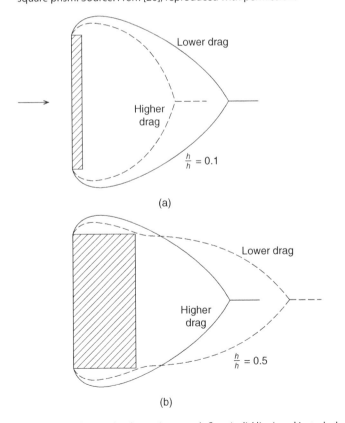

Figure 4.24 Separation layers in smooth flow (solid line) and in turbulent flow (interrupted line). Source: After [23].

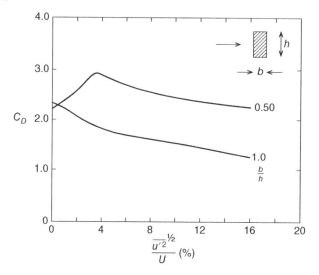

Figure 4.25 Dependence of drag coefficient on turbulence intensity. Source: After [23].

diverted in several directions. Also, the passage of such a flow along a surface sets up boundary-velocity gradients. And three-dimensionality is clearly inherent in turbulent flows. Although the general equations for fluid flow remain available for application, in structural engineering practice most aerodynamic studies rely partially or fully on experiment.

4.6.1 Cases Retaining Two-Dimensional Features

The success of the 2-D flow models discussed in the previous section has in a few cases been considerable because some actual flows retain certain 2-D features, at least to a first approximation. Consider, for example, the case of a long rod of square cross section in an air flow with uniform mean velocity normal to one face. Except near the ends of the rod, the mean flow may, in some cases, be considered for practical purposes as 2-D. However, the effects associated with flow fluctuations are not identical in different strips, the differences between events that take place at any given time increasing with separation distance. This is shown in Figure 4.26 for the pressure difference between centerlines of top and bottom faces of the rod under both laminar and turbulent approaching flow. It is observed that the three-dimensionality of the flow manifests itself through spanwise loss of correlation r_{AB} between pressure differences (measured respectively between point A' at section A and point B' of section B), this correlation loss being accentuated when turbulence is present in the oncoming flow. From this example one may infer that fluctuating phenomena, including vortex shedding, cannot normally be expected to be altogether uniform along the entire length of a cylindrical body, even if the flow has uniform mean speed and the body is geometrically uniform. The animation of Figure 4.27, based on wind tunnel measurements in turbulent boundary-layer flow, clearly demonstrates the imperfect spatial coherence of pressures on a low-rise structure. Investigations reported in [24] were among the first to account explicitly for the imperfect spatial coherence of aerodynamic pressures on low-rise structures.

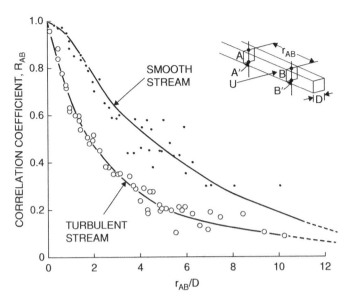

Figure 4.26 Spanwise correlation of the fluctuating pressure difference across the center line of a long square cylinder for flow normal to a face. Source: From [20], reproduced with permission.

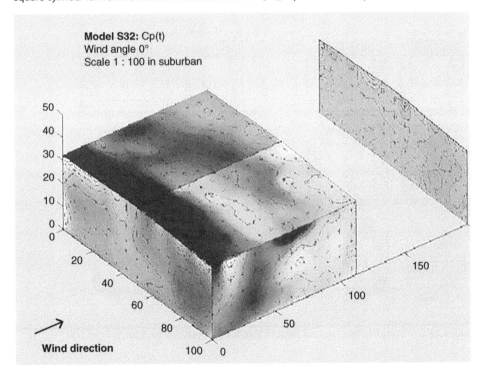

Figure 4.27 Fluctuating wind pressure model for 100 ft × 200 ft × 32 ft building in suburban terrain; gable roof with 1/24 slope. Source: Based on 1 : 100 model scale boundary-layer wind tunnel simulation, University of Western Ontario; animation created by Dr. A. Grazini. Mean wind speed normal to end walls. Note asymmetry of pressures with respect to vertical plane containing ridge line. (Video available at https://www.nist.gov/wind).

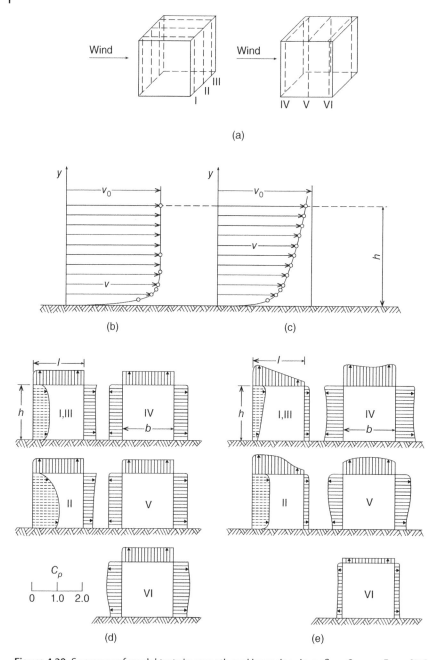

Figure 4.28 Summary of model tests in smooth and boundary-layer flow. Source: From [25].

In practice, mean flow conditions upwind of tall slender structures are usually not uniform; indeed, in the atmospheric boundary layer the mean flow velocity increases with height. Also, certain structures (e.g., stacks) are not geometrically uniform. These important features – in addition to the incident turbulence – further decrease the coherence of vortices shed in the wake of structures.

4.6.2 Structures in Three-Dimensional Flows: Case Studies

The complexities of wind flow introduced by the geometries of typical structures and by the characteristics of the terrain and obstacles upstream emphasize the need to carry out detailed studies of wind pressures experimentally using wind tunnel models and simulation. Wind flows around buildings are prime examples of 3-D flows that cannot be described acceptably by 2-D models. In order to give some idea of the type of results so obtained and to emphasize the important roles of the boundary-layer velocity profile and of the turbulence in such results, a few examples are cited below.

The existence of significant differences between drag or pressure coefficients measured in uniform and boundary-layer flow was first pointed out by Flachsbart in 1932 [25]. Figure 4.28b and c show the respective mean wind speed profiles, and Figure 4.28d and e show pressure coefficient measurement results for wind normal to a building face (Figure 4.28a). As shown in Chapter 5, a large number of large- and full-scale measurements have been made in the intervening years, owing to the need to assess uncertainties in data obtained in conventional wind tunnels.

Figures 4.29 and 4.30 are classic representations by Baines [26] of pressure distributions for structures under laminar and shear flows. Far more detailed measurement

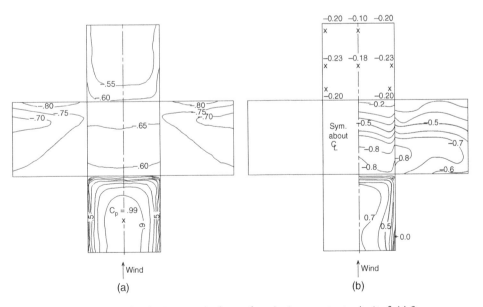

Figure 4.29 (a) Pressure distributions on the faces of a cube in a constant velocity field. Source: From [26]. (b) Pressure distributions on the faces of a cube in a boundary-layer velocity field. Source: From [26].

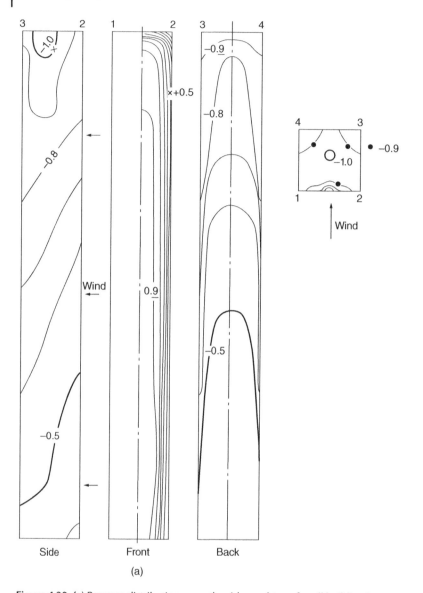

Figure 4.30 (a) Pressure distributions over the sides and top of a tall building in a constant velocity field. Source: From [26]. (b) Pressure distributions over the sides and top of a tall building in a boundary-layer velocity field. Source: From [26].

results, including data on fluctuating pressures, are available in modern databases containing results of wind tunnel measurements (NIST/UWO [27], TPU [28]), as well as in reports on large- and full-scale measurements (e.g., [29–32]).

Load on secondary structural members (e.g., joists) are determined by the algebraic sums of external and internal pressures acting on them. Figure 4.31 depicts the ideal case in which (a) the building is hermetically sealed, so that the internal pressure is

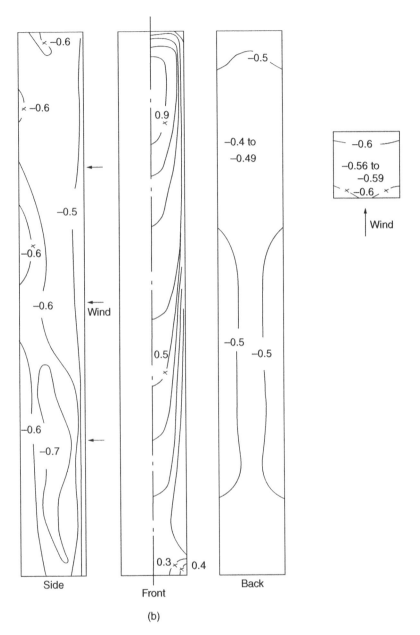

Figure 4.30 (Continued)

not affected by the external wind flow, (b) the building has openings on the windward side only, in which case wind induces positive internal pressures, (c) the building has openings on the leeward side, in which case wind induces internal suctions, and (d) the building has openings on both the windward and leeward sides, in which case induces internal pressures that may be either positive or negative. Wind-tunnel data on internal

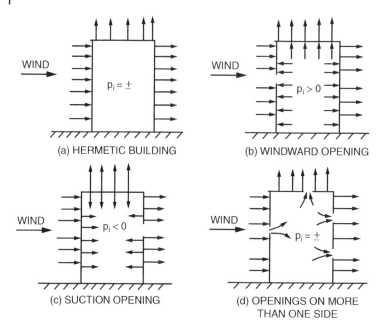

Figure 4.31 Mean internal pressures in buildings with various opening distributions. Source: From [33], with permission from ASCE.

pressures are reported in [34–38]. Recent measurements of internal pressures on a large-scale model of an industrial building, and comparisons with values specified in the ASCE 7-16 Standard [39], are reported in [40].

References

1 Roshko, A. (1993). Perspectives on bluff body aerodynamics. *Journal of Wind Engineering and Industrial Aerodynamics* 49: 79–100.

2 Schluenzen, K. H., ed., Computational Wind Engineering 2014 (CWE 2014). *Proceedings of the Sixth International Symposium on Computational Wind Engineering. Hamburg, Germany*, 2014.

3 Schuster, D. M., "The Expanding Role of Applications in the Development and Validation of CFD at NASA," in *Computational Fluid Dynamics 2010: Proceedings of the Sixth International Conference on Computational Fluid Dynamics, ICCFD6, St Petersburg, Russia, on July 12–16, 2010*, A. Kuzmin, ed., Berlin, Heidelberg: Springer Berlin Heidelberg, 2011, pp. 3–29.

4 Batchelor, G.K. (1967). *An Introduction to Fluid Dynamics*. Cambridge: Cambridge University Press.

5 Centre Scientifique et Technique du Bâtiment (1980). *Aérodynamique*. Nantes, France: Centre Scientifique et Technique du Bâtiment.

6 Nathan, A.M. (2008). The effect of spin on the flight of a baseball. *American Journal of Physics* B76: 119–124.

7 Bénard, H. (1908). Formations de centres de gyration à l'arrière d'un obstacle en movement. *Comptes rendus de l'Académie des Sciences Paris* 147: 839–842.

8 von Kármán, T. (1911). Über den Mechanismus des Widerstandes, den ein bewegter Körper in einer Flüssigkeit erfährt. *Nachrichten von der Gesellschaft der Wissenschaften zu Göttingen, Mathematisch-Physikalische Klasse* 509–517.

9 Gerrard, J.H. (2006). The mechanics of the formation region of vortices behind bluff bodies. *Journal of Fluid Mechanics* 25: 401–413.

10 Nakamura, Y. (1993). Bluff body aerodynamics and turbulence. *Journal of Wind Engineering and Industrial Aerodynamics* 49: 65–68.

11 Strouhal, V. (1878). Über eine besondere Art der Tonerregung. *Annalen der Physik* 241: 216–251.

12 Shih, W.C.L., Wang, C., Coles, D., and Roshko, A. (1993). Experiments on flow past rough circular cylinders at large reynolds numbers. *Journal of Wind Engineering and Industrial Aerodynamics* 49: 351–368.

13 Wooton, L.R. and Scruton, C. (1971). Aerodynamic stability. In: *The Modern Design of Wind-Sensitive Structures* (ed. A.R. Collins), 65–81. London: Construction Industry Research and Information Association.

14 ASCE Task Committee (1961). Wind Forces on structures. *Transactions on ASCE* 126: 1124–1198.

15 Pao, H.P. and Kao, T.W. (1976). On vortex trails over Ocean Islands. *Atmospheric Science, Meteorological Society of the Republic of China, Taiwan* 3: 28–38.

16 Roshko, A. (1955). On the wake and drag of bluff bodies. *Journal of the Aeronautical Sciences* 22: 124–132.

17 Roshko, A. (1961). Experiments on the flow past a circular cylinder at very high Reynolds number. *Journal of Fluid Mechanics* 10: 345–356.

18 Scruton, C., Rogers, E.W.E., Menzies, J.B., and Scorer, R.S. (1971). Steady and unsteady wind loading of buildings and structures [and discussion]. *Philosophical Transactions of the Royal Society of London Series A, Mathematical and Physical Sciences* 269: 353–383.

19 Güven, O., Farell, C., and Patel, V.C. (1980). Surface-roughness effects on the mean flow past circular cylinders. *Journal of Fluid Mechanics* 98: 673–701.

20 Vickery, B.J. (1966). Fluctuating lift and drag on a long cylinder of square cross section in a smooth and turbulent flow. *Journal of Fluid Mechanics* 25: 481–494.

21 Lee, B.E. (1975). The effect of turbulence on the surface pressure field of a square prism. *Journal of Fluid Mechanics* 69: 263–282.

22 So, R.M.C. and Savkar, S.D. (1981). Buffeting forces on rigid circular cylinders in cross flows. *Journal of Fluid Mechanics* 105: 397–425.

23 Laneville, A., Gartshore, I. S., and Parkinson, G. V., "An explanation of some effects of turbulence on bluff bodies," In: *Proceedings, Fourth International Conference, Wind Effects on Buildings and Structures*, Cambridge University Press, Cambridge, 1977.

24 Stathopoulos, T., Davenport, A.G., and Surry, D. (1981). Effective wind loads on flat roofs. *Journal of the Structural Division* 107: 281–298.

25 Flachsbart, O. (1932). Winddruck auf geschlossene und offene Gebäude. In: *Ergebnisse der Aerodynamischen Versuchsanstalt zu Göttingen* (ed. I.V. Lieferung, L. Prandtl and A. Betz), 128–134. Munich and Berlin: Verlag von R. Oldenbourg.

26 Baines, W. D., "Effects of velocity distribution on wind loads and flow patterns on buildings," *Proceedings, Symposium No. 1, Wind Effects on Buildings and Structures, held at the National Physical Laboratory, England, UK*, in 1963, published by HMSO London in 1965.

27 NIST/UWO. NIST/UWO aerodynamic database [Online]. Available: https://www .nist.gov/wind.

28 TPU. TPU aerodynamic database [Online]. Available: http://wind.arch.t-kougei.ac.jp/ system/eng/contents/code/tpu.

29 Levitan, M.L., Mehta, K.C., Vann, W.P., and Holmes, J.D. (1991). Field measurements of pressures on the Texas Tech building. *Journal of Wind Engineering and Industrial Aerodynamics* 38: 227–234.

30 Richards, P.J. and Hoxey, R.P. (2008). Wind loads on the roof of a 6m cube. *Journal of Wind Engineering and Industrial Aerodynamics* 96: 984–993.

31 Richards, P.J. and Hoxey, R.P. (2012). Pressures on a cubic building – Part 1: full-scale results. *Journal of Wind Engineering and Industrial Aerodynamics* 102: 72–86.

32 Richards, P.J. and Hoxey, R.P. (2012). Pressures on a cubic building – Part 2: quasi-steady and other processes. *Journal of Wind Engineering and Industrial Aerodynamics* 102: 87–96.

33 Liu, H. and Saathoff, P.J. (1963). Internal pressure and building safety. *Journal of the Structural Division ASCE* 108: 223–224.

34 Holmes, J. D., "Mean and fluctuating internal pressures induced by wind," in the *Fifth International Conference*, Fort Collins, CO, pp. 435–450, 1980.

35 Liu, H. (1982). Internal pressure and building safety. *Journal of the Structural Division* 108: 2223–2234.

36 Saathoff, P.J. and Liu, H. (1983). Internal pressure of multi-room buildings. *Journal of Engineering Mechanics* 109: 908–919.

37 Harris, R.I. (1990). The propagation of internal pressures in buildings. *Journal of Wind Engineering and Industrial Aerodynamics* 34: 169–184.

38 Vickery, B.J. (1994). Internal pressures and interactions with the building envelope. *Journal of Wind Engineering and Industrial Aerodynamics* 53: 125–144.

39 ASCE, "Minimum design loads for buildings and other structures (ASCE/SEI 7-16)," in ASCE Standard ASCE/SEI 7-16, Reston, VA: American Society of Civil Engineers, 2016.

40 Habte, F., Chowdhury, A.G., and Zisis, I. (2017). Effect of wind-induced internal pressure on local frame forces of low-rise buildings. *Engineering Structures* 143: 455–468.

5

Aerodynamic Testing

5.1 Introduction

To date, testing remains the predominant means of obtaining aerodynamic data usable for the design of engineering structures. It is well established that, for most applications, the testing has to be performed in flows simulating the main features of atmospheric flows.

A rigorous simulation of atmospheric flows would require that the non-dimensional form of the equations of fluid motion and their attendant boundary conditions be the same in the prototype and at model scale. This is not possible in practice, owing primarily to the violation of the Reynolds number similarity requirement and the impossibility of rigorously simulating turbulent atmospheric flows. Wind tunnel testing is therefore an art that requires consideration of the errors inherent in imperfect simulations (see Chapters 7 and 12). Attempts to quantify such errors are made by, among other means, performing full-scale aerodynamic measurements, a difficult endeavor owing to large uncertainties in the prototype wind flow that are often encountered in practice.

The purpose of this chapter is to discuss similarity requirements (Section 5.2), describe aerodynamic testing facilities used for civil engineering purposes (Section 5.3), consider the dependence of the aerodynamic response of wind tunnel models upon Reynolds number and the turbulence characteristics of simulated atmospheric boundary layer flows (Section 5.4), discuss blockage effects (Section 5.5), and describe and comment on wind effects based on High Frequency Force Balance (HFFB) measurements (Section 5.6) and on pressure measurements (Section 5.7). Aeroelastic testing, including testing of suspended-span bridges, is discussed in Part III of the book. For a rich source of useful information see [1].

5.2 Basic Similarity Requirements

5.2.1 Dimensional Analysis

Basic similarity requirements can be determined from dimensional analysis. For engineering structures, it may be assumed that the aerodynamic force F on a body is a function of flow density ρ, flow velocity U, a characteristic dimension D, a characteristic

Wind Effects on Structures: Modern Structural Design for Wind, Fourth Edition. Emil Simiu and DongHun Yeo.
© 2019 John Wiley & Sons Ltd. Published 2019 by John Wiley & Sons Ltd.

frequency n, and the flow viscosity μ. The following relation governing dimensional consistency then holds:

$$F \stackrel{d}{=} \rho^\alpha U^\beta D^\gamma n^\delta \mu^\varepsilon \tag{5.1}$$

where α, β, γ, δ, ε are exponents to be determined. Each of the quantities ρ, U, D, n, μ can be expressed dimensionally in terms of the three fundamental quantities: mass M, length L, and time T, so Eq. (5.1) can be written as

$$\frac{ML}{T^2} \stackrel{d}{=} \left(\frac{M}{L^3}\right)^\alpha \left(\frac{L}{T}\right)^\beta (L)^\gamma \left(\frac{1}{T}\right)^\delta \left(\frac{M}{LT}\right)^\varepsilon \tag{5.2}$$

(for the dimensions of the viscosity follow see Section 4.1.2). Dimensional consistency requires that

$$
\begin{aligned}
M: & \quad 1 = \alpha + \varepsilon \\
L: & \quad 1 = -3\alpha + \beta + \gamma - \varepsilon \\
T: & \quad -2 = -\beta - \delta - \varepsilon
\end{aligned}
\tag{5.3}
$$

from which there follows, for example, that

$$
\begin{aligned}
\alpha &= 1 - \varepsilon \\
\beta &= 2 - \varepsilon - \delta \\
\gamma &= 2 - \varepsilon + \delta
\end{aligned}
\tag{5.4}
$$

Substitution of these relations in Eq. (5.1) yields

$$F \stackrel{d}{=} \rho^{1-\varepsilon} U^{2-\varepsilon-\delta} D^{2-\varepsilon+\delta} n^\delta \mu^\varepsilon \tag{5.5}$$

or

$$F \stackrel{d}{=} \rho U^2 D^2 \left(\frac{Dn}{U}\right)^\delta \left(\frac{\mu}{\rho U D}\right)^\varepsilon \tag{5.6}$$

meaning that the dimensionless force coefficient $F/(\rho U^2 D^2)$ is a function of the dimensionless ratios Dn/U and $\mu/(\rho U D)$ (or of their reciprocals).

Generally, an equation involving n physical variables can be written in terms of $p = n - k$ dimensionless parameters constructed from those original variables, where k is the number of physical dimensions involved in the equation. This statement is a form of the Buckingham π theorem. In the preceding example, $n = 5$ (Eq. [5.1]), $k = 3$ (i.e., M, L, and T) and, as indicated following Eq. (5.6), $p = 2$.

In some wind engineering problems (e.g., the vibrations of suspended bridges) the aerodynamic forces are also functions of the acceleration of gravity, g. By introducing g^ζ into Eq. (5.1) it can easily be shown that the force is also a function of the non-dimensional ratio U^2/Dg, called the *Froude number*. The non-dimensional ratio $\rho U D/\mu = U D/\nu$ is the well-known *Reynolds number* and $\nu = \mu/\rho$ is the kinematic viscosity of the fluid (Section 4.1.2). The parameter nD/U is called the reduced frequency, and its reciprocal is the reduced velocity. If the frequency n being considered is the vortex shedding frequency, the reduced frequency is the *Strouhal number* (Section 4.4). If n is replaced by the Coriolis parameter (Section 1.2), the reduced velocity is called the *Rossby number*.

5.2.2 Basic Scaling Considerations

Similarity requires that the reduced frequencies and the Reynolds numbers be the same in the laboratory and in the prototype. This is true regardless of the nature of the frequencies involved (e.g., vortex shedding frequencies, natural frequencies of vibration, frequencies of the turbulent components of the flow), or of the densities being considered (e.g., fluid density, density of the structure). For example, if the reduced frequency is the same in the prototype and in the laboratory (i.e., at model scale), applying this requirement to the vortex shedding frequency n_v and to the fundamental frequency of vibration of the structure n_s we have

$$\left(\frac{n_v D}{U}\right)_p = \left(\frac{n_v D}{U}\right)_m \tag{5.7}$$

and

$$\left(\frac{n_s D}{U}\right)_p = \left(\frac{n_s D}{U}\right)_m \tag{5.8}$$

where the indexes m and p stand for model and prototype, respectively.

It follows from Eqs. (5.7) and (5.8) that

$$\left(\frac{n_s}{n_v}\right)_p = \left(\frac{n_s}{n_v}\right)_m \tag{5.9}$$

This is also true of the ratios of all other relevant quantities (lengths, densities, velocities). Thus, for the density of the structure and the density of the fluid it must be the case that

$$\left(\frac{\rho_s}{\rho_{air}}\right)_p = \left(\frac{\rho_s}{\rho_f}\right)_m \tag{5.10}$$

where ρ_f is the density of the fluid in the laboratory. For the same reason

$$\left(\frac{U(z_1)}{U(z_2)}\right)_p = \left(\frac{U(z_1)}{U(z_2)}\right)_m \tag{5.11}$$

where z_1 and z_2 are heights above the surface. In particular, if in the prototype the velocities conform to a power law with exponent $\bar{\alpha}$, it follows from Eq. (5.11) that in the laboratory the velocities must conform to the power law with the same exponent $\bar{\alpha}$. To see this, Eq. (5.11) is re-written as follows:

$$\left(\frac{z_1}{z_2}\right)_p^{\bar{\alpha}} = \left(\frac{U(z_1)}{U(z_2)}\right)_m \tag{5.12}$$

Since $(z_1/z_2)_p = (z_1/z_2)_m$ by virtue of geometric similarity, it follows from the preceding equation that similarity is satisfied if

$$\left(\frac{z_1}{z_2}\right)_m^{\bar{\alpha}} = \left(\frac{U(z_1)}{U(z_2)}\right)_m \tag{5.13}$$

Since there are three fundamental requirements concerning mass, length, and time, three fixed choices of scale can be made. This choice determines all other scales. For example, let the *length scale*, the *velocity scale*, and the *density scale* be denoted by $\lambda_L = D_m/D_p$, $\lambda_U = U_m/U_p$, and $\lambda_\rho = \rho_m/\rho_p$. The reduced frequency requirement

$$\left(\frac{nD}{U}\right)_p = \left(\frac{nD}{U}\right)_m \tag{5.14}$$

controls the *frequency scale*, λ_n, for all pertinent test frequencies. From Eq. (5.14) it follows immediately that $\lambda_n = \lambda_U/\lambda_L$. The time scale λ_T is the reciprocal of λ_n.

In principle, for similarity between prototype (i.e., full-scale) and laboratory flows to be achieved, the respective Reynolds numbers $Re = UD/\nu$ must be the same. This requirement is referred to as *Reynolds number similarity*. In aerodynamic facilities for testing models of structures the fluid being used is air at atmospheric pressure, and Reynolds number similarity is unavoidably violated.

5.3 Aerodynamic Testing Facilities

To achieve similarity between the model and the prototype, it is in principle necessary to reproduce at the requisite scale the characteristics of atmospheric flows, that is, (i) the variation of the mean wind speed with height, and (ii) the turbulence characteristics. The purpose of this section is to describe facilities intended to do so, including facilities designed to simulate thunderstorm and tornado winds. Also described in this section are facilities used for full- or large-scale tests of special structures such as lamp posts, and for providing data on wind-driven rain intrusion and on snow deposition.

In long wind tunnels, a boundary layer with a depth of 0.5–1 m develops naturally over a rough floor in test sections with lengths of the order of 20 m in length (Figures 5.1–5.3). In such tunnels, as well as in tunnels with considerably shorter test sections (e.g., 5–10 m), the depth of the boundary layer is increased above these values by placing at the test section entrance passive devices such as spires (e.g., Figure 5.3), grids, barriers, fences, singly or in combination, some of which are illustrated subsequently. The height of long tunnels may be adjusted to achieve a zero-pressure gradient streamwise, which owing to energy losses associated with flow friction at the walls and internal friction due to turbulence would otherwise not occur.

The following procedure for the design of spires with the configuration of Figure 5.5 was proposed in [4][1]:

1) Select the desired boundary-layer depth, δ.
2) Select the desired shape of the mean velocity profile defined by the power law exponent, α.
3) Obtain the height h of the spires from the relation.

$$h = \frac{1.39\delta}{1 + \alpha/2} \tag{5.15}$$

4) Obtain the width b of the spire base from Figure 5.6, in which H is the height of the tunnel test section.

---

1 The base dimension of the triangular splitter plate in Figure 5.5 is $h/4$; the lateral dimension is $h/4$. The lateral spacing between the spires is $h/2$. The width of the tunnel need not be an integral multiple of $h/2$.

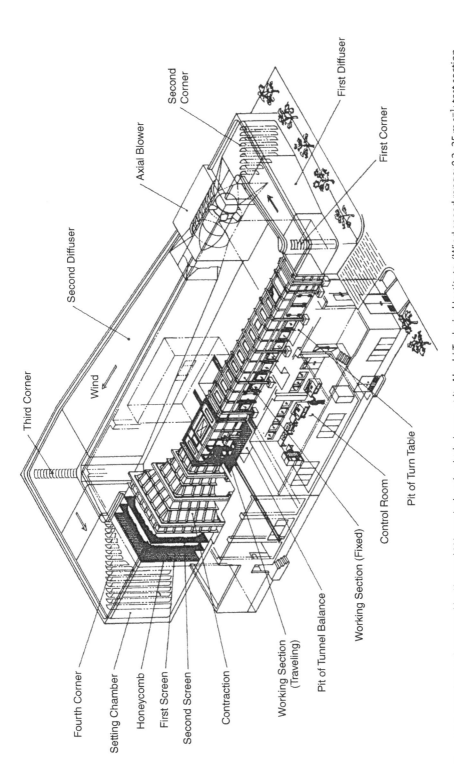

Figure 5.1 Wind tunnel operated by Kawasaki Heavy Industries, Ltd., Japan, at its Akashi Technical Institute. (Wind speed range: 0.2–25 m s^{-1}; test section dimensions: 2.5 × 3 × 20 m.) Source: From: [2], with permission from ASCE.

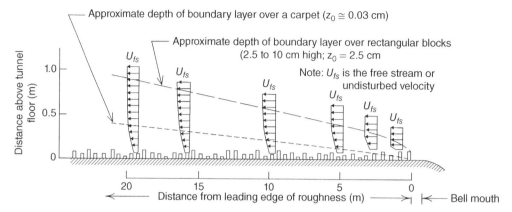

Figure 5.2 Development of boundary layer in a long wind tunnel. Source: After [3].

Figure 5.3 Wind tunnel, Colorado State University. Model and turntable are in the foreground and spires are in the background. Source: Courtesy of Professor B. Bienkiewicz.

The desired mean wind profile occurs at a distance $6h$ downstream from the spires. According to [4, 5], the wind tunnel floor downwind of the spires should be covered with roughness elements, for example, cubes with height k such that

$$\frac{k}{\delta} = \exp\left[\frac{2}{3}\ln\left(\frac{D}{\delta}\right) - 0.1161\left(\frac{2}{C_f} + 2.05\right)^{1/2}\right] \tag{5.16}$$

where D is the spacing of the roughness elements,

$$C_f = 0.136\left(\frac{\alpha}{1+\alpha}\right)^2 \tag{5.17}$$

and α is the exponent of the power law describing the mean wind speed profile. According to [4, 5], Eqs. (5.16) and (5.17) are valid in the range $30 < \delta D^2/k^3 < 2000$. Some laboratories have adopted the system proposed in [4], others have used other methods for designing their flow management system (see, e.g., Figure 5.4).

Figure 5.4 Boundary-layer wind tunnel, University of Florence, Prato, Italy. Source: Courtesy of Professor Claudio Borri.

Figure 5.5 A proposed spire configuration. Source: Reprinted from [4], with permission from Elsevier.

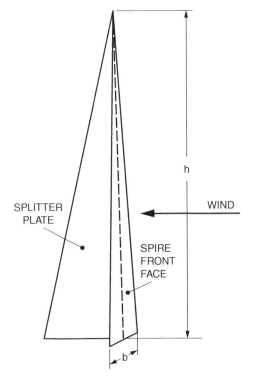

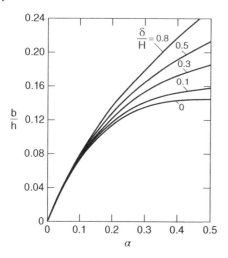

Figure 5.6 Proposed graph for obtaining spire base width. Source: Reprinted from [4], with permission from Elsevier.

Various aerodynamic testing facilities are described in the following.

National Aeronautical Establishment, National Research Council of Canada. A short wind tunnel with 9 m × 9 m cross section, designed for aeronautical applications, has occasionally been used for civil engineering purposes, and is shown in Figure 5.7. The drawback of this facility from a civil engineering point of view is that the test section is too short to allow the flow to develop features with an acceptable resemblance to those of the atmospheric boundary layer.

Figure 5.7 Spire and roughness arrays in a short wind tunnel. Source: Courtesy of the National Aeronautical Establishment, National Research Council of Canada.

Figure 5.8 Interior view of IBHS Research Center with full-scale specimens, placed on the 16.8 m diameter turntable with a surface area of 220 m². The 105-fan array with 300 hp motors is located on the left side of the picture. Source: Courtesy of the Institute for Business & Home Safety.

IBHS Research Center. Figure 5.8 shows an outside and inside view of the Institute for Business & Home Safety (IBHS) Research Center in South Carolina, a multi-peril facility capable of testing structures subjected to realistic Category 1, 2, and 3 hurricanes, extra-tropical windstorms, thunderstorm frontal winds, wildfires, and hailstorms. One purpose of the test performed on the two buildings shown in Figure 5.8 was to offer a vivid illustration of the benefits of robust construction by contrasting, in a video, the good performance of the stronger of the two buildings and the collapse of the weaker building.

Florida International University Wall of Wind Experimental Facility. The Wall of Wind (WoW) is powered by twelve 4.9 m diameter fans and is capable of testing in up to 70 m s^{-1} (157 mph) wind speeds (Figures 5.9 and 5.10). The test section is 6.1 m × 4.3 m, and the turntable diameter is 4.9 m. Testing can be performed at scales approximately twice as large, and Reynolds numbers approximately five times as large, as in facilities such as, for example, the wind tunnel in Figure 5.1. As can be seen in Figure 5.10, the spires and floor roughness elements for the simulation are similar to those used in typical wind tunnels. The facility can be used for destructive testing and for the simulation of water intrusion due to wind-driven rain.

University of Florida (UF) Boundary Layer Wind Tunnel. The University of Florida's major aerodynamic testing facility is its boundary layer wind tunnel, with a 6 m wide, 3 m high, and 40 m long test section, and a 16 m s^{-1} maximum flow speed (Figure 5.11). The floor roughness elements, which help to simulate various surface exposures, are automated and individually controlled. This feature allows fine tuning of the boundary layer at the test section, and rapid reconfiguring for efficient testing using multiple exposures.

Tornado Simulator, Iowa State University (ISU). Basic ideas on facilities for tornado simulation were developed in [6] and [7], among others. The ISU tornado simulator

Figure 5.9 Twelve-fan wall of wind, Florida International University. Source: Courtesy of Professor A. Gan Chowdhury.

Figure 5.10 Twelve-fan wall of wind, Florida International University: view of test section. Source: Courtesy of Professor A. Gan Chowdhury.

Figure 5.11 University of Florida boundary-layer wind tunnel. Source: Courtesy of Professor K. R. Gurley.

is a modern version of the facility described in [6], and is shown schematically in Figure 5.12 [8].

WindEEE Dome. The Wind Engineering, Energy and Environment (WindEEE) Dome [9, 10] is an innovative hexagonal wind tunnel that allows for atmospheric boundary layer simulations over extended areas and complex terrain, and of tornadoes, downbursts, and microbursts (Figures 5.13 and 5.14).

For the atmospheric boundary layer simulation mode (Figures 5.13a and 5.14), the test section is 14 m wide, 3.8 m high, and 25 m long, and the maximum flow velocity is 35 m s^{-1}. The tornado simulation mode (Figures 5.13b) allows the modeling of category F0–F3 tornado flows with vortex diameters of up to 4.5 m, translation speeds of up to 2 m s^{-1}, and flow velocities of up to 25 m s^{-1}. The downburst/microburst simulation mode (Figure 5.13c) can achieve flows with up to 2 m s^{-1} translation speeds and 30 m s^{-1} velocities.

One of the six walls shown in Figure 5.13 has four rows of 15 independently adjustable fans each, used to simulate the atmospheric boundary layer flow. The other five walls have each eight fans at their base. For the tornado simulation mode, directional vanes are placed in front of each of those fans. The angle of orientation of the vanes can be adjusted to impart the desired swirl ratio to the flow (i.e., the ratio between the tangential velocity and the radial velocity in the vortex). Six large fans placed in the upper chamber (Figure 5.13) produce an updraft shown schematically in Figure 5.13b. For details on various capabilities of the WindEEE facility, including measurement capabilities, see [9, 10].

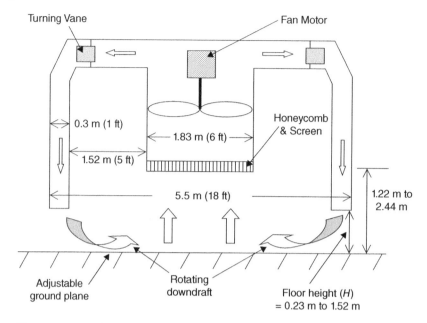

Figure 5.12 Iowa State University tornado simulator. Source: Courtesy of Professor P. Sarkar.

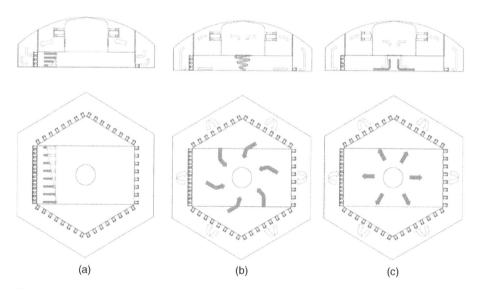

(a) (b) (c)

Figure 5.13 Schematic cross section: (a) Atmospheric boundary-layer simulation mode; (b) tornado flow simulation mode; (c) downburst/microburst simulation mode. Source: Courtesy Professor H. Hangan.

Figure 5.14 View of test section. Source: Courtesy Professor H. M. Hangan.

Politecnico di Milano, Milan, Italy. The test section of its large-scale aerodynamic testing facility is 13.85 m wide, 3.85 m high, and 35 m long, and the maximum wind speed is 16 m s^{-1}.[2]

Centre Scientifique et Technique du Bâtiment (CSTB), Nantes, France. The test section of its large-scale boundary-layer wind tunnel (Figures 5.15 and 5.16) is 4 m wide, 1.7–3.5 m high, and 15 m long, and the maximum wind speed is 30 m s^{-1}. Note in Figures 5.15 and 5.16 that the passive flow management devices being used are different depending upon type of application. Like other prominent laboratories, CSTB

Figure 5.15 Test section of boundary-layer wind tunnel. Source: Photo Florence Joubert; courtesy of CSTB.

―――――――
2 No picture available at the time of printing.

Figure 5.16 Test section of boundary-layer wind tunnel. Source: Courtesy of CSTB. Note that for this application the flow management devices placed at the entrance to the test section are radically different from the typical spires.

operates large facilities for testing: wind-driven rain intrusion (Figure 5.17); roofing (Figure 5.18); snow deposition (Figure 5.19); and other applications.

Technical University Eindhoven (TUE). The TUE boundary-layer wind tunnel test section is 27 m long, 3 m wide, and 2 m high. Wind speeds can be as high as 30 m s^{-1}. The wind tunnel is designed for build environment, maritime, sports, vehicle aerodynamics, air quality, and wind energy applications. Both open and closed circuit modes are feasible (Figure 5.20). Measurement equipment includes 3-D Laser Doppler Anemometry.

Figure 5.17 Wind-driven rain intrusion test. Source: Courtesy of CSTB.

Figure 5.18 Roofing test. Source: Courtesy of CSTB.

Figure 5.19 Snow deposition test. Source: Courtesy of CSTB.

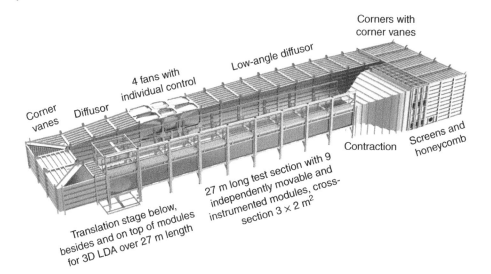

Figure 5.20 TUE boundary layer wind tunnel. Source: Courtesy of Professor B. Blocken.

5.4 Wind Tunnel Simulation of Atmospheric Boundary Layers

5.4.1 Effect of Type of Spires and Floor Roughness Elements

Figure 5.21 [11] shows the mean velocity and the longitudinal and vertical turbulence intensity profiles at (i) 6.1 m and (ii) 18.3 m downwind of the test section entrance, for flows obtained by using three different types of spires, the wind floor being covered by staggered 12.7 mm cubes spaced 50.8 mm apart. In Figure 5.21 the boundary-layer thickness δ, the mean wind speed at elevation δ, and the power law exponent α are denoted by delta, Uinf, and EXP, respectively. It was assumed in the study that the mean flow with power law exponent $\alpha = 0.16$ at station $x = 6.1$ m and $\alpha = 0.29$ at station $x = 18.3$ m are approximately representative of open terrain and suburban terrain, respectively.

Some modelers adopt a geometric scale equal to the ratio between the boundary-layer thickness measured in the laboratory and values z_g of Table 2.4, even though the latter are nominal, rather than physically significant. The use of this geometric scaling criterion for the simulations of Figure 5.21 yielded the geometric scales $\delta / z_g = 0.75/274 = 1/365$ for the flow with open exposure ($\alpha = 0.16$), and 1/400 for the flow with built-up terrain exposure ($\alpha = 0.29$). The respective measured longitudinal turbulence intensities at 50 m above ground are 0.07 and 0.15, versus about 0.15 and 0.225, estimated using Eq. (2.56) for atmospheric boundary-layer flows. As expected, the discrepancy between the longitudinal turbulence intensity in the wind tunnel and the target value in the atmosphere is more severe at the station $x = 6.1$ m, which would correspond to the fetch available in a typical short wind tunnel.

Figure 5.22 [11] shows spectra of the longitudinal velocity fluctuations measured at station $x = 18.3$ m, and elevation $z/\delta = 0.05$ in the three flows described in Figure 5.21. For $nz/U(z) = 1.0$, the spectra corresponding to two of the three types of spires differ from each other by a factor greater than two.

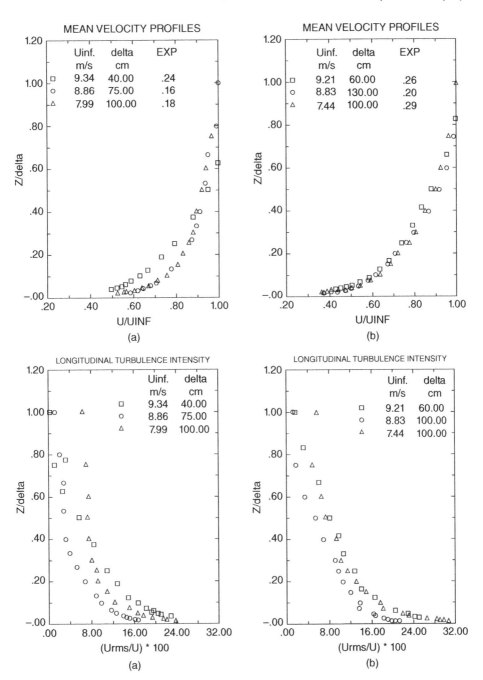

Figure 5.21 Wind tunnel flow features at (a) 6.1 m and (b) 18.3 m downwind of spires, obtained by using three types of spire configurations. Source: Reprinted with permission from [11].

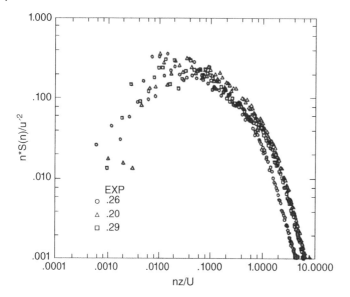

Figure 5.22 Spectra of longitudinal velocity fluctuations measured at 18.3 m downwind of spires. Source: Reprinted with permission from [11].

That wind tunnels with different flow management devices can result in flows with different properties and, hence, in different aerodynamic pressures on bodies immersed in those flows was confirmed by a round robin set of tests reported in [12] (https://www.nist.gov/wind). The tests were conducted by six reputable wind tunnels in the US, Canada, France, and Japan on a model of an industrial building with both open and suburban terrain exposure. Coefficients of variation (CoV) of wind effects determined on the basis of the test results differed significantly from laboratory to laboratory and were found to be as high as 40%.

5.4.2 Effect of Integral Scale and Turbulence Intensity

It is assumed in current practice (see, e.g., ASCE 49-12 Standard [13] and ASCE 7-16 Standard [14]) that wind tunnel flows are satisfactory if, in addition to the mean wind profiles, they reproduce the longitudinal turbulence intensity and, to some degree at least, the longitudinal integral scale of turbulence typical of atmospheric boundary-layer flows. This section discusses the extent to which this assumption is warranted.

Integral Scale and Turbulence Intensity. Some laboratories assume that the integral length is a valid characterization of turbulence for wind tunnel testing purposes. In principle, the geometric scale of the simulation should be consistent with the relation

$$\frac{D_m}{D_p} = \frac{(L_u^x)_m}{(L_u^x)_p} \tag{5.18}$$

where the indexes m and p stand for model and prototype, respectively. However, the usefulness of Eq. (5.18) is questionable for three reasons. First, estimates of integral lengths are typically highly uncertain. Second, the small-scale turbulence transports into the separation bubble free flow particles with large momentum, thus promoting flow reattachment and strongly affecting pressure distributions near separation points [15]; the integral scale is not a significant factor in this phenomenon. Third, integral turbulent scales similar to those occurring in the atmosphere are not achievable in typical conventional wind tunnels at geometric scales used for the simulation of wind effects on low-rise buildings (e.g., $1:50$–$1:100$). This is the case because the size of the eddies associated in the atmosphere with low-frequency flow fluctuations is too large to be accommodated at such scales in wind tunnels with test-section widths of the order of 1.5–3 m. These fluctuations contribute most of the turbulence intensity in atmospheric boundary-layer flows. For these reasons the ASCE 7-16 considerably relaxes the requirement inherent in Eq. (5.18). In addition, it follows from the low-frequency fluctuation deficit in conventional wind tunnels that equal simulated and prototype turbulence intensities may not produce similar aerodynamic effects because the respective flows have different frequency content.

Compensating for Missing Low-Frequency Fluctuations. The effect of the low-frequency fluctuation deficit in conventional boundary-layer wind tunnel tests at geometric scales of the order of $1:100$ can be compensated for by assuming that the energy of those fluctuations is concentrated at frequencies close to or equal to zero. Since zero-frequency (infinite-period) velocity fluctuations are in effect constant velocities, this assumption entails adding to the aerodynamic pressures measured in the wind tunnel, via post-processing, a constant pressure

$$p_d = \frac{1}{2}\rho \overline{C}_p U_{def}^2 \tag{5.19}$$

In Eq. (5.19), ρ is the air density, $\overline{C}_p$ is the mean pressure coefficient measured in the wind tunnel, and U_{def}^2 is the estimated area under the spectral density function of the low-frequency contributions not reproduced in the wind tunnel. This approach is conservative because it implies perfect spatial coherence of the pressures that would be induced by the missing fluctuation components, when in reality that coherence is imperfect. For an alternative approach see [16].

5.4.3 Effects of Reynolds Number Similarity Violation

In principle, for similarity between prototype and wind tunnel flows to be achieved, the respective Reynolds numbers must be the same. This requirement is referred to as *Reynolds number similarity*. In aerodynamic facilities for testing models of structures the fluid being used is air at atmospheric pressure, and Reynolds number similarity is unavoidably violated.

The aerodynamic behavior of the bodies depends upon whether the boundary layers that form on the curved surfaces are laminar or (partially or fully) turbulent. Since boundary layers occurring at high Reynolds numbers are turbulent, it is logical to attempt the reproduction of full-scale flows around smooth cylinders by changing laminar boundary layers into turbulent ones. This can be done by providing the surface with roughness elements [17].

According to [18] the thickness e of the roughness element should satisfy the relations $Ue/v > 400$ and $e/D < 0.01$, where U is the mean speed, v is the kinematic viscosity, and D is the characteristic transverse dimension of the object. For the tower shown in plan in Figure 5.23, the roughness was achieved by fixing onto the surface of the 1/200 model 32 equidistant vertical wires. Three sets of experiments are reported in [18], in which the surface of the cylinder was (i) smooth, (ii) provided with 0.6 mm wires ($e/D = 7 \times 10^3$), and (iii) provided with 1 mm wires, respectively. It was found that the highest mean and peak pressures were more than twice as high on the smooth model than on the models provided with wires. The differences between pressures on the model and with 0.6 mm and the model with 1 mm wires were small. The influence of the roughness on the magnitude of the mean pressures at 20 m (full scale) below the top of the building is shown in Figure 5.23, in which

$$\overline{C}_p = \frac{\overline{p} - p_r}{\frac{1}{2}\rho U_r^2} \tag{5.20}$$

where $\overline{p}$ is the measured mean pressure, p_r is the static reference pressure, U_r is the mean speed at the top of the building, and ρ is the air density.

Unlike bodies with rounded shapes, bodies with sharp edges have fixed separation points (Figure 4.18), whose separation at the edges is independent of Reynolds number. It has therefore been hypothesized that flows around such bodies are similar at full scale and in the wind tunnel, even if Reynolds number similarity is violated. However, in the wind tunnel friction forces are larger in relation to inertial forces than at full scale.

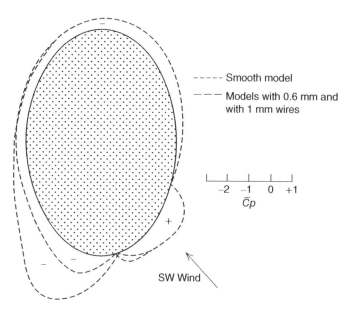

Figure 5.23 Influence of model surface roughness on pressure distribution. Source: Courtesy of Cebtre Scentifique et Technique du Bâtiment, Nantes, France.

This affects the local vorticity at edges and corners in the wind tunnel, resulting in local pressures typically weaker than at full scale. Examples are shown in Section 5.4.4.

5.4.4 Comparisons of Wind Tunnel and Full-Scale Pressure Measurements

Figure 5.24 shows that the negative peak pressures measured at a corner of a low-rise building can be significantly stronger at full scale than in the wind tunnel.

Additional comparisons of pressures on the Texas Tech building and its wind tunnel models tested at Colorado State University and the University of Western Ontario were published in [20]. Figure 5.25 shows that wind tunnel measurements are acceptable for the wall pressures but inadequate for the roof corner.

Figure 5.26a and b show comparisons between wind tunnel and full-scale measurements of pressures at the Commerce Court tower, Toronto. The wind tunnel values were provided at the design stage and are represented by open circles. The solid lines join average values of estimates derived from full-scale measurements; the shaded areas indicate the standard deviation of the full-scale estimates (in Figure 5.26 the notation RMSM denotes the root mean square value about the mean). Note that fluctuating pressures attributable to fluctuating lift differ at some points significantly in the wind tunnel from their full-scale counterparts.

For some tall buildings, the loss of high-frequency velocity fluctuations content in the laboratory can also reduce the strength of the resonant fluctuations induced on the model by the oncoming flow.

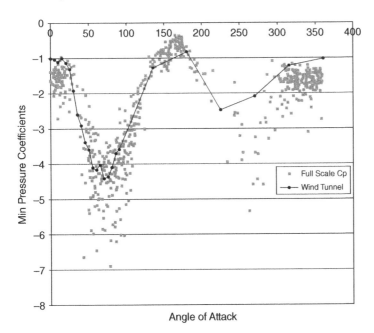

Figure 5.24 Minimum pressure coefficients at building corner, eave level, Texas Tech University experimental building, full-scale and wind tunnel measurements. Source: From [19].

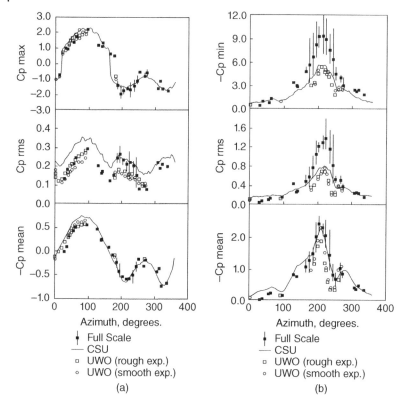

Figure 5.25 Wind pressure coefficients on the Texas Tech Experimental Building – full scale and wind tunnel measurements: (a) wall pressures; (b) corner roof pressures. Source: Reprinted from [20], with permission from Elsevier.

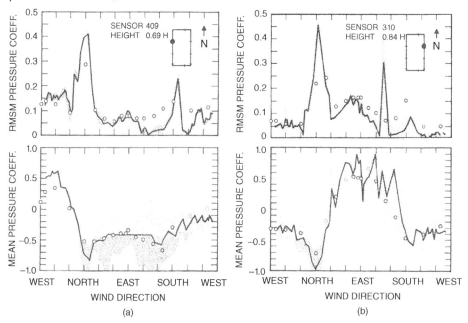

Figure 5.26 Pressures measured on (a) west wall at 20.6 m from NW corner at 46th floor and (b) east wall, at 20.6 m from NE corner at 50th floor, Commerce Court Tower. Source: Reprinted from [21], with permission from Elsevier.

5.5 Blockage Effects

A body placed in a wind tunnel will partially obstruct the passage of air, causing the flow to accelerate. This effect is called blockage. If the blockage is substantial, the flow around the model, and the model's aerodynamic behavior, are no longer representative of prototype conditions.

Corrections for blockage depend upon the body shape, the nature of the aerodynamic effect of concern (i.e., whether drag, lift, Strouhal number, and so forth), the characteristics of the wind tunnel flow, and the relative body/wind tunnel dimensions. Basic studies of blockage are summarized in [22], which contains a bibliography on this topic. For drag measured in closed wind tunnels it is concluded in [22] that the following approximate relation may be used for the great majority of model configurations in all flows, including boundary-layer flows:

$$C_{D_c} = \frac{C_D}{1 + K \ S/C} \tag{5.21}$$

where C_{D_c} is the corrected drag coefficient, C_D is the drag coefficient measured in the wind tunnel, S is the reference area for the drag coefficients C_{D_c} and C_D, and C is the wind tunnel cross-sectional area. The ratio S/C is called the blockage ratio. The coefficient K has been determined only for a limited number of situations. For example, for a bar with rectangular cross section spanning the entire height of a wind tunnel with nominally smooth flow, K was determined to depend upon the ratio a/b as shown in Figure 5.27, where a and b are the dimensions of the along-wind and across-wind sides of the rectangular cross section, respectively.

In practice, it may be assumed that for 2% blockage ratios the blockage corrections are about 5%, and that to a first approximation the blockage correction is proportional to the blockage ratio [22].

For a basic study of blockage effects on bluff-body aerodynamics, see [24].

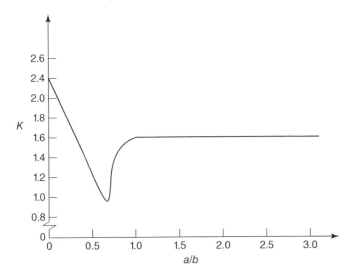

Figure 5.27 Blockage correction factor K for two-dimensional prism ratio a/b in nominally smooth flow [23].

5.6 The High-Frequency Force Balance

The HFFB approach uses rigid test models supported at the base by a high-frequency force (i.e., a rigid) balance. The balance allows measurements of strains proportional to the base bending moments, shears and torsional moments, and experiences very small deformations that render the model motions negligibly small (Figure 5.28).

The HFFB approach is applicable primarily to buildings with approximately straight-line fundamental modal shapes in sway along the principal axes of the building. The expression for the base moment generated by the wind load in the x-direction is

$$M_{b,x}(t) = \int_0^H w_x(z,t)z\,dz \qquad (5.22)$$

where H = building height, $w_x(z, t)$ = wind loading parallel to the x-direction per unit height, and z = elevation above ground. Assuming that the fundamental modal shape is a straight line, the generalized force in the x-direction is also given by right-hand side of Eq. (5.22). Owing to this coincidence, measurement of the base moment yields the generalized force $Q_{x1}(t)$:

$$Q_{x1}(t) = \int_0^H w_x(z,t)(z/H)\,dz \qquad (5.23)$$

where z/H is the fundamental modal shape. The estimation of the fundamental frequency of vibration from the analysis of the structure, and the specification of the damping ratio, then allow the approximate estimation of the dynamic response (see Chapter 11). Similar statements apply to the generalized force in the y-direction [25].

While the generalized aerodynamic torsional moment has the expression

$$Q_{\phi1}(t) = \int_0^H T(z,t)\varphi_{T1}(z)\,dz \qquad (5.24)$$

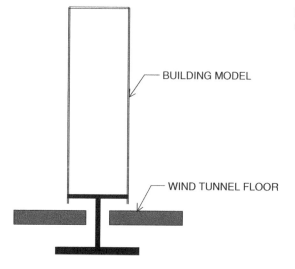

Figure 5.28 Schematic of force-balance model.

BUILDING MODEL

WIND TUNNEL FLOOR

where $T(z, t)$ is the aerodynamic torsional moment per unit height, and $\varphi_{T1}(z)$ is the fundamental mode of vibration in torsion, the base aerodynamic torsional moment measured in the wind tunnel is

$$Q_{\phi 1,\mathrm{HFFB}}(t) = \int_0^H T(z,t)dz. \tag{5.25}$$

Since $\varphi_{T1}(z) \neq 1$, the measured base torsional moment cannot be a substitute for the fundamental generalized torsional moment $Q_{\varphi 1}(t)$. In addition, the HFFB approach provides no information on the contribution of higher modes of vibration to the response.

If the fundamental modes of vibration in the x and y directions do not vary linearly with height, the measured base bending moments are inadequate substitutes for the expressions of the respective modal generalized forces. Corrections accounting for the actual modal shapes can be applied, but they depend upon the distribution of the wind pressures, which until the 1990s could not be obtained by measurements and was therefore generally unknown, especially for buildings affected by aerodynamic interference effects. The corrections, and the corresponding approximations of the generalized torques and moments, therefore depended upon educated guesses concerning the wind pressure distribution. In the 1980s, 1990s, and even the first years of the 2000s, the design of tall buildings was based on the HFFB approach that, in spite of its limitations, was a step forward with respect to earlier practices.

The HFFB procedure has two advantages: it is relatively fast and inexpensive, and it is compatible with the presence of architectural details that may render difficult the use of pressure taps in some cases. The procedure is convenient for use in preliminary studies of aerodynamic alternatives for which only qualitative results are required.

5.7 Simultaneous Pressure Measurements at Multiple Taps

Figure 5.29 shows a model with the large number of pressure taps for which simultaneous pressure measurements are enabled by modern electronic scanning systems. In contrast, Figure 5.30 shows typical tap locations for models subjected to tests compatible with the capabilities available in the late 1970s, on the basis of which ASCE 7 Standard provisions were developed in the 1980s. In addition of the fact that the spatial resolution of the pressure taps is two orders of magnitude higher in modern practice than in the 1970s, the quality of the inferences based on the models with large numbers of taps is due to the fact that, unlike their 1970s predecessors, all data obtained by electronic scanning systems can be recorded and therefore allow transparent post-processing.

A widely used simultaneous pressure measuring system is the Electronic Pressure Scanning System developed by Scanivalve Corporation (www.scanivalve.com) (Figure 5.31). A pressure measuring system includes an Electronic Pressure Scanning Module (e.g., ZOC33 with 64 pressure sensors), a Digital Service Module (e.g., DSM4000, which can service up to eight Electronic Pressure Scanning Modules, i.e., up to 512 sensors, and contains an embedded computer, RAM memory, and a hard disk drive), a pressure calibration system, auxiliary instrumentation to regulate supply of clean, dry air, and data acquisition software.

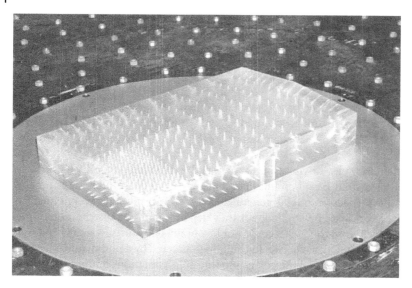

Figure 5.29 Building model in wind tunnel. Source: From [26].

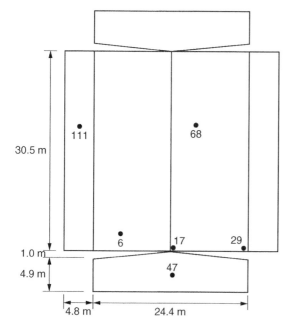

Figure 5.30 Pressure tap arrangement in typical 1970s tests. Source: After [27].

The connection between the Electronic Pressure Scanning Module and the pressure taps is made through plastic tubes. A test model with tubes connecting the pressure taps to the scanning module is shown in Figure 5.32. Tube characteristics must conform to requirements assuring that no significant distortion of pressures acting at the taps occurs [28, 29].

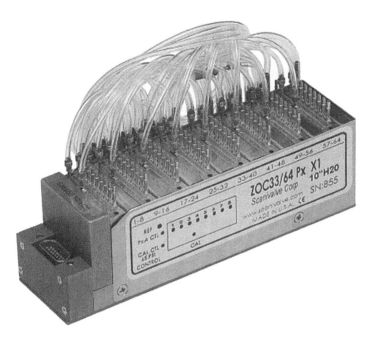

Figure 5.31 View of electronic multi-channel pressure scanning system (www.scanivalve.com).

Figure 5.32 Tubes installed on a small-scale test building.

References

1 Kopp, G. A. (ed.), *Large-Scale and Full-Scale Laboratory Test Methods for Examining Wind Effects on Buildings*, Frontiers in the Built Environment [series], frontiersin .org, [Online]. Available: www.frontiersin.org/research-topics/4739/large-scale-and-full-scale-laboratory-test-methods-for-examining-wind-effects-on-buildings, 2018.

2 Marshall, R.D. (1984). Wind tunnels applied to wind engineering in Japan. *Journal of Structural Engineering* 110: 1203–1221.

3 Davenport, A. G. and Isyumov, N., "The application of the boundary-layer wind tunnel to the prediction of wind loading," *in Proceedings of the International Research Seminar on Wind Effects on Buildings and Structures, Vol. 1*, p. 221. Copyright, Canada, University of Toronto Press, 1968.

4 Irwin, P.A. (1981). The design of spires for wind simulation. *Journal of Wind Engineering and Industrial Aerodynamics* 7: 361–366.

5 Wooding, R.A., Bradley, E.F., and Marshall, J.K. (1973). Drag due to regular arrays of roughness elements of varying geometry. *Boundary-Layer Meteorology* 5: 285–308.

6 Ward, N.B. (1972). The exploration of certain features of tornado dynamics using a laboratory model. *Journal of the Atmospheric Sciences* 29: 1194–1204.

7 Davies-Jones, R.P. (1973). The dependence of core radius on swirl ratio in a tornado simulator. *Journal of the Atmospheric Sciences* 30: 1427–1430.

8 Haan, F.L., Sarkar, P.P., and Gallus, W.A. (2008). Design, construction and performance of a large tornado simulator for wind engineering applications. *Engineering Structures* 30: 1146–1159.

9 Refan, M. and Hangan, H. (2016). Characterization of tornado-like flow fields in a new model scale wind testing chamber. *Journal of Wind Engineering and Industrial Aerodynamics* 151: 107–121.

10 Refan, M., Hangan, H., and Wurman, J. (2014). Reproducing tornadoes in laboratory using proper scaling. *Journal of Wind Engineering and Industrial Aerodynamics* 135: 136–148.

11 Cermak, J.E. (1982). Physical modeling of the atmospheric boundary layer in long boundary-layer tunnels. In: *Wind Tunnel Modeling for Civil Engineering Applications, Proceedings of the international workshop on wind tunnel modeling criteria and techniques in civil engineering applications, Gaithersburg, MD, USA, April 1982*, 1st ed. (ed. T.A. Reinhold), 97–125. Cambridge, UK: Cambridge University Press.

12 Fritz, W.P., Bienkiewicz, B., Cui, B. et al. (2008). International comparison of wind tunnel estimates of wind effects on low-rise buildings: test-related uncertainties. *Journal of Structural Engineering* 134: 1887–1890.

13 ASCE, "Wind tunnel testing for buildings and other structures (ASCE/SEI 49-12)," in *ASCE Standard ASCE/SEI 49-12*, Reston, VA: American Society of Civil Engineers, 2012.

14 ASCE, "Minimum design loads for buildings and other structures (ASCE/SEI 7-16)," in *ASCE Standard ASCE/SEI 7-16*, Reston, VA: American Society of Civil Engineers, 2016.

15 Li, Q.S. and Melbourne, W.H. (1995). An experimental investigation of the effects of free-stream turbulence on streamwise surface pressures in separated and reattaching flows. *Journal of Wind Engineering and Industrial Aerodynamics* 54–55: 313–323.

16 Mooneghi, M.A., Irwin, P.A., and Chowdhury, A.G. (2016). Partial turbulence simulation method for predicting peak wind loads on small structures and building appurtenances. *Journal of Wind Engineering and Industrial Aerodynamics* 157: 47.

17 Szechenyi, E. (1975). Supercritical Reynolds number simulation for two-dimensional flow over circular cylinders. *Journal of Fluid Mechanics* 70: 529–542.

18 Gandemer, J., Barnaud, G., and Biétry, J., "Études de la tour D.M.A. Partie I, Étude des efforts dûs au vent sur les façades," Centre Scientifique et Technique du Bâtiment, Nantes, France, 1975.

19 Long, F., Uncertainties in pressure coefficients derived from full and model scale data, report to the National Institute of Standards and Technology, Wind Science and Engineering Research Center, Texas Tech University

20 Tieleman, H.W. (1992). Problems associated with flow modelling procedures for low-rise structures. *Journal of Wind Engineering and Industrial Aerodynamics* 42: 923–934.

21 Dalgliesh, A. (1975). Comparisons of model full-scale wind pressures on a high-rise building. *Journal of Wind Engineering and Industrial Aerodynamics* 1: 55–66.

22 Melbourne, W.H. (1982). Wind tunnel blockage effects and correlations. In: *Wind Tunnel Modeling for Civil Engineering Applications*, 1st ed. (ed. T.A. Reinhold), 197–216. Cambridge, UK: Cambridge University Press.

23 Courchesne, J. and Laneville, A. (1979). A comparison of correction methods used in the evaluation of drag coefficient measurements for two-dimensional rectangular cylinders. *Journal of Fluids Engineering* 101: 506–510.

24 Utsunomiya, H., Nagao, F., Ueno, Y., and Noda, M. (1993). Basic study of blockage effects on bluff bodies. *Journal of Wind Engineering and Industrial Aerodynamics* 49: 247–256.

25 Tschanz, T. and Davenport, A.G. (1983). The base balance technique for the determination of dynamic wind loads. *Journal of Wind Engineering and Industrial Aerodynamics* 13: 429–439.

26 Ho, C. E., Surry, D., and Moorish, D., NIST/TTU Cooperative Agreement – Windstorm Mitigation Initiative: Wind Tunnel Experiments on Generic Low Buildings, Alan G. Davenport Wind Engineering Group, The University of Western Ontario, 2003

27 Davenport, A. G., Surry, D., and Stathopoulos, T., "Wind loads on low-rise buildings," Final report on phase I and II, BLWT-SS8-1977, University of Western Ontario, London, Ontario, Canada, 1977.

28 Irwin, P.A., Cooper, K.R., and Girard, R. (1979). Correction of distortion effects caused by tubing systems in measurements of fluctuating pressures. *Journal of Wind Engineering and Industrial Aerodynamics* 5: 93–107.

29 Kovarek, M., Amatucci, L., Gillis, K. A., Potra, F. A., Ratino, J., Levitan, M. L., and Yeo, D., "Calibration of dynamic pressure in a tubing system and optimized design of tube configuration: a numerical and experimental study," NIST TN 1994, National Institute of Standards and Technology, Gaithersburg, MD, 2018. https://www.nist.gov/wind.

6

Computational Wind Engineering

6.1   Introduction

Computational Fluid Dynamics (CFD) is a vast field aimed at describing fluid flows using numerical methods. Computational Wind Engineering (CWE) is a CFD subfield whose main objective is to produce descriptions of aerodynamic wind effects on the built environment. In particular, descriptions are sought for use in the structural design of buildings and other structures. It is symptomatic that, while addressing recent CWE accomplishments, state-of-the-art surveys [1–3] mention few if any applications to structural design practice. This is because, to date, with rare exceptions [4], structural designers cannot rely on CWE with the degree of confidence required to ensure the safety of structures whose failure may result in loss of life. However, CWE is increasingly being used in such applications as the evaluation of pedestrian comfort in zones of intensified wind speeds (see Chapter 15) and the estimation of wind effects on solar collectors in solar power plants [5].

In a number of cases CWE can provide solutions that may be used for preliminary design purposes if backed by proper validation – see the *UK Design Manual for Roads and Bridges* (BD 49/01) [6], the Eurocode (prEN 1991-1-4) [7], and the *Architectural Institute of Japan Guidebook* [8]. Currently, CWE research is aimed at creating tools allowing the development of aerodynamic data usable for structural design even in the absence of closely related ad-hoc experimental validation.

The purpose of this chapter is to present a brief compendium of selected information on CWE modeling, numerical issues, and verification and validation procedures, with a view to acquainting wind and structural engineers with the CWE vocabulary and facilitating dialogue between wind and structural engineers on the one hand and CWE professionals on the other.

It is shown in Chapter 12 that uncertainties in the aerodynamic pressures have considerably less weight in the global uncertainty budget than do uncertainties in the wind speeds; for this reason, their effect on the estimates of overall effects of the flow on the structure are less severe than is the case in automotive or aeronautics applications.

The mathematical model used in CWE simulations consists of the governing equations of the flow (Section 6.2). The governing equations need to be discretized, and grids within a computational domain are generated for implementing the discretization (Section 6.3). The requisite initial and boundary conditions are considered in Section 6.4. Numerical solutions for the flow as represented by the discretized computational model are briefly discussed in Section 6.5. Section 6.6 concerns numerical

Wind Effects on Structures: Modern Structural Design for Wind, Fourth Edition. Emil Simiu and DongHun Yeo.
© 2019 John Wiley & Sons Ltd. Published 2019 by John Wiley & Sons Ltd.

stability issues. Section 6.7 summaries turbulence models. Section 6.8 is a concise introduction to verification and validation (V&V), and uncertainty quantification (UQ). Section 6.9 considers the role of wind tunnel testing and CWE prospects in the simulation of aerodynamic effects. Section 6.10 briefly discusses best practice guidelines (PBG).

6.2 Governing Equations

Fluid flows are described by the equation of continuity (conservation of mass) and the Navier–Stokes equations (conservation of momentum). CWE only considers incompressible air flow (see Chapter 4), for which the equation of continuity is

$$\frac{\partial U_i}{\partial x_i} = 0 \tag{6.1}$$

where U_i are the velocity components in the x_i directions in a Cartesian coordinate system ($i = 1, 2, 3$). Einstein notation is used in Eq. (6.1) and subsequent equations.

The Navier–Stokes equations are

$$\frac{\partial U_i}{\partial t} + U_j \frac{\partial U_i}{\partial x_j} = -\frac{1}{\rho} \frac{\partial p}{\partial x_i} + v \frac{\partial^2 U_i}{\partial x_j \partial x_j} + f_i \tag{6.2}$$

where p is the pressure, v is the fluid kinematic viscosity, and f_i is the vector representing body forces (e.g., the gravity or the pressure-gradient force). The non-dimensional form of Eqs. (6.1) and (6.2) highlights the dependence of the flow on the Reynolds number Re:

$$\frac{\partial U_i^*}{\partial x_i^*} = 0 \tag{6.3}$$

$$\frac{\partial U_i^*}{\partial t^*} + U_j^* \frac{\partial U_i^*}{\partial x_j^*} = -\frac{\partial p^*}{\partial x_i^*} + \frac{1}{Re} \frac{\partial^2 U_i^*}{\partial x_j^* \partial x_j^*} + f_i^* \tag{6.4}$$

where the non-dimensional variables based on reference length (L_{ref}) and velocity (U_{ref}) are defined as:

$$x_i^* = \frac{x_i}{L_{ref}}, \quad U_i^* = \frac{U_i}{U_{ref}}, \quad t^* = \frac{t}{L_{ref}/U_{ref}} = \frac{tU_{ref}}{L_{ref}}, \quad p^* = \frac{p}{\rho U_{ref}^2}, \quad f^* = \frac{f_i}{U_{ref}^2 L_{ref}^{-1}} \tag{6.5}$$

6.3 Discretization Methods and Grid Types

Discretization of the governing equations (Eqs. [6.1] and [6.2]) in CWE is commonly performed using finite difference, finite volume, or finite element methods (FDM, FVM, or FEM, respectively). All methods discretize the computational domain using grids and approximate the governing partial differential equations by systems of algebraic equations. FDM, typically restricted to simple geometries, uses Taylor series or polynomial fitting to approximate at each grid point the derivatives that appear in the governing equations. FVM, the most commonly used discretization technique, solves the integral form of the governing equations in a domain subdivided into small contiguous control

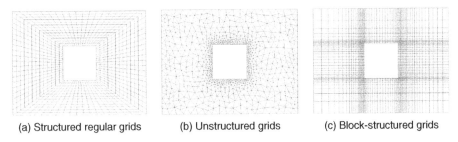

(a) Structured regular grids   (b) Unstructured grids   (c) Block-structured grids

Figure 6.1 Types of grid.

volumes. The method defines the control volume boundaries, rather than the computational nodes, and the values of the variables are approximated at the cell faces from the values at the control volume centers. FEM is similar to FVM, but uses weight functions aimed at minimizing approximation errors [9]. Its main advantage is that it is readily applicable to flows with complex geometries.

Through the grid generation process, the computational domain is composed of by a large number of cells consisting of nodes (vertexes) and lines joining adjacent nodes, thus defining a grid (also called mesh). Grids can be structured or unstructured. Structured grids (Figure 6.1a) are defined as families of grid lines such that lines of a single family do not cross each other, and lines of a family cross lines of other families only once [10]. Unstructured grids (Figure 6.1b) usually consist of triangles or quadrilaterals in two dimensions and tetrahedra and hexahedra in three dimensions, typically in irregular patterns. The generation of unstructured grids can be automated in computational domains with any level of geometric complexity. However, unstructured grids require more computational memory and entail higher costs than structured grids.

For parallel computing, structured grids can be based on a multi-block approach, in which a domain with complex geometries is decomposed into multiple blocks (zones) with simple geometries (Figure 6.1c). Interfaces between blocks should be located in regions in which the flow characteristics (e.g., pressure and velocity gradient) are not rapidly changing. Unstructured grids are typically decomposed into zones by an algorithm embedded in a mesh generation program.

6.4 Initial and Boundary Conditions

Simulations are of two generic types: (i) steady-state simulations, applied to equilibrium problems, and (ii) marching simulations, applied to transient problems. In equilibrium problems, the governing equations are solved once to determine the time-independent solution. In marching problems, the equations are solved at each time step, starting from the initial conditions, to determine the time-dependent solution as it advances in time. Appropriate initial and boundary conditions in conjunction with the governing equations are required for constructing a well-posed mathematical model of the flow.

6.4.1 Initial Conditions

For time-dependent simulations, initial values are generally imposed in the computational domain. The most effective initial conditions are solutions of the fully developed flow obtained from previous simulations. Results from steady-state simulations can be employed to expedite the turbulence development in transient flow simulations.

6.4.2 Boundary Conditions

Boundary conditions (BC) are typically defined in terms of boundary values of the unknown field and their derivatives. BC commonly used in CWE applications are listed next.

Dirichlet boundary conditions assign at the boundary a constant ϕ_0 value for the variable ϕ:

$$\phi = \phi_0. \tag{6.6}$$

If ϕ is a pressure or a velocity, Eq. (6.6) describes a constant pressure or a constant velocity field at the boundary condition, respectively.

Von Neumann boundary conditions assign at the boundary a constant gradient of the variable ϕ:

$$\frac{\partial \phi}{\partial n} = \phi_0 \tag{6.7}$$

where n is normal to the boundary.

Convective boundary conditions (also called non-reflective BC) approximate the variable ϕ at a boundary near which the flow is convective but exhibits no diffusive effects, that is, for an upstream reference velocity $\mathbf{U}_{\text{ref}}$

$$\frac{\partial \phi}{\partial t} + \mathbf{U}_{\text{ref}} \cdot \nabla \phi = 0. \tag{6.8}$$

Periodic boundary conditions (also called cyclic BC) approximate cyclically repeating behavior as follows:

$$\phi(t)|_B = \phi(t)|_{B+L} \tag{6.9}$$

where B represents a boundary and L is the characteristic length of periodicity.

No-slip wall boundary conditions are applied to viscous flow bounded by a solid wall where the flow velocity relative to the wall vanishes, that is, for a stationary wall

$$U_P = U_\perp = 0 \tag{6.10}$$

where U_P and $U_\perp$ are the tangential and normal components of the velocity vector, respectively. This boundary condition is typically used near the wall when the grids in that region are fine enough to resolve the flow throughout the viscous sublayer (i.e., for $z^+ < 1$, where $z^+ = u_* z / v$, z is the direction normal to wall and u_* is the friction velocity.).

Slip (or inviscid) wall boundary conditions model a zero-shear solid wall (i.e., no friction at the interface of fluid and structure). Thus, the velocity component normal to the wall is zero:

$$U_\perp = 0 \tag{6.11}$$

and the gradients normal to the wall of the velocity components are assumed to be zero:

$$\nabla U_P \cdot \mathbf{n} = \nabla U_\perp \cdot \mathbf{n} = 0. \tag{6.12}$$

This can be used for a wall above which viscous effects are negligible or for a far boundary field that influences negligibly the flow physics of interest.

Symmetry Boundary Conditions are employed on a plane when the flow is assumed to be symmetric with respect at that plane. Thus, there is no fluxes across the plane, meaning that the velocity normal to the boundary is zero:

$$U_{\perp} = 0. \tag{6.13}$$

In addition, the gradient of the velocity tangent to the symmetry plane in the direction normal to that plane is zero:

$$\nabla U_p \cdot \mathbf{n} = 0, \tag{6.14}$$

which means that the shear stress is zero, but the normal stress is not zero ($\nabla U_{\perp} \cdot \mathbf{n} \neq 0$) on the symmetry plane, which is not the cases in the no-slip and slip wall boundary condition. Another requirement is zero gradient of all scalar quantities ϕ_s normal to the symmetry plane:

$$\nabla \phi_s \cdot \mathbf{n} = 0. \tag{6.15}$$

6.5   Solving Equations

For CWE applications modeled by nonlinear partial differential equations, matrix equations are solved by iterative methods in which the initial solution is assumed, the equation is linearized, and the solution is improved by repeating the process until an acceptable solution is obtained. More details of the solutions of the systems of equations are provided in [10, Chapter 5].

In incompressible flow, a difficulty arises in the solution of the governing equations, since no independent equation for the pressure is available. The conservation of momentum equations contain pressure gradient terms and, in combination with the continuity equation, can be used to determine the pressure field as a function of time and space using methods discussed in [10, Chapter 7], [11] and [12].

6.6   Stability

Numerical approximations to the governing equations may exhibit unstable behavior, that is, they may magnify errors that occur as a result of discretization. Stability is assured by satisfying the Courant–Friedrichs–Lewy (CFL) condition, which requires that the distance traveled by a fluid element per time step not be larger than the distance between adjacent grid points [13, 14]. In 1-D simulations, the CFL condition is

$$\text{CFL} = \frac{U \, \Delta t}{\Delta x} \leq C_{\max} \tag{6.16}$$

where U and Δx are the flow velocity and the grid size in the x streamwise direction, respectively, Δt is the chosen time step, and $C_{\max}$ is the upper bound of the CFL number, which is less than unity and can vary depending on numerical schemes employed for solving the equations. If $C_{\max} = 0.8$ is chosen, the largest time step used in the simulation is estimated as

$$\Delta t_{\max} = 0.8 \frac{\Delta x}{U} \tag{6.17}$$

The CFL condition can be extended to 3-D simulations as follows:

$$\text{CFL} = \max \left(\frac{U}{\Delta x}, \frac{V}{\Delta y}, \frac{W}{\Delta z} \right) \Delta t \leq C_{\max} \tag{6.18a}$$

or

$$\text{CFL} = \left(\frac{U}{\Delta x} + \frac{V}{\Delta y} + \frac{W}{\Delta z} \right) \Delta t \leq C_{\max} \tag{6.18b}$$

The corresponding largest time step can be estimated by Eq. (6.18a) or, more conservatively, by Eq. (6.18b).

6.7 Turbulent Flow Simulations

6.7.1 Resolved and Modeled Turbulence

A turbulent flow consists of turbulent motions over broad range of length and time scales, as illustrated by the energy spectrum $E(\kappa)$ per unit of wave number κ in Figure 6.2, from energy-containing eddies to energy-dissipation eddies. The smallest scales of turbulent flow, associated with energy-dissipation eddies [15], are:

$$l_\eta \equiv \left(\frac{v^3}{\varepsilon} \right)^{1/4} \quad \text{(length)}$$

$$\tau_\eta \equiv \left(\frac{v}{\varepsilon} \right)^{1/2} \quad \text{(time)}$$

$$u_\eta \equiv (v\varepsilon)^{1/4} \quad \text{(velocity)} \tag{6.19a,b,c}$$

where v is the kinematic viscosity and ε is the rate of energy dissipation of the turbulent kinetic energy k, defined as $k = (1/2)\overline{u_i u_i}$. For details on the energy spectrum see Section 2.4.3.

Strategies for the simulation of turbulence motions depend on the extent to which eddy motions are resolved on the one hand and modeled empirically on the other (Figure 6.2). Direct Numerical Simulation (DNS) resolves all turbulent scales and uses no turbulence modeling (Section 6.7.2). Large Eddy Simulation (LES) resolves the large-scale turbulent eddies and models the small-scale eddies (Section 6.7.3). In steady Reynolds-Averaged Navier–Stokes Simulation (RANS[1]) all turbulent eddies are modeled. Unsteady Reynolds-Averaged Navier–Stokes simulation (URANS) models all turbulent eddies but resolves low-frequency motions associated with unsteadiness in the mean flow, such as vortex-shedding (Section 6.7.4). Hybrid RANS/LES employs the RANS approach near walls and LES in regions far from the walls (Section 6.7.5). Simulation costs increase as the resolved part of the simulation increases. The resolved and modeled parts in each turbulence model are illustrated in Figure 6.2.

6.7.2 Direct Numerical Simulation (DNS)

DNS is the most reliable approach to the simulation of turbulent flows. It consists of solving the discretized governing equations of the fluid motion by explicitly resolving all scales of turbulence down to the dissipation scale, without resorting to empirical

---

1 Reynolds-Averaged Navier-Stokes Simulation is also referred to as Reynolds-Averaged Numerical Simulation.

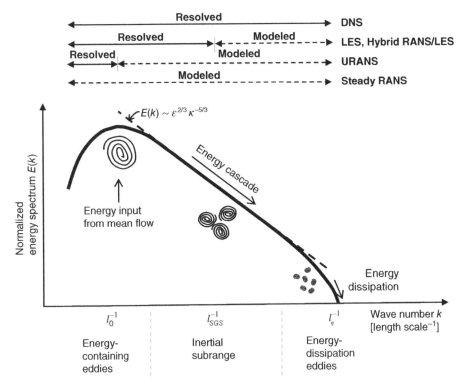

Figure 6.2 Turbulence spectrum.

turbulence modeling. DNS must satisfy the following conditions. First, the smallest resolved scales must be in the order of the dissipation scales, that is, in the order of 1 mm for atmospheric boundary layer (ABL) flow. Second, the dimensions of the computational domain (L_x, L_y, L_z in the x, y, z directions) must be significantly larger than (i) the largest scales of the turbulent flow (the scales can be in the order of hundreds of meters for ABL flow, and (ii) the characteristic length of the structure for signature turbulence [11, 16]. In addition, the domain must be sufficiently large to reduce the blockage effect to an acceptable level (e.g., 2–5% blockage ratio).

Under the assumption that in the of energy-dissipating range the eddies can be resolved by four-point grids in each direction ($\Delta x = \Delta y = \Delta z = l_\eta/4$), the number of cells can be estimated as the ratio of the volume of the computational domain to the volume of a cell, that is,

$$N_{xyz} = \frac{L_x L_y L_z}{(l_\eta/4)^3} \tag{6.20}$$

Assuming for a full-scale simulation that $L_x = 1000$ m, $L_y = L_z = 100$ m, and $l_\eta = 0.001$ m, N_{xyz} is in the order of 10^{18}. It can be shown that the corresponding minimum number of time-steps in a simulation with turnover time T_0 over which the largest eddies, with scale l_0, break down into eddies with dissipation scales l_η, is in the order of

$$N_t = \frac{T_0}{\Delta t} = \frac{l_0/k^{1/2}}{l_\eta/(4k^{1/2})} = 4\frac{l_0}{l_\eta} \tag{6.21}$$

where the square root of the turbulent kinetic energy has the dimension of a velocity. For $L_x = 1000$ m, $L_y = L_z = 100$ m, $l_0 = 150$ m, and $l_\eta = 0.001$ m, $N_t = 6 \times 10^5$, and the computational cost of the DNS simulation for this example is commensurate with $N_{xyz} N_t \approx 10^{23}$.

For boundary layer flows near a wall, the first in a direction normal to the wall should be located at a distance $z^+ \approx 1$ from the wall; there should be 3–5 cells in the direction normal to the wall up to $z^+ < 10$. The grid sizes should be $\Delta x^+ \approx 10$–15 in the direction of the tangent to the wall, and $\Delta y^+ \approx 5$ in the cross-stream direction ($\Delta x^+ = u_* \Delta x / \nu$ and $\Delta y^+ = u_* \Delta y / \nu$) [17]. Therefore, the grid sizes are inversely proportional to friction velocity u_* and, therefore, to the Reynolds number of the flow.

Using current computer technology, DNS can only be applied to practical problems for which the Reynolds numbers are low. For CWE applications, time, and memory requirements for DNS simulations are prohibitive to date. It has been estimated that DNS simulations may become feasible for the analysis of common engineering problems by 2050–2080 [18, 19].

6.7.3 Large Eddy Simulations (LES)

LES resolves the time-averaged and unsteady motions of large-scale turbulent eddies, and models small, subgrid-scale (SGS) eddies. The large-scale eddies contain most of the energy of the flow and have the largest contribution to the Reynolds stress tensor τ_{ij}:

$$\tau_{ij} = \overline{u_i u_j} \qquad (i, j = 1, 2, 3) \tag{6.22}$$

where u_i is the fluctuating velocity component in the i-th direction, the subscripts 1, 2, and 3 represent the x, y, and z directions, respectively, and the overbar denotes time-averaging. The size of the small eddies to be modeled is determined by the filter width Δ_{SGS}. The small eddies are approximately isotropic and do not depend upon the characteristics of large-scale flow.

The velocity field for the unfiltered motion can be written as

$$\mathbf{U}(\mathbf{x}, t) = \tilde{\mathbf{u}}(\mathbf{x}, t) + \mathbf{u}'_{SGS}(\mathbf{x}, t) \tag{6.23}$$

where $\tilde{\mathbf{u}}(\mathbf{x}, t)$ is the velocity in the filtered motion and $\mathbf{u}'_{SGS}(\mathbf{x}, t)$ is the sub-filtered turbulent velocity. The filtered velocity can be obtained using explicit filter functions (e.g., top-hat or Gaussian filter function [20]) or through an implicit filtering process by grid scales. While the former approach is used for fundamental turbulence studies, the latter is commonly used in applications. The filtering approach attenuates small eddies whose sizes are smaller than Δ_{SGS}, and leaves the large and intermediate-scale eddies unchanged. Figure 6.3 illustrates the spatially filtered velocity as affected by the filter width [21].

To resolve the motion of large and intermediate-scale eddies, LES uses the governing equations based on filtered variables:

$$\frac{\partial \tilde{u}_i}{\partial x_i} = 0 \tag{6.24}$$

$$\frac{\partial \tilde{u}_i}{\partial t} + \frac{\partial \tilde{u}_i \tilde{u}_j}{\partial x_j} = -\frac{1}{\rho} \frac{\partial \tilde{p}}{\partial x_i} + \nu \frac{\partial^2 \tilde{u}_i}{\partial x_i \partial x_j} - \frac{\partial \tau_{ij}^R}{\partial x_j} + \tilde{f}_i \tag{6.25}$$

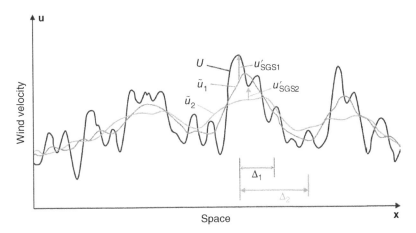

Figure 6.3 Illustration of unfiltered velocity field $U(x)$ and resolved velocity field $\widetilde{u}_i(x)$ based on filter width Δ_i ($i = 1, 2$) [21]. Reprinted from Figure 2.4 of Large-Eddy Simulation in Hydraulics, W. Rodi, G. Constantinescu, and T. Stoesser, 2013, CRC Press [20], with permission from Taylor & Francis.

where $\widetilde{f}_i$ is the filtered external force vector per unit mass, and the SGS stress tensor, called residual stress tensor, is

$$\tau_{ij}^R = \widetilde{U_i U_j} - \widetilde{u}_i \widetilde{u}_j. \tag{6.26}$$

The SGS stress can be decomposed into an isotropic and a deviatoric part

$$\tau_{ij}^R = \frac{1}{3}\delta_{ij}\tau_{kk}^R + \tau_{ij}^r \tag{6.27}$$

where δ_{ij} is the Kronecker delta and $k = 1, 2, 3$. Substituting Eq. (6.27) into Eq. (6.25), the LES governing equations become

$$\frac{\partial \widetilde{u}_i}{\partial t} + \frac{\partial \widetilde{u}_i \widetilde{u}_j}{\partial x_j} = -\frac{1}{\rho}\frac{\partial \widetilde{p}^*}{\partial x_i} + \nu\frac{\partial^2 \widetilde{u}_i}{\partial x_j \partial x_j} - \frac{\partial \tau_{ij}^r}{\partial x_j} + \widetilde{f}_i \tag{6.28}$$

where

$$\widetilde{p}^* = \overline{p} + \frac{1}{3}\rho\delta_{ij}\tau_{kk}^R. \tag{6.29}$$

Closure of Eq. (6.28) requires the development of SGS models, that is, models of the deviatoric SGS stress τ_{ij}^r. The models predict effects of the SGS stresses on the resolved motion whose length scales depend upon the filter width Δ_{SGS}. For uniform grids with mesh size Δ, $\Delta_{\text{SGS}} = \Delta$. For non-uniform grids, $\Delta_{\text{SGS}} = (\Delta x \Delta y \Delta z)^{1/3}$, for example.

The widely used Smagorinsky SGS model [22] approximates the deviatoric SGS stress τ_{ij}^r by assuming the validity of Boussinesq's eddy viscosity hypothesis [23], according to which the deviatoric part of the Reynolds stress is proportional to the strain rate tensor of the filtered (resolved) velocities, $\widetilde{S}_{ij} \equiv (1/2)(\partial\widetilde{u}_i/\partial x_j + \partial\widetilde{u}_j/\partial x_i)$, that is,

$$\tau_{ij}^r = -2\nu_{t,\text{SGS}}\widetilde{S}_{ij} \tag{6.30}$$

where $\nu_{t,\text{SGS}}$ is the kinematic eddy viscosity to be modeled under the assumption that the eddy viscosity is proportional to a typical length scale l_{SGS} and a velocity scale

q_{SGS} of the flow. The SGS turbulent eddy viscosity in the Smagorinsky model can be expressed as

$$\nu_{t,SGS} \propto l_{SGS}q_{SGS}$$
$$= (C_s\Delta_{SGS})^2|\widetilde{S}| \tag{6.31}$$

where the characteristic length and velocity scales are $(C_s \; \Delta_{SGS})$ and $(C_s\Delta_{SGS} | \widetilde{S} |)$, respectively, the Smagorinsky constant C_s varies, depending upon the flow, between 0.1 and 0.2 [24], and $| \widetilde{S} |= (2\widetilde{S}_{ij}\widetilde{S}_{ij})^{1/2}$. This model has been widely used on account of its simplicity and computational efficiency. However, the use of a constant value for C_s makes it difficult to predict accurately complex flows. For example, $C_s \approx 0.17$, as determined for isotropic homogeneous turbulence [25], should be decreased for flow with strong mean shear, especially near a wall [26], in order to reduce the amount of dissipation introduced by the SGS model and the resulting spurious SGS stresses [11]. For this reason, in the Smagorinsky model a near-wall correction is required to capture the near-wall effects.

To address the shortcomings of the Smagorinsky model, dynamic SGS models have been proposed for non-isotropic flows [27, 28], in which the model parameter is automatically reduced near the wall from its value for isotropic flow. Improved SGS models still need to be developed for complex geometry and highly anisotropic flow applications.

Reliable LES simulations require sufficiently fine spatial and temporal scales. The grid sizes should be $l_\eta \ll \Delta_{SGS} \ll l_0$ (see Figure 6.2). The computational domain size required for LES simulations is the same as for DNS. To resolve flows in the wall region, the typical requisite grid sizes close to the wall are $\Delta x^+ \approx 50$ in the along-wall (streamwise) direction and $\Delta y^+ \leq 15$ in the cross-stream direction; in the normal-to-wall direction the first grid point from the wall should be at $z^+ \approx 1$ while at least three grid points in the viscous region ($1 \leq z^+ \leq 10$) and 30–50 grid points within the boundary layer are required [20]. The total number of grid points for wall-resolving LES is smaller than for DNS [29], but it is still prohibitively expensive, particularly for high Reynolds number flows over wall-mounted structures. Approaches to reducing the computational cost include using wall-layer models (called Wall-Modeled LES or WMLES) [30], or using hybrid RANS-LES methods [31], are discussed in Section 6.7.5.

6.7.4 Reynolds-Averaged Navier–Stokes Simulation (RANS)

RANS are a primary approach for practical turbulent flow simulations owing to their simplicity and relatively low computational cost. RANS simulates the averaged fields of turbulent flows by solving the Reynolds-averaged Navier–Stokes equations.

In RANS, the flow field is divided by Reynolds decomposition into a mean flow field and a fluctuating field. For example, the flow velocity can be expressed as:

$$\mathbf{U}(\mathbf{x}, t) = \overline{\mathbf{u}}(\mathbf{x}, t) + \mathbf{u}'_{RANS}(\mathbf{x}, t) \tag{6.32}$$

where $\overline{\mathbf{u}}$ is the time-averaged velocity and $\mathbf{u}'_{RANS}$ is the fluctuating component. Steady RANS, based on time-averaging, is used to simulate time-independent flow. URANS, based on ensemble-averaging, simulates time- and space-dependent flow. It has been noted that, "while all turbulent flows are unsteady, not every unsteadiness is turbulence."

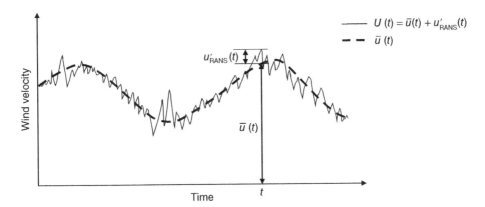

Figure 6.4 Illustration of Reynolds decomposition ($U(t) = \bar{u}(t) + u'_{RANS}(t)$).

For example, in flows with a large-scale periodicity due to vorticity shed in the wake of a structure, that periodicity would be suppressed by time-averaging, but is preserved under ensemble-averaging. URANS is applied to such flows (Figure 6.4).

The equations of Reynolds-averaged flow field are derived by applying the ensemble-averaging operation (Eq. [6.32]) to the governing equations (Eqs. [6.1] and [6.2]). Using the decomposition and noting that $\bar{U}_i = \bar{u}$, the governing equations for URANS are derived as

$$\frac{\partial \bar{u}_i}{\partial x_i} = 0 \tag{6.33}$$

$$\frac{\partial \bar{u}_i}{\partial t} + \frac{\partial}{\partial x_j}\bar{u}_i\bar{u}_j = -\frac{1}{\rho}\frac{\partial \bar{p}}{\partial x_i} + v\frac{\partial^2 \bar{u}_i}{\partial x_i \partial x_j} - \frac{\partial \tau_{ij}}{\partial x_j} + \bar{f}_i \tag{6.34}$$

where τ_{ij} is the Reynolds stress tensor:

$$\tau_{ij} \equiv \overline{u'_{RANSi}\, u'_{RANSj}} = \overline{U_i U_j} - \bar{u}_i \bar{u}_j. \tag{6.35}$$

Equation (6.35) accounts for momentum flux generated by all turbulent fluctuations, while the residual stresses in LES (Eq. [6.26]) exclude the contribution of resolved turbulent fluctuations. Note that the first term in the left-hand side of Eq. (6.34) does not exist in the steady RANS governing equations.

The URANS governing equations cannot be solved because the Reynolds stresses are unknown. To close the system, it is required that the Reynolds stresses be approximated in terms of the averaged quantities. Under the Boussinesq approximation (see Eq. [6.30] in LES) the Reynolds stress tensor is

$$\tau_{ij} = -2v_{t,RANS}\bar{S}_{ij} \tag{6.36}$$

where $\bar{S}_{ij} \equiv (1/2)(\partial \bar{u}_i/\partial x_j + \partial \bar{u}_j/\partial x_i)$ is the rate of strain tensor of averaged flow field and $v_{t,RANS}$ is the RANS turbulent eddy viscosity to be modeled, similar to Eq. (6.31) in LES, as:

$$v_{t,RANS} \propto l_{RANS}\, q_{RANS}$$
$$= C_\mu l_{RANS}\, q_{RANS}. \tag{6.37}$$

In Eq. (6.37), l_{RANS} and q_{RANS} are the typical length and velocity scales of a turbulent flow, respectively and C_μ is a non-dimensional constant determined in a calibration procedure.

A broad selection of closure models of the Reynolds stresses is available [32], including linear eddy viscosity models, nonlinear eddy viscosity models, and Reynolds stress models. Among linear eddy viscosity models, the SST (Shear Stress Transport) model [33] and the Spalart–Allmaras (S–A) model [34] are considered, capable of predicting reliably flows around bluff bodies with strong adverse pressure gradients and massive flow separation. For example, SST uses the k-ω model [32] for boundary-layer (or inner layer) flows and the k-ε model [35, 36] for low shear layer (or outer layer) flows. A blending function is employed for the transition between the two models. For details see [37].

The spatial and temporal requirements for RANS simulations are much less demanding than for DNS and LES. However, RANS simulations should have sufficiently fine grids to capture the change of the averaged flow field, especially for near-wall regions characterized by high velocity-gradient flow. The RANS models typically have two options for the treatment of near-wall flow: (i) resolving the flow (called low-Re model) and (ii) using wall functions (high-Re model). In the flow near the wall, low-Re RANS models generally require grid resolutions as fine as LES in the direction normal to the wall, but much coarser grids in the wall-tangential streamwise and across-stream directions than LES. The increase in aspect ratios of cells near the wall can therefore lead to a substantial reduction in the total number of cells. For the high-Re RANS models using typical wall functions, the grid closest to a wall should be located in the log layer beyond the viscous sublayer (e.g., $30 < z^+ < 500$ where the upper limit depends on the Reynolds number of the flow), so that the wall functions can bridge the gap between the near-wall and the fully turbulent flow region. This option can save considerable computational time due to the alleviated grid requirement, but the performance can be poor, especially for flows around bluff bodies, since wall functions are generally developed for relatively simple flows, such as flows over flat plates.

6.7.5 Hybrid RANS/LES Simulation

URANS models typically perform unsatisfactorily for massively separated flows characterized by large turbulence scales [16]. Such flows can be better simulated by LES. However, LES simulations of high Reynolds number flows over wall-mounted structures are still challenging, owing to the prohibitive grid requirements for near-wall regions. To alleviate the near-wall grid resolution problem in massively separated flows, hybrid RANS/LES models have been proposed, for example [31, 38]. These models work in the RANS mode for near-wall flow regions, and transition to the LES model for regions away from the wall. The near-wall flow is simulated by a less accurate but computationally more efficient RANS, and large turbulent eddies from massively separated flow are resolved by LES at manageable computational cost.

Detached Eddy Simulations (DES) [31] are the most widely used hybrid RANS/LES model for flows over wall-mounted structures at high Reynolds numbers, including

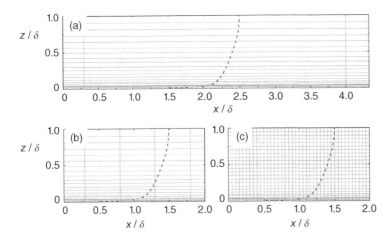

Figure 6.5 Types of grid in boundary layers. The dashed line represents the velocity profile [39].
Source: Reproduced with permission of ANNUAL REVIEWS.

flows over bluff bodies of interest in wind engineering. DES is a non-zonal type[2] of model that modifies the original RANS model and includes a transition from RANS to LES. The DES S–A [31] and DES SST hybrid models [37] are based on the S–A RANS [34] and the SST RANS model [33], respectively.

The computational cost of DES is much lower than for LES, but is still higher than RANS. Grid generation strategy is much more complicated for DES than for RANS or LES. As reported in [16, 39, 40], the DES model induces "grey areas" in which the flow is not adequately solved by either pure RANS or pure LES. In those areas the turbulence energy modeled in RANS may not be adequately transferred to LES-resolved turbulence energy. This effect, called Modeled-Stress-Depletion (MSD), may cause premature separation due to inadequate grid spacing (Grid-Induced Separation, or GIS) [39]. As shown in Figure 6.5, the grid sizes in boundary-layer flow are assumed to be $\Delta x \approx \Delta y \gg \Delta z$ (i.e., $\Delta_{max} = \max [\Delta x, \Delta y, \Delta z] = \Delta x$), where the x, y, and z directions are the along-wind, the across-wind, and the normal-to-wall direction, respectively. In the DES formulation, for grids with $\Delta_{max}/\delta > 1$ (see Figure 6.5a), the RANS mode is activated in the whole boundary layer. If $\Delta_{max}/\delta < 1$ (Figure 6.5b, c), the switch from RANS to LES is activated within the boundary layer thickness. This causes unphysical behaviors associated with MSD and GIS. Updated versions of DES, called Delayed DES (DDES) [41] and Improved DDES (IDDES) [42], have been proposed that attempt to improve upon DES. For grids with $\Delta_{max}/\delta < 0.5$ to 1 (Figure 6.5b), DDES and IDDES prevent LES mode activation. For grids with $\Delta_{max}/\delta \ll 1$ (Figure 6.5c), IDDES fully enables the LES mode, except for wall modeling, which is performed in the RANS mode, as in Wall-Modeled LES. DES performance depends upon type of grid as represented in Figure 6.5. In particular, it has been observed that in some instances the DES performance does not necessarily improve if the grid size is reduced [39, 43].

---

2 Another approach to hybrid RANS/LES models is a zonal model, with distinct zones occupied by pure RANS and pure LES, and discontinuous solutions at interfacing boundaries. See details in [16].

(a) 2-D SST RANS ($C_D = 0.78$)

(b) 2-D SST URANS ($C_D = 1.73$)

(c) 3-D SST URANS ($C_D = 1.24$)

(d) S-A DES (coarse grid, $C_D = 1.16$)

(e) S-A DES (fine grid, $C_D = 1.26$)

(f) SST DES (fine grid, $C_D = 1.28$)

S-A: Spalart-Allmaras
SST: Shear Stress Transport

Figure 6.6 Vorticity iso-surfaces around a circular cylinder ($Re = 5 \times 10^4$, experimental drag coefficient $C_D = 1.15$–1.25) [39]. Source: Reproduced with permission of *Annual Reviews*.

6.7.6 Performance of Turbulence Models

Figure 6.6 shows visualizations of resolved vortical flow structures around a circular cylinder simulated using various turbulence models [39]. As expected, 2-D (steady) RANS cannot predict the vortex shedding (Figure 6.6a). Note that even 3-D steady RANS cannot accurately predict the averaged flow characteristics in such unsteady and separated flows [44]. Two-dimensional URANS allows the simulation of 2-D large eddies associated with vortex shedding, but does not capture 3-D flow structures (Figure 6.6b). 3-D URANS captures 3-D flow structures, but cannot resolve smaller flow structures using finer grids (Figure 6.6c) [45]. DES predicts 3-D flow structures up to finer scales than URANS (Figures 6.6d and e); with a sufficiently fine grid it can resolve fine flow structures in the separated flow region. Figures 6.6e and f show that the performance of DES in the LES region does not depend significantly upon the choice of its RANS models (i.e., whether S–A or SST) [39].

6.8 Verification and Validation. Uncertainty Quantification

The credibility of CWE simulations depends upon the quality of the physical modeling, the competence of the analysts performing the simulations, the simulations' verification and validation (V&V), and the UQ of the simulation results [46]. The analyst's depth of understanding of the modeling details and of the simulation results plays a decisive role in the simulation process.

V&V consists of procedures required for assessing the accuracy of simulation results. Uncertainty quantification is aimed at identifying, characterizing, and estimating quantitatively the factors in the analysis that affect the accuracy of the simulation results [46, 47]. The amount of research into V&V and UQ is vast [48–53]. This section is limited to introducing the reader to a few salient features of their respective procedures. Section 6.8.1 briefly discusses sources of inaccuracy in CWE simulations. Section 6.8.2 is a summary description of V&V aims and procedures. Section 6.8.3 is concerned with UQ.

6.8.1 Sources of Inaccuracy in CWE Simulations

Although, as shown in Section 6.8.3, errors and uncertainties are distinct concepts, it will be convenient in this section to refer to both as errors. Errors arising in CWE simulation results are typically of four types [11, 46]: (i) physical modeling, (ii) discretization, (iii) iteration, and (iv) programing/user errors.

Physical modeling errors, δ_{model}, are differences between the behavior of the real physical object and its model counterpart:

$$\delta_{model} = p_{model} - p_{real} \tag{6.38}$$

where p_{model} and p_{real} are the respective response values of interest (e.g., velocity or pressure). They arise from approximations of complex behavior in the governing equations (e.g., approximations inherent in turbulence models), effects of computational domain size and boundary conditions, and assumptions on fluid properties (e.g., constant air density and temperature).

Discretization errors, δ_h, are differences between the exact analytical solution of a mathematical model and the exact solution of the model's discretized counterpart:

$$\delta_h = p_h - p_{model} \tag{6.39}$$

where p_h is the response calculated from the discretized model. Discretization error should be estimated for every new type of grid, solution scheme, or application. Among the sources associated with numerical errors, the discretization errors are usually the largest and their estimation is the most challenging [46].

Iterative errors, δ_{it}, are differences between the exact and computed solutions of the discretized equations,

$$\delta_{it} = p_{comp} - p_h \tag{6.40}$$

where p_{comp} is the solution obtained from a computing machine, which may entail round-off errors and convergence errors inherent in iterative methods. Round-off errors resulting from low precision in computer calculations can affect the stability of the solutions. In simulations with a stable scheme and negligible round-off error accumulation, the round-off errors are usually very small compared to other errors [11]. The iteration–convergence errors are present because a linearized system of discretized equations is typically solved iteratively. In general, iterative errors are at least one or two orders of magnitude lower than the discretization errors [11]. However, if a flow solver uses implicit time integration for unsteady simulations, a loose iterative convergence criteria at each time step may lead to significant influence on accuracy of the numerical solution [64].

Programing/user errors, caused by mistakes or bugs in the software, can be classified into two types [46]: critical errors by which the software cannot execute a simulation or generate reasonable results, and less critical but still non-negligible errors due to dormant software faults that may not be easily identified by code verification. User errors are due to blunders or mistakes from users in input preparation for simulation and in post-processing for output data analysis. Human errors generally are not easily detected, especially when large-scale simulations of complex systems are performed.

6.8.2 Verification and Validation

The objective of verification and validation (V&V) is to establish the credibility of a computational model by assessing the degree of accuracy of the simulation results [46]. The philosophy, definition, and procedure of V&V on modeling and simulation have been developed in practicing communities, such as AIAA (American Institute of Aeronautics and Astronautics) [50], ASME (American Society of Mechanical Engineers) [48, 49], and DOE (Department of Energy) [53].

Verification is the process of determining that a computational model accurately represents the underlying mathematical model and its solution [49]. Validation is the process of determining the degree to which a model is an accurate representation of the real world from the perspective of the intended uses of the model [50]. V&V processes start with determining the intended uses of the computational model. The accuracy requirements for the responses of interest are determined accordingly. Verification process addresses the correct implementation of a numerical model in a code and the estimation of numerical errors in solutions of discretized equations. Model validation process employs the verified simulation results and relevant experimental data and assesses the predictive capacity of the model. If the agreement between model predictions and experimental outcomes satisfies the accuracy requirement, the V&V processes end. A successful V&V can claim that the accuracy of the computational model is adequate for the intended use of the model. Otherwise, the V&V processes are repeated by updating the model and, if necessary, carrying out additional experiments until the agreement is acceptable. For details see [49]. Note that the documentation of the V&V activities and results serves not only for justifying the current intended use, but also for providing information/experience for potential future uses.

Verification is limited to estimating numerical errors and is not concerned with the accuracy of physical modeling. The verification process can be divided into code verification and solution verification. Code verification addresses the correct implementation of the numerical algorithm in the computer code by evaluating the error for a known highly accurate solution referred to as verification benchmark. Code verification in grid-based simulations can be performed by a systematic discretization convergence test (e.g., [54]) and its convergence to a benchmark solution. Practical approaches have been developed, for example, the method of manufactured solutions [55], to generate analytical solutions required for code verification. Code verification is usually performed by code developers/vendors, but should also be performed for specific applications by CWE users of commercial/open-source codes [46, 48].

After the code verification is completed, solution verification is conducted. The solution verification deals with (i) correctness of the input and output data for a particular solution of a problem of interest and (ii) numerical accuracy (error/uncertainty estimation) for the simulated solution in the discretized time and space domain [46]. Numerical solutions and the errors inherent in them (δ_h in Eq. [6.39] and δ_{it} in Eq. [6.40]) estimated in the solution verification process are considered in the validation process. For typical CWE problems, numerical errors can be estimated *a posteriori*, for example, by using multiple simulations with different grid resolutions [47]. Solution verification should be performed by CWE users and be required by structural engineers who use CWE simulation results for structural design.

The interest of CWE users in V&V lies in validation of a computational model for the intended use. The validation process assesses the accuracy of the computational model by comparison with experimental data, quantifies predictive uncertainty in interpolation or extrapolation of the model, and evaluates the acceptability of the model for the intended use [46, 56].

6.8.3 Quantification of Errors and Uncertainties

Error and uncertainty are often used interchangeably. In particular, this is the case in Chapters 7 and 12 for applications unrelated to CWE. However, in the AIAA V&V guide for CFD [50] errors are defined as recognizable deficiencies in all phases or activities of modeling and simulation that are not due to lack of knowledge, whereas uncertainties are defined as potential deficiencies in any phase or activity of the modeling process that are due to lack of knowledge.

Errors can be classified as acknowledged errors and unacknowledged errors. Acknowledged errors can be identified and eliminated (e.g., round-off errors, discretization errors, iterative errors). Unacknowledged errors cannot be found or removed (e.g., programming errors, improper use of the CWE code). Uncertainties can be classified as aleatory and epistemic. Aleatory (irreducible) uncertainties are associated with inherent randomness (e.g., input parameters of a model). Epistemic (reducible) uncertainties are related to a lack of knowledge of (or information on) a physical model. For details, see [46].

The ASME V&V approach [48] provides quantitative evaluations of uncertainties in simulation results by comparison with their counterparts in experiments, and employs concepts and definitions of error and uncertainty borrowed from metrology [57].

6.9 CWE versus Wind Tunnel Testing

Wind tunnel testing is currently an indispensable tool used (i) to obtain aerodynamic or aeroelastic data on special structures for which no such data are available, and (ii) to improve standard provisions. Its drawbacks include the following: (i) its first costs and the maintenance costs are high; (ii) testing is time-consuming; (iii) it typically entails violation of the Reynolds number and of other similarity criteria applicable to certain types of special structures (e.g., air-supported structures); and (iv) it is not consistently reliable (see Appendix F for high-rise building and [58] for low-rise building testing).

As computer technology and numerical techniques have evolved, the prospect of performing CWE simulations has become increasingly attractive, given their following potential advantages: (i) ready availability; (ii) relatively low initial and maintenance costs; (iii) relatively fast turnover times; (iv) less restrictive model scale limitations; (v) capability to solve multi-physics problems (e.g., wind-structure interaction or rain-wind scenarios), and (vi) as is also the case for wind tunnel simulations, the fact that errors and uncertainties affecting the estimation of aerodynamic effects have significantly less weight than their wind climatological counterparts – see Chapter 12. However, CWE is not yet accepted as a structural design tool because, typically, its results cannot be used confidently. Wind tunnel testing and, whenever possible, full-scale measurements, will still be required for validation purposes until CWE will have evolved into a fully reliable independent tool.

6.10 Best Practice Guidelines

Using CWE for selected applications requires the development of appropriate mathematical models, computational grids and domains, spatial and temporal discretization schemes, solvers, turbulence models, boundary conditions, and convergence criteria, capable of being successfully subjected to rigorous V&V procedures. Best practice guidelines can facilitate the use of such development, and cover general applications [59] as well as specific fields, such as urban environmental wind [60, 61], nuclear power plants (e.g., nuclear reactor safety application [62], dry cask application [63]), and structural loads on buildings [8]. Best practice guidelines cover a limited number of simulations. Therefore, it is recommended that V&V procedures be applied to simulations that deviate in any significant aspect from existing simulations covered by the guidelines.

References

1 Tamura, Y. and Phuc, P.V. (2015). Development of CFD and applications: monologue by a non-CFD-expert. *Journal of Wind Engineering and Industrial Aerodynamics* 144: 3–13.

2 Blocken, B. (2014). 50 years of computational wind engineering: past, present and future. *Journal of Wind Engineering and Industrial Aerodynamics* 129: 69–102.

3 Dagnew, A.K. and Bitsuamlak, G.T. (2013). Computational evaluation of wind loads on buildings: a review. *Wind & Structures* 16: 629–660.

4 Michalski, A., Kermel, P.D., Haug, E. et al. (2011). Validation of the computational fluid-structure interaction simulation at real-scale tests of a flexible 29 m umbrella in natural wind flow. *Journal of Wind Engineering and Industrial Aerodynamics* 99 (4): 400–413.

5 Andre, M., Mier-Torrecilla, M., and Wüchner, R. (2015). Numerical simulation of wind loads on a trough parabolic solar collector, using lattice Boltzmann and finite element methods. *Journal of Wind Engineering and Industrial Aerodynamics* 146: 185–194.

6 HE, "Volume I: Highway structures: approval procedures and general design in Design manual for roads and bridges (DMRB)," Highways England (HE), 2001.

7 CEN, "Eurocode 1: Actions on structures - Part 1–4: Gernal actions - Wind actions," in *EN 1991-1-4*, ed: European Committee for Standardization (CEN), 2005.

8 AIJ, "Guidbook of recommendation for loads on buildings 2: Wind-induced response and load estimation/Practical guides of CFD for wind resistant design," Tokyo, Japan: Architectural Institute of Japan, 2017, p. 434.

9 Donea, J. and Huerta, A. (2003). *Finite Element Methods for Flow Problems*, 1st ed. Chichester, UK: Wiley.

10 Ferziger, J.H. and Peric, M. (2002). *Computational Methods for Fluid Dynamics*, 3rd ed. New York: Springer Verlag.

11 Zikanov, O. (2010). *Essential Computational Fluid Dynamics*, 1st ed. Hoboken, New Jersey: Wiley.

12 Issa, R.I. (1986). Solution of the implicitly discretised fluid flow equations by operator-splitting. *Journal of Computational Physics* 62: 40–65.

13 Courant, R., Friedrichs, K., and Lewy, H. (1967). On the partial difference equations of mathematical physics. *IBM Journal* 11: 215–234.

14 Courant, R., Friedrichs, K., and Lewy, H. (1928). Über die partiellen Differenzengleichungen der mathematischen Physik. *Mathematische Annalen* 100: 32–74.

15 Kolmogorov, A. N., "The local structure of turbulence in incompressible viscous fluid for very large Reynolds numbers," in *Dokl. Akad. Nauk SSSR*, 1941, pp. 299–303.

16 Sagaut, P., Deck, S., and Terracol, M. (2013). *Multiscale and Multiresolution Approaches in Turbulence*, 2nd ed. London: Imperial College Press.

17 Tucker, P.G. (2014). *Unsteady Computational Fluid Dynamics in Aeronautics*, 1st ed. Dordrecht, the Netherlands: Springer.

18 Spalart, P.R. (2000). Strategies for turbulence modelling and simulations. *International Journal of Heat and Fluid Flow* 21: 252–263.

19 Voller, V.R. and Porté-Agel, F. (2002). Moore's law and numerical modeling. *Journal of Computational Physics* 179: 698–703.

20 Sagaut, P. (2006). *Large Eddy Simulation for Incompressible Flows*, 3rd ed. Berlin, Germany: Springer-Verlag Berlin Heidelberg.

21 Rodi, W., Constantinescu, G., and Stoesser, T. (2013). *Large-Eddy Simulation in Hydraulics*. London, UK: CRC Press.

22 Smagorinsky, J. (1963). General circulation experiments with the primitive equations. I. The basic experiment. *Monthly Weather Review* 91: 99–164.

23 Boussinesq, J. (1877). Essai sur la théorie des eaux courantes. *Mémoires présentés par divers savants à l'Académie des Sciences* 23: 1–680.

24 Porté-Agel, F., Meneveau, C., and Parlange, M.B. (2000). A scale-dependent dynamic model for large-eddy simulation: application to a neutral atmospheric boundary layer. *Journal of Fluid Mechanics* 415: 261–284.

25 Lilly, D.K. (1967). The representation of small-scale turbulence in numerical simulation experiments. In: *IBM Scientific Computing Symposium on Environmental Sciences* (ed. H.H. Goldstein), 195–210. New York: Yorktown Heights.

26 Deardorff, J.W. (1971). On the magnitude of the subgrid scale eddy coefficient. *Journal of Computational Physics* 7: 120–133.

27 Germano, M., Piomelli, U., Moin, P., and Cabot, W.H. (1991). A dynamic subgrid-scale eddy viscosity model. *Physics of Fluids A: Fluid Dynamics* 3: 1760–1765.

28 Lilly, D.K. (1992). A proposed modification of the germano sub-grid closure method. *Physics of Fluids* 4: 633.

29 Choi, H. and Moin, P. (2012). Grid-point requirements for large eddy simulation: Chapman's estimates revisited. *Physics of Fluids* 24: 011702.

30 Piomelli, U. and Balaras, E. (2002). Wall-layer models for large-eddy simulations. *Annual Review of Fluid Mechanics* 34: 349–374.

31 Spalart, P. R., Jou, W. H., Strelets, M., and Allmaras, S. R., "Comments on the feasibility of LES for wings, and on a hybrid RANS/LES approach," in *Advances in*

DNS/LES, 1st AFOSR International Conference on DNS/LES, Ruston, LA, 1997, pp. 137–147.

32 Wilcox, D. C., Turbulence Modeling for CFD: DCW Industries, 2006.

33 Menter, F., "Zonal two equation k-ω turbulence models for aerodynamic flows," in *23rd Fluid Dynamics, Plasmadynamics, and Lasers Conference,* ed: American Institute of Aeronautics and Astronautics, 1993.

34 Spalart, P. R. and Allmaras, S. R., "A one-equation turbulence model for aerodynamic flows," in *30th Aerospace Sciences Meeting and Exhibit,* Reno, NV, 1992, pp. 1–22.

35 Jones, W.P. and Launder, B.E. (1972). The prediction of laminarization with a two-equation model of turbulence. *International Journal of Heat and Mass Transfer* 15: 301–314.

36 Launder, B.E. and Sharma, B.I. (1974). Application of the energy-dissipation model of turbulence to the calculation of flow near a spinning disc. *Letters in Heat and Mass Transfer* 1: 131–137.

37 Menter, F.R., Kuntz, M., and Langtry, R. (2003). Ten years of industrial experience with the SST turbulence model. In: *Turbulence, Heat and Mass Transfer 4* (ed. K. Hanjalic, Y. Nagano and M. Tummers), 625–632. Begell House, Inc.

38 Speziale, C.G. (1998). Turbulence modeling for time-dependent RANS and VLES: a review. *AIAA Journal* 36: 173–184.

39 Spalart, P.R. (2009). Detached-eddy simulation. *Annual Review of Fluid Mechanics* 41: 181–202.

40 Spalart, P.R., Deck, S., Shur, M.L. et al. (2006). A new version of Detached-Eddy Simulation, resistant to ambiguous grid densities. *Theoretical and Computational Fluid Dynamics* 20: 181.

41 Strelets, M., "Detached eddy simulation of massively separated flows," in 39th Aerospace Sciences Meeting and Exhibit, Reno, NV, 2001.

42 Gritskevich, M.S., Garbaruk, A.V., Schütze, J., and Menter, F.R. (2012). Development of DDES and IDDES formulations for the k-ω shear stress transport model. *Flow, Turbulence and Combustion* 88: 431–449.

43 Ke, J. and Yeo, D., "RANS and hybrid LES/RANS simulations of flow over a square cylinder," Presented at the 8th International Colloquium on Bluff Body Aerodynamics and Applications, Boston, MA, 2016. https://www.nist.gov/wind.

44 Iaccarino, G., Ooi, A., Durbin, P.A., and Behnia, M. (2003). Reynolds averaged simulation of unsteady separated flow. *International Journal of Heat and Fluid Flow* 24: 147–156.

45 Shur, M., Spalart, P.R., Squires, K.D. et al. (2005). Three-dimensionality in Reynolds-Averaged Navier-Stokes solutions around two-dimensional geometries. *AIAA Journal* 43: 1230–1242.

46 Oberkampf, W. L. and Roy, C. J., *Verification and Validation in Scientific Computing,* Cambridge, UK: Cambridge University Press, 2010.

47 Roache, P.J. (1997). Quantification of uncertainty in computational fluid dynamics. *Annual Review of Fluid Mechanics* 29: 123–160.

48 ASME, "Standards for verification and validation in computational fluid dynamics and heat transfer," in *ASME V&V 20-2009,* New York, NY: American Society of Mechanical Engineers, 2009.

49 ASME, "Guide for verification and validation in computational solid mechanics," in ASME V&V 10-2006, New York, NY: American Society of Mechanical Engineers, 2006.

50 AIAA, "Guide for the verification and validation of computational fluid dynamics simulations," AIAA-G-077-1998, American Institute of Aeronautics and Astronautics, Reston, Virginia, 1998.

51 NASA, "NASA handbook for models and simulations: An implementation guide for NSA-STD-7009," NASA-HDBK-7009, National Aeronautics and Space Administration, Washington, DC, 2013.

52 Kaizer, J. S., "Fundamental Theory of Scientific Computer Simulation Review," NUREG/KM-0006, Nuclear Regulatory Commission, Washington, DC, 2013.

53 Pilch, M., Trucano, T., Moya, J., Froehlich, G., Hodges, A., and Peercy, D., "Guidelines for Sandia ASCI verification and validation plans – Content and Format Version 2.0," SAND2000-3101, Sandia National Laboratory, Albuquerque, NM, 2001.

54 Roache, P.J. (1994). Perspective: a method for uniform reporting of grid refinement studies. *Journal of Fluids Engineering* 116: 405–413.

55 Roache, P.J. (2002). Code verification by the method of manufactured solutions. *Journal of Fluids Engineering* 124: 4–10.

56 Oberkampf, W.L. and Trucano, T.G. (2008). Verification and validation benchmarks. *Nuclear Engineering and Design* 238: 716–743.

57 JCGM, "International vocabulary of metrology – basic and general concepts and associated terms" JCGM 200:2012 (JCGM 200:2008 with minor corrections), Joint Committee for Guides on Metrology, 2012.

58 Fritz, W.P., Bienkiewicz, B., Cui, B. et al. (2008). International comparison of wind tunnel estimates of wind effects on low-rise buildings: test-related uncertainties. *Journal of Structural Engineering* 134: 1887–1890.

59 Casey, M. and Wintergerste, T. (2000). *ERCOFTAC Special Interest Group on Quality and Trust in Industrial CFD: Best Practice Guidelines*. Brussels, Belgium: ERCOFTAC (European Research Community on Flow, Turbulence and Combustion).

60 Blocken, B. (2015). Computational fluid dynamics for urban physics: importance, scales, possibilities, limitations and ten tips and tricks towards accurate and reliable simulations. *Building and Environment* 91: 219–245.

61 Franke, J., Hellsten, A., Schlünzen, H., and Carissimo, B., "Best practice guideline for the CFD simulation of flows in the urban environment," COST Action 732, COST Brussels, Belgium, 2007.

62 Menter, F., "CFD Best Practice Guidelines for CFD Code Validation for Reactor-Safety Applications," EVOL– ECORA – D01, European Commission, 5th EURATOM Framework Programme, 2002.

63 Zigh, G. and Solis, J., "Computational Fluid Dynamics Best Practice Guidlines for Dry Cask Applications," NUREG-2152, Nuclear Regulatory Commission, Washington, DC, 2013.

64 Eça, G. Vaz, L., and Hoekstra, "Iterative errors in unsteady flow simulations: Are they really negligible?," Presented at the 20th Numerical Towing Tank Symposium (NuTTS 2017), Wageningen, The Netherlands, 2017.

7

Uncertainties in Wind Engineering Data

7.1 Introduction

Structural design for wind is affected by errors and uncertainties[1] in the measurement and modeling of the micrometeorological, wind climatological, and aerodynamic factors that determine the wind load.

Uncertainty quantification is a complex task on which research is ongoing. Owing to insufficient information and data, it is in many cases necessary to estimate uncertainties not only on the basis of measurements and statistical theory, but also by making use of subjective assessments, inferences from past practice, and simplified structural reliability methods (see Appendix E).[2] To provide context on the use of the uncertainties discussed in this chapter, Section 7.2 presents a simple statistical framework, originally developed in [1], that relates uncertainty estimates to the development of safety factors with respect to wind loads, called wind load factors. The wind load factor specified in the pre-2010 versions of the ASCE 7 Standard is larger than unity (ASCE: American Society of Civil Engineers). The 2010 and 2016 versions of the Standard specify a wind load factor equal to unity, and to make up for this change, specify far longer mean recurrence intervals (MRIs) of the design wind speeds than their pre-2010 counterparts (e.g., 700 years in lieu of 50 years). Section 7.3 discusses the uncertainties considered in this chapter. These are used in Chapter 12 to define wind load factors and mean recurrence intervals of design wind effects.

7.2 Statistical Framework for Estimating Uncertainties in the Wind Loads

The peak wind effect is a random variable: it varies from realization to realization. The following approximate expressions commonly hold for the expectation and coefficient of variation (CoV, i.e., ratio of the standard deviation to the expectation) of the peak

---

1 For convenience, the term "uncertainties" also applies to errors and uncertainties as defined in Chapter 6.
2 The use of far more elaborate and rigorous methods than those developed so far for civil engineering purposes is required by NASA and the Department of Energy for a wide variety of applications. Such methods, which are beyond the scope of this chapter, are discussed in NASA's Handbook for models and simulations available at https://standards.nasa.gov/standard/nasa/nasa-hdbk-7009, and in other documents mentioned in Chapters 6 and 12.

Wind Effects on Structures: Modern Structural Design for Wind, Fourth Edition. Emil Simiu and DongHun Yeo.
© 2019 John Wiley & Sons Ltd. Published 2019 by John Wiley & Sons Ltd.

wind effect p_{pk} (e.g., pressure, force, moment) with an $\overline{N}$-year mean recurrence interval:

$$\overline{P}_{pk}(\overline{N}) \approx a\,\overline{E}_z\,\overline{K}_d\,\overline{G}(\theta_m)\,\overline{C}_{p,pk}(\theta_m)\overline{U}^2(z_{ref},\overline{N}) \tag{7.1}$$

$$\mathrm{CoV}[p_{pk}(\overline{N})] \approx \{\mathrm{CoV}^2(E_z) + \mathrm{CoV}^2(K_d) + \mathrm{CoV}^2[G(\theta_m)] + \mathrm{CoV}^2[C_{p,pk}(\theta_m)]$$
$$+ 4\mathrm{CoV}^2[U(z_{ref},\overline{N})]\}^{1/2} \tag{7.2}$$

In Eq. (7.1) the factor a is a constant that depends upon the type of wind effect, and the overbar denotes expectation. E_z is a surface exposure factor defined by the wind profile and specified in the ASCE Standard; the subscript z denotes height above the surface. The aerodynamically most unfavorable wind direction is denoted by θ_m. K_d is a wind directionality reduction factor that accounts for the fact that the direction θ_m and the direction of the largest directional wind speeds typically do not coincide. The peak aerodynamic coefficient $C_{p,pk}(\theta_m)$ depends upon the area being considered, which can be as small as a roof tile or as large as an entire building. Once this dependence is taken into account, for rigid structures the gust response factor G is unity. Flexible structures experience dynamic effects that depend on both wind engineering and structural engineering features. The factor G that characterizes dynamic effects is considered in Chapter 12. $U(z_{ref},\overline{N})$ is the wind speed with an $\overline{N}$-year MRI, estimated from largest wind speed data regardless of direction. The uncertainty in the wind speed $U(z_{ref},\overline{N})$ is due to measurement, micrometeorological, and probabilistic modeling errors, and to the limited size of the data sample on which the estimation is based. According to approximate estimates similar to those of [1], $\mathrm{CoV}(E_z) \approx 0.16$, $\mathrm{CoV}[C_{p,pk}(\theta_m)] \approx 0.12$ and, on the basis of wind speed data at seven locations not exposed to hurricane winds, $0.09 < \mathrm{CoV}[U(\overline{N}=50\,\mathrm{years})] < 0.16$. Research reported in [2] suggests that $\mathrm{CoV}(K_d) \approx 0.10$.

Derivation of Eqs. (7.1) and (7.2). Consider the product $p = xy$ of two random variables x and y with means $\overline{x}, \overline{y}$, fluctuations about the mean x', y', and variances $\overline{x'^2}, \overline{y'^2}$. Then

$$p = \overline{p} + p'$$
$$= (\overline{x} + x')(\overline{y} + y')$$
$$= \overline{x}\overline{y} + \overline{x}y' + \overline{y}x' \tag{7.3}$$

$$\overline{p} = \overline{x}\,\overline{y} \tag{7.4}$$

$$p'^2 = \overline{x}^2 y'^2 + \overline{y}^2 x'^2 + 2\overline{x}\,\overline{y}x'y' \tag{7.5}$$

$$\overline{p'^2} = \overline{x}^2\overline{y'^2} + \overline{y}^2\overline{x'^2} + 2\overline{x}\,\overline{y}\overline{x'y'} \tag{7.6}$$

If x', y' are independent, the last term in Eq. (7.6) vanishes, and

$$\mathrm{CoV}^2(p) = \mathrm{CoV}^2(y) + \mathrm{CoV}^2(x) \tag{7.7}$$

However, if $x \equiv y$, as in the case of the square of the wind speeds, it follows from Eq. (7.6) that $\overline{p'^2} = 4\overline{x}^2\overline{x'^2}$; hence the factor 4 in Eq. (7.2). It is easy to see that Eqs. (7.4) and (7.7) can be extended to a product of any number of mutually independent variables.

The larger the individual uncertainties in the factors that determine the wind loading, the larger the overall uncertainty in the wind effect $p_{pk}(\overline{N})$ being considered, and the larger the requisite wind load factor. For example, for a site at which wind speed data are obtained from weather balloon measurements (or from wind speeds at locations with poorly defined surface roughness conditions, and/or from a short extreme wind speed data record), the overall uncertainty in the wind effect and therefore the corresponding wind load are greater than for a site at which the wind speed measurements are more reliable. Similarly, uncertainties in the measurement of aerodynamic pressures can be large if obtained in wind engineering laboratories that use inadequate simulation and measurement techniques.

Equations (7.1) and (7.2) make it possible to consider the effects of individual uncertainties *collectively*, rather than *in isolation*, and enable the estimation of the uncertainty in the overall wind effect as a function of individual uncertainties. This allows a rational allocation of resources when considering the reduction of any individual uncertainty. For example, when using public databases of pressure coefficients, the lack of data directly applicable to a building with a particular set of dimensions requires the use of interpolations. This can result in errors as large as 15%, say. The reduction of such errors would require the development of databases with larger sets of model dimensions. However, if the 15% error in the pressure coefficient resulted in an error in the estimation of the design wind effect of only 5%, say, the expensive development of a database with higher resolution might in practice be considered unnecessary (see Section 12.4.2).

Structural engineers have pointed out that wind engineering laboratory reports do not provide any indication on the requisite magnitude of the wind load factor (see Appendix F), or of augmented design mean recurrence intervals, consistent with the uncertainties specific to the project at hand. Equations (7.1) and (7.2), or similar estimates, make it possible to depart from the notion that "one wind load factor fits all." They enable a differentiated approach that accounts, albeit approximately, for the explicit dependence of the wind load factor on individual uncertainties, which may differ for some structures from their typical values. The wind engineering laboratory can therefore help to achieve safe structural designs by providing, in addition to point estimates, uncertainty estimates of relevant aerodynamic and wind climatological features.

7.3 Individual and Overall Uncertainties

As noted in Section 7.1, uncertainty quantification is typically difficult or impossible to achieve rigorously, and must therefore be based wholly or in part on subjective assessments based on consensus among informed professionals, in addition to being based on measurements, physical considerations, and statistical methods.

7.3.1 Uncertainties in the Estimation of Extreme Wind Speeds

Large-Scale Extratropical Storms and Thunderstorms. It is reasonable to assume that the distributions of extreme wind speeds in large-scale extratropical storms and thunderstorms are Extreme Value Type I, with parameters that differ at the same site for the two types of storm. It is therefore possible to estimate the respective uncertainties by accounting for (i) measurement errors and (ii) sampling errors in the estimation of wind speeds for each of the two types of storm. For design wind speeds with specified mean recurrence intervals, sampling errors may be determined by using, for example, Eq. (3.9).

If the terrain exposure around the anemometer tower is open, measurement errors may be assumed to be relatively small, that is, in the order of 5%, say. However, if the terrain around the tower is built up, the conversion of wind speeds measured at the site to standardized wind speeds (i.e., wind speeds averaged over a specified time interval, e.g., 3 s, at a specified elevation, e.g., 10 m, above terrain with open exposure), the errors can be considerably larger (see [18]). Errors are likely to be even larger if wind speed measurements are performed using weather balloon data.[3]

Hurricanes. Hurricane wind speeds used for structural design are obtained by simulations that involve: the physical modeling of the hurricane wind flow at high altitudes (Section 1.3.1 and Eq. [1.4]); observations of pressure defects, radii of maximum rotational wind speeds, and storm translation speeds and directions (see Section 3.2.3); probabilistic models based on observations; empirical methods for transforming wind speeds at high altitudes into surface wind speeds; and calibration of the physical and probabilistic models against the rare available direct measurements of hurricane wind speeds, or against inferences on hurricane wind speeds based on observed hurricane wind damage to buildings and other structures. Added to the uncertainties inherent in the physical and probabilistic models used in the simulations are statistical uncertainties due to the relatively small number of hurricane events at various locations on the Gulf and Atlantic coasts. In particular, available observations may not include the occurrence of abrupt changes of direction of the hurricane translation velocity, resulting in the possible failure of engineering models to predict high wind speeds and/or storm surge. The lack of such observations might explain why, according to the ASCE 7 Standard, estimated design wind speeds in the New York City area are the same as, for example, in Arizona or western Massachusetts, or the failure to predict hurricane Sandy's severity [4]. Since rigorous estimates of uncertainties in hurricane wind speeds are in practice not possible, it is typically necessary to resort to engineering judgment. It is argued in [5] that theoretical models of natural phenomena such as hurricanes or earthquakes, while useful, should be superseded by prudent risk management considerations that weigh the relatively modest additional costs of conservative design against the costs of potential catastrophic failures. Even though, in spite of efforts reported in [6] and [7], the rigorous estimation of uncertainties in hurricane wind speeds is difficult if not impossible in the current state of the art, it is definitely the case that these uncertainties are greater than their counterparts for extratropical storms (note that the estimated uncertainties are considerably smaller in [6] than in [7]).

7.3.2 Uncertainties in the Estimation of Exposure Factors

Exposure factors represent ratios between squares of the wind speeds at various elevations over suburban terrain or water surfaces and their counterparts at 10 m above the open terrain. Wind profiles within cities, especially city centers, cannot be described in general terms, and are simulated in wind tunnels that reproduce to scale the built environment, as required, for example, in [8]. Wind tunnel simulations for locations with surface exposure difficult to define tend to reduce the uncertainty in the exposure factor.

---

3 Useful information on uncertainties inherent in weather balloon measurements could be obtained by performing such measurements at a location where reliable surface observations are available.

7.3.3 Uncertainties in the Estimation of Pressure Coefficients

Errors in the laboratory estimation of pressure coefficients are due to: (i) the violation of the Reynolds number in wind tunnels and, to a lesser extent, in large-scale aerodynamic facilities; (ii) differences between simulated and full-scale atmospheric boundary-layer flows; (iii) laboratory measurement errors; (iv) the estimation of pressure coefficient time history peaks; (v) the duration of the pressure coefficient record; and (vi) possible errors due to blockage (Chapter 5).

(i) Reynolds Number Effects. Wind tunnel simulations of aerodynamic pressures are typically performed at geometric scales in the order of $1 : 50$–$1 : 500$ and velocity scales of about $1 : 4$, say. Since in wind tunnels commonly used for structural engineering applications the fluid is air, that is, the same as for the prototype, Reynolds number similarity is typically violated by a factor in the order of 100–1000. In some large-scale aerodynamic facilities, the geometric and the velocity scales are in the order of $1 : 10$–$1 : 50$ and $1 : 1$–$1 : 2$, respectively, so that the Reynolds number is violated by a factor in the order of $1 : 10$–$1 : 100$.

The violation of the Reynolds number can be especially consequential for aerodynamic pressures on bodies with rounded shapes. As shown in Chapters 4 and 5, this is the case because, at the high Reynolds numbers typical of wind flows around buildings, the boundary layer that forms at the surface of the body is typically turbulent. The turbulent fluctuations transport particles with large momentum from the free flow into the boundary layer, thus helping the boundary-layer flow to overcome negative pressure gradients, and causing flow separation to occur farther downstream, thus reducing the drag on the body with respect to its value at lower Reynolds numbers. A remedial measure commonly used in wind tunnel simulations is to force the boundary-layer flow to be turbulent by rendering the body surface rougher. However, the resulting flow still differs from the high Reynolds number flow. This contributes to increasing the uncertainty in the pressure coefficients.

It has been argued that the violation of the Reynolds number is not consequential for flows around bodies with sharp corners, since for such bodies flow separation occurs at the corners, regardless of Reynolds number. This argument is not necessarily borne out by comparisons between full-scale and wind tunnel measurements. This has been shown in [9], which reported that peak negative pressure coefficients measured in the wind tunnel can underestimate their prototype counterparts by as much as a 25% (see Section 5.4.4). In such cases corrections of wind tunnel data, based on comparisons between full-scale and laboratory, are warranted. A systematic effort to develop such corrections remains to be performed. Positive pressure coefficients measured in the wind tunnel appear to be adequate, however.

(ii) Errors in the Simulation of Atmospheric Boundary Layer (ABL) Flows. Wind tunnel simulations of ABL profiles and turbulence are largely empirical (see Chapter 5). They depend upon the length of the test section, the type of roughness used to retard the flow near the wind tunnel floor, and the geometry of, and distance between, the spires placed at the entrance into the test section to help transform uniform flows into shear flows. Such simulations can achieve flows bearing at least a qualitative resemblance between simulated and prototype flows.

Differences between wind tunnel flows can result in significant differences between the respective pressure coefficient measurements. An international round-robin test reported in [10] showed that the coefficients of variation of the peak pressure coefficients measured in six reputable wind tunnel laboratories were as high as 10–40%. On the other hand, after the elimination of suspected outliers from results of tests performed by 12 laboratories, the respective measurements of pressures on a square cylinder were considered to be acceptable provided that the wind profiles and the turbulence intensities did not differ significantly from laboratory to laboratory [11].

For wind tunnel tests performed at relatively large geometric scales (e.g., 1 : 100 for low-rise buildings, rather than, say, 1 : 500 for tall buildings), an additional simulation problem arises: the inability to simulate in the wind tunnel the low-frequency portion of the longitudinal velocity spectra (see, e.g., [12]).

(iii) Uncertainties Associated with Measurement Equipment. A significant contributor to pressure measurement errors is the calibration of dynamic pressures in tubing systems connecting models to sensors. The pressure waves propagating inside a thin, circular tube distort the aerodynamic pressures on the model owing to the acoustic and visco-thermal effects brought about by fluid action on the tube [13]. According to [14], uncertainties associated with measurement equipment are typically approximately 10%.

(iv) Statistical Estimation of Pressure Coefficient Peaks. Appendix C describes a powerful peaks-over-threshold method that estimates peak pressure coefficients and their probability distributions. An alternative method is discussed in the following.

Let the pressure coefficient record $C_p(t)$ for any given direction θ have length T and be divided into a number n of subintervals ("epochs") of length T/n. The peak value of the pressure coefficient in any one epoch i ($i = 1, 2,\ldots, n$) (i.e., over any one subinterval of length T/n), denoted by $C_{p,pk\,i}(T/n)$, forms a data sample of size n. It is assumed that the epochs are sufficiently large that their respective peaks are independent, and that the data are identically distributed. Experience has shown that, typically, the data $C_{p,pk\,i}(T/n)$ are best fitted by a Type I Extreme Value (EV I) cumulative distribution function (see Eqs. [3.4] and [3.5]):

$$P\left[C_{p,pk}\left(\frac{T}{n}\right)\right] = \exp\left\{-\exp\left[-\frac{\left(C_{p,pk}\left(\frac{T}{n}\right) - \mu\right)}{\sigma}\right]\right\}, \tag{7.8}$$

where $P[C_{p,pk}(T/n)]$ is the probability that the variate $C_{p,pk}(T/n)$ is not exceeded during any one epoch of length T/n. The probability $F_r[C_{p,pk}\,(T/n)]$ that the variate $C_{p,pk}\,(T/n)$ is not exceeded during the 1st epoch, *and* the 2nd epoch, $\ldots$, *and* the rth epoch, is

$$F_r\left[C_{p,pk}\left(\frac{T}{n}\right)\right] = \left\{P\left[C_{p,pk}\left(\frac{T}{n}\right)\right]\right\}^r = \exp\left\{-r\exp\left[-\frac{\left(C_{p,pk}\left(\frac{T}{n}\right) - \mu\right)}{\sigma}\right]\right\}$$

$$\tag{7.9}$$

Inversion of Eq. (7.9) yields

$$C_{p,pk}\left(\frac{T}{n}\right)\Big|_{Fr} = (\mu + \sigma \ln r) - \sigma \ln(-\ln F_r) \tag{7.10}$$

Equation 7.9 shows that F_r is an EV I cumulative distribution function with location parameter $\mu + \sigma \ln r$ and scale parameter σ (see Eqs. [3.4]–[3.6]). The expectation of the largest $C_{p,pk}(T/n)|_{F_r}$ values over r epochs, denoted by $\overline{C}_{p,pk}(T/n, r)$, is

$$\overline{C}_{p,pk}\left(\frac{T}{n}, r\right) = (\mu + \sigma \ln r) + 0.5772\, \sigma \qquad (7.11)$$

(see Eq. [3.5a]). It follows from Eqs. (7.10) and (7.11) that

$$-\ln[-\ln F_r(\overline{C}_{p,pk})] = 0.5772 \qquad (7.12)$$

hence

$$F_r(\overline{C}_{p,pk}) = \exp[-\exp(-0.5772)] = 0.5704 \qquad (7.13)$$

Equation (7.13) may be interpreted as follows. Given a large number of realizations, in 57% of the cases the observed peak will be lower, and in 43% of the cases it will be larger than the expected value. The parameters μ and σ can be estimated from the sample of data $C_{p,pk\,i}(T/n)$ ($i = 1, 2, \ldots, n$) by using, for example, the BLUE estimator or the method of moments (Section 3.3.3).

In applications, design peak pressures are currently estimated by substituting in Eq. (7.10) estimated values for the "true" values of the parameters μ and σ, and assuming the probability $F_r = 0.78$ or 0.8 (as specified in [15, p. 22]), rather than $F_r = 0.5704$. The use of the probability $F_r = 0.8$, rather than $F_r = 0.5704$, is an instance of double counting, since it increases in Eq. (7.1) the pressure (or force) coefficient *above* its expected value, while also accounting in Eq. (7.2) for the deviation of the pressure from its expected value [16].

It has been argued that the use of the 0.78 or 0.8 value of F_r is consistent with storm durations in excess of 1 hour (e.g., 3 hours). However, if a storm duration longer than 1 hour were assumed, the expected peak corresponding to it should be estimated directly by using in Eq. (7.9) a value of r consistent with that duration. Also, the assumption that storm durations are longer than one hour would be at variance with U.S. standard practice, which follows the convention of 1-hour storm durations.

For a thorough study of peaks of time series of pressure coefficients, see [17].

(v) Estimation of Pressure Coefficient Peaks from Short Records. In some applications the available records are short. This is the case, for example, for pressure measurements performed in large aerodynamic facilities, where operation time is expensive.

Example 7.1 Consider a $T = 90$-second long record of pressure coefficients at the tap of a roof on a model with length scale 1 : 8 and velocity scale 1 : 2. The length of the prototype counterpart of the record is obtained from the condition

$$T_p = \left(\frac{L_p}{L_m}\right)\left(\frac{U_m}{U_p}\right) T_m = 8 \times \left(\frac{1}{2}\right) \times 90 \text{ seconds} = 360 \text{ seconds}.$$

Let $n = 16$. The prototype length of each subinterval is then $T_p/16 = 360/16 = 22.5$ seconds.

For the 360-second prototype record being considered, the mean and standard deviation of the sample consisting of the peak pressures of the 16 subintervals (epochs) are assumed to be $|E[C_{p,pk}(T/n)]| = 4.72$ and $SD[C_{p,pk}(T/n)] = 0.75$, respectively, to which there correspond the estimated Type I Extreme Value distribution parameters (Eq. [3.5]):

$$\sigma\left(\frac{T}{n}\right) = \sqrt{\frac{6}{\pi}}\, \text{SD}\left(\frac{T}{n}\right) = 0.78 \times 0.75 = 0.585 \quad \text{and}$$

$$\mu\left(\frac{T}{n}\right) = \overline{C}_{p,pk}\left(\frac{T}{n}\right) - 0.5772\ \sigma\left(\frac{T}{n}\right) = 4.72 - 0.5772 \times 0.585 = 4.38.$$

The estimated means of the peak $C_{p,pk}(T/16 = 22.5\,\text{s}, r)$ for $r = 16$ and $r = 160$ (i.e., for a 360-s and a 3600-s long prototype record) are

$$\left|\overline{C}_{p,pk}\left(\frac{T}{16}, r = 16\right)\right| = \mu + \sigma \ln r + 0.5772\sigma = 4.38 + 0.585 \ln 16$$

$$+ 0.5772 \times 0.585 = 6.34$$

(Eq. [7.13]) and

$$\left|\overline{C}_{p,pk}\left(\frac{T}{16}, r = 160\right)\right| = \mu + \sigma \ln r + 0.5772\sigma = 4.38 + 0.585 \ln(160)$$

$$+ 0.5772 \times 0.585 = 7.70.$$

The standard deviations of the sampling errors in the estimation of the mean peak $\overline{C}_{p,pk}(T/16 = 22.5\,\text{seconds}, r)$ can be obtained from Eq. (3.9). Note that in both cases the sample size is $n = 16$.

7.3.4 Uncertainties in Directionality Factors

According to a study reported in [2], uncertainties in the directionality factors may be assumed to be typically in the order of 10%.

References

1 Ellingwood, B., Galambos, T. V., MacGregor, J. G., and Cornell, C. A., "Development of a probability-based load criterion for American National Standard A58," NBS Special Publication 577, National Bureau of Standards, Washington, DC, 1980.

2 Habte, F., Chowdhury, A., Yeo, D., and Simiu, E. (2015). Wind directionality factors for nonhurricane and hurricane-prone regions. *Journal of Structural Engineering* 141: 04014208.

3 Panofsky, H.A. and Dutton, J.A. (1984). *Atmospheric Turbulence: Models and Methods for Engineering Applications*, 1e. Wiley.

4 Yeo, D., Lin, N., and Simiu, E. (2014). Estimation of hurricane wind speed probabilities: application to New York City and other coastal locations. *Journal of Structural Engineering* 140: 04014017.

5 Emanuel, K. (2012). Probable cause: are scientists too cautious to help us stop climate change? *Foreign Policy*, Nov. 9, 2012. https://foreignpolicy.com/2012/11/09/probable-cause.

6 Vickery, P.J., Wadhera, D., Twisdale, L.A. Jr., and Lavelle, F.M. (2009). U.S. hurricane wind speed risk and uncertainty. *Journal of Structural Engineering* 135: 301–320.

7 Coles, S. and Simiu, E. (2003). Estimating uncertainty in the extreme value analysis of data generated by a hurricane simulation model. *Journal of Engineering Mechanics* 129: 1288–1294.

8 ASCE, "Wind tunnel testing for buildings and other structures (ASCE/SEI 49-12)," in *ASCE Standard ASCE/SEI 49-12*, Reston, VA: American Society of Civil Engineers, 2012.

9 Long, F., "Uncertainties in pressure coefficient derived from full and model scale data," Report to the National Insititue of Standards and Technology, Wind Science and Engineering Research Center, Department of Civil Engineering, Texas Technical University, Lubbock, TX, 2005.

10 Fritz, W.P., Bienkiewicz, B., Cui, B. et al. (2008). International comparison of wind tunnel estimates of wind effects on low-rise buildings: test-related uncertainties. *Journal of Structural Engineering* 134: 1887–1890.

11 Hölscher, N. and Niemann, H.-J. (1998). Towards quality assurance for wind tunnel tests: a comparative testing program of the Windtechnologische Gesellschaft. *Journal of Wind Engineering and Industrial Aerodynamics* 74–76: 599–608.

12 Mooneghi, M.A., Irwin, P.A., and Chowdhury, A.G. (2016). Partial turbulence simulation method for predicting peak wind loads on small structures and building appurtenances. *Journal of Wind Engineering and Industrial Aerodynamics* 157: 47.

13 Irwin, P.A., Cooper, K.R., and Girard, R. (1979). Correction of distortion effects caused by tubing systems in measurements of fluctuating pressures. *Journal of Wind Engineering and Industrial Aerodynamics* 5: 93–107.

14 Diaz, P. S. Q., "Uncertainty analysis of surface pressure measurements on low-rise buildings," M.S. thesis, Civil Engineering, University of Western Ontario, London, Ontario, Canada, 2006.

15 ISO, "Wind Actions on Structures" in *ISO 4354*, Geneva, Switzerland: International Standards Organization 2009.

16 Simiu, E., Pintar, A.L., Duthinh, D., and Yeo, D. (2017). Wind load factors for use in the wind tunnel procedure. *ASCE-ASME Journal of Risk and Uncertainty in Engineering Systems, Part A: Civil Engineering* 3. https://www.nist.gov/wind.

17 Gavanski, E., Gurley, K.R., and Kopp, G.A. (2016). Uncertainties in the estimation of local peak pressures on low-rise buildings by using the Gumbel distribution fitting approach. *Journal of Structural Engineering* 142: 04016106.

18 Masters, F.J., Vickery, P.J., Bacon, P., and Rappaport, E.N. (2010). Toward objective, standardized intensity estimates from surface wind speed observations. *Bulletin of the American Meteorological Society* 91: 1665–1681.

Part II

Design of Buildings

8

Structural Design for Wind

An Overview

This chapter starts with a brief history of approaches to the design of structures for wind (Section 8.1). It then presents an overview of two design procedures based on recently developed technology allowing the simultaneous measurement of pressure time histories at large numbers of taps placed on wind tunnel models.[1] Both procedures depend on "big data" processing and entail iterative computations (including dynamics calculations) that, once the wind climatological and aerodynamic data are provided by the wind engineer, are most effectively performed by the structural engineer. The first of these procedures is called Database-Assisted Design (DAD) and is discussed in Section 8.2. DAD uses recorded time series of randomly varying pressure coefficients to determine, by rigorously accounting for dynamic and directional effects, peak demand-to-capacity indexes (DCIs) with specified mean recurrence intervals (MRIs) for any desired number of structural members (for details on DCIs see Chapter 13). DAD can be applied to buildings regardless of the complexity of their shape. (Examples of buildings with complex shapes are the CCTV building, the Shanghai World Financial Center, and the Burj Khalifa tower.) The second procedure, discussed in Section 8.3, uses time series of measured pressure coefficients only for the computation of the aerodynamic and inertial forces acting at the building floor levels, following which it determines static wind loads used to calculate design DCIs with specified MRIs. If the resulting DCIs are close to their counterparts produced by the DAD procedure, those loads can be regarded as equivalent static wind loads (ESWLs). It can be inferred from Chapter 14 that, unlike DAD, the procedure for determining ESWLs is typically applicable only to buildings with relatively simple geometries (e.g., buildings with rectangular shape in plan). Section 8.4 briefly compares the DAD and ESWL procedures; in particular, it discusses the verification of ESWL results against benchmark values obtained by DAD.

8.1 Modern Structural Design for Wind: A Brief History

Modern structural design for wind emerged in the 1960s as a synthesis of the following developments:

- Modeling of the neutrally stratified atmospheric boundary layer flow, including (i) the variation of wind speeds with height above the ground as functions of upwind surface roughness, and (ii) the properties of atmospheric turbulence.

---

1 These procedures may be inapplicable in the rare cases in which the configuration of the building models does not allow the placement of pressure taps.

Wind Effects on Structures: Modern Structural Design for Wind, Fourth Edition. Emil Simiu and DongHun Yeo.
© 2019 John Wiley & Sons Ltd. Published 2019 by John Wiley & Sons Ltd.

- Probabilistic modeling of extreme wind speeds.
- Modeling of pressures induced on a face of a rectangular building by atmospheric flow normal to that face.
- Frequency-domain modeling of the dynamic along-wind response produced by atmospheric flow normal to a building face.

The increase of wind speeds with height above ground was first reported by Helmann in 1917 [1]. The aerodynamic effects of turbulent shear flows were first researched by Flachsbart in 1932 [2] (Figure 4.31).[2] Flachsbart's work influenced the approach to the 1933 tests of the Empire State Building reported by Dryden and Hill [5]. Probabilistic models of extreme values for geophysical applications were developed by Gumbel in the 1940s [6]. A pioneering approach to the analytical estimation of the dynamic response of bodies immersed in turbulent flow was developed by Liepmann in 1952 [7]. A synthesis of these developments was first achieved in the 1960s by Davenport [8, 9], a University of Bristol student of the eminent engineer Sir Alfred Pugsley. However, that synthesis could not account for wind effects induced by vorticity shed in the wake of the structure, by winds skewed with respect to a building face or affected by the presence of neighboring buildings, or for aeroelastic behavior. Specialized wind tunnels were therefore developed in the 1960s with a view to simulating the atmospheric boundary layer flow and its aerodynamic, dynamic, and aeroelastic effects on structures.

During the 1970s wind tunnel techniques were not sufficiently developed to allow the accurate determination of wind effects for structural design purposes. Information on wind effects was based in large part on non-simultaneous pressures measured at typically small numbers of taps (e.g., six taps for a model that currently accommodates hundreds of pressure taps – see Figures 5.29 and 5.30), with unavoidable errors that can be significant.

An improvement in the capability to determine wind effects was achieved in the late 1970s with the development of the high frequency force balance (HFFB) [10]. The HFFB approach, used in conjunction with frequency-domain analyses, is applied to tall buildings designed to have no unfavorable aeroelastic response under realistic extreme wind loading – that is, in practice, to all well-designed tall buildings. HFFB provides time histories of the effective (aerodynamic and dynamic) base moments induced by the wind loads. Its chief drawback is that it provides no information on the distribution of the wind loads with height, since that distribution cannot be inferred from the base moments or shears (see, e.g., [11]). The loading information needed to calculate the demand-to-capacity ratios therefore depended largely on guesswork, especially for buildings influenced aerodynamically by neighboring structures. Nevertheless, the HFFB approach can be useful in the preliminary phase of the design process, for the rapid if only qualitative aerodynamic assessment of building configurations, orientations, and aerodynamic features. The HFFB approach is also useful for buildings with facade configurations that do not allow the effective placing of pressure taps.

From the 1990s on, the development of the pressure scanner (see Section 5.7) has radically changed the approach to structural design for wind and has rendered the HFFB approach largely obsolete. The pressure scanner allows the simultaneous measurement

---

2 Flachsbart was dismissed by the Nazi authorities for refusing to divorce his Jewish wife [3] and was therefore unable to complete his research. Some of his results were re-discovered independently by Jensen in the 1960s [4].

of pressures at as many as hundreds of taps and, therefore, the capture of the pressures' variation as a function of time and spatial separation. To exploit this new measurement technology two computer-intensive procedures have been developed, which are used in conjunction with time-domain analyses: Database-assisted Design and Equivalent Static Wind Loads (ESWL), applicable, like the HFFB procedure, to tall buildings designed to have no unfavorable aeroelastic response under realistic extreme wind loading. Introductions to DAD and ESWL procedures are presented in Sections 8.2 and 8.3, respectively.

8.2 Database-Assisted Design

DAD is a computer-intensive technique based on the full use of aerodynamic pressure data for structural design purposes. It provides benchmark values against which results of procedures based on ESWLs can be assessed. DAD uses time-domain methods, which are typically more straightforward, transparent, and effective than their frequency-domain counterparts.

Structural design for wind uses two types of wind engineering data: (i) time series of pressure coefficients on a structure measured simultaneously at multiple taps, and (ii) wind climatological data at the building site. The task of the wind engineering laboratory is to deliver these data as well as estimates of the uncertainties inherent in them.

The tasks of the structural engineer are the following:

1. Select the structural system, and determine the structure's preliminary member sizes based on a simplified model of the wind loading (e.g., a static wind loading taken from standard provisions). The structural design so achieved is denoted by D_0.
2. For the design D_0, determine the system's mechanical properties, including the modal shapes, natural frequencies of vibration, and damping ratios, as well as the requisite influence coefficients; and develop on their basis a dynamic model of the structure. P-Δ and P-δ effects can be accounted for by using, for example, the geometric stiffness matrix (Chapter 9).
3. From the time histories of simultaneously measured pressure coefficients, determine the time histories of the randomly varying aerodynamic loads induced at all floor levels by directional mean wind speeds $U(\theta)$ for a sufficient number of speeds U (e.g., $20\,\mathrm{m\,s^{-1}} < U \leq 80\,\mathrm{m\,s^{-1}}$, say) and directions θ ($0° \leq \theta < 360°$). The reference height for the mean wind speeds is typically assumed to be the height of the structure (Chapter 10).
4. For each of the directional wind speeds defined in task 3, perform the dynamic analysis of the structure D_0 to obtain the time histories at floor k of (i) the inertial forces induced by the respective aerodynamic loads and (ii) the effective wind-induced loads $F_k[U(\theta), t]$ applied at the structure's center of mass. The lateral loading determined in this task consists of the three components acting along the principal axes x, y, and the torsional axis ϑ (Chapter 11).
5. For each cross section m of interest, use the appropriate influence coefficients to determine time series of the DCIs induced by the combination of factored gravity loads and effective wind loads obtained in task 4. The DCIs are the left-hand sides of the design interaction equations, and are typically used to size members subjected to

more than one type of internal force. For example, the interaction equations for steel members subjected to flexure and axial forces are [12]:

$$\text{If} \quad \frac{P_r}{\phi_p P_n} \geq 0.2, \quad \frac{P_r}{\phi_p P_n} + \frac{8}{9}\left(\frac{M_{rx}}{\phi_m M_{nx}} + \frac{M_{ry}}{\phi_m M_{ny}}\right) \leq 1.0, \tag{8.1}$$

$$\text{If} \quad \frac{P_r}{\phi_p P_n} < 0.2, \quad \frac{P_r}{2\phi_p P_n} + \left(\frac{M_{rx}}{\phi_m M_{nx}} + \frac{M_{ry}}{\phi_m M_{ny}}\right) \leq 1.0, \tag{8.2}$$

In Eqs. (8.1) and (8.2), P_r and P_n are the required and available tensile or compressive strength; M_{rx} and M_{nx} are the required and available flexural strength about the strong axis; M_{ry} and M_{ny} are the required and available flexural strength about the weak axis; ϕ_p and ϕ_m are resistance factors.[3] The required strengths are based on combinations of wind and gravity effects specified in the applicable code. A similar, though simpler expression for the DCI, is applied to shear forces. Additional material on DCIs is provided in Chapter 13.

6. For each cross section m of interest, construct the response surfaces of the peak combined effects being sought as functions of wind speed and direction; that is, for each of the directional wind speeds considered in task 3, determine the corresponding peak of the DCI time series (e.g., Eqs. [8.1] and [8.2]), and construct from the results so obtained the peak DCI response surface. The response surfaces are properties of the structure, dependent upon its aerodynamic and mechanical characteristics, but independent of the wind climate. They provide for each cross section of interest the peak DCIs as functions of wind speed and direction. Response surfaces are also constructed for peak inter-story drift ratios and peak accelerations. For details, see Chapter 13.

7. Use the information contained in the response surfaces and the matrices of directional wind speeds at the site to determine, by accounting for wind directionality, the *design DCIs*, that is, the peak DCIs with the specified MRI $\overline{N}$ for the cross sections of interest. For each cross section m the steps required for this purpose are:

   (i) In the directional wind speed matrix $[U_{ij}]$, where i and j denote the storm number identifier and the wind direction, respectively, replace the entries U_{ij} by the peak DCIs $\text{DCI}_m^{pk}(U = U_i, \theta = \theta_j)$ taken from the response surface for the cross section m.

   (ii) Transform the matrix $[\text{DCI}_m^{pk}(U_i, \theta_j)]$ so obtained into the vector $\{\max_j[\text{DCI}_m^{pk}(U_i, \theta_j)]\}^T$ where T denotes transpose, by disregarding in each row i all entries lower than $\max_j[\text{DCI}_m^{pk}(U_i, \theta_j)]$.

   (iii) Rank-order the quantities $\max_j[\text{DCI}_m^{pk}(U_i, \theta_j)]$ and use non-parametric statistics in conjunction with the mean annual rate of storm arrival λ, to obtain the design DCIs, that is, the quantities $\text{DCI}_m^{pk}(\overline{N})$ (Chapter 13 and Section A.8). Similar operations are performed for inter-story drift ratios and accelerations.

If, for the member being considered, the design DCI is approximately unity, the design of that member is satisfactory from a strength design viewpoint. If the uncertainties in the wind velocity and/or the aerodynamic data are significantly larger than their

---

3 Some indexes used in Eqs. (8.1) and (8.2) are used elsewhere in this book in different contexts, in which they are clearly defined.

typical values on which code requirements are based, the design MRIs will exceed the MRIs specified in, for example, the ASCE 7-16 Standard, and can be determined as in Section 12.5 [13].

In general, the preliminary design D_0 does not satisfy the strength and/or serviceability design criteria. The structural members are then re-sized to produce a modified structural design D_1. This iterative process continues until the final design is satisfactory. If necessary, to help satisfy serviceability criteria, motion mitigation devices such as Tuned Mass Dampers are used (Chapter 16).

Tasks 2 through 7 are repeated as necessary until the design DCIs are close to unity, to within serviceability constraints. Each iteration entails a re-sizing of the structural members consistent with the respective estimated design DCIs.

Features of interest of the DAD approach are summarized next:

- The wind engineer performs wind engineering tasks, and the structural engineer performs structural engineering tasks. The wind engineer's tasks are to provide the requisite aerodynamic and wind climatological data, with the respective uncertainty estimates. These data are used by the structural engineer to determine the stochastic aerodynamic loading and perform the dynamic analyses required to obtain the effective wind-induced loading, as well as all the subsequent operations resulting in the structure's final design. Included in these operations is the estimation of the design DCIs, inter-story drift ratios, and building accelerations with the respective specified MRIs, consistent with the uncertainties in the aerodynamic and the wind climatological data (Chapter 13). This division of tasks is efficient and establishes clear lines of accountability for the wind engineer and for the structural engineer. The structural engineer's role in designing structures for wind thus becomes similar to the role of the designer of structures for seismic effects, whose tasks include performing the requisite dynamic analyses.
- DAD allows higher modes of vibration and any modal shape to be rigorously accounted for.
- Wind effects with specified MRIs obtained by accounting for wind directionality are determined by the structural engineer rigorously and transparently, as functions of the properties of the structure inherent in the final structural design.
- The aerodynamics and wind climatological data provided by the wind engineer, as well as the operations performed by the structural engineer, can be recorded and documented in detail, allowing the full development of Building Information Modeling (BIM) for the structural design for wind [14]. This feature enables ready traceability and detailed scrutiny of the data by the project stakeholders.
- Owing to currently available computational capabilities the requisite tasks can be readily performed in engineering offices.
- Combined wind effects, including DCIs, induced by wind loads acting on all building facades as well as by wind-induced torsion, are determined automatically by using specialized software. The software can be accessed via links provided at the end of this chapter.

Typically, satisfactory designs for strength, that is, designs resulting, to within serviceability constraints, in DCIs close to unity, require more than one iteration, owing to possibly significant successive changes in member sizes and in the structure's dynamic properties. As noted earlier, once the aerodynamic and wind climatological data, as well as estimates of the respective uncertainties, are provided by the wind engineer, the calculations – including all dynamic calculations – are performed by the structural engineer. This eliminates unnecessary, time-consuming interactions, required in earlier practices, between the wind engineering laboratory and the structural engineering office.

8.3 Equivalent Static Wind Loads

The ESWL procedure presented in this book is a variant of DAD and, like DAD, requires the wind engineer to provide wind climatological data at the building site, time series of pressure coefficients measured simultaneously at multiple taps, and measures of uncertainties inherent in those data. As in the case of DAD, once these tasks are completed, the ESWL-based design process is fully the responsibility of the structural engineer. The ESWL procedure, which by definition yields design DCIs that approximate their benchmark counterparts determined by DAD, is typically applicable to structures with simple geometries.

The structural engineer's tasks 1–4 are identical to their counterparts for DAD. The subsequent tasks are performed for each of the wind speeds and directions considered in task 3, as follows:

4a. Determine the static loads $F_{kx,p}^{\text{ESWL}}(U, \theta)$, and acting at the mass center of floor k ($k = 1, 2, \ldots, n_f$) in the direction of the building's principal axes x, y, and about the torsional axis ϑ, where the subscript p ($p = 1, 2, \ldots, p_{\max}$) identifies distinct wind loading cases WLC$_p$ associated with superpositions of the three EWSL loads, and $p_{\max}$ is a function of the number n_{pit} of points in time (pit) used to obtain the peak effects of interest [15]. This task is described in detail in Chapter 14.

5. For each cross section m of interest, calculate the internal forces used to determine its DCI, and substitute their expressions into the expressions for the DCIs (e.g., Eq. [8.1]). This task requires the use of (i) the static wind loads determined in task 4a, (ii) the influence coefficients, and required to calculate the wind-induced internal forces, and (iii) the factored gravity loads and the respective influence coefficients. For example, is the internal force induced at cross section m by a unit force acting in direction x at the center of mass of floor k. The wind-induced internal forces are denoted by. Their expression is

$$f_{m,p}^{\text{ESWL}}(U, \theta) = \sum_{k=1}^{n_f} r_{mk,x} F_{kx,p}^{\text{ESWL}}(U, \theta) + \sum_{k=1}^{n_f} r_{mk,y} F_{ky,p}^{\text{ESWL}}(U, \theta) + \sum_{k=1}^{n_f} r_{mk,\vartheta} F_{k\vartheta,p}^{\text{ESWL}}(U, \theta)$$

(8.3)

6. The corresponding DCIs, denoted by DCI$_{m,p}$, are obtained by substituting the calculated internal forces into the expressions for the DCIs. For design purposes only the largest of these DCIs is of interest, that is,

$$\text{DCI}_m^{\text{RS,ESWL}}(U, \theta) = \max_p(\text{DCI}_{m,p}^{\text{ESWL}}(U, \theta))$$

(8.4)

The surface representing, for each cross section m of interest, the dependence of its demand-to-capacity index DCI$_m$ upon wind speed U and direction θ, is called the *response surface* for the cross section m. (The superscript RS denotes "response surface.")

7. Use the response surfaces constructed in task 6, the climatological wind speed matrix at the building site $[U_{ij}]$, and the non-parametric statistical procedure described in detail in Chapter 13, to determine the design peak DCIs with the specified $\overline{N}$-year MRI. As was also noted for the DAD procedure, depending upon the uncertainties in the aerodynamic and climatological wind speed data as determined by the wind engineering laboratory, the design MRI may have to differ from the value specified, for example, in the ASCE 7-16 Standard, in which case it can be determined as indicated in Chapter 12.

If the design DCIs determined in task 7 differ significantly from unity, the structure's members are re-sized to create a new design D_1. Tasks 2–7 are then performed on that

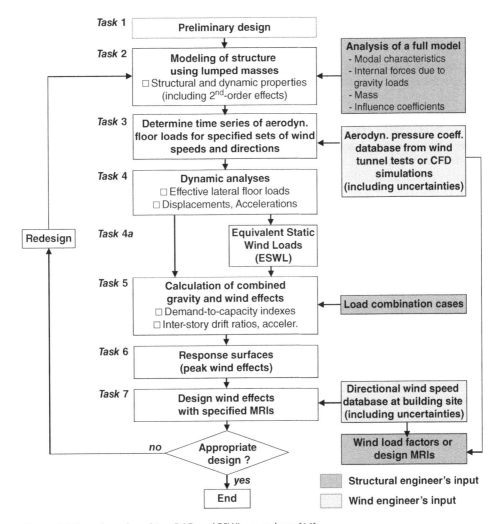

Figure 8.1 Flowchart describing DAD and ESWL procedures [16].

design. This process is iterated until a structural design is achieved for which, in each structural member, the design DCI is close to unity, to within serviceability constraints. As is the case for DAD, all calculations are automated. The requisite software can be accessed via the link provided at the end of this chapter.

8.4 DAD versus ESWL

The ESWL procedure has the same useful features listed for the DAD procedure in Section 8.2. It has been argued that, at least for the time being, some structural engineers may prefer performing the design for wind by using ESWLs. However, since both the DAD and the ESWL procedures are automated, the amount of labor required on the part of the structural engineer is the same regardless of which procedure is used. In addition, it is worth noting that while design for seismic loads was originally based on static seismic loads, structural engineering culture has evolved to the point where this is no longer necessarily the case. Design for wind is expected to undergo a similar evolution.

Given the substitution of static loads for the actual stochastic loads, it is appropriate to verify the extent to which the ESWL procedure actually results in structural designs approximately equivalent to those produced by DAD. This is achieved by comparing DCIs induced by ESWL and DAD (see Chapter 18). The use of peak DCIs obtained by DAD as benchmarks against which DCIs induced by ESWL can be verified is justified by the superior accuracy inherent in the DAD procedure.

A flowchart describing the sequence of operations leading to the final structural design by the DAD and the ESWL procedures is shown in Figure 8.1 [16]. The software DAD_ESWL version 1.0, a detailed user's manual [17], and a tutorial with detailed examples [18], are available for the two procedures at https://www.nist.gov/wind.

References

1 Hellmann, G., *Über die Bewegung der Luft in den untersten Schichten der Atmosphäre*: Königlich Preussischen Akademie der Wissenschaften, 1917.

2 Flachsbart, O. (1932). Winddruck auf offene und geschlossene Gebäude. In: *Ergebnisse der Aerodynamischen Versuchanstalt zu Göttingen* (ed. L.L. Prandtl and A. Betz), 128–134. Munich: R. Oldenbourg Verlag.

3 Plate, E., Personal comminication, 1995.

4 Jensen, M. and Franck, N. (1965). *Model Scale Tests in Turbulent Wind*. Copenhagen: Danish Technical Press.

5 Dryden, H.L. and Hill, G.C. Wind pressure on a model of the empire state building. *Bureau of Standards Journal of Research* 10 (4): 493–523. Research Paper 545, April 1933.

6 Gumbel, E.J. (1958). *Statistics of Extremes*. New York: Columbia University Press.

7 Liepmann, H.W. (1952). On the application of statistical concepts to the buffeting problem. *Journal of the Aeronautical Sciences* 19: 793–800.

8 Davenport, A.G. (1961). The application of statistical concepts to the wind loading of structures. *Proceedings of the Institution of Civil Engineers* 19: 449–472.

9 Davenport, A.G. (1967). Gust loading factors. *Journal of the Structural Division, ASCE* 93: 11–34.

10 Tschanz, T. (1982). Measurement of total dynamic loads using elastic models with high natural frequencies. In: *Wind Tunnel Modeling for Civil Engineering Applications* (ed. T.A. Reinhold), 296–312. Cambridge: Cambridge University Press.

11 Chen, X. and Kareem, A. (2005). Validity of wind load distribution based on high frequency force balance measurements. *Journal of Structural Engineering* 131: 984–987.

12 ANSI/AISC, "Specification for Structural Steel Buildings," in *ANSI/AISC 360–10*, Chicago, Illinois: American Institute of Steel Construction, 2010.

13 Simiu, E., Pintar, A.L., Duthinh, D., and Yeo, D. (2017). Wind load factors for use in the wind tunnel procedure. *ASCE-ASME Journal of Risk and Uncertainty in Engineering Systems, Part A: Civil Engineering* 3: 04017007. (https://www.nist.gov/wind).

14 ARUP. (June 14, 2017). *Building Information Modelling (BIM)*. Available: www.arup.com/services/building_modelling

15 Yeo, D. (2013). Multiple points-in-time estimation of peak wind effects on structures. *Journal of Structural Engineering* 139: 462–471. (https://www.nist.gov/wind).

16 Park, S., Simiu, E., and Yeo, D., "Equivalent static wind loads vs. database-assisted design of tall buildings: An assessment," *Engineering Structures* (submitted). https://www.nist.gov/wind.

17 Park, S. and D. Yeo, *Database-Assisted Design and Equivalent Static Wind Loads for Mid- and High-Rise Structures: Concepts, Software, and User's Manual*, NIST Technical Note 2000, National Institute of Standards and Technology, Gaithersburg, MD, 2018 (Available: https://doi.org/10.6028/NIST.TN.2000).

18 Park, S. and Yeo, D., *Tutorial for DAD and ESWL: Examples of High-Rise Building Designs for Wind*, NIST Technical Note 2001, National Institute of Standards and Technology, Gaithersburg, MD, 2018 (Available: https://doi.org/10.6028/NIST.TN.2001).

9

Stiffness Matrices, Second-Order Effects, and Influence Coefficients

For structures with linearly elastic material behavior, structural, and dynamic analyses can be performed by using stiffness matrices (Section 9.1) and accounting as necessary for second-order effects (e.g., via geometric stiffness matrices, Section 9.2). Influence coefficients, representing wind effects of interest induced by unit loads acting at mass centers along the structure's principal axes, and unit torsional moments about the centers of mass are considered in Section 9.3.

Stiffness matrices, geometric stiffness matrices, and influence coefficients can be determined by using finite element software. Second-order effects can be determined by a variety of methods other than the geometric stiffness matrix method, including the simple moment amplification method [1]. Software and user manuals described and accessible via links provided in Chapters 17 and 18 contain modules that perform the requisite calculations.

9.1 Stiffness Matrices

To define the stiffness matrix of the linearly elastic structural system of a building with n_f floors, consider the flexibility matrix

$$[a] = \begin{bmatrix} [x_{i,1x}\ x_{i,2x} \cdots x_{i,n_f x}] & [x_{i,1y}\ x_{i,2y} \cdots x_{i,n_f y}] & [x_{i,1\vartheta}\ x_{i,2\vartheta} \cdots x_{i,n_f \vartheta}] \\ [y_{i,1x}\ y_{i,2x} \cdots y_{i,n_f x}] & [y_{i,1y}\ y_{i,2y} \cdots y_{i,n_f y}] & [y_{i,1\vartheta}\ y_{i,2\vartheta} \cdots y_{i,n_f \vartheta}] \\ [\vartheta_{i,1x}\ \vartheta_{i,2x} \cdots \vartheta_{i,n_f x}] & [\vartheta_{i,1y}\ \vartheta_{i,2y} \cdots \vartheta_{i,n_f y}] & [\vartheta_{i,1\vartheta}\ \vartheta_{i,2\vartheta} \cdots \vartheta_{i,n_f \vartheta}] \end{bmatrix} \quad (9.1)$$

which consists of nine component sub-matrices represented in the right-hand side of Eq. (9.1) by their respective ith rows ($i = 1, 2, \ldots, n_f$); for example, the entry denoted in Eq. (9.1) by $[y_{i,\,1x}, y_{i,\,2x}, \cdots y_{i,n_f x}]$ represents the matrix

$$\begin{bmatrix} y_{1,1x} & y_{1,2x} \cdots y_{1,n_f x} \\ y_{2,1x} & y_{2,2x} \cdots y_{2,n_f x} \\ & \vdots \\ y_{n_f,1x} & y_{n_f,2x} \cdots y_{n_f,n_f x} \end{bmatrix} \quad (9.2)$$

The size of matrix $[a]$ is $3n_f \times 3n_f$. The terms of matrix $[a]$ are displacements in the x or y direction or torsional rotations about the mass center of floor i ($i = 1, 2, \ldots, n_f$) due to a unit horizontal force in the x or y direction or a unit torsional moment about the mass center of floor j ($j = 1, 2, \ldots, n_f$), and can be obtained by using standard structural analysis

Wind Effects on Structures: Modern Structural Design for Wind, Fourth Edition. Emil Simiu and DongHun Yeo.
© 2019 John Wiley & Sons Ltd. Published 2019 by John Wiley & Sons Ltd.

programs. (For example, the term $y_{1,2x}$ is the y displacement of the mass center of floor 1 due to a unit horizontal force acting at the mass center of floor 2 in direction x.)

The stiffness matrix of the system is the inverse of the matrix $[a]$:

$$[k] = [a]^{-1}, \tag{9.3}$$

As follows from Eq. (9.3), the product $[k][a]$ is the identity matrix. In the matrix $[k]$, for example, the restoring force $k_{1x,2y}$ represents the horizontal force in the x direction at the mass center of floor 1, induced by a unit horizontal displacement in the y direction at the mass center of floor 2.

For structures with members of known sizes and properties the matrix $[k]$ is determined by using standard finite element software.

9.2 Second-Order Effects

Wind forces induce horizontal displacements that give rise to overturning moments acting at every floor, equal to the weight of the floor times the floor's horizontal displacements. These overturning moments result in an amplification of the wind effects. The study of second-order effects is concerned with this amplification and its structural and dynamic consequences.

In linear elastic analysis equilibrium is based on the undeformed geometry of the structure. In elastic, geometrically nonlinear analysis, equilibrium is based on the deformed geometry of the structure, while the material behavior is assumed to be elastic; in inelastic, geometrically nonlinear analysis the equilibrium is based on the deformed geometry and the material behavior is assumed to be inelastic [1]. In this book, unless otherwise indicated, the structural behavior is assumed to be elastic. The analysis includes both chord rotation effects due to sway at the member ends (i.e., P-Δ effects), and member curvature effects (i.e., P-δ effects) [2]. Both effects are illustrated in Figure 9.1.

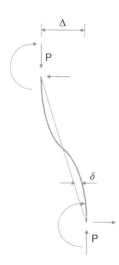

In Chapter 18 second-order effects are determined by the *geometric stiffness method*, in which the total displacements of the structure are obtained by subtracting from the stiffness matrix $[k]$ a geometric matrix $[k_g]$ developed as shown in [1]. The resultant matrix, denoted by $[k_s]$ and henceforth referred to as the *effective matrix*, is "softer" than the matrix $[k]$, and replaces the latter in calculations of the structural response to wind and gravity loading, including calculations of influence coefficients and dynamic response. In the geometric stiffness method, the variation of transverse displacements along the member's length is commonly approximated by a cubic polynomial. An example of the derivation of terms of the matrix $[k_g]$ for a two-dimensional beam-column member with six degrees of freedom is shown in [2]. As is the case for the matrix $[k]$, for structures with known member sizes and properties the matrix $[k_g]$ can be calculated by using standard finite element software [3]. This approach has limitations noted in [2], which appear not to be significant for

Figure 9.1 P-Δ (member chord) and P-δ (member curvature) effects.

tall buildings subjected to wind loads. It is suggested, however, that the validity of this statement be the object of further research, and that an alternative approach to the estimation of second-order effects be considered if necessary.

9.3 Influence Coefficients

Influence coefficients are used in conjunction with wind and gravity loads acting on the structure to determine internal forces, displacements, and accelerations induced by those loads. Consider the aerodynamic wind load time series $F_{kx}[U(\theta), t]$, $F_{ky}[U(\theta), t]$, $F_{k\vartheta}[U(\theta), t]$, induced along the principal axes and in torsion by wind with mean speed U and direction θ at reference height z_{ref}, acting at the center of mass of floor k ($k = 1, 2, \ldots, n_f$). The time series of the internal forces denoted by $f_m[U(\theta), t]$ induced by those load time series at a cross section m can be written as the sum

$$f_m[U(\theta), t] = \sum_{k=1}^{n_f} \{ r_{mk,x} F_{kx}[U(\theta), t] + r_{mk,y} F_{ky}[U(\theta), t] + r_{mk,\vartheta} F_{k\vartheta}[U(\theta), t] \} \quad (9.4)$$

where the influence coefficients $r_{mk,x}$, $r_{mk,y}$, $r_{mk,\vartheta}$ are internal forces induced at cross section m by a unit load acting at the center of mass of floor k along the axes x and y and around the vertical axis. Similar relations apply to displacements and accelerations, and to gravity loads. For any specified wind speed and direction, the wind load time series are computed from time histories of pressure coefficients provided by the wind engineer (see Chapter 10).

References

1 ANSI/AISC (2010). *Steel Construction Manual*, 14th ed. Chicago, IL: American Institute of Steel Construction.
2 White, D.W. and Hajjar, J.F. (1991). Application of second-order elastic analysis in LRFD: research to practice. *Engineering Journal, American Institute of Steel Construction* 28: 133–148.
3 Park, S. and Yeo, D. (2018). Second-order effects on wind-induced structural behavior of high-rise buildings. *Journal of Structural Engineering* 144: doi: 10.1061/(ASCE)ST.1943-541X.0001943. https://www.nist.gov/wind.

10

Aerodynamic Loads

Main Structure, Secondary Members, and Cladding

10.1 Introduction

Aerodynamic loads are based on time series of aerodynamic pressure coefficients measured simultaneously at multiple taps on the surfaces of the wind tunnel building model. Two main cases are considered in this chapter. In the first case, the objective is to determine, for specific structures, aerodynamic loads at the center of mass of each floor, on main members, on secondary members (e.g., purlins and girts), and on cladding. In the second case, the objective is to develop standard provisions on pressure coefficients. In both cases details of the procedures for determining the loading differ to some extent depending upon whether the pressure taps are placed in orthogonal patterns, as is the case for the National Institute of Standards and Technology/University of Western Ontario (NIST/UWO) database https://www.nist.gov/wind [1], or in non-orthogonal patterns, as in the Tokyo Polytechnic University (TPU) database [2].

Section 10.2 discusses orthogonal and non-orthogonal tap placement patterns and the determination of individual tap tributary areas. Section 10.3 is concerned with the determination of aerodynamic loads at floor levels, and on main members, secondary members, and cladding. Section 10.4 describes a method used to develop standard provisions on pressure coefficients as functions of areas contained within specified zones. Section 10.5 concerns wind-driven rain intrusion.

10.2 Pressure Tap Placement Patterns and Tributary Areas

Pressure taps may be placed in orthogonal or non-orthogonal patterns. Examples of orthogonal pressure tap patterns are shown Figure 10.1, which shows rectangular tributary areas of pressure taps represented by cross symbols.

A non-orthogonal pattern is shown in Figure 10.2a, in which circles indicate pressure tap locations. Individual tap tributary areas are conveniently calculated using Voronoi diagrams [5]. The diagrams can be derived from Delaunay triangulation [6], which connects a given set of point taps to form triangles that: (i) do not overlap, (ii) cover the entire interior space, and (iii) do not have any tap within the triangle's circumcircle. The corresponding Voronoi diagram is created by drawing perpendicular bisectors to the sides of the triangles. Regions formed from these bisectors contain one tap each and bound an area containing points that are closer to that tap than to any other tap. The

Wind Effects on Structures: Modern Structural Design for Wind, Fourth Edition. Emil Simiu and DongHun Yeo.
© 2019 John Wiley & Sons Ltd. Published 2019 by John Wiley & Sons Ltd.

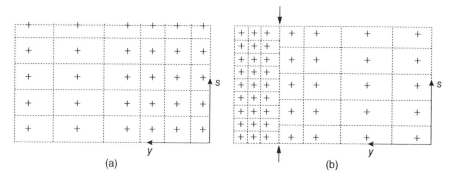

Figure 10.1 Rectangular tap tributary areas: (a) simple tap array; (b) tap array with varying tap density [3].

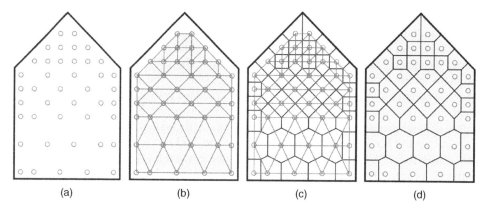

Figure 10.2 Example of non-orthogonal pattern of pressure tap placement and of tributary area assignments [4]. Source: With permission from ASCE.

Voronoi MATLAB function [7] can generate both Delaunay triangulation and Voronoi diagrams. Figure 10.2b connects the taps using Delaunay triangulation. Figure 10.2c shows how a Voronoi diagram can be derived from the Delaunay triangulation. Figure 10.2d shows the Voronoi diagram; the bounded area created around a tap is the tributary area of that tap.

10.3 Aerodynamic Loading for Database-Assisted Design

Pressure data on the structure's envelope are provided as time series of non-dimensional pressure coefficients C_p, typically based on the hourly mean wind speed V_H at the building roof height H

$$C_p = \frac{p}{\frac{1}{2}\rho V_H^2} \tag{10.1}$$

where p is the net pressure relative to the atmospheric pressure and ρ is the air density (1.225 kg m^{-3} for $15°C$ air at sea level).

From the similarity requirement for the reduced frequency nD/V, where n is the sampling frequency and D is a characteristic dimension of the structure, it follows that the prototype time interval $\Delta t_p = 1/n_p$ is

$$\Delta t_p = \frac{D_p}{D_m} \frac{V_m}{V_p} \Delta t_m \tag{10.2}$$

where the subscripts p and m stand for prototype and model, respectively, D_m/D_p is the geometric scale, V_m/V_p is the velocity scale, and Δt_m is the reciprocal of the sampling frequency n_m at model scale.

Calculations of aerodynamic pressure coefficients based on pressure measurements at taps placed on wind tunnel models require:

1) The creation of virtual pressure taps at each edge of the model surface. The time series of the pressure coefficients at those taps are obtained by extrapolation from the time histories at the outermost and next to outermost pressure taps (Figure 10.3a). This operation is necessary because actual pressure taps cannot be placed at the structure's edges.

2) The generation of a mesh for interpolations between time series of pressure coefficients measured at actual taps or estimated at virtual taps (Figure 10.3b). Each mesh element has dimensions $\Delta B \times \Delta H$, where $\Delta B = B/(2n_B)$, $\Delta H = H/(2N)$, B is the building width, H is the building height (including, for buildings with parapets, the parapet height, in which case the height of the uppermost mesh element is equal to the height of the parapet), n_B is the number of pressure taps in each pressure tap row,

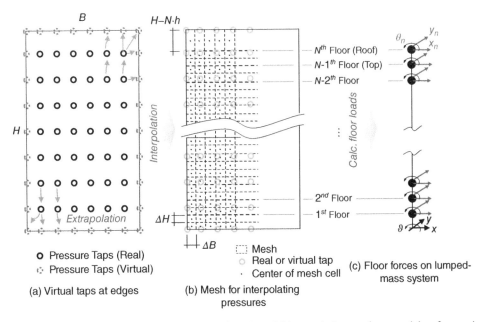

(a) Virtual taps at edges

(b) Mesh for interpolating pressures

(c) Floor forces on lumped-mass system

Figure 10.3 (a) Actual and virtual pressure tap locations, (b) Interpolation mesh on model surface and points of application of wind forces obtained by interpolation at centers of mesh cells (h = floor height), (c) Wind forces applied at floor-level lumped masses.

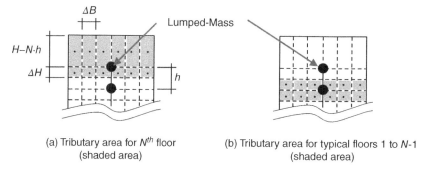

(a) Tributary area for N^{th} floor
(shaded area)

(b) Tributary area for typical floors 1 to N-1
(shaded area)

Figure 10.4 Tributary areas for calculation of floor wind loads.

and N is the number of floors. Software described and applied in Chapter 18 offers the option of carrying out the interpolations by any of three methods supported by MATLAB. An alternative method is described in [8].

3) For multi-story buildings, time series of floor wind loads are applied at the floor centers of mass. They are determined as functions of time series of pressure coefficients obtained by interpolation, of the respective tributary areas (Figure 10.4), and of the mean wind speeds at the elevation of the top of the building. The wind loads consist, at each floor, of forces acting along the two principal directions of the structure, and a torsional moment (Figure 10.3c).

It is typically assumed that pressure coefficients do not depend significantly upon Reynolds number and are therefore identical for the model and the prototype. However, for the design of cladding, a more conservative approach may be adopted to account for the fact that wind tunnel simulations may underestimate peak suctions, as shown in Figure 5.24.

10.4 Peaks of Spatially Averaged Pressure Coefficients for Use in Code Provisions

10.4.1 Pressures Within an Area A Contained in a Specified Pressure Zone

Standards specify pressures applicable to areas of various sizes A contained in specified zones (e.g., middle, edge, or corner zones of roofs or walls) within which it is assumed for practical design purposes that the pressures are uniform. Except for an area A covering the entire area of the zone being considered, the number of areas of specified size A within a zone exceeds unity. To develop standard provisions on pressure coefficients, the following steps are required [9]:

1) Identify all areas of size A within the zone (see Section 10.4.2).
2) Determine the tributary area B_l of each tap l contained in the zone.
3) For each of the areas A, determine its intersections, a_l, with the tap tributary areas B_l. For example, let the four rectangles shown in Figure 10.5 represent tap tributary areas B_1, B_2, B_3, B_4, and let the area A of interest be the shaded area of Figure 10.5. The intersections of area A with the areas B_l (l = 1, 2, 3, 4) are denoted by a_l (l = 1, 2, 3, 4).

4) For each wind direction θ_j, and for each of the areas of size A within the zone being considered, obtain the time history

$$p(A, t, \theta_j) = \sum_l p_l(t, \theta_j)\frac{a_l}{A} \qquad (10.3a)$$

$$\sum_l a_l = A \qquad (10.3b)$$

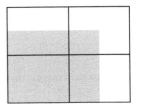

Figure 10.5 Intersection of four pressure tap tributary areas (cells) with shaded area A [9].

where $p_l(t, \theta_j)$ is the time history of the pressure induced by wind with direction θ_j at the tap contained in area B_l.

5) Determine, for each of the areas A and for each of the directions θ_j, the peak of the time history $p(A, t, \theta_j)$ using, for example, the procedure described in Appendix C. (Alternatively, the procedure described in https://www.nist.gov/wind may be used. This requires the partitioning of the record into equal segments and the creation of a sample of peak values consisting of the peak of each of the segments.) The largest of all those peaks is the pressure being sought for codification purposes.

6) Divide that pressure by the dynamic pressure $(1/2)\rho\,\max_j[U^2(z_{ref}, \theta_j)]$ to obtain the corresponding pressure coefficient $C_p(A, t)$.

For compliance with ASCE 7 Standard requirements, the pressure coefficients $C_p(A, t)$ are re-scaled to be consistent with 3-second peak gust wind speeds, and are reduced via multiplication by a directionality reduction factor (see Section 13.5).

10.4.2 Identifying Areas A Within a Specified Pressure Zone

Pressure Taps with Rectangular Tributary Areas. The summation process in Eq. (10.3a) is simplest when the cells representing the tributary areas of the taps are rectangular (Figure 10.1a). Special consideration must be given to areas A in edge and corner zones since such areas generally do not coincide with cell boundaries (see, e.g., Figure 10.5), and to cases in which grids of different densities merge, as indicated by arrows in Figure 10.1b.

To see how various areas of size A are determined within a specified zone with area larger than A, consider the six-cell zone with orthogonal tap placement shown in Figure 10.6 [9]. We seek the number of distinct rectangles with areas A within that zone. The areas A may consist of one cell, or of rectangular conterminous aggregates of two, three, four, or six cells. There are six possible rectangular areas consisting of one cell each. The cell on the upper left corner is denoted Aa. To Aa is added a cell in the downward y-direction; the two-cell rectangle so obtained is denoted Aa2. With this step, the lower boundary of the zone is reached; therefore, no additional cell can be added in direction y. To the cell selected in step Aa, one cell is added in direction s, rightward. The two-cell rectangle so obtained is denoted Aa3,1. In a next step, denoted Aa3,2, an additional cell is added, again in direction s, rightward. Thus, two additional rectangles have been created in step Aa3. With step Aa3,2 the rightmost boundary of the zone has been reached, so further expansion in the direction s is not possible. Next, step Ab consists of adding to the cell selected in step Aa via expansion in both

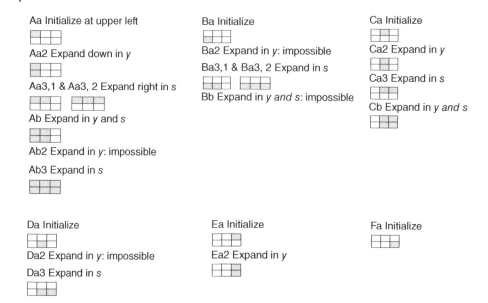

Figure 10.6 Six-cell zone with orthogonal tap placement [9].

directions, y downward and s rightward. Thus, a rectangle consisting of four cells is created. Expansion in the direction y downward is attempted in step Ab2, but is not possible. Step Ab3 consists of expanding in the s direction rightward, which results in a six-cell rectangle. All possibilities of expansion from the single cell selected in step Aa being exhausted, one proceeds to the next initial cell direction rightward (step Ba). The procedure is repeated until all possible initial cells have been used. Figure 10.6 shows six rectangles formed by one cell, seven rectangles formed by two cells, two rectangles formed by three cells, two rectangles formed by four cells, and one rectangle formed by six cells, for a total of eighteen rectangles. If the cells are rectangles of unit area, for the zone of area 6 the following numbers of pressure time series result: 6 with area $A = 1$; 7 with area $A = 2$; 2 with area $A = 3$; 2 with area $A = 4$; and 1 with area $A = 6$. In this example, areas A have aspect ratios ranging from 1 to 3. We need to calculate the peak average pressure coefficient for each of the 6 one-cell areas; for each of the seven two-cell areas; and so forth.

To limit the number of combinations for large zones, the aspect ratio of the rectangles formed by the aggregation of cells is limited to four at most. This aspect ratio covers many practical units of components and cladding, and allows consideration of long, narrow zones along the edges of roofs and walls. The number of area combinations increases very quickly with the size of the grid, for example, a 19×8 grid produces 4290 areas of aspect ratio ≤ 4, whereas a subset 16×7 of the same grid produces 2351 such areas.

If the zone being studied overlaps areas of different tap density, the coarser density is used overall, and full and partial cell areas in the high-density region are combined as needed. Figure 10.7a shows a portion of a zone with two grid densities. To conform with the coarser grid, the two rows of three cells at the bottom of the figure are transformed into two rectangles each (Figure 10.7b).

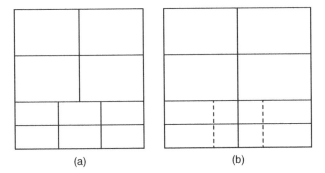

(a)   (b)

Figure 10.7 Combination of areas with different tap densities.

Example 10.1 Results are shown for Building 7 (open country exposure) of the NIST/UWO database (data set jp1, https://www.nist.gov/wind [1]. The building (Figure 10.8a) was modeled at a 1 : 100 scale, and data were collected for 100 seconds at 500 Hz; its full-scale width, length and height are 12.2 m (40 ft), 19.1 m (62.5 ft) and 12.2 m (40 ft), respectively, and its roof slope is 4.8°. The peak averaged pressure coefficients were re-scaled to be consistent with ASCE 7-10 Standard [10] 3-second peak gust wind speeds. Figure 10.8b shows that ASCE 7-10 specifications, in which peak wind pressure coefficients are denoted by (GC_p), underestimate negative pressures over almost all of the areas within the corner zone by factors of up to 2.3. These results, and a thorough study in [11], confirm the finding that negative pressures specified in the ASCE 7-10 Standard tend to be strongly unconservative.

Pressure Taps with Polygonal Tributary Areas. To produce intersections of tap tributary areas with rectangular areas A contained within a specified zone, each building façade and roof surface is swept in small discrete steps by overlaid rectangles with area A. The first set of rectangles with area A have sides equal to the horizontal and vertical distances between adjacent taps, that is, the smallest possible useful rectangles. In the subsequent sets, the sizes of the rectangles are progressively increased horizontally, vertically, and both horizontally and vertically, until the largest rectangle is determined by the dimensions of the facade. Step-wise offsetting of each of those sets of rectangles by amounts equal to the smallest distances between taps ensures that no rectangular area A for which the averaged pressure coefficient needs to be calculated is missed.

Example 10.2 Consider the wall represented in Figure 10.2. Let the smallest horizontal and vertical distances between taps be 2 m. Two sets of rectangular areas A are shown in the figure: a set consisting of a 2×2 m grid (Figure 10.9a), and a set consisting of a 2×4 m grid (Figure 10.9c). Figure 10.9b shows 2×2 m rectangles with 1 m offset in the x and y directions. Figure 10.9d shows rectangles with dimensions 2×4 m and 1 m offset in the x direction. In Figure 10.9, all the shaded areas contain pressures. As the rectangle's borders cross outside a building surface, partial or incomplete elements created by such borders are neglected. They are represented in Figure 10.9 as blank areas within the façades. Table 10.1 lists grids with minimum sizes $A = 2 \times 2$ m and maximum sizes $A = 3 \times 3$ m, with no offset, with 1 m offsets, and with 2 m offsets.

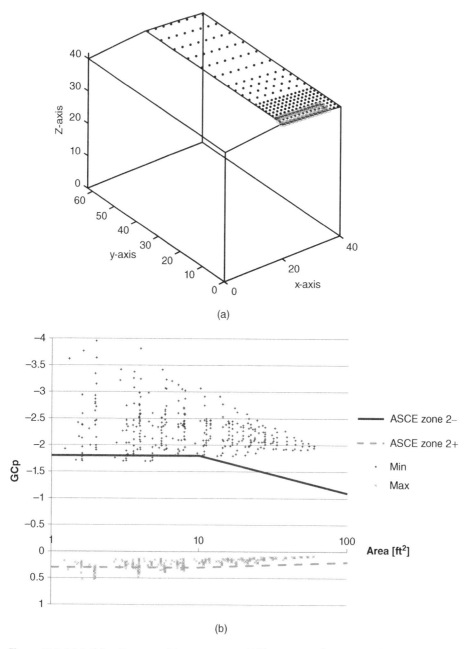

(a)

(b)

Figure 10.8 (a) Building 7 corner, with pressure taps (1 ft² = 0.0929 m²); (b) Peak of averaged pressure time histories. Source: After [9].

The method just described was programmed using MATLAB [7] to process buildings available in TPU's low-rise building pressures database, specifically case numbers 13–108 [2]. The wind tunnel tests of low-rise buildings without eave were performed at a length scale of 1/100, velocity scale of 1/3 (i.e., a 3/100 time scale), for suburban terrain. At a reference height of 0.1 m, the turbulence intensity was 0.25 and the test wind

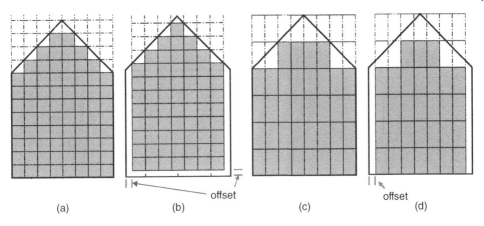

(a)   (b)   (c)   (d)

Figure 10.9 Example of superposed rectangular surfaces with areas $A = 2\,\text{m} \times 2$ and $A = 2\,\text{m} \times 4\,\text{m}$, with no offsets and with 1 m offsets offsets [4]. Source: With permission from ASCE.

Table 10.1 Grid areas and offset combinations.

Grid size		Offset	
x (m)	*y* (m)	*x* direction	*y* direction
2	2	0	0
2	2	0	1
2	2	1	0
2	2	1	1
2	3	0	0
2	3	0	1
2	3	0	2
2	3	1	0
2	3	1	1
2	3	1	2
3	2	0	0
3	2	0	1
3	2	1	0
3	2	1	1
3	2	2	0
3	2	2	1
3	3	0	0
3	3	0	1
3	3	0	2
3	3	1	0
3	3	1	1
3	3	1	2
3	3	2	0
3	3	2	1
3	3	2	2

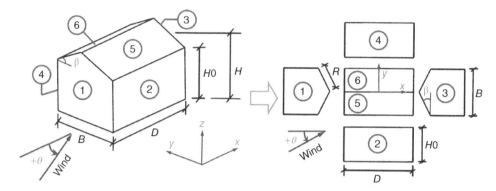

Figure 10.10 TPU low-rise building showing geometric parameters. Source: From [2]. Courtesy of Professor Y. Tamura, Tokyo Polytechnic University.

velocity was $7.4\,\mathrm{m\,s^{-1}}$, which corresponds to a $22\,\mathrm{m\,s^{-1}}$ mean hourly wind speed at a 10 m height in full scale. Wind pressure time-history data were recorded at 500 Hz for 18 seconds, or $18 \times 100/3$ seconds = 10-minute full scale. An example of such a building is shown in Figure 10.10.

TPU's aerodynamic database incorporates a moving average calculation for the pressure time series data. Denoting the data sampling interval by Δt and the net pressure above ambient at tap i at time t by $p_i(t, \theta)$, TPU defines the pressure denoted by $p(i, t, \theta)$ at tap i at time t as

$$p(i, t, \theta) = \mathrm{avg}[p_i(t - \Delta t, \theta), p_i(t, \theta), p_i(t + \Delta t, \theta)] \tag{10.4}$$

An example of a building from the TPU database is shown in Figure 10.10.

Consider building TP-1 (case 61 of the database, $B = 16\,\mathrm{m}$, $D = 24\,\mathrm{m}$, $H0 = 12\,\mathrm{m}$, roof slope 4.8°, see Figure 10.10). Figure 10.11 shows the Voronoi diagram applied to that building; the pressure taps are indicated by circles, bounded by the polygons that define the tributary areas. With the tributary areas in place, the overlaid rectangle/offset combinations can then be specified. The smallest overlaid rectangle was chosen as $2 \times 2\,\mathrm{m}$, based on the minimum 2 m tap spacing; the largest was chosen to be $7 \times 7\,\mathrm{m}$. The rectangles were incremented from $2 \times 2\,\mathrm{m}$ up to $7 \times 7\,\mathrm{m}$ by increments of 0.5 m, and were offset in increments of 0.5 m in the x and y directions.

Based on these rectangle/offset combinations, the total number of combinations is 9801, each rectangle having an aspect ratio of 3.5 or less. ASCE 7-16 Commentary limits the aspect ratio of areas relevant to the design of components and cladding to 3. The process by which peaks of average pressures are calculated as functions of areas within code-specified zones involves the use of Boolean algebra and the MATLAB function Polybool, and is repeated for all available tested wind directions θ: 0°, 15°, 30°, 45°, 60°, 75°, and 90°. For additional details on the method and its application to the assessment of ASCE 7 Standard provisions, see [4], in which it is noted that the TPU tap and wind direction resolution are lower than their NIST/UWO counterparts, particularly for buildings for which ASCE 7 Standard edge zones and corner zones have small dimensions. Nevertheless, no significant differences were found between wind loads on main structural members based on [1] on the one hand and [2] on the other [12].

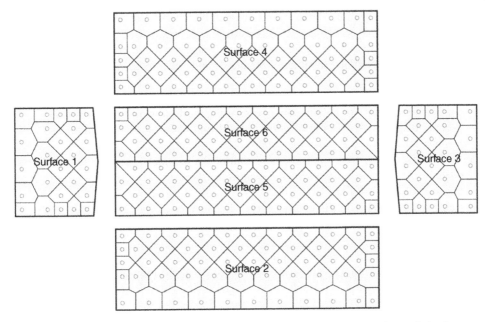

Figure 10.11 Tributary areas achieved using the Voronoi diagram [4]. Source: With permission from ASCE.

10.5 Aerodynamic Pressures and Wind-Driven Rain

Recent advances in the area of wind-driven rain water intrusion include the development of full-scale testing under conditions simulating (i) atmospheric boundary layer hurricane force winds and (ii) up to 760 mm h^{-1}. rain simulated by continuous spraying of water through a plumbing system with spray nozzles [13]. The frontal area of the wind and wind-driven-rain field simulated in [13] exceeded 30 m^2. Measurements were performed of the amount of water intruded through nailed and through self-adherent heavy and light secondary water barriers, and of internal and external aerodynamic pressures induced by the wind flow. Tests of specimens with different slopes showed that the severity of the intrusion increases as the roof slope decreases.

Additional testing described in [14] was conducted using records of tropical cyclone wind-driven rain data as a basis for the development of the target parameters considered in the simulation, including raindrop size distribution. The results of the tests were used to propose enhancements to simplified test protocols specified in current standards. For additional material on rain water intrusion due to directly impinging rain drops and surface runoff, see [15] and references quoted therein.

References

1 Ho, T., Surry, D., and Morrish, D. *NIST/TTU Cooperative Agreement/ Windstorm Mitigation Initiative: Wind Tunnel Experiments on Generic Low Buildings.* BLWT-SS20–2003, Boundary Layer Wind Tunnel Laboratory, University of Western Ontario, London, Canada, 2003.

2 Tamura, Y. "Aerodynamic Database of Low-Rise Buildings." Global Center of Excellence Program, Tokyo Polytechnic University, Tokyo, Japan, 2012.

3 Main, J. A. and Fritz, W. P. *Database-Assisted Design for Wind*. NIST Building Science Series 180, National Institute of Standards and Technology, Gaithersburg, MD, 2006.

4 Gierson, M.L., Phillips, B.M., Duthinh, D., and Ayyub, B.M. (2017). Wind-pressure coefficients on low-rise building enclosures using modern wind-tunnel data and Voronoi diagrams. *ASCE-ASME Journal of Risk and Uncertainty in Engineering Systems, Part A: Civil Engineering* 3: 04017010.

5 Voronoi, G. (1908). New applications of continuous parameters to the theory of quadratic forms. *Journal of Pure and Applied Mathematics* 133 (133): 97–178. doi:10.1515/crll.1908.133.97.

6 Delaunay, B. (1934). On the empty sphere." In Memory of Georges Voronoi. *Bulletin of the USSR Academy of Sciences, Section: Mathematics and Natural Sciences* 6: 793–800.

7 MATLAB. MATLAB documentation 2014b. The Mathworks Inc., 2014.

8 Uematsu, Y., Kuribara, O., Yamada, M. et al. (2001). Wind-induced dynamic behavior and its load estimation of a single-layer latticed dome with a long span. *Journal of Wind Engineering and Industrial Aerodynamics* 89: 1671–1687.

9 Duthinh, D., Main, J. A., and Phillips, B. M. *Methodology to Analyze Wind Pressure Data on Components and Cladding of Low-Rise Buildings*. NIST TN 1903, National Institute of Standards and Technology, Gaithersburg, MD, 2016. https://www.nist .gov/wind.

10 ASCE. "Minimum design loads for buildings and other structures (ASCE/SEI 7-10)," in *ASCE Standard ASCE/SEI 7-10*, Reston, VA: American Society of Civil Engineers, 2010.

11 Gavanski, E., Gurley, K.C., and Kopp, G.A. (2016). Uncertainties in the estimation of local peak pressures on low-rise buildings by using the Gumbel distribution fitting approach. *Journal of Structural Engineering* 142 (11): 04016106.

12 Hagos, A., Habte, F., Gan Chowdhury, A., and Yeo, D. (2014). Comparisons of two wind tunnel pressure databases and partial validation against full-scale measurements. *Journal of Structural Engineering* 149: 04014065.

13 Bitsuamlak, G.T., Gan Chowdhury, A., and Sambare, D. (2009). Application of a full-scale testing facility for assessing wind-driven-rain intrusion. *Building and Environment* 44: 2430–2441.

14 Baheru, T., Gan Chowdhury, A., Bitsuamlak, G. et al. (2014). Simulation of wind-driven rain associated with tropical storms and hurricanes using the 12-fan Wall of Wind. *Building and Environment* 76: 18–29.

15 Baheru, T., Gan Chowdhury, A., Pinelli, J.-P., and Bitsuamlak, G. (2014). Distribution of wind-driven rain deposition on low-rise buildings: direct impinging raindrops versus surface runoff. *Journal of Wind Engineering and Industrial Aerodynamics* 133: 27–38.

11

Dynamic and Effective Wind-Induced Loads

11.1 Introduction

Unlike seismic loads, which consist of forces of inertia, wind loads consist of sums of applied aerodynamic forces *and* forces of inertia. Rigid structures are by definition structures for which wind-induced forces of inertia are negligible. Flexible structures are defined as structures for which the wind-induced forces of inertia are significant.

The forces of inertia are due to resonant amplification effects. A well-known example of resonant amplification is the effect on a bridge of a military formation marching in lock-step at a frequency equal or close to the bridge's fundamental frequency of vibration. The effects of successive steps are additive: a first step causes a deflection whose maximum is reached when the second step strikes. The second step causes an additional deflection and subsequent steps keep adding to the response. The randomly fluctuating wind loading can be represented as a sum of harmonic components (see Appendices B and D). Wind-induced resonant amplification effects are caused by harmonic loading components with frequencies equal or close to the natural frequencies of vibration of the structure.

The forces of inertia are yielded by dynamic analyses based on second-order ordinary differential equations of motion, in accordance with Newton's second law. The analyses can be performed by solving the equations of motion in the frequency domain or in the time domain. The use of the frequency domain approach was predominant in the 1960s, primarily because it does not require the direct solution of the differential equations; instead, the latter are converted to algebraic equations via Fourier transformation (see Appendix D).

The development of pressure scanners allows the simultaneous wind tunnel measurement of pressure time histories at large numbers of taps mounted on the external surfaces of rigid building models. Inherent in the measurements is phase information on pressure fluctuations and, therefore, information on the extent to which the pressures acting at different points on the structure are in or out of phase; that is, the extent to which those pressures are mutually coherent (see Figure 4.27 for an illustrative animation). It is currently a routine task to obtain, via simple weighted summations of pressure time histories, the time histories of the wind-induced forces acting on the structure (see Chapter 10). Once those time histories are determined, it is again a routine matter to solve numerically the equations of wind-induced motion of the structure in the time domain.

Wind Effects on Structures: Modern Structural Design for Wind, Fourth Edition. Emil Simiu and DongHun Yeo.
© 2019 John Wiley & Sons Ltd. Published 2019 by John Wiley & Sons Ltd.

The purpose of this chapter is to present the basic theory that governs the multi-degree-of-freedom behavior of structural systems assumed to be linearly elastic. Section 11.2 discusses the simple case of the linear single-degree-of-freedom system. The multi-degree-of-freedom case is considered in Section 11.3. The solution of the structure's equations of motion yields the forces of inertia induced by the wind loading, as well as the structure's displacements and accelerations. Section 11.4 concerns, for any specified direction of the wind speed, the determination of the corresponding effective wind loads, defined as the sum of the aerodynamic and inertial loads.

In the High Frequency Force Balance (HFFB) approach dynamic response calculations are performed partly by the structural engineer and partly by the wind engineer. This practice is left over from the late 1970s, when dynamic calculations were performed in the frequency domain to avoid computations involving the solution of differential equations of motion. The drawbacks of this practice include: (i) difficulties in the estimation of combined wind effects, (ii) the lack of information on the distribution of the wind loads with height, which prevents the realistic determination of wind effects in structural members, (iii) the impossibility of determining the dynamic response in higher modes of vibration, and (iv) the need to resort to correction factors to compensate, with varying degrees of approximation, for the errors due to the assumptions that the shape of the fundamental modes of vibration in sway are linear and that the shape of the fundamental torsional mode is independent of height. These drawbacks are especially significant for buildings affected aerodynamically by neighboring buildings. The advances in computational capabilities achieved in the twenty-first century render the HFFB approach obsolete – in the sense in which, for example, the moment distribution method is obsolete. This is the case for detailed, final design purposes, although the use of the HFFB approach for rapid, preliminary design purposes remains warranted.

11.2 The Single-Degree-of-Freedom Linear System

The system of Figure 11.1 consists of a particle of mass M concentrated at point B of a member AB with linear elastic behavior and negligible mass. The particle is subjected to a force $F(t)$.

The displacement $x(t)$ of the mass m is opposed by (i) a restoring force $-kx$ supplied by the elastic spring inherent in the member AB and (ii) a damping force $-c\, dx/dt \equiv c\dot{x}^1$ where k is the system's *stiffness* (i.e., the magnitude of the restoring force corresponding to a unit displacement x of the mass M) and c is the damping coefficient. Both k and c are assumed to be constant. The inverse of the system's stiffness k is referred to as the *flexibility* of the system (i.e., the system's displacement corresponding to a unit restoring force).

Newton's second law states that the product of the particle's mass by its acceleration, $M\ddot{x}$, is equal to the total force applied to the particle. The equation of motion of the system is then

Figure 11.1 Single-degree-of-freedom system.

$$M\ddot{x} = -c\dot{x} - kx + F(t) \tag{11.1}$$

---

1 Here and elsewhere in the book the dot denotes differentiation with respect to time, that is, $\dot{x} \equiv dx/dt$.

With the notations $n_1 = \sqrt{k/M}/(2\pi)$ and $\zeta_1 = c/(2\sqrt{kM})$ where n_1 denotes the frequency of vibration of the oscillator[2] and ζ_1 is the damping ratio (i.e., the ratio of the damping c to the critical damping $c_{cr} = 2\sqrt{kM}$ beyond which the system's motion would no longer be oscillatory), Eq. (11.1) becomes

$$\ddot{x} + 2\zeta_1(2\pi\, n_1)\dot{x} + (2\pi\, n_1)^2 x = \frac{F(t)}{M} \tag{11.2}$$

For structures, ζ_1 is typically small (in the order of 1%). We note for future reference that the product of the system's stiffness and flexibility is $k \times (1/k) = 1$, and that the system's kinetic energy and strain energy are $T = \frac{1}{2} M \dot{x}^2$ and $V = \int kx\, dx = \frac{1}{2} kx^2$.

An alternative derivation of the equation of free vibrations of the undamped and unforced system (i.e., of the system with $c = 0$ and $F(t) \equiv 0$) can be obtained from the system's Lagrangian:

$$L = T - V \tag{11.3}$$

where T is the total kinetic energy and V is the potential energy (e.g., strain energy) of the system.

For the system under consideration,

$$L = \frac{1}{2}M\dot{x}^2 - \frac{1}{2}kx^2 \tag{11.4}$$

From Lagrange's equations

$$\frac{d}{dt}\left(\frac{\partial L}{\partial \dot{q}_i}\right) - \frac{\partial L}{\partial q_i} = 0 \tag{11.5}$$

where the generalized coordinate $q_i \equiv x\ (i = 1)$ it follows that

$$M\ddot{x} + kx = 0 \tag{11.6}$$

11.3 Time-Domain Solutions for 3-D Response of Multi-Degree-of-Freedom Systems

In general, the dynamic response to wind of flexible buildings with linearly elastic behavior entails translational motions (sway) along their principal axes and torsional motions about the building's elastic center. The torsional motions are due to the eccentricity of the aerodynamic and inertial forces with respect to the elastic center. An example of torsional deformations induced by wind is shown in Figure 11.2.

The system's equations of free vibration are obtained by following steps analogous to those that led, for the single degree-of-freedom system, to Eq. (11.6).

Figure 11.2 Torsional deformation of Meyer–Kiser building in 1926 Miami hurricane. Source: From [1].

---

2 The quantity $2\pi n$ is called circular frequency and is commonly denoted by ω.

11.3.1 Natural Frequencies and Modes of Vibration

The total kinetic energy of a structure with n_f masses (e.g., n_f floors) is

$$T = \frac{1}{2}\sum_{n=1}^{n_f}(m_n\dot{x}_n^2 + m_n\dot{y}_n^2 + I_n\dot{\vartheta}_n^2) \tag{11.7}$$

where x_n, y_n are the displacements of the mass m_n in the x and y directions, respectively, ϑ_n is the torsional rotation of the nth mass about its elastic center, and n_f is the total number of masses.

The total strain energy of the system is

$$V = \frac{1}{2}\{q\}^T[k]\{q\}, \tag{11.8}$$

$$\{q\}^T = \{x_1, x_2, \dots, x_{n_f}, y_1 y_2, \dots, y_{n_f}, \vartheta_1, \vartheta_2, \dots, \vartheta_{n_f}\} \tag{11.9}$$

where T denotes tranpose, and $[k]$ is the stiffness matrix (see Section 9.1).

For the freely vibrating structure, the displacements of the mass center and the torsional rotation about the mass center at the elevation z_i of the ith floor form a vector $\{w(t)\}$ of dimension $3n_f$. Its terms are denoted as follows:

$$w_1(t) = x_1(t), w_2(t) = x_2(t), \dots, w_{n_f}(t) = x_{n_f}(t);$$

$$w_{n_f+1}(t) = y_1(t), w_{n_f+2}(t) = y_2(t), \dots, w_{2n_f}(t) = y_{n_f}(t);$$

$$w_{2n_f+1}(t) = \vartheta_1(t), w_{2n_f+2}(t) = \vartheta_2(t), \dots, w_{3n_f}(t) = \vartheta_{n_f}(t). \tag{11.10}$$

The equations of motion of the undamped, freely vibrating system

$$[M]\{\ddot{w}(t)\} + [k]\{w(t)\} = \{0\} \tag{11.11}$$

are obtained from the Lagrange equations (Eq. [11.5]). In Eq. (11.11) $[M]$ is a diagonal matrix of the floor masses (for sway motions) or mass moments of inertia (for torsional motions). Equation (11.11) are coupled owing to the cross-terms of the matrix $[k]$. Assume solutions of the form:

$$\{w(t)\} = \{A\}\cos(\omega t + \varphi) \tag{11.12}$$

where $\{A\}$ is a vector to be defined subsequently. Substitution of these solutions in Eq. (11.11) yields

$$(-[M]\omega^2 + [k])\{A\} = \{0\} \tag{11.13}$$

Equation (11.13) is a system of linear homogeneous equations in the unknowns A_1, $A_2, \dots, A_{n_f}$, that is,

$$(k_{11} - M_1\omega^2)A_1 + \quad k_{12}A_2 + \dots \quad \dots + \quad k_{1n_f}A_{3n_f} = 0$$
$$k_{21}A_1 + (k_{22} - M_2\omega^2)A_2 + \dots \quad \dots + \quad k_{2\,n_f}A_{3n_f} = 0$$
$$k_{3n_f1}A_1 + \quad k_{3n_f2}A_2 + \dots \quad \dots + (k_{3n_f3n_f} - M_{3n_f}\omega^2)A_{3n_f} = 0$$

$$\tag{11.14}$$

For Eq. (11.14) to have non-zero solutions, the determinant of the coefficients of the unknowns $\{A\}$ must vanish. This condition yields a $3n_f$-degree equation in ω^2 called the

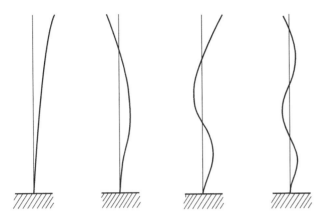

Figure 11.3 First four normal modes of a cantilever beam.

characteristic equation. Its $3n_f$ roots are called *eigenvalues*. The rank-ordered frequencies $\omega_1 < \omega_2 < \cdots < \omega_{3n_f}$ are called the system's *natural frequencies of vibration*. To each of the $3n_f$ eigenvalues there corresponds an eigenvector with $3n_f$ components obtained from Eq. (11.14), which defines a *natural* (or *normal*) *mode of vibration*. The eigenvectors corresponding to the $3n_f$ eigenvalues $\{\omega\}$ form a $3n_f \times 3n_f$ matrix $[\phi]$. For $i \neq j$ the vectors $\{\phi_i\}$ and $\{\phi_j\}$ can be shown to be mutually orthogonal with respect to mass or mass moments of inertia weighting, that is

$$\sum_{k=1}^{3n_f} \phi_{ik}\phi_{jk}M_k = 0 \qquad (i \neq j) \tag{11.15}$$

The free vibrations, with their normal modal shapes and associated frequencies, are properties of the structural system, independent of the loads. The first four normal modes along one of the principal axes of a continuous cantilever beam are shown in Figure 11.3.

11.3.2 Solutions of Equations of Motion of Forced System

The equations of motion of the forced system are

$$[M]\{\ddot{w}(t)\} + [k]\{w(t)\} = \{F(t)\} \tag{11.16}$$

where $\{F(t)\}$ is the vector of the wind forces (torsional moments) with components $F_{x1}(t), F_{x2}(t), \ldots, F_{xn_f}(t), F_{y1}(t), F_{y2}(t), \ldots, F_{yn_f}(t), M_{\theta 1}(t), M_{\theta 2}(t), \ldots, M_{\theta n_f}(t)$ acting at the centers of mass of floors $1, 2, \ldots, n_f$. The variables $w(t)$ can be written as

$$\{w(t)\} = [\phi]\{\xi(t)\}, \tag{11.17}$$

where $[\phi]$ is the matrix consisting of the $3n_f$ eigenvectors $\{\phi_j\}$ $(j = 1, 2, \ldots, 3n_f)$, and the coefficients $\{\xi(t)\}$, called *generalized coordinates*, indicate what fraction of each mode enters the given deflection pattern. Substitution of Eq. (11.17) into Eq. (11.16) yields

$$[M][\phi]\{\ddot{\xi}(t)\} + [k][\phi]\{\xi(t)\} = \{F(t)\} \tag{11.18}$$

Premultiplication of Eq. (11.18) by $[\phi]^T$, where the superscript T denotes transpose, yields:

$$[\phi]^T[M][\phi]\{\ddot{\xi}(t)\} + [\phi]^T[k][\phi]\{\xi(t)\} = [\phi]^T\{F(t)\} \tag{11.19}$$

Owing to the orthogonality of the eigenvectors, Eq. (11.19), to which modal viscous damping terms proportional to the modal damping ratios ζ_m are added, can be written as

$$M_m\ddot{\xi}_m(t) + 2M_m\omega_m\zeta_m\dot{\xi}(t) + M_m\omega_m^2\xi(t) = \{[\phi]^T\{F(t)\}\}_m \qquad (m = 1, 2, \dots, 3n_f) \tag{11.20}$$

In Eq. (11.20), the quantities $M_m = [[\phi]^T[M][\phi]]_m$ and the quantities in right-hand side of Eq. (11.20) are called the *m*-th mode *generalized masses* and *generalized forces*, respectively. It follows from the unforced equation of motion of the system that $M_m\omega_m^2 = [[\phi]^T[k][\phi]]_m$. Once Eq. (11.20) are solved numerically, the physical coordinates $\{w(t)\}$ (i.e., the coordinates $x_1(t)$, $x_2(t)$, ..., $x_{n_f}(t)$, $y_1(t)$, $y_2(t)$, ..., $y_{n_f}(t)$, $\vartheta_1(t)$, $\vartheta_2(t)$, ..., $\vartheta_{n_f}(t)$ are given by Eq. (11.17), which can be written as

$$\{w(t)\} = \sum_{m=1}^{m_{max}} \phi_m\xi_m(t) \tag{11.21}$$

where m_{max} is the highest mode that contributes significantly to the response. Accelerations $\{\ddot{w}(t)\}$ are obtained by differentiating Eq. (11.21) twice, the second derivatives of the generalized coordinates being known once Eq. (11.20) are solved. The requisite numerical calculations are performed using software that outputs directly the natural frequencies and modes of vibration of the structure and the forces of inertia induced by the wind loading being considered.

The total time-dependent wind-induced forces acting on the structure consist of the sums of the applied aerodynamic forces and the inertial forces associated with the structure's dynamic response. If tuned mass dampers are used to reduce the magnitude of the dynamic response, they can be viewed as additional masses connected to the structure by springs and damping-producing devices; for details see Chapter 16.

11.4 Simultaneous Pressure Measurements and Effective Wind-induced Loads

One of the inputs to Eq. (11.20) is the vector $\{F(t)\}$ of the applied aerodynamic loads (Eq. [11.16]). Figure 11.4 is an example of the placement of taps used to obtain time histories of simultaneously wind-induced pressure coefficients.

The *applied* aerodynamic forces and torsional moments induced by wind with speeds $U(z_{ref}, \theta_w)$, where θ_w is the wind direction and z_{ref} is the reference height, act at the locations of the pressure taps and are obtained from measured time histories of aerodynamic force coefficients. Their resultants acting at the mass centers of each floor or group of floors are obtained as indicated in Section 10.3 and are added algebraically to their *inertial* counterparts, thus yielding the *effective* wind-induced lateral forces and torsional moment at the center of mass of each floor. These are used in conjunction with influence coefficients (Section 9.3) to determine internal forces and their weighted combinations; this forms the basis on which the building's structural members are sized, as shown in subsequent chapters.

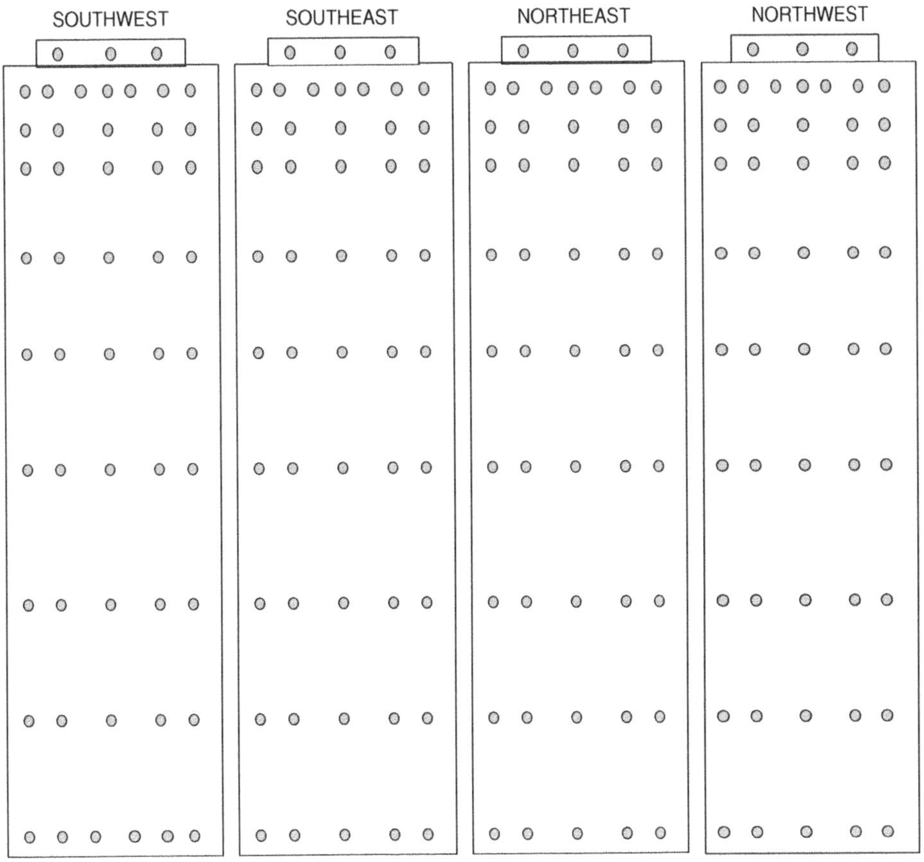

Figure 11.4 Example of pressure tap arrangement on the facades of a building model. Source: Courtesy of Dr. I. Venanzi, University of Perugia, and Dr. G. Bartoli, University of Florence.

Reference

1 Schmit, F.E. (1926). The Florida hurricane and its effects. *Engineering News Record* 97: 624–627.

12

Wind Load Factors and Design Mean Recurrence Intervals

12.1   Introduction

A 2004 landmark report by Skidmore Owings and Merrill (Appendix F) noted the absence in wind engineering laboratory reports of information or guidance pertaining to wind load factors. The latter depend upon uncertainties in the micrometeorological, aerodynamic and wind climatological parameters that govern structural design. These uncertainties can differ in some cases from those on which standard provisions are based. The purpose of this chapter is to discuss the development for such cases of appropriate wind load factors (or, if wind load factors are specified to be equal to unity, of appropriately augmented mean recurrence intervals (MRI) of design wind effects – see Section 12.5).

Attempts to develop design criteria applicable to structural *systems* have been unsuccessful owing in large part to difficulties arising in the reliability analysis of statically indeterminate structures. For this reason, strength design criteria are generally focused on individual structural *members* (see Appendix E). In modern codes, factors assuring that probabilities of failure are acceptably low differ according to whether they apply to loads or resistances, and are called load and resistance factors, respectively (hence the term "load and resistance factors design," or LRFD). Load factors depend upon the type of load (e.g., dead, live, snow, wind loads), and are defined as the quantities by which *nominal* loads or load effects need to be multiplied to obtain the *design* loads. Their magnitude is so calibrated that the resulting structural designs are comparable to proven designs based on past practices.

The calibration is of necessity imperfect owing to the variety of materials, construction techniques and design procedures used in past practices. However, a feature of past practices that was preserved in LRFD is the choice of MRIs of nominal wind effects, which are approximately 50 or 100 years, a choice largely based on engineering judgment and experience.

The load factor that multiplies the nominal $\overline{N}$-year wind load is called the $\overline{N}$-year *wind load factor* and is denoted by $\gamma_w(\overline{N})$. In pre-2010 versions of the ASCE 7 Standard the wind load factor was specified to be approximately 1.6. In the 2010 and 2016 versions, to simplify the design process the wind load factor was specified to be unity. However, to compensate for the reduction of the load factor from 1.6 to 1, and achieve design wind effects approximately equal to those implicit in pre-2010 ASCE 7 Standard requirements, the MRIs of the design wind speeds associated with a wind load factor equal to unity were augmented from 50 or 100 to 700 and 1700 years, respectively. In addition,

Wind Effects on Structures: Modern Structural Design for Wind, Fourth Edition. Emil Simiu and DongHun Yeo.
© 2019 John Wiley & Sons Ltd. Published 2019 by John Wiley & Sons Ltd.

wind speeds with a 3000-year MRI are currently specified for structures classified as essential, such as, for example, fire and police stations.

Uncertainties in quantities that determine wind-induced effects on rigid structures were discussed in Chapter 7. For flexible structures uncertainties in the dynamic features of the structure come into play as well, and are discussed in Section 12.2. The definition of the wind load factor is introduced in Section 12.3, which also discusses the calibration of the wind load factor with respect to past practice. Section 12.4 provides examples of the dependence of the wind load factor upon uncertainties specific to particular design situations. The examples show that the uncertainty in the wind speeds dominates the other individual uncertainties. Section 12.5 concerns the use of augmented design MRIs in lieu of products of wind load factors larger than unity by nominal wind loads or wind effects.[1]

12.2   Uncertainties in the Dynamic Response

The dynamic response of the structure depends upon its dynamic properties (natural frequencies, modal shapes, and damping ratios). The uncertainty in the dynamic response can in principle be estimated approximately by performing Monte Carlo simulations of the response, based on assumed probability distributions of the structure's dynamic properties. In practice, the estimation of the uncertainty must be performed largely on the basis of engineering judgment and experience by accounting for, among others, the cracking behavior of reinforced concrete and the behavior of joints in some types of steel structures. It is suggested that for flexible structures the assumption $\mathrm{CoV}(G) \approx 0.12$ used in the development of the wind load factor specified in earlier versions of the ASCE 7 Standard is reasonable for preliminary calculations based on Eq. (7.2).

---

1 It was mentioned in Chapter 7 that NASA and the Department of Energy require the use of far more elaborate approaches to uncertainty quantification than are currently available for civil engineering purposes – see, for example, https://standards.nasa.gov/standard/nasa/nasa-hdbk-7009, which provides "technical information, clarification, examples, processes, and techniques to help institute good modeling and simulation practices in … NASA. As a companion guide to NASA-STD-7009, the Handbook provides a broader scope of information than may be included in a Standard and promotes good practices in the production, use, and consumption of NASA modeling and simulation products. NASA-STD-7009 specifies what a modeling and simulation activity shall or should do (in the requirements) but does not prescribe how the requirements are to be met, which varies with the specific engineering discipline, or who is responsible for complying with the requirements, which depends on the size and type of project. A guidance document, which is not constrained by the requirements of a Standard, is better suited to address these additional aspects and provide necessary clarification." As indicated in [1], the NASA Jet Propulsion Laboratory at the California Institute of Technology "is pursuing Quantification of Margins and Uncertainty (QMU) technology to enable certification of models and simulations for extrapolation to poorly-testable … conditions,…, and provides a formalism for establishing credibility of a 'digital twin' that would predict system performance under difficult-to-test conditions." Among the tools used in QMU are Sandia's Sierra Mechanics and Multiphysics tools on models and simulations, Sandia's DAKOTA uncertainty analysis tool, the ASME V&V 10-2006 Guide for Verification and Validation in Computational Solid Mechanics, the AIAA Guide for the Verification of Computational Fluid Dynamics Simulation, and Department of Energy and Defense Guidelines and Recommended Practices. These tools and recommended practices are outside the scope of this chapter, but it may be anticipated that adapting them or their principles for civil engineering applications will be considered in the future.

12.3 Wind Load Factors: Definition and Calibration

The design peak wind effect with a 50-year MRI is defined as

$$p_{pk\,des}(\overline{N} = 50\text{ years}) \approx \overline{p}_{pk}(\overline{N} = 50\text{ years})\{1 + \beta\,\text{CoV}[p_{pk}(\overline{N} = 50\text{ years})]\} \quad (12.1)$$

where $\overline{p}_{pk}(\overline{N} = 50\text{ years})$ is the expectation, and $\text{CoV}[p_{pk}(\overline{N} = 50\text{ years})]$ is the coefficient of variation, of the peak wind effect p_{pk} with a 50-year MRI. For codification purposes the factor β is determined by calibration with respect to past practice and consensus among expert practitioners; the value $\beta \approx 2$ suggested in [2] appears to be reasonable and is adopted here for illustrative purposes. The quantity

$$\gamma_w(\overline{N} = 50\text{ years}) = 1 + \beta\,\text{CoV}[p_{pk}(\overline{N} = 50\text{ years})] \quad (12.2)$$

is the wind load factor by which the nominal expected peak wind effect with MRI $\overline{N} = 50$ years must be multiplied to yield the design peak wind effect. Therefore,

$$p_{pk\,des}(\overline{N} = 50\text{ years}) \approx \gamma_w(\overline{N} = 50\text{ years})\,\overline{p}_{pk}(\overline{N} = 50\text{ years}). \quad (12.3)$$

Example 12.1 For a rigid building, with the notations of Eq. (7.2), let $\text{CoV}(E_z) \approx 0.12$, $\text{CoV}(K_d) \approx 0.1$, $\text{CoV}[G(\theta_m)] \approx 0.05$, say, where θ_m defines the most unfavorable wind direction, $\text{CoV}[C_{p,pk}(\theta_m)] \approx 0.12$, and $\text{CoV}[U(\overline{N} = 50\text{ years})] \approx 0.12$ (see [2]). It follows from Eq. (7.2) that $\text{CoV}[p_{pk}(\overline{N})] = 0.315$ and, with $\beta = 2.0$, $\gamma_w \approx 1.63$, as calculated in Eq. (12.4):

$$\gamma_w = 1 + 2\,(0.12^2 + 0.1^2 + 0.05^2 + 0.12^2 + 4 \times 0.12^2)^{1/2} \approx 1.63 \quad (12.4)$$

This is, approximately, the value adopted in the ASCE 7-05 Standard [3, 4].

The following excerpt from [2, pp. 6, 7] illustrates the problems arising in the calibration of the factor β with respect to past practice:

> …reliability with respect to wind … loads appears to be relatively low compared to that for gravity loads,…, at least according to the methods used for structural safety checking in conventional design.[2] These are methods which are simplified representations of real building behavior and they have presumably given satisfactory performance in the past. It was decided to propose load factors for combinations involving wind … loads that will give calculated β values which are comparable to those existing in current practice, and not to attempt to raise these values to those for gravity loads by increasing the nominal loads or the load factors for wind … loading. Based on the information given here the profession may well feel challenged (1) to justify more explicitly (by analysis or test) why current simplified wind … calculations may be yielding conservative estimates of loads, resistances or safety; (2) to justify why current safety levels for gravity loads are higher than necessary if indeed this is true; (3) to explain why lower safety levels are appropriate for wind … vis-à-vis gravity loads, or (4) to agree to raise the wind … loads or load factors to achieve a similar reliability as that inherent in gravity loads. While the authors feel that arguments can be cited in favor and against all four options, they decided that this report is not the appropriate forum for what should be a profession-wide debate.

---

2 According to those methods the factor β for gravity loads is 3.0, rather than 2.0 [2].

12.4 Wind Load Factors vs. Individual Uncertainties

Equations (7.1, 7.2, 12.2 and 12.3) show that the uncertainty in the peak wind effect and, therefore, the magnitude of the wind load factor, depend upon the individual uncertainties that appear in the right-hand side of Eq. (7.2). This section examines, for various cases of interest, the degree to which the influence of an individual uncertainty on the magnitude of the wind load factor is significant. Except as otherwise noted, the individual uncertainties being considered are assumed to be those of Example 12.1.

12.4.1 Effect of Wind Speed Record Length

Assume that the length of the record of the largest yearly wind speeds to which there corresponds the value $\text{CoV}[U(\overline{N}=50 \text{ years})] \approx 0.12$ is 30 years. That value is due to measurement and sampling errors for which the CoVs are assumed to be 0.07 and 0.1, respectively. It was seen that to the CoVs of the uncertainties considered in Example 12.1, there corresponds a wind load factor $\gamma_w \approx 1.63$. Assume now that the record length on the basis of which the sampling errors in the estimation of the 50-year speed was estimated was only 10 years, as may be the case for remote locations for which few reliable meteorological measurements are available. Since the standard deviation of the sampling error is approximately proportional to the reciprocal of the square root of the sample size (see Eq. 3.9), the coefficient of variation characterizing the sampling errors may be assumed to be $\sqrt{3}$ times larger than for the case in which the record length is 30 years. Therefore, $\text{CoV}[U(\overline{N})] \approx [0.07^2 + (0.1 \times \sqrt{3})^2]^{1/2} \approx 0.187$. Instead of $\gamma_w \approx 1.6$, it follows from Eq. (12.2) that the estimated wind load factor is

$$\gamma_w = 1 + 2\,(0.12^2 + 0.1^2 + 0.05^2 + 0.12^2 + 4 \times 0.187^2)^{1/2} \approx 1.85 \tag{12.5}$$

The ratio between the wind load factors based on the 10-year wind speed record and the 30-year record of wind speeds, all other uncertainties being unchanged, is approximately 1.14. This is in part a consequence of the multiplication of $\text{CoV}[U(\overline{N})]$ by the factor 4 (see Eq. 7.2), owing to which the contribution to the wind load factor of the uncertainty in the wind speed dominates the contributions of the other individual uncertainties.

12.4.2 Effect of Aerodynamic Interpolation Errors

Large sets of aerodynamic pressure data used for database-assisted design cannot cover all possible model dimensions and roof slopes. For this reason, interpolations based on databases with limited numbers of models are typically necessary in the design process. According to calculations reported in [5], such interpolations entail errors that, depending upon the number of models in the database, can have CoVs as large as 0.15, say. Accounting for this CoV in the expression for the load factor

$$\gamma_w = 1 + 2\,[0.12^2 + 0.1^2 + 0.05^2 + (0.12^2 + 0.15^2) + 4 \times 0.12^2]^{1/2} \approx 1.70 \tag{12.6}$$

rather than 1.63; that is, the increase in the estimated value of the wind load factor in this example is approximately 5%. This result suggests that the number of models in large aerodynamic databases does not necessarily have to be increased unless the CoVs of the interpolation errors in the estimation of the pressure coefficients exceed 15%, say.

12.4.3 Number of Pressure Taps Installed on Building Models

The lower the number of taps placed on the model, the larger will be the errors in the estimation of the wind effects. Figures 5.29 and 5.30 show the vast difference between the numbers of taps typically used before and after the development of pressure scanners to determine wind loads. For strength design purposes, useful assessments of the extent to which the number of pressure taps installed on the building model is adequate by modern standards can be made by comparing base shears and moments obtained by high-frequency force balance measurements to their counterparts based on pressure time histories at the taps [6]; or, in some cases, by comparing wind effects based on all the available taps on the one hand and on, say, half the number of taps on the other.

12.4.4 Effect of Reducing Uncertainty in the Terrain Exposure Factor

Ad-hoc wind tunnel testing that reproduces to scale the built environment of the structure being designed has the advantage of reducing the uncertainty in the terrain exposure factor. Because the wind tunnel simulation of the atmospheric flow is imperfect, the reduction may be relatively modest, from $CoV(E_z) = 0.12$ (as in [2]) to $CoV(E_z) = 0.05$, say. For a rigid structure, this would result in a less than 3% reduction of the wind load factor from $\gamma_w \approx 1.63$ (see Eq. [12.4]) to $\gamma_w \approx 1.59$:

$$\gamma_w = 1 + 2(0.05^2 + 0.10^2 + 0.05^2 + 0.12^2 + 4 \times 0.12^2)^{1/2} \approx 1.59 \tag{12.7}$$

12.4.5 Flexible Buildings

Flexible structures experience dynamic effects that may be expressed in terms of a dynamic response factor, G. According to [2], typically $CoV(G) \approx 0.12$. This value is based on early studies of uncertainties in the along-wind response [7]. In some instances, natural frequencies, modal shapes and modal damping ratios are dependent upon factors that are difficult to quantify and on which relatively few reliable data exist. For example, estimates of the extent to which cracking of concrete influences the structure's stiffness characteristics may still be affected by significant uncertainties. For these reasons the coefficients of variation of the uncertainty in the dynamic effects may be larger than 0.12.

Non-zero values of $CoV(G)$ increase the coefficient of variation of the peak wind effect and will therefore result in wind load factors larger than their rigid structure counterparts. This explains the quest by structural engineers for ad-hoc wind load factors applicable to tall buildings (see Appendix F). Assuming, for example, that $CoV(G) = 0.12$ and that the other individual uncertainties affecting the wind load factor have the values used in Eq. (12.4),

$$\gamma_w = 1 + 2(0.12^2 + 0.1^2 + 0.12^2 + 0.12^2 + 4 \times 0.12^2)^{1/2} \approx 1.67 \tag{12.8}$$

rather than 1.63 for the rigid structure case. If, in addition, the length of the wind speed record is 10 years and $CoV[U(\overline{N})] = 0.187$, as in Section 12.4.1, $\gamma_w = 1.88$.

12.4.6 Notes

1) Except for the uncertainty in the wind speed, individual uncertainties in the quantities that determine wind effects typically have relatively small or negligible effects

on the magnitude of the wind load factors. This fact should be considered before significant resources are devoted to efforts to reduce these uncertainties.

2) The magnitude of the wind load factor can be affected significantly by uncertainties in the wind speeds that are larger than those typically assumed in standards. This is especially true of hurricane wind speeds, for which estimates of uncertainties are difficult to determine reliably.

3) Wind load factors are larger for flexible buildings than for rigid buildings, and the joint effect of uncertainties in the dynamic response and of larger than typical uncertainties in the wind speeds can result in large increases in the wind load factors. Standard provisions on the wind tunnel procedure should clearly indicate this fact.

4) It was shown that typical uncertainties in pressure coefficients obtained in wind tunnel tests have relatively minor effects on the magnitude of the wind load factor. This suggests that the use of Computational Wind Engineering simulations to obtain estimates of pressure coefficients should be acceptable for practical purposes as long as the CoVs of the uncertainties in those estimates are lower than, say, 15%.

12.5 Wind Load Factors and Design Mean Recurrence Intervals

ASCE 7-05 Standard and earlier versions specified a typical MRI of the design wind speed $\overline{N} = 50$ years, and a wind load factor $\gamma_w \approx 1.6$. Later versions instead specify no wind load factor (i.e., a wind load factor $\gamma_w = 1$), and an augmented MRI $\overline{N}_1$ of the design wind speed such that the design wind loads are approximately the same in the earlier and the current standards. Since wind effects determined in accordance with conventional provisions of the ASCE 7 Standard are proportional to the square of the wind speeds U, this condition yields the relation

$$U^2(\overline{N}_1) = \gamma_w U^2(\overline{N}). \tag{12.9}$$

For $\overline{N} = 50$ years (100 years) and $\gamma_w \approx 1.6$, this relation was assumed to yield $\overline{N}_1 = 700$ years (1700 years). These values correspond to typical probability distributions of extreme wind speeds. Since those probabilities can depend fairly strongly on geographical location, the values $\overline{N}_1$ of the MRIs of the design wind speeds and wind effects may turn out to differ, in some cases significantly, from 700 or 1700 years.

Example 12.2 Let the mean $E(U)$ and the standard deviation $SD(U)$ of the extreme annual wind speed sample be 59 and 6.41 mph, respectively. The 50- and 700-years wind speeds are estimated by Eq. (3.7b) to be 75.6 and 88.9 mph, respectively. The design wind effect is

$$p_{pk\ des} = c\gamma_w U^2(\overline{N} = 50 \text{ years})$$
$$= cU^2(\overline{N}_1 \text{ years})$$

where c is a coefficient that reflects the relation between wind effect and the square of the wind speed. Therefore, from Eq. (12.9),

$$U(\overline{N}_1 \text{ years}) = \gamma_w^{1/2} U (50 \text{ years})$$

where $\overline{N}$ can be estimated using Eqs. (3.5) and (3.7). For $\gamma_w = 1.6$, $U(\overline{N}_1$ years) $= 95.6$ mph (rather than 88.9 mph) and $\overline{N}_1 \approx 2700$ years, rather than $\overline{N}_1 = 700$ years as specified in the ASCE 7-16 Standard. For $\gamma_w = 1.5$, $U(\overline{N}_1$ years) $= 92.6$ mph and $\overline{N}_1 \approx 1500$ years.

This example shows that, like the wind load factors, the MRIs of the design wind effects should be specified by accounting for the wind climate statistics and the specific uncertainties in the micrometeorological, wind climatological, aerodynamic, directionality, and dynamic features of the structure of interest. In light of this example it appears that the validity of the neat correspondence suggested in [8, figure 3] between MRIs and factors β (see Eq. [12.3]) is not warranted.

References

1 Peterson, L. D., "Quantification of margins and uncertainties for model-informed flight system qualification," presented at the NASA Thermal and Fluids Analysis Workshop, Hampton, VA, 2011, https://kiss.caltech.edu/workshops/xTerra/presentations1/peterson.pdf.

2 Ellingwood, B., Galambos, T. V., MacGregor, J. G., and Cornell, C. A., "Development of a probability-based load criterion for American National Standard A58," NBS Special Publication 577, National Bureau of Standards, Washington, DC, 1980. https://www.nist.gov/wind.

3 ASCE, "Minimum design loads for buildings and other structures," in *ASCE Standard ASCE/SEI 7-05*, Reston, VA: American Society of Civil Engineers, 2005.

4 Simiu, E., Pintar, A.L., Duthinh, D., and Yeo, D. (2017). Wind load factors for use in the wind tunnel procedure. *ASCE-ASME Journal of Risk and Uncertainty in Engineering Systems, Part A: Civil Engineering* 3: doi:10.1061/AJRUA6.0000910. https://www.nist.gov/wind.

5 Habte, F., Chowdhury, A.G., Yeo, D., and Simiu, E. (2017). Design of rigid structures for wind using time series of demand-to-capacity indexes: application to steel portal frames. *Engineering Structures* 132: 428–442.

6 Dragoiescu, C., Garber, J., and Kumar, K. S., "The use and limitation of the pressure integration technique for predicting wind-induced responses of tall buildings," in *European and African Conference on Wind Engineering*, Florence, Italy, 2009, pp. 181–184.

7 Vickery, B. J., "On the reliability of gust loading factors," in *Technical Meeting Concerning Wind Loads on Buildings and Structures*, Washington, DC, 1970. https://www.nist.gov/wind.

8 McAllister, T.P., Wang, N., and Ellingwood, B.R. Risk-informed mean recurrence intervals for updated wind maps in ASCE 7-16. *Journal of Structural Engineering* 144 (5): 06018001. doi:10.1061/(ASCE)ST.1943-541X.0002011.

13

Wind Effects with Specified MRIs: DCIs, Inter-Story Drift, and Accelerations

13.1 Introduction

Wind directionality is accounted for in different ways depending upon whether the wind climatological data consist of directional or non-directional wind speeds, as defined in Section 3.2.2.

If the design is based on directional wind speeds, as is commonly the case for designs using wind tunnel test results, wind effects with specified mean recurrence intervals (MRIs) are determined by accounting explicitly for the dependence of both the wind speeds and the wind effects upon direction. For the database-assisted design approach this requires:

1) The use of matrices of directional wind speeds $[U_{ij}]$ provided by the wind engineering laboratory, where U_{ij} is the mean wind speed at top of building in storm event i ($i = 1, 2, ..., n_s$) from direction $\theta = \theta_j$ ($j = 1, 2, ..., n_d$), based on a sample of measured or simulated directional wind speeds. The number n_s of storms for which wind speeds are available in the matrix $[U_{ij}]$ must be sufficiently large to allow the reliable estimation by non-parametric statistics of wind effects with the required MRI. If, as is the case for the ASCE 7-10 and ASCE 7-16 Standards, the MRI is 700 years or larger, Monte Carlo simulations are used to meet this requirement (see [1], and Sections 3.3.7 and A.8).

2) The development, for each type of wind effect of interest (e.g., base shear, base moment, internal force, peak demand-to-capacity index [DCI], displacement, acceleration) of time series $R(U, \theta, t)$, representing the dependence of the wind effect R upon the wind speed U, the direction θ and the time t. The length T of the time series $R(U, \theta, t)$ is equal to the length of the time series of pressure coefficients provided by the wind engineering laboratory. However, the peak value of $R(U, \theta, t)$, that is, $\max_t(R(U, \theta, t))$, henceforth denoted as $R^{pk}(U, \theta)$, can be determined for time series with any specified length $T_1 > T$ by using, for example, the procedure described in Section 7.3.3 (v), or the procedure in Appendix C. The *response surface* is the three-dimensional plot of $R^{pk}(U, \theta)$ as a function of wind speed U and direction θ (Section 13.2).

3) The transformation of the directional wind speed matrix $[U_{ij}]$ into the matrix $[R^{pk}(U_{ij})]$. This is accomplished by substituting the quantities $R^{pk}(U_{ij})$ for the wind velocities U_{ij} in the matrix $[U_{ij}]$ (Sections 13.3.1 and 13.3.2).

Wind Effects on Structures: Modern Structural Design for Wind, Fourth Edition. Emil Simiu and DongHun Yeo.
© 2019 John Wiley & Sons Ltd. Published 2019 by John Wiley & Sons Ltd.

4) The transformation of the matrices $[R^{pk}(U_{ij})]$ into vectors $\{R_i\} = \max_j(R^{pk}(U_{ij}))$ by disregarding in each windstorm i all wind effects $R^{pk}(U_{ij})$ lower than the largest wind effect, $\max_j(R^{pk}(U_{ij}))$, occurring in that storm (Section 13.3.3).

5) The application to the n_s rank-ordered quantities $\{R_i\}$ of the non-parametric statistical estimation procedure in Section A.9 for regions with one or two types of windstorm, to obtain the wind effects $R(\overline{N})$ with the specified MRI (Section 13.4).

If the design is based on non-directional wind speeds, which is the case if directional wind speed data are not available, the design is based on pressure coefficients and wind speeds with the respective most unfavorable directions, which typically do not coincide. It follows from the assumed linear dependence of the mean wind loads upon the square of the non-directional wind speeds that the MRI of the wind loads is the same as the MRI of the wind speeds. However, a correction factor smaller than unity, called wind directionality factor, is applied to the wind effect to account approximately for the non-coincidence of the most unfavorable pressure and wind directions (Section 13.5).

Material on DCIs and on inter-story drift and accelerations is provided in Sections 13.6 and 13.7, respectively.

A method for estimating directionality effects developed in the 1970s, and still being used by some wind engineering laboratories, is described in Section B.6, which discusses the reasons why the method is impractical and prone to yielding inadequate estimates of the wind effects being sought. In addition, that method is viewed by structural engineers as lacking transparency, as indicated in Appendix F. An alternative method proposed in [2] is also being used by some laboratories, in spite of the fact that it can yield unconservative results.

In practical applications, operations covered by this chapter can be performed by using software for which links are provided in Chapters 17 and 18.

13.2 Directional Wind Speeds and Response Surfaces

Once the wind engineering laboratory provides the requisite aerodynamic and wind climatological data as affected by terrain exposure at the structure's site, the structural engineer's first step toward determining peak wind effects $R(\overline{N})$, where $\overline{N}$ denotes the specified MRI, is to develop *response surfaces*, that is, three-dimensional plots representing the dependence of peak wind effects $R^{pk}(U_{ij})$ upon wind speed and direction. A response surface is constructed for each wind effect of interest. An example of response surface for a peak DCI (involving the axial force and bending moment at a given member cross section) is shown in Figure 13.1.

In general, owing to nonlinearities inherent in resonance effects and/or column instability, the ordinates of the DCI response surfaces are not proportional to the squares of the wind speeds. The wind effect of interest must therefore be determined separately for each wind speed and direction.

The response surfaces are properties of the structure independent of the wind climate. As shown subsequently, they are used in a simple non-parametric statistical procedure that yields peak wind effects with any specified MRI.

Response surfaces for DCIs are developed as follows. Consider the time series of the effective forces $F_{kx}(U,\theta,t)$, $F_{ky}(U,\theta,t)$, $F_{k\vartheta}(U,\theta,t)$ (t = time) induced by wind with

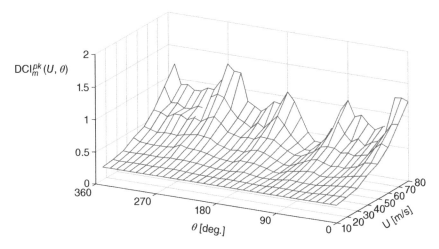

Figure 13.1 Response surface for the peak demand-to-capacity index of a cross section m as a function of wind speed and direction.

speed U from direction θ, and acting at the center of mass of floor k ($= 1, 2, ..., n_f$) in the directions of the building's principal axes and in torsion, respectively. The internal forces $f_m(U, \theta, t)$ induced by the effective wind forces at any given cross section m can be written as

$$f_m(U, \theta, t) = \sum_{k=1}^{n_f} r_{mk,x} F_{kx}(U, \theta, t) + \sum_{k=1}^{n_f} r_{mk,y} F_{ky}(U, \theta, t) + \sum_{k=1}^{n_f} r_{mk,\vartheta} F_{k\vartheta}(U, \theta, t)$$

(13.1)

where $r_{mk,x}$, $r_{mk,y}$, $r_{mk,\vartheta}$ are influence coefficients (see Section 9.3). Equation (13.1) is then used to obtain time series of demand-to-capacity DCIs at cross sections m, denoted by $\text{DCI}_m(U, \theta, t)$, in which it is recalled that effects of factored gravity loads are also accounted for. A similar approach is used for displacements and accelerations.

13.3 Transformation of Wind Speed Matrix into Vectors of Largest Wind Effects

13.3.1 Matrix of Largest Directional Wind Speeds

In the following we focus on the $\text{DCI}_m^{pk}(U_{ij})$ induced in cross section m by the wind speed U_{ij} acting in storm i from direction j at the building site. Similar approaches can be used for any other wind effects.

Consider, for illustrative purposes, the 3×4 matrix of wind speeds (in m s^{-1})

$$[U_{ij}] = \begin{bmatrix} 34 & \mathbf{45} & 32 & 44 \\ 37 & 39 & 36 & \mathbf{51} \\ 42 & 44 & 35 & \mathbf{46} \end{bmatrix}$$

(13.2)

at the site of the structure. (See Section 3.2.3 for wind speed data that are available or that can be developed by simulation from such data.) Under the convention inherent in the notation U_{ij}, the 3×4 matrix corresponds to three storm events and four wind directions, that is, $i = 1, 2, 3$ and $j = 1, 2, 3, 4$. For example, the wind speed that occurs in the second storm event from the third direction is $U_{23} = 36\ \text{m s}^{-1}$. (The entries in the wind speed matrix could be, for example, mean hourly speeds at the elevation of the top of the structure with direction j over terrain with suburban exposure.) In the matrix of Eq. (13.2) the largest wind speeds in each of the three storms are indicated in bold type.

13.3.2 Transformation of Matrix $[U_{ij}]$ into Matrix of Demand-to-Capacity Indexes $[\text{DCI}_m^{pk}(U_{ij})]$

Transform the matrix $[U_{ij}]$ into the matrix $[\text{DCI}_m^{pk}(U_{ij})]$ by substituting for the quantities U_{ij} the ordinates $\text{DCI}_m^{pk}(U_{ij})$ of the cross section m response surface. Assume that these quantities are DCIs, and that the result of this operation is the matrix

$$[\text{DCI}_m^{pk}(U_{ij})] = \begin{bmatrix} 0.70 & \mathbf{1.02} & 0.80 & 0.68 \\ 0.83 & 0.83 & \mathbf{1.01} & 0.91 \\ \mathbf{1.07} & 0.98 & 0.96 & 0.74 \end{bmatrix} \tag{13.3}$$

13.3.3 Vector $\{\text{DCI}_{m,i}\} = \{\max_j(\text{DCI}_m^{pk}(U_{ij}))\}$

The directional wind effects induced by the wind speeds occurring in storm i depend upon the wind direction j. It is only the largest of those wind effects, that is, $\text{DCI}_{m,i} = \max_j (\text{DCI}_m^{pk}(U_{ij}))$ ($i = 1, 2, 3$), that are of interest from a design viewpoint. These largest wind effects, shown in bold type in Eq. (13.3), form a vector $\{1.02, 1.01, 1.07\}^T$, where T denotes transpose. Note that $\text{DCI}_{m,i}$ is not necessarily induced by the speed $\max_j(U_{ij})$. For example, $\text{DCI}_{m,3} = 1.07$ is not induced by the speed $\max_j(U_{3j}) = U_{34} = 46\ \text{m s}^{-1}$, but rather by the speed $U_{31} = 42\ \text{m s}^{-1}$.

The components of the vector $\{\text{DCI}_{m,i}\}$ constitute the sample of the largest peak wind effects occurring in each of the n_s storm events (in this example $i = 1, 2, 3$; $n_s = 3$). The estimation of the response with any specified MRI is based on this sample, used in conjunction with the mean annual rate of occurrence of the storms (see Section 13.4).

13.4 Estimation of Directional Wind Effects with Specified MRIs

The peak wind effects $\text{DCI}_m(\overline{N})$, where $\overline{N}$ denotes the specified MRI in years, could in principle be determined by using parametric statistics. This would entail the fitting of a cumulative distribution function (CDF) to the sample $\text{DCI}_{m,i}$ ($i = 1, 2, \ldots, n_s$). The variate DCI_m with an MRI N_f, where N_f is the number of average time intervals between successive storms, corresponds in the example of Section 13.3 to a CDF ordinate $P = 1 - 1/N_f$. However, the designer is interested in the variate DCI_m with an MRI $\overline{N}$ in *years*, rather than in average time intervals between successive storms. Since the mean annual rate of storm arrival is λ, $\overline{N} = N_f\ /\ \lambda$. For example, if the storms being considered are tropical cyclones, it is typically the case that $\lambda < 1$ storm/year, so $\overline{N} > N_f$. The converse is true for the case $\lambda > 1$ storm/year. Therefore, the variate DCI_m with an MRI $\overline{N}$, DCI_m $(\overline{N})$, corresponds to the ordinate $P = 1 - 1/(\lambda \overline{N})$ of the CDF fitted to the data sample $DCI_{m,i}$ $(i = 1, 2, \ldots, n_s)$.

A drawback of parametric statistics for this type of application would be that few studies have been performed on, and little is known about, the best fitting types of probability distribution of the various wind effects (as opposed to wind speeds). If, as is the case for the ASCE 7-16 Standard, the MRIs of interest are large (e.g., 300–3000 years), the uncertainty inherent in the choice of the best fitting type of probability distribution may entail significant probabilistic modeling errors. It is therefore prudent to use the non-parametric approach.

The application of non-parametric statistics requires the development by the wind engineering laboratory of synthetic directional wind speed samples from measured directional wind data. The development entails three phases. In the first phase the measured directional wind speeds are processed by the wind engineer so that they are micrometeorologically consistent with the wind speeds used in the development of the directional aerodynamic pressure time series. In the second phase the directional wind speed data so obtained are fitted to Extreme Value Type I distributions (see Chapters 3 and Appendix A), which are widely accepted as appropriate for the probabilistic description of extreme wind speeds. A probability distribution is fitted to the wind speeds from each direction j. When doing so, it may be assumed for practical purposes that wind speeds from different directions are for practical purposes mutually independent, provided that the respective azimuths do not differ by less than $10°$, say. In the third phase the Extreme Value Type I distributions are used to develop by Monte Carlo simulation (see Section A.8) the requisite large sets of directional extreme wind speed data [1]. These sets are provided to the structural engineer by the wind engineering laboratory.

The structural engineer can then use the simulated extreme wind speed data as input to software subroutines for the estimation of wind effects with specified MRIs. This approach is implemented in the software presented in Chapters 17 and 18, which uses the procedure of Section A.9.1 for regions with a single type, or Section A.9.2 for regions with two types of storm hazard (e.g., synoptic storms and thunderstorms).

13.5 Non-Directional Wind Speeds: Wind Directionality Reduction Factors

If dynamic effects are negligible, design wind loads $W_{std}(\overline{N})$ are typically based in standards on:

- non-directional sets of pressure or force coefficients $C_p = \max_j(C_{pj})$, where C_{pj} is the peak directional force or pressure coefficient corresponding to wind direction j ($j = 1, 2, \ldots, j_{max}$; e.g., $j_{max} = 16$)
- wind speeds with an $\overline{N}$-year MRI, $U(\overline{N})$, estimated from the non-directional wind speeds data $U_i = \max_j(U_{ij})$, where U_{ij} is the largest directional wind speed from direction j during storm event i, defined for the appropriate terrain exposure, height above ground, and averaging time

- a factor K_d that accounts for directionality effects

Therefore,

$$W_{std}(\overline{N}) = a \, K_d \, C_p \, U^2(\overline{N}) \tag{13.4}$$

where a is a constant. The wind directionality factor K_d is defined as the ratio of calculated wind effects $W_{dir}(\overline{N})$ and $W_{nd}(\overline{N})$ that account and do not account for wind directionality, respectively.

$$K_d \approx \frac{W_{dir}(\overline{N})}{W_{nd}(\overline{N})} \tag{13.5}$$

where the numerator and the denominator are estimated, respectively, from the data

$$W_{i\,dir} = a \, \max_j (C_{pj} U_{ij}^2) \tag{13.6}$$

and

$$W_{i\,nd} = a \, \max_j (C_{pj}) \, (\max_j (U_{ij}))^2 \tag{13.7}$$

($i = 1, 2,\ldots, n_s$; the indexes "dir" and "nd" stand for "directional" and "non-directional"). It is clear that, typically, $W_{i\,nd} > W_{i\,dir}$ and that $K_d < 1$.

Example 13.1 Consider the directional wind speed matrix $[U_{ij}]$ of Eq. (13.2). Assume that the directional aerodynamic coefficients C_{pj} are 0.7, 0.8, 1.2, and 0.6 for directions $j = 1, 2, 3$, and 4, respectively. It can be easily verified that the entries in Table 13.1 are smaller for column (2) than for column (1).

For the simplified estimated value of $W_{std}(\overline{N})$ to be reasonably correct, it is required that the directionality factor in Eq. (13.4) be approximately equal to the ratio $W_{dir}(\overline{N})/W_{nd}(\overline{N})$. According to the ASCE 7 Standard this is the case for typical buildings if $K_d = 0.85$, reflecting the fact that the climatologically and aerodynamically most unfavorable wind directions typically do not coincide. Calculations reported in [3] indicate that the use of this value in design is typically reasonable, although for hurricane-prone regions it is prudent to use the value $K_d = 0.9$.

Table 13.1 Comparison of non-directional and directional wind load estimates.

i	(1) $\max_j (C_j) \max_j (U_{ij}^2)$ (m² s⁻²) (non-directional)	(2) $\max_j [C_j \, U_{ij}^2]$ (m² s⁻²) (directional)
1	$1.2 \times 45^2 = 2430$	$0.8 \times 45^2 = 1620 \ (j = 2)$
2	$1.2 \times 51^2 = 3121$	$1.2 \times 36^2 = 1561 \ (j = 3)$
3	$1.2 \times 46^2 = 2539$	$0.8 \times 44^2 = 1549 \ (j = 2)$

13.6 Demand-to-Capacity Indexes

This section is a brief presentation of material on DCIs for steel and reinforced concrete buildings. The DCI_m is a measure of the degree to which the strength of a structural cross section m is adequate. In general, the index is defined as a ratio or sum of ratios of the required internal forces to the respective available capacities. A DCI larger than unity indicates that the design of the cross section being considered is inadequate.

The general expression for the DCIs used in design is

$$DCI^{PM}(t) = f\left(\frac{P_u(t)}{\phi_p P_n(t)}, \frac{M_u(t)}{\phi_m M_n(t)}\right) \leq 1 \tag{13.8}$$

$$DCI^{VT}(t) = f\left(\frac{V_u(t)}{\phi_v V_n(t)}, \frac{T_u(t)}{\phi_t T_n(t)}\right) \leq 1 \tag{13.9}$$

where the symbols P, M, V, and T represent compressive or tensile strength, flexural strength, shear strength, and torsional strength, respectively; the subscripts u and n indicate required and available strength, respectively; and ϕ_i resistance factors ($i = p, m, v, t$, corresponding to axial, flexural, shear, and torsional strength, respectively). The available strength is specified by the AISC Steel Construction Manual [4] for steel structures, and the ACI Building Code Requirements for Structural Concrete [5], or other documents. For details on the application of [4] and [5] in the context of this book see [6, 7].

13.7 Inter-Story Drift and Floor Accelerations

The approach to determining wind effects with specified MRIs considered in Sections 13.2–13.4 is applicable, in particular, to inter-story drift ratios and floor accelerations.

The time-series of the inter-story drift ratios at the kth story, $d_{k,x}(t)$ and $d_{k,y}(t)$, corresponding to the x- and y-principal axis of the building, are (Figure 13.2):

$$d_{k,x}(t) = \frac{[x_k(t) - D_{k,y}\vartheta_k(t)] - [x_{k-1}(t) - D_{k-1,y}\vartheta_{k-1}(t)]}{h_k} \tag{13.10a}$$

$$d_{k,y}(t) = \frac{[y_k(t) + D_{k,x}\vartheta_k(t)] - [y_{k-1}(t) + D_{k-1,x}\vartheta_{k-1}(t)]}{h_k} \tag{13.10b}$$

Figure 13.2 Position parameters at floor k for inter-story drift and accelerations.

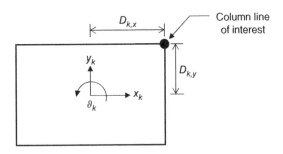

where $x_k(t)$, $y_k(t)$, and $\vartheta_k(t)$ are the displacements and rotation at the mass center at the kth floor, $D_{k,x}$ and $D_{k,y}$ are distances along the x- and y-axis from the mass center on the kth floor to the point of interest on that floor, and h_k is the kth story height.

The time-series of the resultant acceleration at floor k, $a_{k,r}(t)$ is yielded by the expression:

$$a_{k,r}(t) = \sqrt{[\ddot{x}_k(t) - D_{k,y}\ddot{\vartheta}_k(t)]^2 + [\ddot{y}_k(t) + D_{k,x}\ddot{\vartheta}_k(t)]^2} \tag{13.11}$$

where accelerations $\ddot{x}_k(t)$, $\ddot{y}_k(t)$, and $\ddot{\vartheta}_k(t)$ of the mass center at the kth floor pertain to the x-, y-, and ϑ- (i.e., rotational) axis, and $D_{k,x}$ and $D_{k,y}$ are the distances along the x- and y- axis from the mass center to the point of interest on the kth floor (Figure 13.2).

References

1 Yeo, D. (2014). Generation of large directional wind speed data sets for estimation of wind effects with long return periods. *Journal of Structural Engineering* 140: 04014073. https://www.nist.gov/wind.

2 Simiu, E. and Filliben, J.J. (2005). Wind tunnel testing and the sector-by-sector approach to wind directionality effects. *Journal of Structural Engineering* 131: 1143–1145. https://www.nist.gov/wind.

3 Habte, F., Chowdhury, A., Yeo, D., and Simiu, E. (2015). Wind directionality factors for nonhurricane and hurricane-prone regions. *Journal of Structural Engineering* 141: 04014208.

4 ANSI/AISC (2010). *Steel Construction Manual*, 14 ed. American Institute of Steel Construction.

5 ACI (2014). *Building Code Requirements for Structural Concrete (ACI 318-14) and Commentary*. Farmington Hills, MI: American Concrete Institute.

6 Yeo, D., Database-assisted design for wind: Concepts, software, and example for of high-rise reinforced concrete structures, NIST Technical Note 1665, National Institute of Standards and Technology, Gaithersburg, MD, 2010. https://www.nist.gov/wind.

7 Park, S., Yeo, D., and Simiu, E., Database-assisted design and equivalent static wind loads for mid- and high-rise structures: concepts, software, and user's manual, NIST Technical Note 2000, National Institute of Standards and Technology, Gaithersburg, MD, 2018. https://www.nist.gov/wind.

14

Equivalent Static Wind Loads

14.1   Introduction

This chapter presents a procedure for determining equivalent static wind loads (ESWLs) on mid- and high-rise buildings. A similar procedure is presented in [1] and [2] and is demonstrated in a case study in Chapter 18.

The tasks performed by using ESWLs, commonalities and differences between those tasks and the tasks performed by using Database-Assisted Design (DAD), and comparative ESWL and DAD features, were considered in Sections 8.3 and 8.4. Section 14.2 describes a procedure for determining ESWLs. It follows from the description of that procedure that ESWL-based designs are typically limited to buildings with simple geometries. For structures with complex geometries, risk-consistent designs require the use of the more computer-intensive – and more accurate – DAD procedure.

Like DAD, the ESWL procedures presented in this chapter and in [1, 2] are user-friendly, transparent, readily subjected to effective public scrutiny, and easily integrated into Building Information Modeling (BIM) systems. Also, like DAD, ESWL renders obsolete the High Frequency Force Balance (HFFB) practice wherein analyses of wind-induced dynamic effects are performed by the wind engineer in the absence of information of the distribution of the wind loads with height. In contrast to HFFB, ESWL allows iterative structural designs to be readily performed with no time-consuming back-and-forth interactions between the wind and the structural engineer. For structures with relatively simple shapes, wind effects calculated by using ESWL approximate reasonably closely their DAD counterparts. The latter may serve as reliable benchmarks against which ESWL calculations can be verified. However, the ESWL procedure can be less effective if wind speeds from a direction that is unfavorable from a structural point of view are dominant. Also, for structures with complex shapes the ESWL procedure may be inapplicable.

14.2   Estimation of Equivalent Static Wind Loads

Earlier approaches to the estimation of ESWLs are described in [3–5]. This section describes an approach to structural design that, typically, induces in structural members DCIs approximately equal to their counterparts obtained by using DAD. As noted earlier, like other ESWL approaches, the approach presented in this section is applicable only to structures with relatively simple shapes.

Wind Effects on Structures: Modern Structural Design for Wind, Fourth Edition. Emil Simiu and DongHun Yeo.
© 2019 John Wiley & Sons Ltd. Published 2019 by John Wiley & Sons Ltd.

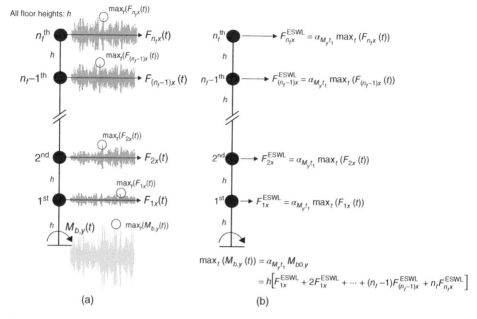

Figure 14.1 Lumped mass structure with fluctuating wind loads acting in the direction x. (a) fluctuating wind loads in DAD and (b) equivalent static wind loads in ESWL. Typically, owing to imperfect spatial correlations, peak wind-induced loads at different floors do not occur at the same times.

We first consider the simple though physically unrealistic case in which the wind loads are assumed to act on the structure only in the direction x of a principal axis of the building. We denote the effective (i.e., aerodynamic plus dynamic) randomly fluctuating load at floor k by $F_{kx}(t)$, where $k = 1, 2, \dots n_f$ (see Figure 14.1). Assume for the sake of simplicity that all floors have height h. The sum of the moments of the loads $\max_t(F_{kx}(t))$ with respect to the building base is

$$M_{b0,y} = h[\max_t (F_{1x}(t)) + 2 \max_t (F_{2x}(t)) + \dots + n_f \max_t (F_{n_fx}(t))] \qquad (14.1)$$

Owing to the imperfect spatial correlation between any pair of time-dependent floor loads, the peak of the actual base moment $M_{b,y}(t)$ induced by the effective loads $F_{kx}(t)$ is

$$\max_t(M_{b,y}(t)) < M_{b0,y} \qquad (14.2)$$

Denote by F_{kx}^{ESWL} the ESWL acting at the floor k. In order for the static loads F_{kx}^{ESWL} to produce a peak base moment $\max_t(M_{b,y}(t))$, the peak floor loads, $\max_t(F_{kx}(t))$, are multiplied by a reduction coefficient $\alpha_{M_y t_1}$ such that

$$F_{kx}^{\text{ESWL}} = \alpha_{M_y t_1} \max_t(F_{kx}(t)) \qquad (14.3a)$$

$$\alpha_{M_y t_1} = \frac{\max_t(M_{b,y}(t))}{M_{b0,y}} \qquad (14.3b)$$

where t_1 is the time of occurrence of the peak base moment $\max_t(M_{b,y}(t))$. The ESWLs F_{kx}^{ESWL} determined as described here are acceptable for design purposes if they induce in each structural member DCIs approximately equal to the peak DCIs induced by the fluctuating loads.

The equivalence of static and fluctuating forces must apply to the internal forces $f_m(t)$ at all cross sections m within the structure, where

$$f_m(t) = r_{m1,x} F_{1x}(t) + r_{m2,x} F_{2x}(t) + \dots + r_{mn_f,x} F_{n_fx}(t) \tag{14.4}$$

$m = 1, 2, \dots, m_{max}$ identifies the cross section being considered, and $r_{mk,x}$ ($k = 1, 2, \dots, n_f$) are influence coefficients; that is, the loads F_{kx}^{ESWL} must satisfy the system of equations

$$\max_t(f_m(t)) = r_{m1,x} F_{1x}^{ESWL} + r_{m2,x} F_{2x}^{ESWL} + \dots + r_{mn_f,x} F_{n_fx}^{ESWL}$$
$$(m = 1, 2, \dots, m_{max}; k = 1, 2, \dots, n_f) \tag{14.5}$$

Since $m_{max} > n_f$, Eq. (14.5) cannot be satisfied exactly and, in certain cases, even approximately. In reality, loads induced by wind with given velocity $U(\theta)$ do not act along direction x only, as was assumed for simplicity in Eqs. (14.1–14.5). Rather, they act simultaneously along the structure's principal axes x and y, and about the vertical torsional axis, ϑ. In addition, during any one storm, the structure is subjected to winds from all directions θ, with each of the velocities $U(\theta)$ inducing three simultaneous loads along the axes x, y, and about the axis ϑ. It is shown in Chapter 18 that if directional wind effects are accounted for, equations analogous to Eq. (14.5) can in practice be satisfied to within an approximation in the order of 10% or less. This is attributed to the fact that, for some wind directions, those equations overestimate, while for other directions they underestimate the wind effects being sought. However, if the extreme wind climate is dominated by winds with direction unfavorable from a structural point of view, for some members the approximation may be in the order of 20% or more.

If a member experiences effects of three simultaneous fluctuating loads, an approximate estimate of the peak of the combined effects induced in the member by those loads can be obtained by the following approach. Three wind loading cases (WLCs) are considered. In the first WLC, denoted by WLC1, the peak effect induced by the first load, called the WLC1 principal load, is added to the effects induced by the second and third loads, called WLC1 companion loads, at the time t_1 of occurrence of that peak. In the second (third) WLC case, denoted by WLC2 (WLC3), the peak effect induced by the second (third) load, called the WLC2 (WLC3) principal load, is added to the effects induced by the first and third (second) loads, called WLC2 (WLC3) companion loads, at the time t_2 (t_3) of occurrence of that peak. Of the three WLCs, only the WLC producing the largest wind effect is retained for design purposes.

By applying this approach to the problem at hand, we have

$$F_{kx}^{ESWL} = \alpha_{M_y,t_1}^{princ} \max_t(F_{kx}(t)) \tag{14.6}$$

where

$$\alpha_{M_y,t_1}^{princ} = \frac{\max_t(M_{b,y}(t))}{M_{b0,y}} = \frac{M_{b,y}(t_1)}{M_{b0,y}} \tag{14.7}$$

t_1 is the time of occurrence of the peak of $M_{b,y}(t)$, and $M_{b0,y}$ is the base moment induced by the loads $\max_t(F_{kx}(t))$ ($k = 1, 2, \dots, n_f$). The superscript "princ" indicates that the reduction factor α_{M_y,t_1}^{princ} applied to the loads $\max_t(F_{kx}(t))$ acting in the x direction corresponds to the peak value of the base moment $M_{b,y}$. We rewrite Eq. (14.6) in the form

$$F_{kx}^{ESWL} = \alpha_{M_y,t_1}^{princ} \left\{ \begin{array}{c} \max_t(F_{n_f x}(t)) \\ \vdots \\ \max_t(F_{2x}(t)) \\ \max_t(F_{1x}(t)) \end{array} \right\} \tag{14.8}$$

For the companion loads we have

$$F_{ky}^{ESWL} = \alpha_{M_x,t_1}^{comp} \left\{ \begin{array}{c} \max_t(F_{n_f y}(t)) \\ \vdots \\ \max_t(F_{2y}(t)) \\ \max_t(F_{1y}(t)) \end{array} \right\} \tag{14.9}$$

where

$$\alpha_{M_x,t_1}^{comp} = \frac{M_{b,x}(t_1)}{M_{b0,x}} \tag{14.10}$$

and

$$F_{k\vartheta}^{ESWL} = \alpha_{M_\vartheta,t_1}^{comp} \left\{ \begin{array}{c} \max_t(F_{n_f \vartheta}(t)) \\ \vdots \\ \max_t(F_{2\vartheta}(t)) \\ \max_t(F_{1\vartheta}(t)) \end{array} \right\} \tag{14.11}$$

where

$$\alpha_{M_\vartheta,t_1}^{comp} = \frac{M_{b,\vartheta}(t_1)}{M_{b0,\vartheta}} \tag{14.12}$$

The procedure just described is based on the "point-in-time" (PIT) estimator of the peak of a sum of random time series. A similar but more reliable estimator was developed in [6] and is based on the "multiple points-in-time" (MPIT) estimator, illustrated in Figure 14.2. The MPIT approach makes use of rank-ordered peaks in each time series of base moments and base torsion. Let the number of largest values of time series $M_{b,x}$ be $n_{pit} = 4$ (see the upper four 'circle' symbols in Figure 14.2a). Denote the times of occurrence of these values by t_j ($j = 1, 2, 3, 4$). The moments $M_{b,x}(t_j)$, called principal components, are combined with the values $M_{b,y}(t_j)$ and $M_{b,\vartheta}(t_j)$ (see 'x' symbols in Figure 14.2b and c), called companion components. The same procedure is used for the lowest (negative) values of $M_{b,x}$. Next, the procedure is used for the $n_{pit} = 4$ peak positive values and the peak negative values of $M_{b,y}$, and finally for $M_{b,\vartheta}$. The total number of WLCs is then $4 \times 2 \times 3 = 24$. It is shown in the case study of Chapter 18 that the accuracy of the estimated DCIs (i.e., the degree to which the DCIs obtained by ESWL are

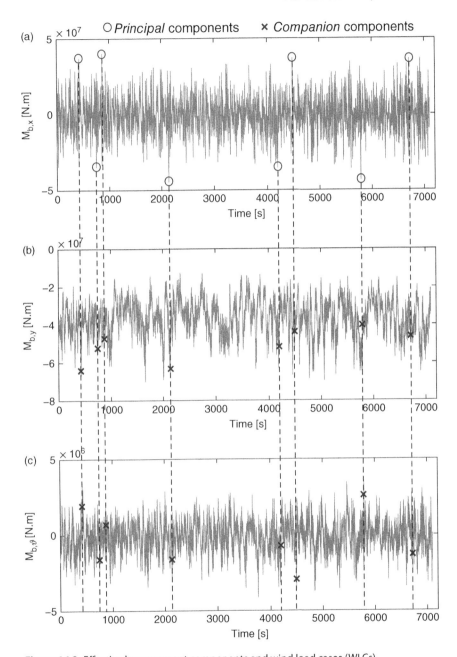

Figure 14.2 Effective base moment components and wind load cases (WLCs).

close to the peak DCIs obtained by DAD) improves as the number n_{pit} of points in time increases. For additional details that further explain the accuracy of ESWLs estimated by the approach presented in this chapter, see Sections 8.3 and 8.4.

DAD and ESWL computations can be performed by using, respectively, the DAD and the ESWL option of the DAD_ESWL version 1.0 software (see Chapter 18).

References

1 Park, S., Yeo, D., and Simiu, E., "Database-assisted design and equivalent static wind loads for mid- and high-rise structures: concepts, software, and users manual," NIST Technical Note 2000, National Institute of Standards and Technology, Gaithersburg, MD, 2018. https://www.nist.gov/wind.

2 Park, S., Simiu, E., and Yeo, D., "Equivalent static wind loads vs. database-assisted design of tall buildings: An assessment," *Engineering Structures*, (submitted). https://www.nist.gov/wind.

3 Boggs, D. and Lepage, A. (2006). *Wind Tunnel Methods*, vol. 240. Special Publication, 125–142. Farmington Hills, MI: American Concrete Institute.

4 Garber, J., Browne, M. T. L., Xie, J., and Kumar, K.S. "Benefits of the pressure integration technique in the design of tall buildings for wind." *12th International Conference on Wind Engineering (ICWE12)*, Cairns. 2007.

5 Huang, G. and Chen, X. (2007). Wind load effects and equivalent static wind loads of tall buildings based on synchronous pressure measurements. *Engineering Structures* 29: 2641–2653.

6 Yeo, D. (2013). Multiple points-in-time estimation of peak wind effects on structures. *Journal of Structural Engineering* 139: 462–471. https://www.nist.gov/wind.

15

Wind-Induced Discomfort in and Around Buildings

15.1   Introduction

It is required that structures subjected to wind loads be sufficiently strong to perform adequately from a structural safety viewpoint. For tall buildings the designer must also take into account wind-related serviceability requirements, meaning that structures should be so designed that their wind-induced motions will not cause unacceptable discomfort to the building occupants.

Wind-induced discomfort is also of concern in the context of the serviceability of outdoor areas within a built environment. Certain building and open space configurations may give rise to relatively intense local wind flows. It is the designer's task to ascertain in the planning stage the possible existence of zones in which such flows would cause unacceptable discomfort to users of the outdoor areas of concern. Appropriate design decisions must be made to eliminate such zones if they exist.

The notion of unacceptable discomfort may be defined as follows. In any given design situation, various degrees of wind-induced discomfort may be expected to occur with certain frequencies that depend upon the features of the design and the wind climate at the location in question. The discomfort is unacceptable if these frequencies are judged to be too high. Statements specifying maximum acceptable frequencies of occurrence for various degrees of discomfort are known as comfort criteria. In practice, reference is made to a suitable parameter, various values of which are associated with various degrees of discomfort. In the case of wind-induced structural motions the relevant parameter is the building acceleration at the top floors. In criteria pertaining to the serviceability of pedestrian areas, the parameter employed is an appropriate measure of the wind speed near the ground at the location of concern. It is therefore necessary to assign maximum probabilities of exceedance to the parameters corresponding to various degrees of discomfort.

Verifying the compliance of a design with requirements set forth in a given set of comfort criteria involves two steps. First, an estimate must be obtained of the wind velocities under the action of which the parameter of concern will exceed the critical values specified by the comfort criteria. Second, the probabilities of exceedance of those velocities must be estimated on the basis of appropriate wind climatological information. The design is regarded as adequate if the probabilities so estimated are lower than the maximum acceptable probabilities specified by the comfort criteria.

The development of comfort criteria for the design of tall buildings is discussed in Section 15.2. Comfort criteria for pedestrian areas are considered in Sections 15.3–15.5.

Wind Effects on Structures: Modern Structural Design for Wind, Fourth Edition. Emil Simiu and DongHun Yeo.
© 2019 John Wiley & Sons Ltd. Published 2019 by John Wiley & Sons Ltd.

15.2 Occupant Wind-Induced Discomfort in Tall Buildings

15.2.1 Human Response to Wind-Induced Vibrations

Studies of human response to mechanical vibrations have been conducted predominantly by the aerospace industry. Because the frequencies of vibration of interest in aerospace applications is relatively high (usually 1–35 Hz), the usefulness of these studies to the structural engineer is generally limited. Nevertheless, results obtained for high frequencies have been extrapolated to frequencies lower than 1 Hz [1], as shown in Table 15.1.

Results of experiments aimed at establishing perception thresholds for periodic motions of 0.067–0.2 Hz suggested that about 50% of the subjects reported perception thresholds of 1–0.6% g, respectively [2]. According to [3], for frequencies of 0.1–0.25 Hz perception thresholds vary between 0.6 and 0.3% g, respectively. It is noted in [4] that creaking noises that occur during building motions tend to increase significantly the feeling of discomfort and should be minimized by proper detailing.

Based on the results reported in [2], [5] proposed a simple criterion that limits the average number of 1% g accelerations at the top occupied floor to at most 12 per year. On the basis of interviews with building occupants it was tentatively suggested in [6] that: "The return periods, for storms causing an rms horizontal acceleration at the building top that exceeds 0.5 % g, shall not be less than 6 years. The rms shall represent an average over the 20-min period of the highest storm intensity and be spatially averaged over the building floor."

The first step in verifying the compliance of a design with requirements set forth in comfort criteria is the estimation, for each wind direction, of the speeds that would induce the acceleration levels of interest. Database-assisted design methods can be used for obtaining plots of wind speed versus accelerations for the wind velocities that induce critical building accelerations. An example of such a plot is shown in Figure 15.1. If tuned mass dampers are used to reduce building motions, the accelerations can be estimated by methods mentioned in Chapter 16. The second step is the estimation of the frequency of occurrence of accelerations higher than the critical value specified in the comfort criteria. The frequency may be defined as the mean number of days per year during which the maximum wind speeds exceed the values corresponding to the plot of Figure 15.1. This information can be obtained from wind speed data typically available

Table 15.1 Proposed correspondence between degrees of user discomfort and the accelerations causing them.

Degree of discomfort perceived	Accelerations (as percentages of discomfort from the acceleration of gravity, g)
Imperceptible	$<\frac{1}{2}\% \, g$
Perceptible	$\frac{1}{2} - 1\frac{1}{2}\% \, g$
Annoying	$1\frac{1}{2} - 5\% \, g$
Very Annoying	$5 - 15\% \, g$
Intolerable	$>15\% \, g$

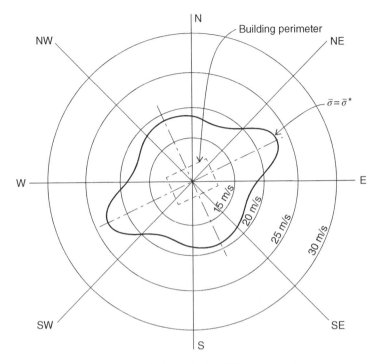

Figure 15.1 Wind speeds inducing critical building accelerations.

in the United States in the public domain (see Section 3.2.3). For details on research concerning human discomfort due to building motions see, for example, [7].

15.3 Comfort Criteria for Pedestrian Areas Within a Built Environment

The problem of wind-induced discomfort in pedestrian areas is not new (Figure 15.2). For the sake of its historical interest, we reproduce in Figure 15.3 a note by the great naturalist Buffon describing the flow changes occurring upwind of a tower, for which it offers a charming but no longer tenable explanation. A translation of the note follows.

On reflected wind
I must report here an observation which it seems to me has escaped the attention of physicists, even though everyone is in a position to verify it. It seems that reflected wind is stronger than direct, and the more so as one is closer the obstacle that reflects it. I have experienced this a number of times by approaching a tower that is about 100 feet high and is situated on the north end of my garden in Montbard. When a strong wind blows from the south, up to thirty steps from the tower one feels strongly pushed, after which there is an interval of five of six steps where one ceases to be pushed and where the wind, which is reflected by the tower, is so to speak in equilibrium with the direct wind. After this, the closer one

Figure 15.2 The Gust. Lithograph by Marlet. Source: Photo Bibliothèque Nationale de France.

approaches the tower, the more the wind reflected by it is violent. It pushes you back much more strongly than the direct wind pushed you forward. The cause of this effect, which is a general one, and can be experienced against all large buildings, against sheer cliffs, and so forth, is not difficult to find. The air in the direct wind acts only with its ordinary speed and mass; in the reflected wind the speed is slightly lower, but the mass is considerably increased by the compression that the air suffers against the obstacle that reflects it, and as the momentum of any motion is composed of the speed multiplied by the mass, the momentum is considerably larger after the compression than before. It is a mass of ordinary air that pushes you in the first case, and it is a mass of air that is once or twice as dense that pushes you back in the second case.

15.3.1 Wind Speeds, Pedestrian Discomfort, and Comfort Criteria

Observations of wind speeds on people and calculations involving the rate of working against the wind suggest that the following degrees of discomfort are induced by wind speeds V at 2 m above ground, averaged over 10 min–1 h: $V = 5 \, \text{m s}^{-1}$: onset of discomfort; $V = 10 \, \text{m s}^{-1}$: definitely unpleasant; $V = 20 \, \text{m s}^{-1}$: dangerous [8]. According to [8], if mean speeds V occur less than 10% of the time, complaints about wind conditions are unlikely to arise. If such speeds occur between 10 and 20% of the time, complaints might arise. For frequencies in excess than 20%, remedial measures are necessary. An alternative set of comfort criteria proposed in [9] is shown in Table 15.2.

à l'Histoire Naturelle.   15

ADDITIONS

A l'Article qui a pour titre : Des
Vents réglés, *page 224.*

I.

Sur le Vent réfléchi, page 242.

Je dois rapporter ici une observation
qui me paroît avoir échappé à l'attention
des Phyficiens , quoique tout le monde
foit en état de la vérifier ; c'eft que le
vent réfléchi eft plus violent que le vent
direct , & d'autant plus qu'on eft plus
près de l'obftacle qui le renvoie. J'en
ai fait nombre de fois l'expérience , en
approchant d'une tour qui a près de
cent pieds de hauteur & qui fe trouve
fituée au nord , à l'extrémité de mon
jardin , à Montbard ; lorfqu'il fouffle
un grand vent du midi , on fe fent for-
tement pouffé jufqu'à trente pas de la
tour ; après quoi, il y a un intervalle de
cinq ou fix pas , où l'on ceffe d'être

16   *Supplément*

pouffé & où le vent , qui eft réfléchi par
la tour , fait , pour ainfi dire , équilibre
avec le vent direct ; après cela , plus on
approche de la tour & plus le vent qui
en eft réfléchi eft violent , il vous repouffe
en arrière avec beaucoup plus de force
que le vent direct ne vous pouffoit en
avant. La caufe de cet effet qui eft gé-
néral , & dont on peut faire l'épreuve
contre tous les grands bâtimens , contre
les collines coupées à plomb , &c. n'eft
pas difficile à trouver. L'air dans le vent
direct n'agit que par fa vîteffe & fa
maffe ordinaire ; dans le vent réfléchi ,
la vîteffe eft un peu diminuée , mais la
maffe eft confidérablement augmentée
par la compreffion que l'air fouffre
contre l'obftacle qui le réfléchit ; &
comme la quantité de tout mouvement
eft compofée de la vîteffe multipliée
par la maffe , cette quantité eft bien plus
grande après la compreffion qu'aupara-
vant. C'eft une maffe d'air ordinaire , qui
vous pouffe dans le premier cas , & c'eft
une maffe d'air une ou deux fois plus
denfe , qui vous repouffe dans le fecond
cas.

Facsimile of note on reflected wind. From *Histoire Naturelle, Générale et Particulière, Contenant les Epoques de la Nature*, Par M. le comte de Buffon, Vol. 13, Intendant du Jardin et du Cabinet du Roi, de l'Académie Française, de celle des Sciences, etc., Tome Treizième, A Paris, De L'Imprimerie Royale, 1778.

Figure 15.3 Facsimile of note on reflected wind. Source: From *Histoire Naturelle, Générale et Particulière, Contenant les Epoques de la Nature, Par M. le Comte de Buffon, Intendant du Jardin et du Cabinet du Roi, de l'Académie Française, de celle des Sciences,* etc., *Tome Treizième. A Paris, De l'Imprimerie Royale, 1778.*

A more elaborate view of pedestrian comfort, that accounts for local climate characteristics other than wind speeds, including thermal characteristics, is discussed in [10].

15.4   Zones of High Surface Winds Within a Built Environment

15.4.1   Wind Effects Near Tall Buildings

As noted in [8], high wind speeds occurring at pedestrian level around tall buildings are
in general associated with the following types of flow:

1) Vortex flows that develop near the ground (Figures 15.4 and 15.5).
2) Corner streams (Figure 15.4).
3) Air flows through ground floor openings connecting the windward to the leeward
   side of a building (Figure 15.4), or cross-flows from the windward side of one building
   to the leeward side of a neighboring building (Figure 15.6).

Table 15.2 Comfort criteria for various pedestrian areas.

Criterion	Area Description	Limiting Wind Speed	Frequency of Occurrence
1	Plazas and Parks	Gusts to about $6 \, \mathrm{m \, s^{-1}}$	10% of the time (about $1000 \, \mathrm{h \, yr^{-1}}$)
2	Walkways and other	Gusts to about $12 \, \mathrm{m \, s^{-1}}$	1 or 2 times a month (about $50 \, \mathrm{h \, yr^{-1}}$) areas subject to pedestrian access
3	All of the above	Gusts to about $20 \, \mathrm{m \, s^{-1}}$	About $5 \, \mathrm{h \, yr^{-1}}$
4	All of the above	Gusts to about $25 \, \mathrm{m \, s^{-1}}$	Less than $1 \, \mathrm{h \, yr^{-1}}$

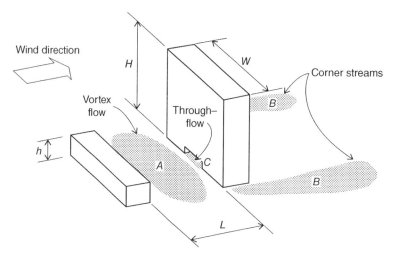

Figure 15.4 Regions of high surface winds around a tall building. Source: By permission of the Director, Building Research Establishment, U.K. Copyright, Controller of Her Britannic Majesty's Stationery Office (HMSO).

The flow visualization of Figure 15.5 was obtained by injecting smoke into the airstream. The flow patterns in the immediate vicinity of the windward face are consistent with the fact that pressures are highest at roughly two-thirds of the height of the taller building; that is, the air flows from zones of higher to zones of lower pressure. Part of the air deflected downward by the building forms a vortex and thus sweeps the ground in a reverse flow (area *A*, marked "vortex flow" in Figure 15.4). Another part is accelerated around the building corners and forms jets that sweep the ground near the building sides (areas *B*, marked "corner streams" in Figure 15.4). If an opening connecting the windward to the leeward side is present at or near the ground level, part of the descending air will be sucked from the zone of relatively low pressures (suctions) on the leeward side. A through-flow will thus sweep the area *C* in Figure 15.4. Through-flows of this type have caused serious discomfort to users of the MIT Earth Sciences Building in Cambridge, Massachusetts, a structure about 20 stories in height. Cross-flows between pairs of buildings are caused by similar pressure differences, as shown in Figure 15.6.

Figure 15.5 Wind flow in front of a tall building (wind blowing from left to right). Source: By permission of the Director, Building Research Establishment, U.K. Copyright, Controller of HMSO.

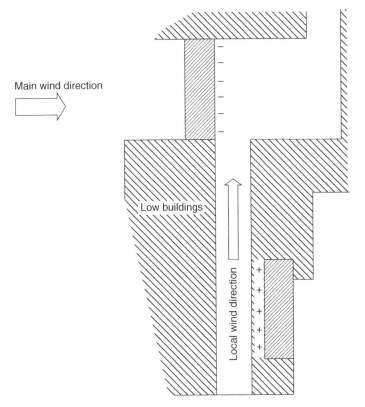

Figure 15.6 Cross-flow between two tall buildings. Source: By permission of the Director, Building Research Establishment, U.K. Copyright, Controller of HMSO.

15.4.2 Wind Speeds at Pedestrian Level in a Basic Reference Case [8]

The pattern of the surface wind flow within a site depends in a complex way upon the relative location, the dimensions, the shapes, and certain of the architectural features of the building of interest, upon the roughness and the topographical features of the terrain around the site, and upon the possible presence near the site of one or several tall buildings. To study the surface flow under conditions significantly different from those depicted schematically in Figures 15.4 and 15.6 it may be necessary to conduct wind tunnel tests or perform Computational Wind Engineering simulations (e.g., [11, 12]), either of which can provide useful, if approximate, information.

However, for suburban built environments that retain a basic similarity with the configurations of Figures 15.4–15.6, and in which the height of the buildings does not exceed 100 m or so, information based on aerodynamic studies reported in [8] is useful for the prediction of surface winds in a wide range of practical situations. The surface winds depend upon the dimensions H, W, L, and h defined in Figure 15.4 and are expressed in terms of ratios V/V_H, where V and V_H are mean speeds at pedestrian level and at elevation H, respectively. In certain applications, it is useful to estimate the ratio V/V_0, where V_0 is the mean speed at 10 m above ground in open terrain. The ratios V/V_0 can be obtained as follows:

$$\frac{V}{V_0} = \frac{V}{V_H}\frac{V_H}{V_0} \tag{15.1}$$

Approximate ratios V_H/V_0 corresponding to suburban built environments suggested in [8] are listed in in Table 15.3.

In the material that follows, the wind direction is assumed to be normal to the building face unless otherwise stated.

Speeds in Vortex Flow. V_A and V_H denote the maximum mean wind speed at pedestrian level in zone A and the mean speed at elevation H, respectively (Figure 15.4). Approximate ratios V_A/V_H are given in Figure 15.7 as functions of W/H for various rations L/H and for the ranges of values H/h shown. The height h corresponded in the model tests to typical heights of suburban buildings (7–16 m). It is noted that as the building becomes slenderer (as the ratio W/H decreases) the ratio V_A/V_H decreases.

Typical examples of the variation of V_A with individual variables are shown in Figure 15.8. If the distance L between the low-rise and high-rise building is small, the vortex cannot penetrate effectively between the buildings and V_A is small. If L is very large or if h is very small, the vortex that forms upwind of the tall building will be poorly organized and weak; V_A will therefore be relatively low. If h approaches the value of H, the taller building will in effect be sheltered and the speed V_A will thus be low.

It is noted that the ratio V_A/V_H is in the order of 0.5 for a range of practical situations. *Speeds in Corner Streams.* Figure 15.9 shows the approximate dependence of the ratio V_B/V_H upon H/h, where V_B is the largest mean speed at pedestrian level in the zones swept by the corner stream, and V_H is the mean speed at elevation H. Examples of the

Table 15.3 Approximate ratios V_H/V_0.

H (m)	20	30	40	50	60	70	80	90	100
V_H/V_0	0.73	0.82	0.89	0.94	0.99	1.04	1.08	1.11	1.14

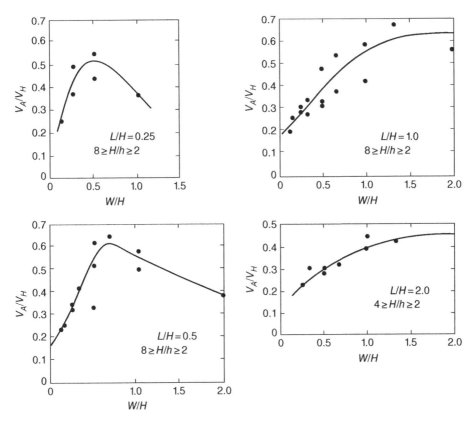

Figure 15.7 Ratios V_A/V_H [8]. Source: By permission of the Director, Building Research Establishment, U.K. Copyright, Controller of HMSO.

variation of V_B with the variables H, L, W, h, and θ are given in Figure 15.10. The speed V_B varies weakly with the angle θ between the mean wind direction and the normal to the building face. However, the orientation of the corner streams and, hence, the position of the point of maximum speed V_B may depend significantly upon the direction of the mean wind.

Information about the wind speed field around the corner of a wide building model ($H = 0.4$ m, $W = 0.4$ m) is given in Figure 15.11. The wind speed decreases rather slowly within a distance from the building corner approximately equal to H. The ratio $Y/(D/2)$, where Y is defined as in Figure 15.11 and D is the building depth, provides an approximate measure of the position of the corner stream. Measured values of this ratio for various values of H and of $W/(D/2)$ are shown in Figure 15.12.

It is noted that the ratio V_B/V_H is approximately 0.95 for a range of practical situations. *Speeds in a Through-Flow.* Let V_C denote the maximum mean wind speed through a ground floor passageway connecting the windward and the leeward side of a building (Figure 15.4). Figure 15.13 shows the approximate dependence of the ratio V_C/V_H upon the parameter H/h. Examples of the variation of V_C with H, W, L, h, and θ are given in Figure 15.14. The data of Figures 15.13 and 15.14 are based on tests in which the entrances to the passageways were sharp-edged. If the edges of the entrance are rounded

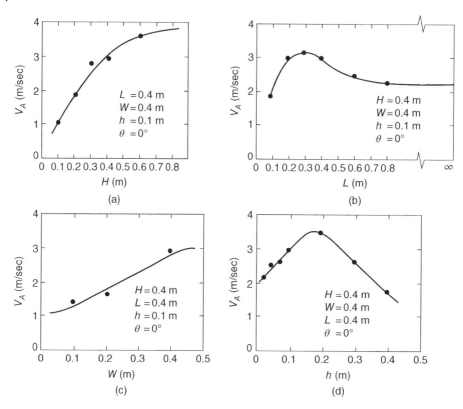

Figure 15.8 Examples of the variation of V_A with individual parameters [8]. Source: By permission of the Director, Building Research Establishment, U.K. Copyright, Controller of HMSO.

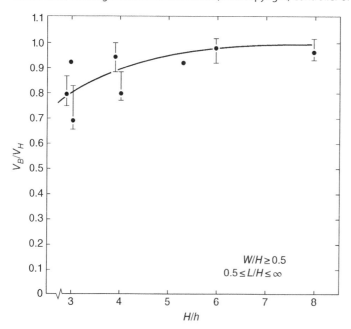

Figure 15.9 Ratios V_B/V_H [8]. Source: By permission of the Director, Building Research Establishment, U.K. Copyright, Controller of HMSO.

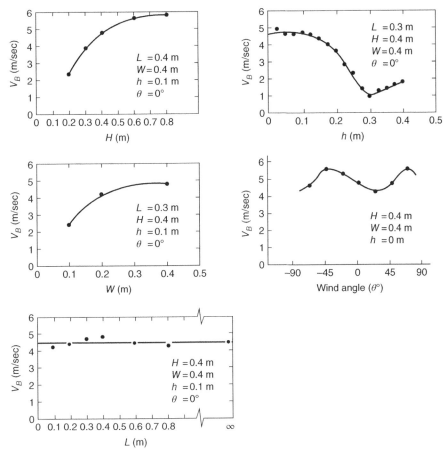

Figure 15.10 Examples of the variation of V_B with individual parameters [8]. Source: By permission of the Director, Building Research Establishment, U.K. Copyright, Controller of HMSO.

to form a bellmouth shape, the speeds V_C are reduced with respect to those data by as much as 25%. It is noted that the ratio V_C/V_H is approximately 1.2 for a range of practical situations.

As noted earlier, the approximate validity of the information provided in Figures 15.7–15.14 is limited to buildings with regular shape in plan and heights of 100 m or less.

15.4.3 Case Studies

Case 1: Model of a Building in Utrecht, The Netherlands [13]. The model of a building with height $H = 80$ m, width $W = 80$ m, depth $D = 22$ m, $H/h = 8.0$, and $L/H = 0.5$, is shown in plan in Figure 15.15. Contours of ratios V/V_H, shown in Figure 15.15 for south and north winds, were obtained in wind tunnel tests [13]. Ratios V_A/V_H and V_B/V_H are about 0.65 at the centerline of the building and 0.90, respectively, versus the values 0.60 and 1.00 from Figures 15.7 and 15.9 – a reasonably good agreement. Note that the vortex flow is asymmetrical and contains regions in which the ratios V/V_H are as high as 0.8.

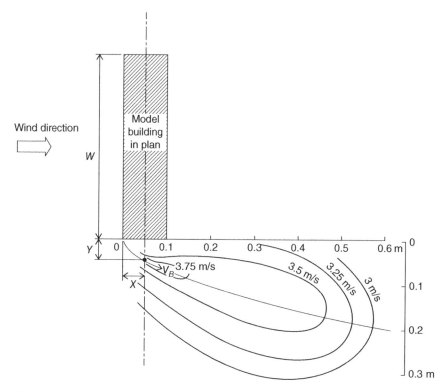

Figure 15.11 Surface wind speed field in a corner stream [8]. Source: By permission of the Director, Building Research Establishment, U.K. Copyright, Controller of HMSO.

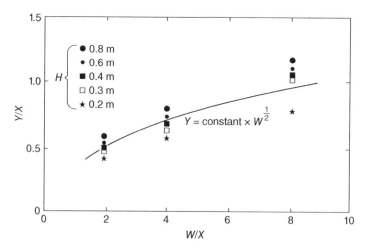

Figure 15.12 Empirical curve Y/X versus W/X [8]. Source: By permission of the Director, Building Research Establishment, U.K. Copyright, Controller of HMSO.

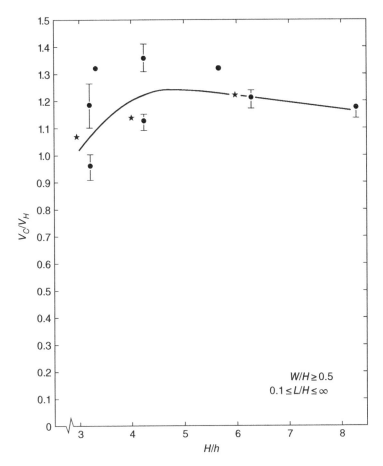

Figure 15.13 Ratios V_C/V_H [8]. Source: By permission of the Director, Building Research Establishment, U.K. Copyright, Controller of HMSO.

Case 2: Model of Place Desjardins, Montreal [14]. Figure 15.16 shows a 1 : 400 model of a design considered for a development in Place Desjardins, Montreal. Tests were conducted only for the predominant wind direction shown in the Figure 15.17. Surface flow patterns were observed by using thread tufts taped to the model surfaces, a wool tuft on the end of a hand-held rod, and a liquid mixture of kerosene-chalk (china clay) sprayed over the horizontal surfaces of the model. As the wind blows over the model, the mixture is swept away from the high-speed zones and accumulates in zones of stagnating flow. After the evaporation of the kerosene, the white accumulations of chalk indicate zones of low speeds while areas that are dark indicate zones surface winds are high. Wind speed measurements were made in the latter zones.

The numbers in Figure 15.17 are ratios of mean wind speeds at the locations shown to the mean speed $V_{(1)}$ at 1.8 m above ground at the northwest corner of the development. The percentages in the figure represent turbulent intensities, and the arrows show the direction of the wind component that was measured by the probe. The quantities not between

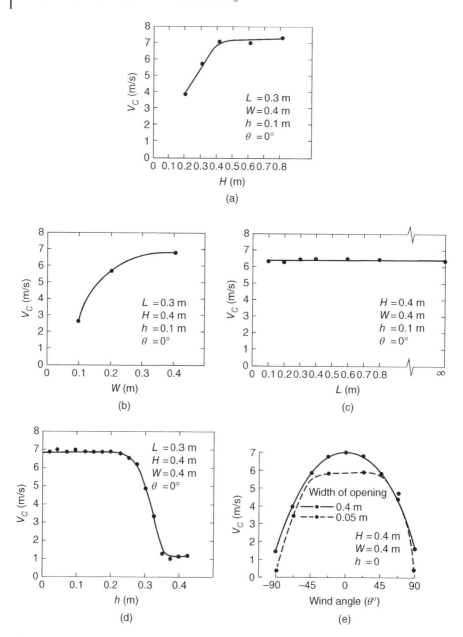

Figure 15.14 Examples of the variation of V_C with individual parameters [8]. Source: By permission of the Director, Building Research Establishment, U.K. Copyright, Controller of HMSO.

parentheses correspond to measurements made in the absence of a projected 50-story tower near the southwest corner of the development. Results of measurements made with the tower in place are shown in parentheses. The presence of the tower changed the ratios of the wind speeds at locations 8 and 10 to the wind speed $V_{(1)}$ from 3.11 to 2.96 to approximately 3.38 and 2.48, respectively.

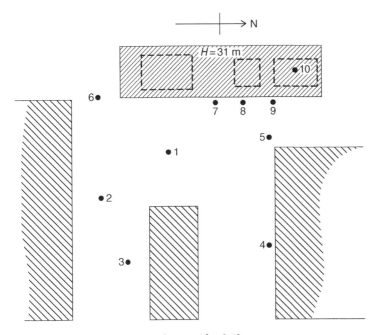

Figure 15.15 Plan view, Case 1. Source: After [13].

The results just listed correspond to the case of the uncovered mall. In the absence of the tower, covering the mall reduced the mean wind speeds by a factor of almost three at location 8; however, there was no reduction at location 10; with the tower in place, it reduced the mean wind speeds by a factor of five at location 8 and a factor of 1.67 at location 10.

Case 3: Commerce Court Plaza, Toronto [15, 16]. A wind tunnel model and a plan view of the Commerce Court project in Toronto are shown in Figures 15.18 and 15.19. Surface flow patterns obtained by smoke visualization are shown for two wind directions in Figures 15.20 and 15.21 [15]. Ratios V/V_H, where V and V_H are mean wind speeds are 2.7 and 240 m above ground, were obtained from measurements in the wind tunnel and, after the completion of the structures, on the actual site. The results of the measurements are shown in Figure 15.22 as functions of wind direction for locations 1 through 7. The agreement between wind tunnel and full-scale values is generally acceptable, although differences of 30%, 50%, and larger can be noted in certain cases.

After the completion of the Commerce Court Plaza, conditions were found to be particularly annoying on windy days for pedestrians walking from the relatively protected zone north of the 32-story tower into the flow funneled through the passageway 2–3 (see Figure 15.19). Wind tunnel tests indicated that the provision of screens at the ground level as shown in Figure 15.23a would result at locations 2, 5, and 6 in reductions of undesirable mean speeds in the order of 40%. However, the placement of screens was rejected for architectural reasons. Instead, potted evergreens about 3 m high were placed as shown in Figure 15.23b. This reduced the mean speeds by about 20% at locations 2, 10% at location 5, and 33% at location 6.

Case 4: Shopping Center, Croydon, England [8]. An office building 75 m tall, 70 m wide, and 18 m deep adjoins a shopping center 75 m long. A passageway 12 m wide and 3.7 m

Figure 15.16 Place Desjardins model. Source: Courtesy of the National Aeronautical Establishment, National Research Council of Canada.

high underneath the building connects the shopping center on the west side of the building to the street on the east side (Figure 15.24). The complex was designed and built without a roof over the shopping mall. After the completion of the building complex, it became apparent that remedial measures were necessary to reduce wind speeds in the passageway and the shopping mall. The ground level wind flow was investigated in the wind tunnel, first for the complex as initially built (with the mall not covered), and then with various arrangements of roofs over the mall and screens within the passageway. Ratios V/V_H measured in the wind tunnel (V and V_H are the mean speeds at 1.8 and 75 m above ground, respectively) are shown in Figure 15.24 for three cases. For the

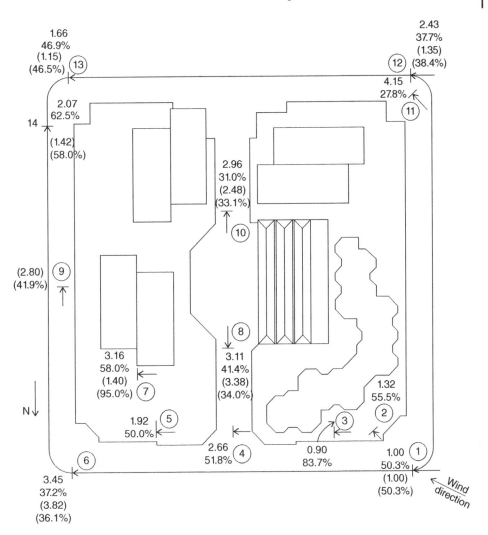

Figure 15.17 Wind speed ratios and turbulence intensities, place Desjardins. Source: Courtesy of the National Aeronautical Establishment, National Research Council of Canada.

complex as first built, the highest values of the ratio V/V_H were 0.68 in the vortex flow zone and 1.01 in the through-flow zone. The provision of a full roof over the mall but of no screens within the passageway reduced considerably pedestrian level speeds caused by west winds. However, with east winds, the flow was trapped under the roof and the wind speeds within the mall were, for this reason, high; as shown in Figure 15.24, the speeds were also high at the east entrance of the passageway. A solid roof close to the tall building followed by a partial roof over the rest of the mall, and a screen obstructing 75% of the passageway area resulted in a significant reduction of surface winds.

The solution actually applied, which proved effective, was to provide (i) a full roof over the entire mall and (ii) screens with 75% blockage in the passageway.

Figure 15.18 Commerce Court model. Source: Reprinted from [15], with permission from Elsevier.

15.5 Frequencies of Ocurrence of Unpleasant Winds

15.5.1 Detailed Estimation Procedure

Let $V_0(V, \theta)$ denote the wind speeds at 10 m above ground in open terrain that induce pedestrian winds V blowing from direction θ at a given location in a built environment. The frequency of occurrence of wind speeds larger than V, denoted by f^V is, approximately,

$$f^V = \sum_{i=1}^{n} f_i^{V_0} \tag{15.2}$$

where $f_i^{V_0}$ are the frequencies of occurrence in open terrain of winds with speeds larger than $V_0(V, \theta_i)$ and the directions $\theta_i - \frac{\pi}{n} < \theta < \theta_i + \frac{\pi}{n}$, where $\theta_i = 2\pi i/n$ ($i = 1, 2,..., n$). In practical applications a 16-point compass is sometimes used, so that $n = 16$.

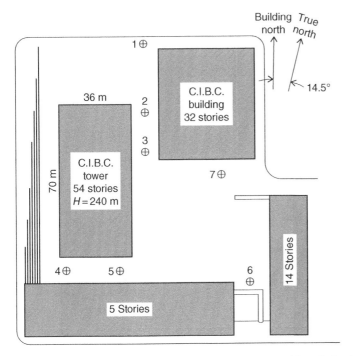

Figure 15.19 Plan view, Commerce Court. Source: Reprinted from [15], with permission from Elsevier.

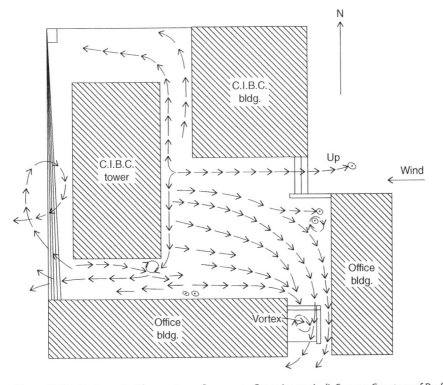

Figure 15.20 Surface wind flow pattern, Commerce Court (east wind). Source: Courtesy of Professor A. G. Davenport.

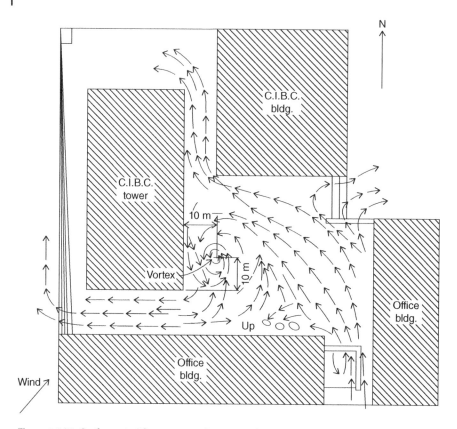

Figure 15.21 Surface wind flow pattern, Commerce Court (southwest wind). Source: Courtesy of Professor A. G. Davenport.

To obtain f^V it is necessary, first, to estimate the values of $V_0(V, \theta_i)$. From wind climatological data it is then possible to estimate the frequencies $f_i^{V_0}$. The speed $V_0(V, \theta_i)$ can be written as

$$V_0(V, \theta_i) = \frac{1}{V/V_H(\theta_i)} \frac{V_0(\theta_i)}{V_H(\theta_i)} V \tag{15.3}$$

The ratios $V_0(\theta_i)/V_H(\theta_i)$ characterize the site micrometeorologically. For standard roughness conditions in open terrain, these ratios depend upon the elevation H and upon the roughness conditions upwind of the site. The ratios $V/V_H(\theta_i)$ are obtained from wind tunnel tests.

Consider, for example, all three-hour interval observations in a year (8 obs day^{-1} × 365 days = 2920 obs), and assume that 58 of these observations represent northnorthwesterly (NNW) winds with speeds in excess of 6 m s^{-1}. The frequency of occurrence of such wind can then be estimated as $f_1^6 = 58/2920 = 2\%$. It is desirable to base frequency estimates on several years of data. In some applications it may be of interest to estimate frequencies for individual seasons or for a grouping of seasons. Also, data for times not relevant from a pedestrian comfort viewpoint (e.g., between 11 p.m.–5 a.m.) may in some cases be eliminated from the data set.

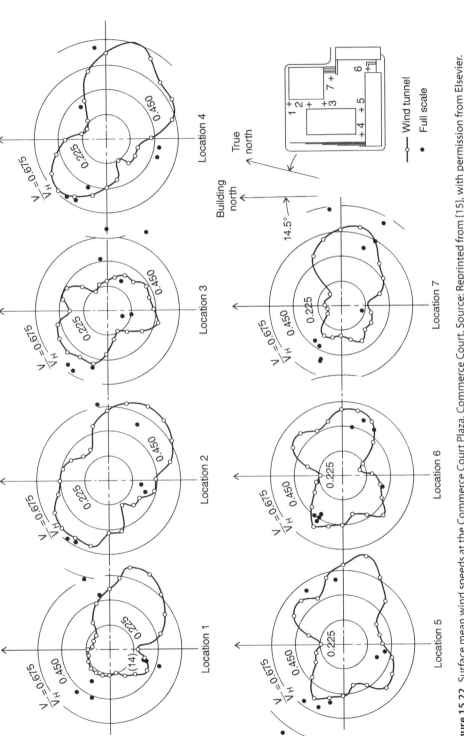

Figure 15.22 Surface mean wind speeds at the Commerce Court Plaza, Commerce Court. Source: Reprinted from [15], with permission from Elsevier.

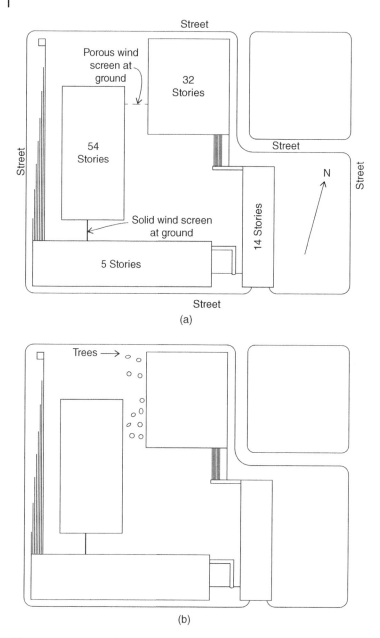

Figure 15.23 Remedial measures at Commerce Court: (a) screens; (b) trees. Source: After [17].

15.5.2 Simplified Estimation Procedure

A simplified version of the procedure just presented is suggested in [8] for built environments similar in configuration to the basic reference case (Figure 15.4). In this version, the aerodynamic information used, rather than being a function of wind direction, is limited to the results given, for example, in Figures 15.7, 15.9 and 15.13. The ratios

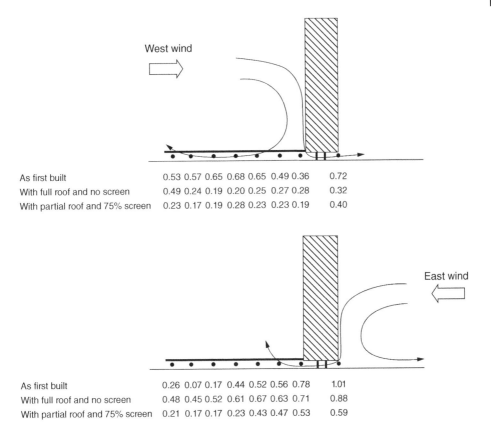

Figure 15.24 Model test results, Croydon [8]. Source: By permission of the Director, Building Research Establishment, U.K. Copyright, Controller of HMSO.

V_H/V_0 of mean wind at elevation H in the in the built environment to mean wind at 10 m above ground in open terrain may be taken from Table 15.2 or determined as shown in Section 2.3.6. The requisite climatological information consists of the of the frequencies of occurrence of all winds with speeds in excess of various values V_0, regardless of their direction. According to [8], this simplified procedure provides generally reliable indications on the serviceability of pedestrian areas in a built environment of the type represented in Figure 15.4, provided that it is used in conjunction with the comfort criteria proposed in [8].

To illustrate the procedure proposed in [8], consider the case of a building complex for which $H = 70$ m, $W = 50$ m, $L = 35$ m, and $h = 10$ m. (For these notations see Figure 15.4.). From Figures 15.7 and 15.9, $V_A/V_H \approx 0.6$ and $V_B/V_H \approx 0.95$, where V_A and V_B are the highest mean speeds in the vortex and corner flow, respectively. For $H = 70$ m, $V_H/V_0 \approx 1.04$ (Table 15.2), so $V_A/V_0 \approx 0.63$ and $V_B/V_0 \approx 1.00$. Given the requisite wind speed data it is possible to estimate the frequencies of winds $V_A > 5$ and $V_B > 5$ m s^{-1}. In order for $V_A > 5$, $V_0 > 5/0.63 = 8$ m s^{-1}. For $V_B > 5$ m s^{-1}, $V_0 > 5/1.00 = 5.00$ m s^{-1}. The frequency of 5 m s^{-1} winds depends upon the local wind climate. If that frequency exceeds 20%, according to the comfort criterion of [8] (see Section 15.2.1), the wind conditions at the site are unacceptable.

References

1 Chang, F.-K. (1973). Human response to motions in tall buildings. *Journal of the Structural Division* 99: 1259–1272.

2 Chen, P.W. and Robertson, L.E. (1972). Human perception thresholds of horizontal motion. *Journal of the Structural Division* 92: 1681–1695.

3 Goto, T., "Human Perception and Tolerance of Motion," *Monograph of Council on Tall Buildings and Urban Habitat*, PC, 1981, pp. 817–849.

4 Reed, J. W., "Wind-induced motion and human discomfort in tall buildings," Research Report No. R71–42, Department of Civil Engineering, MIT, Cambridge, 1971.

5 Feld, L. (1971). Superstructure for 1350 ft. World Trade Center. *Civil Engineering ASCE* 41: 66–70.

6 Hansen, R.J., Reed, J.W., and Vanmarcke, E.H. (1973). Human response to wind-induced motion of buildings. *Journal of the Structural Division* 99: 1589–1605.

7 Lamb, S. and Kwok, K.C.S. (2017). The fundamental human response to wind-induced building motion. *Journal of Wind Engineering and Industrial Aerodynamics* 165: 79–85.

8 Penwarden, A. D. and Wise, A. F. E., "Wind environment around buildings," Building Research Establishment Report, Department of the Environment, Building Research Establishment: Her Majesty's Stationery Office, London, 1975.

9 Apperley, L. W. and Vickery, B. J., "The prediction and evaluation of the ground level environment," in the *Fifth Australian Conference on Hydraulics and Fluid Mechanics*, University of Canterbury, Christchurch, New Zealand, 1974.

10 Wu, H. and Kriksic, F. (2012). Designing for pedestrian comfort in response to local climate. *Journal of Wind Engineering and Industrial Aerodynamics* 104-106: 397–407.

11 Mochida, A. and Lun, I.Y.F. (2008). Prediction of wind environment and thermal comfort at pedestrian level in urban area. *Journal of Wind Engineering and Industrial Aerodynamics* 96: 1498–1527.

12 Llaguno-Munitxa, M., Bou-Zeid, E., and Hultmark, M. (2017). The influence of building geometry on street canyon air flow: validation of large eddy simulations against wind tunnel experiments. *Journal of Wind Engineering and Industrial Aerodynamics* 165: 115–130.

13 Poestkoke, R., "Windtunnelmetingen aan een model van het Transitorium II van de Rijksuniversiteit Utrecht," Report No. TR72110L, National Aerospace Laboratory NLR, The Netherlands, 1972.

14 Standen, N. M., "A wind tunnel study of wind conditions on scale models of place Desjardins, Montreal," Laboratory Technical Report No. LTR-LA-101, National Research Council of Canada, National Aeronautical Establishment, Ottawa, 1972.

15 Isyumov, N. and Davenport, A.G. (1975). Comparison of full-scale and wind tunnel wind speed measurements in the Commerce Court Plaza. *Journal of Wind Engineering and Industrial Aerodynamics* 1: 201–212.

16 Davenport, A. G., Bowen, C. F. P., and Isyumov, N., "A study of wind effects on the Commerce Court project, Part II, wind environment at pedestrian level," Engineering Science Research Report No. BLWT-3-70, University of Western Ontario, Faculty of Engineering Science, London, Canada, 1970.

17 Isyumov, N. and Davenport, A.G., "The ground level wind environment in built-up areas," in *Proceedings of the Fourth International Conference on Wind Effects on Buildings and Structures*. London, 1975. Cambridge University Press, Cambridge, 1977, pp. 403–422.

16

Mitigation of Building Motions

Tuned Mass Dampers

16.1   Introduction

Tuned mass dampers (TMDs) are the most commonly used devices for reducing tall structure accelerations and inter-story drift due to translation and torsion. Generally, TMD effects are not taken into account in strength calculations.

The TMD was invented in 1909 by Frahm and was originally used in mechanical engineering systems. Since the 1970s TMDs have been used to mitigate building motions. Examples of buildings in which TMDs were used include the Citicorp Center, New York City, the John Hancock tower, Boston (equipped with dual TMDs designed to control both drift and torsional motions), and the Taipei 101 tower. For additional examples, see [1]. Basic TMD theory was developed in [2].

TMDs consist of one or more masses in the order of 2% of the total mass of the structure, added to and interacting dynamically with the structure through springs and damping devices. The structure's motion is reduced by the forces of inertia due to the motion of the TMDs. A schematic view of a TMD operating on the top floor of the Citicorp Center building is shown in Figure 16.1. The mass of the TMD consists in this case of a 400-ton concrete block bearing on a thin oil film. The TMD structural stiffness is provided by pneumatic springs that can be tuned to the actual frequency of vibration of the building as determined experimentally in the field. The damping is provided by hydraulic shock absorbers. The system included fail-safe devices to prevent excessive travel of the concrete block [3]. Descriptions and theory applicable to buildings are presented in [1] for various types of TMD, including translation and pendular TMDs placed at or near the top of the building, TMD pairs placed at opposite sides of the top building floor, designed to reduce torsional motions, and TMDs installed at several elevations, tuned to reduce motions in more than one mode of vibration. Dampers that produce forces of inertia due to fluid motion have also been used. Early contributions to the design of TMDs for building motion control were made in [3] and [4]. For recent developments on multi-degree-of-freedom system TMDs under random excitation see [5], which provides comprehensive references.

Wind Effects on Structures: Modern Structural Design for Wind, Fourth Edition. Emil Simiu and DongHun Yeo.
© 2019 John Wiley & Sons Ltd. Published 2019 by John Wiley & Sons Ltd.

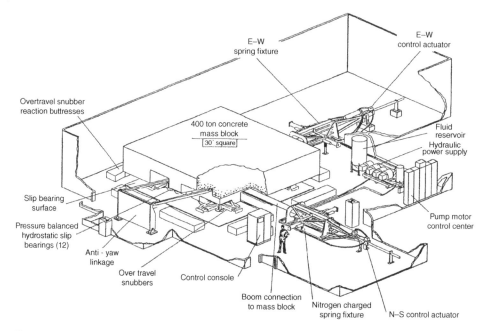

Figure 16.1 Tuned mass damper system, Citicorp Center, New York City. Source: Courtesy of MTS Systems Corp., Minneapolis.

16.2 Single-Degree-of-Freedom Systems

Figure 16.2 is a TMD schematic, in which m, c, k, and m_d, c_d, k_d are the mass, viscous damping coefficient and spring constant of the structure idealized as a single-degree-of-freedom system (SDOF) and of the TMD, respectively.

Assume that the forcing function $p(t)$ in Figure 16.2 is harmonic. The equations of motion of the system are

$$m\,\ddot{x} + c\,\dot{x} + k\,x = p\sin\Omega t + c_d\,\dot{x}_d + k_d\,x_d \tag{16.1}$$

$$m_d(\ddot{x} + \ddot{x}_d) + c_d\dot{x}_d + k_d x_d = 0 \tag{16.2}$$

where x is the displacement of the SDOF system and x_d is the displacement of the TMD with respect to the SDOF system. The solutions of Eqs. (16.1) and (16.2) are harmonic and have amplitudes

$$X = \frac{p}{k}H, \qquad x = \frac{p}{k_d}H_d \tag{16.3a,b}$$

where the dynamic amplification factors (also known as mechanical admittance functions) of the structure and of the TMD are denoted by H and H_d:

$$H = \frac{\sqrt{(\beta_d^2 - \beta^2)^2 + (2\zeta_d\beta\beta_d)^2}}{|D|}, \qquad H_d = \frac{\beta^2}{|D|} \tag{16.4a,b}$$

$$|D| = \{[-\beta_d^2\beta^2\gamma + (1 - \beta^2)(\beta_d^2 - \beta^2) - 4\zeta\zeta_d\beta_d\beta^2]^2$$
$$+ 4[\zeta\beta(\beta_d^2 - \beta^2) + \zeta_d\beta_d\beta(1 - \beta^2(1 + \gamma))]^2\}^{1/2} \tag{16.5}$$

Figure 16.2 Schematic of a damped system equipped with a damped tuned mass damper.

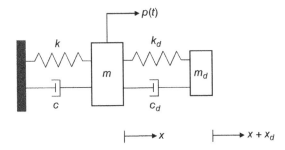

(see, e.g., [2]). In Eqs. (16.4a,b) and (16.5) the following non-dimensional parameters are used:

$$\beta = \frac{\Omega}{\omega} = \frac{\Omega}{\sqrt{k/m}}, \qquad \beta_d = \frac{\omega_d}{\omega} = \frac{\sqrt{k_d/m_d}}{\sqrt{k/m}} \qquad (16.6a,b)$$

$$\zeta = \frac{c}{2\sqrt{km}}, \qquad \zeta_d = \frac{c_d}{2\sqrt{k_d m_d}} \qquad (16.7a,b)$$

An optimal design of TMD should consider the largest acceptable levels of the response of the structure and the TMD, that is, H and H_d. Figure 16.3 shows, for a given set of γ, β_d, and ζ values, and for several values of ζ_d, the dependence of the dynamic amplification factor H upon the non-dimensionalized excitation frequency β. For $\zeta_d = 0$ the amplification factor H of the structure has two separate peaks, as does the amplification factor H_d of the TMD. As ζ_d increases up to $\zeta_{d\,opt}$ (i.e., approximately 0.09, see Eq. [16.9]), the ordinates of the two peaks of the factors H and H_d decrease. As ζ_d increases further, the two peaks of H and H_d merge into one peak. For H, that peak increases as ζ_d approaches unity, whereas for H_d the peak continues to decrease. As shown in Figure. 16.3, if the ratio β of the excitation frequency to the natural frequency ω of the structure is contained in the interval 0.85–1.15, the TMD reduces the response by amounts that depend upon that ratio.

As explained in [1, p. 247], because of the dependence of $|D|$ upon ζ, no analytical expressions can be obtained for the optimal tuning frequency ratio $\beta_{d\,opt}$ and optimal damping $\zeta_{d\,opt}$ as functions of the mass ratio γ. Numerical calculations are therefore resorted to. The reader is referred to [1] for plots of the calculated optimal values of H and H_d as functions of m_d/m for various values of ζ.

The optimal values of the parameters β_d and ζ_d as functions of m_d/m and ζ can be obtained from the following expressions based on curve fitting schemes proposed in [6]:

$$\beta_{d\,opt} = \left(\frac{\sqrt{1 - 0.5\ m_d/m}}{1 + m_d/m} + \sqrt{1 - 2\zeta^2} - 1 \right)$$
$$- (2.375 - 1.034\sqrt{m_d/m} - 0.426\ m_d/m)\zeta\sqrt{m_d/m}$$
$$- (3.730 - 16.903\sqrt{m_d/m} + 20.496\ m_d/m)\zeta^2\sqrt{m_d/m} \qquad (16.8)$$

$$\zeta_{d\,opt} = \sqrt{\frac{3\gamma}{8(1 + m_d/m)(1 - 0.5m_d/m)}}$$
$$+ (0.151\zeta - 0.170\zeta^2) + (0.163\zeta + 4.980\zeta^2)m_d/m \qquad (16.9)$$

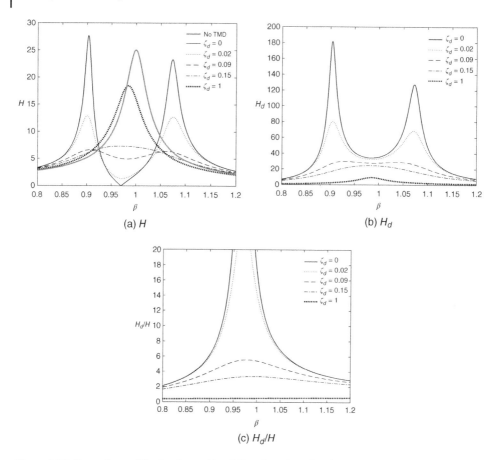

Figure 16.3 Dynamic amplification factor H and H_d as functions β with various values of ζ_d ($m_d/m = 0.03$, $\beta_d = 0.97$, $\zeta = 0.02$).

For a single-degree-of-freedom linear oscillator with no TMD the largest possible value of the mechanical admittance function is

$$H_{SD} = \frac{1}{2\zeta_{SD}\sqrt{1 - \zeta_{SD}^2}} \tag{16.10}$$

where ζ_{SD} denotes the oscillator's damping ratio. For ζ_{SD} in the order of 0.02, say,

$$H_{SD} \approx \frac{1}{2\zeta_{SD}} \tag{16.11}$$

Similarly, the equivalent damping for the mass m provided to the system described by Eqs. (16.1) and (16.2) can be written as

$$\zeta_e = \frac{1}{2H_{opt}} \tag{16.12}$$

Example 16.1 Following [1, p. 251], it is assumed that the damping ratio is $\zeta = 0.02$, and that the dynamic amplification factor H_{opt} and the ratio between the amplitudes of the TMD and the structure are limited by the inequalities

$$H_{opt} < 7 \tag{16.13}$$

$$\frac{H_{d\ opt}}{H_{opt}} < 6 \tag{16.14}$$

that is, $H_{d\ opt} < 42$. From [1, figure 5.28] it follows that for $\zeta = 0.02$ the required ratio $m_d/m \approx 0.03$. From [1, figure 5.29] it follows that Eq. (16.14) is satisfied. The value $\beta_{d\ opt}$, the stiffness k_d, and the optimal damping ratio $\zeta_{d\ opt}$ are then obtained from Eqs. (16.8), (16.6b), and (16.9), respectively. The equivalent damping provided by the TMD is $1/(2H_{opt}) \approx 0.07$.

16.3 TMDs for Multiple-Degree-of-Freedom Systems

Figure 16.4 shows a two-degree-of-freedom (2-DOF) system. The equations of motion of masses m_1, m_2, and m_d are [1]:

$$m_1\ddot{x}_1 + c_1\dot{x}_1 + k_1x_1 - k_2(x_2 - x_1) - c_2(\dot{x}_2 - \dot{x}_1) = p_1 \tag{16.15}$$

$$m_2\ddot{x}_2 + c_2(\dot{x}_2 - \dot{x}_1) + k_2(x_2 - x_1) - k_d x_d - c_d\dot{x}_d = p_2 \tag{16.16}$$

$$m_d\ddot{x}_d + c_d\dot{x}_d + k_d x_d = -m_d\ddot{x}_2 \tag{16.17}$$

Expressing x_1 and x_2 in terms of model shapes and generalized coordinates

$$\begin{Bmatrix} x_1 \\ x_2 \end{Bmatrix} = \begin{bmatrix} \phi_{11} & \phi_{12} \\ \phi_{21} & \phi_{22} \end{bmatrix} \begin{Bmatrix} q_1 \\ q_2 \end{Bmatrix} \qquad \text{or} \qquad \mathbf{x} = \mathbf{\Phi}\ \mathbf{q} \tag{16.18a,b}$$

where $\mathbf{\Phi}$ is the modal matrix and $\mathbf{q}$ is the generalized coordinate vector. Based on the orthogonality of natural modes, Eqs. (16.15) and (16.16) are transformed into the uncoupled equations of a SDOF structure:

$$m_j^*\ddot{q}_j + c_j^*\dot{q}_j + k_j^*q_j = \phi_{j1}p_1 + \phi_{j2}(p_2 + c_d\dot{x}_d + k_d x_d) \qquad \text{for} \quad j = 1, 2 \tag{16.19}$$

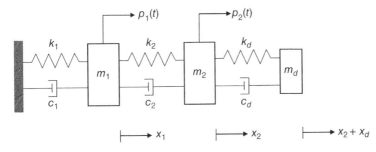

Figure 16.4 Two-DOF system with tuned mass damper.

where the modal mass, stiffness, and damping matrices are defined as

$$m_j^* = \mathbf{\Phi}_j^T \mathbf{M} \mathbf{\Phi}_j; \tag{16.20}$$

$$c_j^* = \mathbf{\Phi}_j^T \mathbf{C} \mathbf{\Phi}_j; \tag{16.21}$$

$$k_j^* = \mathbf{\Phi}_j^T \mathbf{K} \mathbf{\Phi}_j; \tag{16.22}$$

In Eqs. (16.20)–(16.22), the *j*-th modal vector of $\mathbf{\Phi}$ is

$$\mathbf{\Phi}_j = \begin{bmatrix} \phi_{1j} \\ \phi_{2j} \end{bmatrix} \tag{16.23}$$

Consider the case of a TMD designed to control the first modal response (i.e., $j = 1$). If the external forcing frequency is close to $\omega_1 = \sqrt{k_1/m_1}$, the response in the first mode dominates. Equations (16.18a,b) then yield $x_2 \approx \phi_{21} q_1$, and

$$q_1 = \frac{x_2}{\phi_{21}} \tag{16.24}$$

Substitution of Eq. (16.24) into Eq. (16.19) in which $j = 1$ yields

$$m_{1e}^* \ddot{x}_2 + c_{1e}^* \dot{x}_2 + k_{1e}^* x_2 = p_{1e}^* + c_d \dot{x}_d + k_d x_d \tag{16.25}$$

where the equivalent mass, damping, stiffness, and force matrices are

$$m_{1e}^* = \frac{m_1^*}{\phi_{21}^2} \tag{16.26}$$

$$k_{1e}^* = \frac{k_1^*}{\phi_{21}^2} \tag{16.27}$$

$$c_{1e}^* \approx \alpha k_{1e}^* \tag{16.28}$$

$$p_{1e}^* = \frac{\phi_{11} p_1 + \phi_{12} p_2}{\phi_{21}} \tag{16.29}$$

Equation 16.28 is derived under the assumption that damping is proportional to the stiffness [1].

Equations (16.25) and (16.17) have the same form as Eqs. (16.1) and (16.2), respectively. Therefore, with appropriate changes of notation, the solutions discussed in Section 16.2 are also applicable to Eqs. (16.25) and (16.17). For details and a numerical example, see [1].

Reference [5] presents a frequency-domain approach to the optimization of TMDs installed at several levels of multiple-degree-of-freedom structures subjected to wind loads defined by their power spectral densities.

References

1 Connor, J. and Laflamme, S. (2016). Tuned mass damper systems. In: *Structural Motion Engineering*, 199–285. Springer International Publishing.
2 Den Hartog, J.P. (1956). *Mechanical Vibrations*, 4th ed. New York: McGraw-Hill.
3 McNamara, R.J. (1977). Tuned mass dampers for buildings. *Journal of the Structural Division, ASCE* 103: 1785–1798.

4 Luft, R.W. (1979). Optimal tuned mass dampers for buildings. *Journal of the Structural Division, ASCE* 105: 2766–2772.

5 Lee, C.-L., Chen, Y.-T., Chung, L.-L., and Wang, Y.-P. (2006). Optimal design theories and applications of tuned mass dampers. *Engineering Structures* 28: 43–53.

6 Tsai, H.-C. and Lin, G.-C. (1993). Optimum tuned-mass dampers for minimizing steady-state response of support-excited and damped systems. *Earthquake Engineering & Structural Dynamics* 22: 957–973.

17

Rigid Portal Frames

Case Study

17.1 Introduction

Conventional methods for determining wind loads on rigid structural systems, as defined by the "analytical method" of the ASCE 7 Standard [1], involve the use of tables and plots contained in standards and codes. Wind effects determined by such methods can differ from those consistent with laboratory measurements by amounts that can exceed 50% [2, 3]. This is due in part to the severe data storage limitations inherent in conventional standards, in which vast amounts of aerodynamic data varying randomly in time and space are reduced to a far smaller number of enveloping time-invariant data. In addition, for low-rise buildings of the type covered by [1], the specified wind loads, referred to in the standard as "pseudo-loads," do not account for (i) the distance between frames, which affects the spatial coherence of the aerodynamic pressures impinging on the frames, and (ii) the structural system's actual member sizes and, therefore, the influence coefficients used in the structural calculations. Lastly, the ASCE 7 provisions are based on wind tunnel experiments conducted in part between three and four decades ago with obsolete pressure measurement technology, no archived records of pressure measurements, and numbers of building geometries and pressure taps lower by more than one order of magnitude than those of current aerodynamic databases [4]. In contrast, in the DAD approach "pseudo-loads" are replaced by close approximations to the actual loads.

This chapter presents an application of the DAD approach to the design of portal frames, wherein time-domain methods allow wind effects to be calculated by using large numbers of stored time series of measured pressure coefficients, and wind effect combinations are performed objectively and rigorously via summations of time series. The DAD approach accounts naturally for the imperfect spatial coherence of pressures acting at different points of the structure, examples of which are shown in the visualization of Figure 4.27.

Software for the application of the DAD approach to rigid structures was first developed for frames of simple gable roof buildings in [5]. This chapter presents an updated version of this approach and a case study reported in [6] that calculates peak demand-to-capacity indexes (DCIs) directly used by the structural engineer to size structural members of gable roof building frames. The aerodynamic pressure coefficients used in [5] and [6] were taken from the NIST/UWO database [7]. Results based on the NIST/UWO and the Tokyo Polytechnic University (TPU) database

Wind Effects on Structures: Modern Structural Design for Wind, Fourth Edition. Emil Simiu and DongHun Yeo.
© 2019 John Wiley & Sons Ltd. Published 2019 by John Wiley & Sons Ltd.

[8] – the largest available to date – were found in [9] to yield comparable results. Calculations reported in [6] confirmed this conclusion.

Checking the adequacy of a member cross section consists of ascertaining that, subject to possible serviceability and constructability constraints, its DCI is close to unity. If the DCI does not satisfy this condition, the cross section is redesigned. The member properties based on this iteration process can then be used to recalculate the influence coefficients by which revised wind loads are transformed into wind effects, and to check the adequacy of the resulting DCIs.

Since the capacity of members in compression is determined by stability considerations, their DCIs depend nonlinearly upon axial load and are therefore not proportional to the squares of the wind speeds. For this reason, to estimate wind effects with the requisite mean recurrence intervals it is necessary to produce DCI response surfaces (see Section 13.2). The estimation of the peak DCIs from DCI time series can be performed by a multiple-points-in-time method based on observed peak values [10]. An alternative approach to the estimation of peaks, based on rigorous statistical methods, and capable of producing error estimates, is presented in Appendix C.

The peak DCI response surfaces are properties of the structure, independent of the wind climate, and depend upon the structure's terrain exposure, aerodynamic behavior, structural system, and member sizes. The response surfaces are used in conjunction with non-parametric statistics to estimate peak DCIs with any specified mean recurrence intervals (MRIs) (Sections 13.4 and A.9). Since the design MRIs specified in [1] are in the order of hundreds or thousands of years, the use of non-parametric statistics requires the wind speed data sets to be commensurately large. Databases of simulated hurricane wind speeds that meet this requirement are available, see Section 3.2.3, and Monte Carlo simulations can be performed to develop large wind speed data sets from smaller sets of measured data [11].

The results obtained in the case study presented in this chapter confirm the existence of significant errors in the estimation of wind effects by the ASCE 7 Standard envelope procedure. The requisite software and a detailed user's manual are available in [12].

The DAD procedure as used in this chapter is typically applicable to any low- or mid-rise buildings, in addition to simple buildings with gable roofs, portal frames, and bracing parallel to the ridge. Depending upon the preferences of the user, alternative methods for the estimation of time series peaks, the interpolation of results based on buildings with dimensions different from those of the building of interest, and the estimation of secondary effects, may be used in lieu of the methods employed in [6].

17.2 Aerodynamic and Wind Climatological Databases

Aerodynamic databases are developed by wind engineering laboratories and contain time histories of simultaneously measured pressure coefficients at large numbers of taps. Figure 17.1 shows a building model with the locations of the taps. Pressure coefficient time-history databases for one-of-a-kind structures are obtained in ad-hoc wind tunnel tests, rather than from pre-existing databases.

Climatological databases are also developed by wind engineering laboratories. They typically consist of directional or non-directional extreme wind speeds that account for the building's directional terrain exposure, and cover periods in the order of typically

Figure 17.1 Wind tunnel model of an industrial building. Source: Courtesy of the Boundary Layer Wind Tunnel Laboratory, University of Western Ontario.

tens of years of measured data or as many as thousands of years of synthetic data, as well as providing the mean rate of arrival of storm events per year (Section A.6.4). Directional wind speed data U_{ij} ($i = 1, 2,..., n_s$; $j = 1, 2,..., n_d$) are typically presented in the form of $n_s \times n_d$ matrices, in which n_s is the number of storm events and n_d is the number of wind directions (e.g., $n_d = 16$); non-directional wind speed data sets are vectors with components U_i ($i = 1, 2,..., n_s$), where U_i = largest wind speed in storm i, regardless of direction (see Chapter 13). The climatological database considered in the case study presented in this chapter consisted of directional hurricane wind speeds generated by Monte Carlo simulations for 999 storm events and 16 wind directions (Section 3.2.3).

17.3 Structural System

The structural system being considered consists of equally spaced moment-resisting steel portal frames commonly used in low-rise industrial buildings (Figure 17.2). Roof and wall panels form the exterior envelope of the buildings and are attached to purlins and girts supported by the frames. Bracing is provided in the planes of the exterior walls parallel to the ridge. The coupling between frames due to the roof diaphragms is neglected. The purlins and girts are attached to the frames by hinges. The purlins and girts act as bracings to the outer flanges, and the inner flanges are also braced. The following limitations are imposed: (i) The taper should be linear or piecewise linear, and (ii) the taper slope should typically not exceed 15° [13].

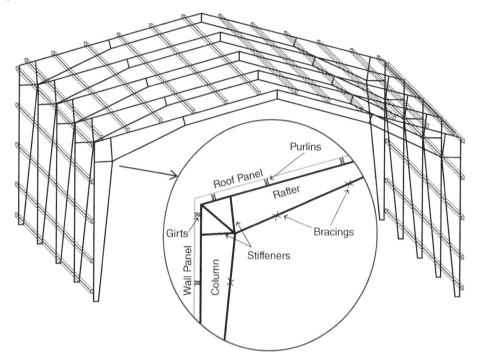

Figure 17.2 Schematic of the structural system.

17.4 Overview of the Design Procedure

The sizing of the structural members requires calculations of the respective peak DCIs. The DCIs pertaining to axial forces and bending moments at any cross sections of the frames are determined using Eqs. (8.1) and (8.2). A similar, simpler equation pertains to shear forces [14].

The wind forces acting along the axis parallel to the ridge and the torsional moment about the structure's elastic center are resisted by secondary bracing members; hence only the wind forces due to winds normal to the building's ridge contribute to the frame DCIs. Therefore, for the application at hand the quantities with subscript y in Eqs. (8.1) and (8.2) need not be considered. The time histories of the internal forces in the expressions for the DCIs are computed as sums of factored load effects due to wind loads and gravity loads. Design for strength requires considering the following five LRFD load combination cases [1]:

Case 1 : 1.4D,
Case 2 : 1.2D + 1.6 L + 0.5L$_r$,
Case 3 : 1.2D + 1.6L$_r$ + 0.5W,
Case 4 : 1.2D + 1.0W + 0.5L$_r$,
Case 5 : 0.9D + 1.0W,

where D, L$_r$, and W denote dead load, roof live load and wind load, respectively. The dead load includes both superimposed dead load and frame self-weight. The superimposed dead load and roof live load are assumed to be uniformly distributed on the roof surface. They impose forces on the frame through the frame-purlin connections in the vertical downward direction. Self-weights are determined by dividing the frames into sufficiently large numbers of elements. The directional wind speeds matrix (see Section 13.3.1) and the mean annual rate of storm arrival were assumed to be those listed for Miami (milepost 1450 in Figure 3.1).

The member capacities are determined as specified in [13, 14]. To comply with AISC requirements on second-order effects, a first-order analysis method can be used that accounts for geometric imperfections [6, 14]. The axial capacity of a member in compression is the smaller of the calculated in-plane and out-of-plane buckling capacities computed by the method of successive approximations described in [15].

Equations (8.1) and (8.2) and their shear force counterpart maintain the phase relationship among the axial force, bending moments and shear force, hence they result in DCIs that rigorously reflect the actual combined wind effects.

The preliminary design of the structure starts with an informed guess as to the structural system's member sizes, that is, with a preliminary design denoted by D_0, to which there corresponds a set of influence coefficients denoted by IC_0. The wind loads applied to this preliminary design are taken from the standard or code being used. For the case study presented here the loads used for the preliminary design were obtained from the ASCE 7 Standard [1].

As performed in [6], the next step is the calculation of the peak DCIs with the specified mean recurrence interval $\overline{N}$ inherent in the design D_0 (see Chapter 13). Unless those DCIs are close to unity, the cross sections are modified. This results in a new design, D_1, for which the corresponding set of influence coefficients, IC_1 is calculated. A new set of DCIs is calculated, based again on the wind loads taken from the standard. The procedure is repeated until a design D_n is achieved such that the effect of using a new set of influence coefficients, IC_{n+1}, is negligible, that is, until the design D_{n+1} is in practice identical to the design D_n. At this point the procedure is applied by using, instead of the ASCE 7 Standard wind pressures, wind pressures based on the time histories of the pressure coefficients taken from the aerodynamics database. This results in a design D_{n+2}, to which there corresponds a set of influence coefficients IC_{n+2} and a new set of DCIs. The cross sections are then modified and the calculations are repeated until the DCIs are close to unity. Typically, this will be the final design D_{final}, although the user may perform an additional iteration to check that convergence of the DCIs to unity has been achieved, to within constructability and serviceability constraints. For the structural system considered in this chapter the approach just described was found to yield the requisite results faster than the alternative approach in which the loads based on the aerodynamic database are used to determine the designs D_1 through D_{n+1}. This is due to the fact that load estimates specified in [1] for the type of structure depicted in Figure 17.1 are less unrealistic than those specified in [1] for other types of structure.

17.5 Interpolation Methods

For the databases with large numbers of data measured on models with different dimensions to be of practical use, simple and reliable interpolation schemes need to be developed that enable the prediction of wind responses for building dimensions not available in the databases. This issue was addressed in, among others, Refs [5, 6, 16].

The interpolation scheme presented in detail in [6] produces responses of the building of interest that, unless the interpolations are performed between buildings with significantly different dimensions, differ from the actual responses by amounts in the order of 5–10%. It is shown in Section 12.4.2 that even larger errors are typically inconsequential from a structural design viewpoint.

17.6 Comparisons Between Results Based on DAD and on ASCE 7 Standard

This section presents results of comparisons between 700-year (i) DAD-based DCIs involving axial forces and moments, denoted by DCI^{PM}, and DCIs involving shear forces, denoted by DCI^V, to their counterparts based on the ASCE 7-10 Standard (Chapter 28). Additional sets of comparisons are reported in [6]. Unless otherwise specified, the assumed frame spacing was 7.6 m. Results are shown for the first interior frame. The frame supports were assumed to be pinned, and all the calculations were conducted for the "enclosed" building enclosure category.

17.6.1 Buildings with Various Eave Heights

For buildings with various eave heights, Figure 17.3 shows ratios of DCI^{PM}s based on DAD to their counterparts based on the ASCE 7-10 Standard, Chapter 28. The buildings had the following dimensions: $B = 24.4$ m, $L = 38.1$ m, roof slope $= 4.8°$, and $H = 4.9$ m, 7.3 m, and 9.8 m.

In most cases represented in Figure 17.3 the DCIs are underestimated by the ASCE 7-10 Standard provisions, especially for suburban exposure.

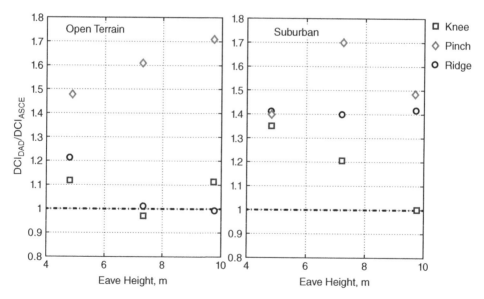

Figure 17.3 DCI_{DAD}/DCI_{ASCE} as a function of eave height.

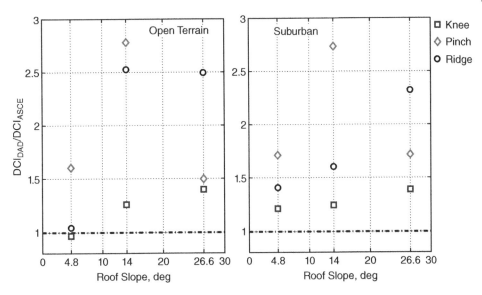

Figure 17.4 DCI_{DAD}/DCI_{ASCE} as a function of roof slope.

17.6.2 Buildings with Various Roof Slopes

For buildings with different roof slopes, Figure 17.4 shows ratios between DCI^{PM}s computed by using DAD and the ASCE 7-10 Standard, Chapter 28. The buildings have the following dimensions: $B = 24.4$ m, $L = 38.1$ m, $H = 7.3$ m, and roof slope $= 4.8, 14.0$, and $26.7°$.

Owing to a strong discontinuity of the pressure coefficient variation at roof slopes of about $22°$, interpolations cannot be performed between wind effects on roofs with slopes lower than $22°$ on the one hand and larger than $22°$ on the other [17].

References

1 ASCE, "Minimum design loads for buildings and other structures (ASCE/SEI 7-16)," in *ASCE Standard ASCE/SEI 7-16*, Reston, VA: American Society of Civil Engineers, 2017.

2 Coffman, B.F., Main, J.A., Duthinh, D., and Simiu, E. (2010). Wind effects on low-rise buildings:databased-assisted design vs. ASCE 7-05 standard estimates. *Journal of Structural Engineering* 136: 744–748.

3 Pierre, L.M.S., Kopp, G.A., Surry, D., and Ho, T.C.E. (2005). The UWO contribution to the NIST aerodynamic database for wind loads on low buildings: part 2. Comparison of data with wind load provisions. *Journal of Wind Engineering and Industrial Aerodynamics* 93: 31–59.

4 Davenport, A. G., Surry, D., and Stathopoulos, T., "Wind loads on low-rise buildings, Part 1. The Boundary Layer Wind Tunnel," University of Western Ontario, London, Ontario, Canada, 1977.

5 Main, J. A. and Fritz, W. P., "Database-Assisted Design for Wind: Concepts, Software, and Examples for Rigid and Flexible Buildings," NIST Building Science Series 170, National Institute of Standards and Technology, Gaithersburg, MD, 2006. https://www.nist.gov/wind.

6 Habte, F., Chowdhury, A.G., Yeo, D., and Simiu, E. (2017). Design of rigid structures for wind using time series of demand-to-capacity indexes: application to steel portal frames. *Engineering Structures* 132: 428–442.

7 NIST. (Dec. 18, 2017). *NIST/UWO aerodynamic database.* Available: https://www.nist.gov/wind.

8 TPU. (Dec. 18, 2017). TPU aerodynamic database. Available: http://wind.arch.t-kougei.ac.jp/system/eng/contents/code/tpu.

9 Hagos, A., Habte, F., Chowdhury, A., and Yeo, D. (2014). Comparisons of two wind tunnel pressure databases and partial validation against full-scale measurements. *Journal of Structural Engineering* 140: 04014065.

10 Yeo, D. (2013). Multiple points-in-time estimation of peak wind effects on structures. *Journal of Structural Engineering* 139: 462–471. https://www.nist.gov/wind.

11 Yeo, D. (2014). Generation of large directional wind speed data sets for estimation of wind effects with long return periods. *Journal of Structural Engineering* 140: 04014073. https://www.nist.gov/wind.

12 Habte, F., Chowdhury, A. G., and Park, S., "The Use of Demand-to-Capacity Indexes for the Iterative Design of Rigid Structures for Wind," NIST Technical Note 1908, National Institute of Standards and Technology, Gaithersburg, MD, 2016. https://www.nist.gov/wind.

13 Kaehler, R. C., White, D. W., and Kim, Y. D., "Frame Design Using Web-Tapered Members," American Institute of Steel Construction, 2011.

14 ANSI/AISC, "Specification for Structural Steel Buildings," in *ANSI/AISC 360-10*, Chicago, IL: American Institute of Steel Construction, 2010.

15 Timoshenko, S. and Gere, J.M. (1961). *Theory of Elasticity Stability*, 2nd ed. McGraw-Hill.

16 Masters, F., Gurley, K., and Kopp, G.A. (2010). Multivariate stochastic simulation of wind pressure over low-rise structures through linear model interpolation. *Journal of Wind Engineering and Industrial Aerodynamics* 98: 226–235.

17 Stathopoulos, T., Personal communication, 2007.

18

Tall Buildings

Case Studies[1]

18.1 Introduction

Tall buildings can be designed by using the Database-Assisted Design (DAD) option, or the related Equivalent Static Wind Loads (ESWL) option, of the DAD_ESWL v. 1.0 software. Both options are available at https://www.nist.gov/wind. A user's manual [1] provides detailed guidance on the use of the software and its application to several examples, including steel and reinforced concrete building examples.

The purpose of this chapter is to introduce the reader to that software and illustrate the application of its two options. Section 18.2 briefly discusses an approach to performing a structure's preliminary design, and outlines the subsequent iterative use of DAD_ESWL to perform the final design. Section 18.3 lists the contributions of the wind engineering laboratory to the design process. Section 18.4 is an introduction to the software. Section 18.5 briefly presents the application of the DAD approach and of the ESWL approach to the structural design of a 47-story steel building. The software is also applicable to the design of mid-rise buildings via the simple device of using as input appropriately large values of the natural frequencies of vibration in the fundamental sway and torsional modes and disregarding higher modes.

18.2 Preliminary Design and Design Iterations

The structural design process starts with the development of a preliminary design. This entails the choice of a structural system for the building being considered (e.g., moment frames), the geometry and morphological features of which must be consistent with architectural and other non-structural design requirements. The member sizes of the preliminary system are initially guessed at by the structural designer on the basis of experience. This will produce a system that, typically, will not meet strength and serviceability requirements. It is therefore advisable to redesign the structural system produced by the structural engineer's educated guesses by using for the wind loading simple models specified, for example, in the ASCE 7 Standard for buildings of all heights. The new design so obtained is referred to here as design D_0.

The structural engineer must check the adequacy of design D_0, that is, whether it satisfies the specified strength and serviceability when subjected to realistic, rather

---

[1] Major contributions to this chapter by Dr. Sejun Park are acknowledged with thanks.

Wind Effects on Structures: Modern Structural Design for Wind, Fourth Edition. Emil Simiu and DongHun Yeo.
© 2019 John Wiley & Sons Ltd. Published 2019 by John Wiley & Sons Ltd.

than simplified wind loads. The information inherent in design D_0, and the data provided by the wind engineering laboratory, are used by the structural engineer in the DAD_ESWL software to determine the members' demand-to-capacity indexes (DCIs), inter-story drift ratios, and accelerations, with the respective specified design mean recurrence intervals (Chapter 13). For strength design, it is required that no member DCI exceed unity, or be significantly less than unity except as required by serviceability constraints. If, as is typically the case for design D_0, these requirements are not satisfied, the members' cross sections need to be modified, and the software is applied iteratively to successive designs D_1, D_2, …, until a satisfactory final design is achieved.

18.3   Wind Engineering Contribution to the Design Process

Realistic wind loads must be based on the following information provided by the wind engineering laboratory:

1) Aerodynamic data, consisting of pressure coefficient time series obtained simultaneously at multiple taps on the façades of the building model, either from ad-hoc wind tunnel tests (or, in the future, by ad-hoc Computational Wind Engineering [CWE] simulations), or from databases such as [2, 3]. The following prototype data are required: Elevation of the reference wind speed (usually the elevation of the top of the building), wind directions, sampling rate, number of sampling points, and coordinates defining the location of the taps on the building façades.

2) Wind climatological data, consisting of q matrices $n_{sq} \times n_{dq}$ of directional wind speed data, and the respective rates of storm arrival, of up to q types of storm (see Section A.9), where, depending upon the wind climate, $q = 1$ (e.g., synoptic storms only), $q = 2$ (e.g., hurricanes and thunderstorms), or $q = 3$ (e.g., hurricanes, nor'easterns, and thunderstorms). The n_{sq} rows correspond to a number n_{sq} of storms (see Sections 13.3.1, 3.2.3, and [4]); the n_{dq} columns correspond to, say, $n_{dq} = 16–36$ wind directions. The matrix entries are mean wind speeds averaged over, say, 30–60 minutes, at the location of the empty (pre-construction) building site and the elevation of the reference wind speed (see item 1)).

3) Measures of uncertainty in the pressure coefficients and the directional wind speeds, to be used in procedures for producing estimates of wind load factors or of augmented design mean recurrence intervals of the wind effects of interest (see Chapters 7 and 12).

The contribution of the wind engineering laboratory to the design process is completed once the information described here is delivered to the structural engineer. The same information is used, with no modification, for the analysis of each of the iterative designs.

18.4   Using the DAD_ESWL Software

For a structure with given mechanical properties, the DAD_ESWL software is used by the structural engineer to determine the effects of interest induced by combinations of

(i) gravity loads and (ii) wind loads based on the aerodynamic and wind climatological information provided by the wind engineering laboratory. This section provides a summary description of the DAD_ESWL software, based on the detailed description available in [1].

18.4.1 Accessing the DAD_ESWL Software

DAD_ESWL v. 1.0 can be accessed via the website https://www.nist.gov/wind. The stand-alone executable version of DAD_ESWL requires installation of **MCRInstaller.exe**, which is available on the main page. The website includes, among others, the input files for the examples described in Section 18.5.

18.4.2 Project Directory and its Contents

It is recommended that a directory named DAD_ESWL, with the structure shown in Figure 18.1, be created for each project on the user's local drive. The directory saves all downloaded files and directories. It is recommended that the executable file for the software, **DAD_ESWL_v1p0.exe**, be included in the project directory.

The *"Aerodynamic_data"* directory contains data files (.MAT format): (i) identifying each of the pressure taps located on the exterior building surfaces, (ii) listing their coordinates, and (iii) containing pressure coefficient time series from wind-tunnel testing (or, in the future, from CWE simulations) corresponding to a sufficient number of directions, to allow the construction of the requisite response surfaces (see Sections 8.2 and 13.2).

The *"Building_data"* directory includes the building's geometric and structural data (members' properties, mass matrix, influence coefficients, internal forces of members induced by gravity loads, and modal dynamic properties). The building's structural data are calculated and prepared in advance by using finite element software, following the user's choice of whether second-order effects are accounted for or disregarded (see Chapter 9). The alternative option of using OpenSees to obtain the building's structural data is available (see [1] for details), in which case the *"OpenSees"* directory is added.

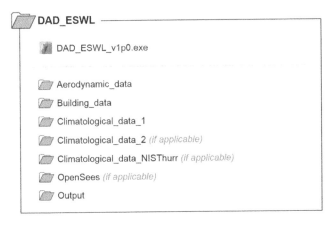

Figure 18.1 Recommended directory structure.

The "*Climatological_data_1*," "*Climatological_data_2*," and "*Climatological_data_3*" directories contain simulated directional wind speed data of up to three distinct types of storm.

The "*Output*" directory contains results of calculations performed by DAD_ESWL.

18.4.3   Software Activation. Graphical User Interface

To run the software, the user double-clicks the **DAD_ESWL_v1p0.exe** file in the project directory. This opens a panel (Figure 18.2) of the Graphical User Interface (GUI) allowing the user to select the type of structure (steel or reinforced concrete). Clicking the button "Start" opens the first of five pages that prompt the user to (i) fill in values of requisite data (e.g., building dimensions, modal periods), (ii) choose between various options (e.g., second-order effects accounted for or disregarded, use of input from FE analyses or OpenSees, use of DAD or ESWL procedure), and (iii), after clicking "Browse" buttons, fill in the respective paths and names of input files used in the calculations to be performed by DAD_ESWL. At the bottom of each of the five pages there is a group of five buttons called *input panel navigator*: **Bldg. modeling**, **Wind loads**, **Resp. surface**, **Wind effects**, and **Results & Plots** (see Figure 18.3). These are activated in succession as the calculations proceed. In addition to the input panel navigator, the five pages contain the following buttons: **Save inputs**, used to save input data, data file paths, and selected options as MAT files for future use in DAD_ESWL; **Open inputs**, used to download the saved input data and allowing empty boxes and unselected options in the input panels to be filled and activated; and **Exit**, which can be clicked at any time to terminate DAD_ESWL.

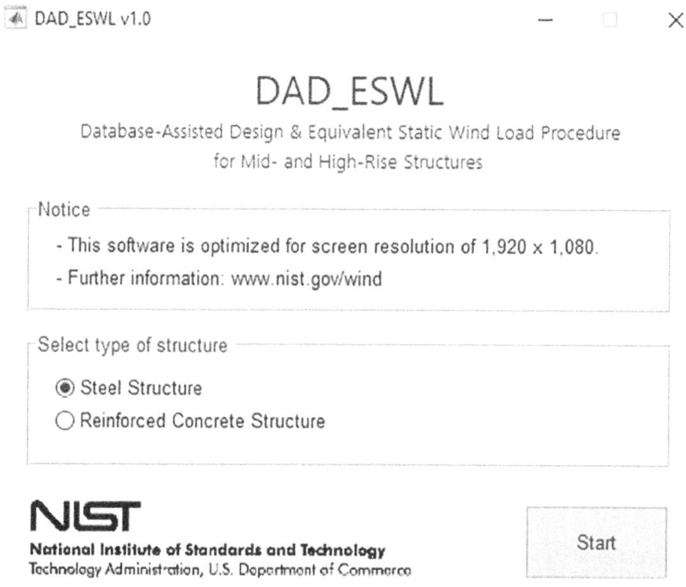

Figure 18.2 Structural type selection panel.

Figure 18.3 Page of "Bldg. modeling."

18.5 Steel Building Design by the DAD and the ESWL Procedures: Case Studies

18.5.1 Building Description

The structure being considered is a 47-story steel building with rigid diaphragm floors, outriggers and belt truss system, and dimensions $40 \times 40 \times 160$ m in depth, width, and height, respectively (Figure 18.4). The structure consists of approximately 2300 columns, 3950 beams, and 2300 diagonal bracings. Columns are of three types: core, external core, and interior columns. Beams are of three types: exterior, internal, and core beams. Diagonal bracings are of two types: core and outrigger bracings. Each type of structural member has the same dimensions for 10 successive floors of the building's lowest 40 floors, and for the seven highest floors. The columns and bracings consist of built-up hollow structural sections (HSS), and the beams consist of rolled W-sections selected from the AISC Steel Construction Manual [5]. The steel grade is ASTM A570 steel, grade 50.

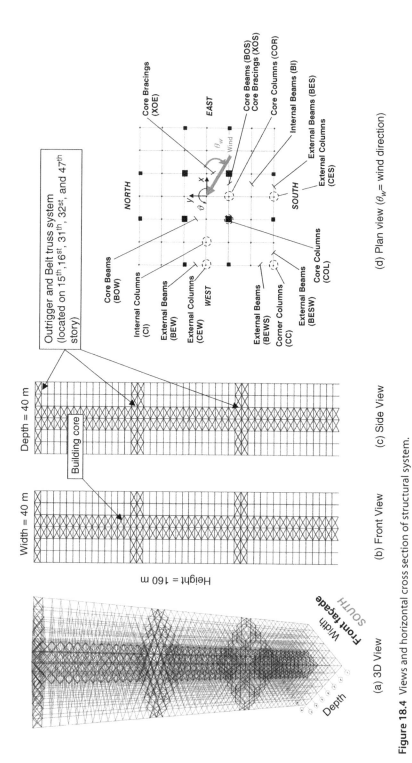

Figure 18.4 Views and horizontal cross section of structural system.

The structure is assumed to be sited in open terrain exposure in South Carolina, near the shoreline of milepost 1950 (for a map showing milepost locations see Figure 3.1). The wind speed data being used are the NIST hurricane data transformed into hourly mean speeds at the elevation of the top of the building at the empty building site. The orientation angle of the building is 270° clockwise from the north, that is, one of the four identical façades of the building faces east. The aerodynamic pressure coefficient time histories are obtained from the Tokyo Polytechnic University (TPU) high-rise building aerodynamic database [6]. Wind direction is defined by the clockwise angle θ_w, with the positive x-axis heading east, and the y-axis heading north (see Figure 18.4d). A total of 60 types of structural members are selected for the final design: six types of column, seven types of beam, and two types of bracing, at the 1st, 17th, 33rd, and 45th floors.

Task 1 (Figure 8.1 in Chapter 8) consists of performing the preliminary design based on, for example, ASCE 7–16 Standard provisions for "buildings of all heights." This task yielded the member sizes listed in Table 18.1.

Table 18.1 Member sizes for the preliminary design denoted by D_0 (in mm), and member nomenclature[a].

Member type	Section ID	Sectional type	Depth	Width	Flange thickness	Web thickness
Bracing	D01–16	Box/Tube	350	350	14	14
	D17–32	Box/Tube	300	300	14	14
	D33–47	Box/Tube	200	200	12	12
Column	Int01–16	Box/Tube	700	700	30	30
	Int17–32	Box/Tube	500	500	24	24
	Int33–47	Box/Tube	300	300	15	15
	Core01–16	Box/Tube	1500	1500	60	60
	Core17–32	Box/Tube	1200	1200	50	50
	Core33–47	Box/Tube	1000	1000	40	40
	ExCore01–16	Box/Tube	1200	1200	50	50
	ExCore17–32	Box/Tube	1000	1000	40	40
	ExCore33–47	Box/Tube	800	800	30	30
Beam	W10X26	I/Wide Flange	261.62	146.56	11.18	6.60

a) D01–16: Diagonal bracing, floors 1–16 and all outriggers and belt trusses.
D17–32: Diagonal bracing, floors 17–32.
D33–47: Diagonal bracing, floors 33–47.
Int01–16: Internal columns, 1–16.
Int17–32: Internal columns, floors 17–32.
Int33–47: Internal columns, floors 33–47.
Core01–16: Core columns, floors 1–16.
Core17–32: Core columns, floors 17–32.
Core33–47: Core columns, floors 33–47.
ExCore01–16: External Core Columns, floors 1–16.
ExCore17–32: External Core Columns, floors 17–32.
ExCore33–47: External Core Columns, floors 33–47.
W10X26: All beams.

18.5.2 Using the DAD and the ESWL Options

This section is a brief summary of salient features of the user's manual in [1], which describes the software in detail.

DAD option. Task 2 (Figure 8.1 and Section 8.2) begins by clicking the button "Start" shown in Figure 18.2, and selecting the "Steel Structure" option. This opens the page shown in Figure 18.3. The page activated by the button "**Bldg. modeling**" contains a "Building information" and a "Structural properties" panel. The user fills in the requisite data (i.e., "No. of stories," "Building height," and so forth) and selects the appropriate option where a choice is offered (i.e., for this example, "Second-order analysis" rather than "Linear," and "Input analysis results from arbitrary FE software" rather than "Use OpenSees,"). The user also clicks the "Browse" buttons and fills in the respective paths and file names containing the results (obtained by FE or OpenSees, depending upon the analyst's choice) available in the "Building_data" directory (Figure 18.1).

Task 3 starts by clicking the button "**Wind loads**" at the bottom of the GUI page shown in Figure 18.5. The user fills in the requisite data (i.e., "Model length scale," "Wind directions," and so forth) in the "Wind tunnel test/CWE data" panel and clicks the "Calculate floor wind loads from pressures measured at taps on building model facade" option in

Figure 18.5 Page "Wind loads."

the "Floor wind loads at model scale [N and N.m]" subpanel. After clicking the "Browse" buttons, the user selects the appropriate files from the "Aerodynamic_data" directory (Figure 18.1). Input files for the pressure coefficient data (Cp_XXXpX.mat), tap identification (tap_loc.mat), and tap coordinates (tap_coord.mat) are provided by the wind engineering laboratory, as indicated in Section 18.3. The user then selects the interpolation method for calculating the floor wind loads (right-hand side of this panel). Clicking the button "Calculate floor wind loads" starts the automatic calculation of the floor wind loads, and activates a pop-up window showing the progress of the computations. The wind pressure information can be checked by clicking the button "Display." The floor wind loading data are saved for each direction in the user-specified directory (in this example, "WL_floors," as shown in Figure 18.5). The "Wind speed range" panel specifies the wind speeds used for the construction of response surfaces discussed in Sections 8.2 and 13.2. In this example, wind speeds from 20 to 80 m s^{-1} in increments of 10 m s^{-1} were used. Finally, the 80% selection that pertains to ASCE 7-16 Standard section 31.4.4 was made in the "Lower limit requirement" panel.

The page opened by clicking the button "**Resp. surface**" contains three panels (Figure 18.6). The "Load combination cases" panel specifies the gravity and wind load

Figure 18.6 Page "Resp. surface."

combinations including the associated load factors. The "Calculation option" panel requires the user to choose between using the DAD and the ESWL approach. The user must specify the length of the initial part of the time series of inertial forces that is discarded in order to eliminate non-stationary effects.

Tasks 4, 5, and 6 require the use of the information provided in the "Response surface" panel, and consist of calculating the ordinates of the response surfaces that yield peak member DCIs, inter-story drift ratios, and accelerations. These are obtained by performing dynamic analyses of the structure D_0 for each of the directional wind speeds with directions entered in the panel "Wind tunnel test/CWE data" and speeds entered in the panel "Wind speed range" of the page "**Wind loads.**" *Task 4* determines, for each of those directional wind speeds, the effective loads consisting of the sums of the aerodynamic and inertial loads. *Task 5* uses the appropriate influence coefficients to determine time series of the DCIs induced by combinations of factored gravity loads and the effective wind loads obtained in Task 4. *Task 6* consists of calculating the ordinates of the response surfaces for peak DCIs, inter-story drift ratios, and accelerations induced in structural members by each of the directional wind speeds considered in Task 4.

Task 7 is executed by attending to the panels "Wind climatological data" and "Design responses for specified MRIs" of Figure 18.7 (page "**Wind effects**"). Typically, the wind

Figure 18.7 Page "Wind effects."

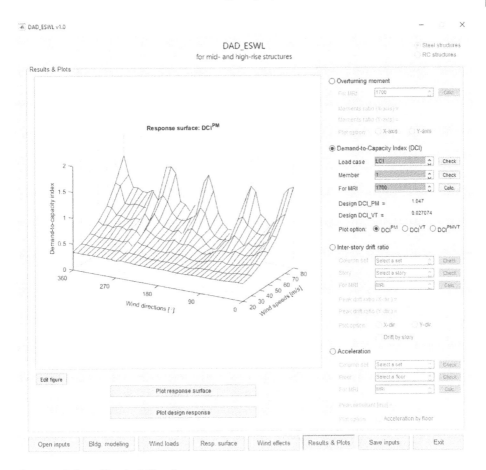

Figure 18.8 Page "Results & Plots."

engineering laboratory provides directional mean wind speed data at the elevation of the top of the building roof at the empty site of the building. The page **"Results & Plots"** allows the user to show the calculated wind effects (i.e., DCIs, inter-story drift ratios, and accelerations) with specified MRIs, as shown in Figure 18.8. Superscripts P, M, V, and T of DCIs stand for axial load (P) and bending moment (M), shear force (V), and torsion (T), respectively. As expected, DCI values for design D_0 were typically inadequate. The design was modified accordingly to yield design D_1; an additional iteration yielded design D_2. Details are provided subsequently.

ESWL option. Tasks 1 through 4 are identical for both ESWL and DAD. However, unlike DAD, ESWL requires completing Task 4a, which consists of calculating Equivalent Static Wind Loads (see Figure 8.1). To do so, in the page **"Resp. surface"** (Figure 18.6) and the panel "Calculation options," the user chooses the option "ESWL (only applicable for DCIs)" and responds to the prompt "No. of Multiple Points-In-Time" (see Section 3.2) in [1]. Tasks 5, 6, and 7 are performed by using the same pages as those used for the DAD option, but with the input required for ESWLs.

DCIs for selected members are listed in Table 18.2 for designs D_0, D_1, and D_2. Member sizes for design D_2 are shown in Table 18.3. The results of the ESWL calculations depend upon the number of points in time n_{pit}. Calculations performed for the example

Table 18.2 DCIs (axial force and bending moments) based on DAD and on ESWL for designs D_0, D_1 and D_2.

Member ID[a]	Method	D_0				D_1				D_2			
		1st	17th	33rd	45th	1st	17th	33rd	45th	1st	17th	33rd	45th
CC	DAD	0.89	1.14	1.51	0.42	0.78	0.76	0.59	0.13	0.83	0.88	0.84	0.87
	ESWL	0.89	1.13	1.51	0.42	0.76	0.74	0.57	0.13	0.83	0.88	0.84	0.86
CEW	DAD	0.62	0.97	1.53	0.37	0.65	0.76	0.68	0.15	0.96	0.80	0.88	0.87
	ESWL	0.62	0.96	1.52	0.37	0.66	0.74	0.66	0.15	0.95	0.80	0.88	0.87
CI	DAD	0.81	1.08	1.64	0.30	0.84	0.85	0.77	0.13	0.83	0.90	1.00	0.80
	ESWL	0.81	1.07	1.62	0.29	0.84	0.84	0.75	0.13	0.83	0.90	0.99	0.79
COL	DAD	1.99	1.62	1.02	0.36	0.79	0.62	0.52	0.15	0.84	0.62	0.90	0.91
	ESWL	2.01	1.61	1.02	0.36	0.76	0.59	0.49	0.10	0.84	0.62	0.89	0.90
CES	DAD	0.78	1.18	1.73	0.39	0.67	0.78	0.69	0.15	0.72	0.90	0.95	0.89
	ESWL	0.78	1.17	1.69	0.38	0.67	0.75	0.66	0.14	0.72	0.90	0.95	0.88
COR	DAD	1.37	1.44	1.57	0.61	1.10	0.92	0.63	0.23	0.56	0.67	0.87	0.78
	ESWL	1.38	1.42	1.55	0.60	1.11	0.88	0.59	0.22	0.56	0.67	0.86	0.78
BESW	DAD	0.66	1.00	1.00	0.66	0.63	0.96	1.00	0.66	0.91	0.98	0.69	0.97
	ESWL	0.64	0.97	0.97	0.64	0.61	0.95	0.97	0.65	0.90	0.98	0.69	0.96
BES	DAD	0.65	0.96	0.93	0.60	0.62	0.93	0.92	0.57	0.62	0.95	0.96	0.83
	ESWL	0.63	0.93	0.91	0.58	0.60	0.92	0.90	0.56	0.61	0.94	0.96	0.83
BI	DAD	0.86	1.40	1.62	1.50	0.85	1.34	1.56	1.42	0.85	0.88	0.95	0.97
	ESWL	0.85	1.39	1.62	1.48	0.85	1.34	1.54	1.40	0.85	0.88	0.93	0.97
BOS	DAD	0.77	0.78	0.81	0.62	0.77	0.77	0.80	0.62	0.77	0.78	0.81	0.90
	ESWL	0.76	0.77	0.79	0.62	0.77	0.77	0.79	0.61	0.77	0.77	0.80	0.90
BEWS	DAD	0.70	1.18	1.26	0.88	0.56	0.95	1.05	0.74	0.87	0.68	0.77	0.75
	ESWL	0.67	1.13	1.24	0.87	0.58	0.95	1.00	0.71	0.85	0.67	0.77	0.75
BEW	DAD	0.69	1.17	1.26	0.79	0.56	0.96	1.02	0.67	0.87	0.69	0.74	0.99
	ESWL	0.67	1.13	1.19	0.77	0.57	0.96	1.01	0.67	0.85	0.68	0.74	0.99
BOW	DAD	0.82	0.87	0.95	0.65	0.76	0.80	0.87	0.63	0.78	0.82	0.89	0.93
	ESWL	0.83	0.86	0.93	0.64	0.77	0.80	0.86	0.63	0.78	0.82	0.89	0.92
XOS	DAD	0.73	0.76	0.79	0.35	0.67	0.70	0.79	0.35	0.69	0.71	0.85	0.80
	ESWL	0.74	0.70	0.73	0.33	0.70	0.64	0.67	0.31	0.69	0.71	0.83	0.80
XOE	DAD	0.82	0.71	1.10	0.44	0.72	0.63	0.86	0.35	0.68	0.62	0.46	0.82
	ESWL	0.83	0.71	1.03	0.42	0.73	0.60	0.80	0.35	0.68	0.60	0.45	0.82

a) CC = corner column; CEW = external column at west side of the building plan; CI = internal column; COL = core column at left side of the core; CES = external column at south; COR = core column at right side of the core; BESW = external beam at southern west; BES = external beam at south; BI = internal beam; BOS = core beam at south; BEWS = external beam at western south; BEW = external beam at west; BOW = core beam at west; XOS = core bracing at south; XOE = core bracing at east. See Figure 18.4d for details.

Table 18.3 Member sizes for design D_2 (in mm), and member nomenclature[a].

Members' type	Section ID		Depth	Width	Flange thickness	Web thickness
Bracing	D01–16	Box/Tube	350	350	14	14
	D17–32	Box/Tube	300	300	14	14
	D33–40	Box/Tube	200	200	12	12
	D41–47	Box/Tube	145	145	9	9
Column	Int01–16	Box/Tube	600	600	35	35
	Int17–32	Box/Tube	400	400	15	15
	Int33–40	Box/Tube	254	254	13	13
	Int41–47	Box/Tube	230	230	10	10
	Core01–10	Box/Tube	1800	1800	100	100
	Core11–20	Box/Tube	1600	1600	80	80
	Core21–30	Box/Tube	1200	1200	50	50
	Core31–40	Box/Tube	565	565	25	25
	Core41–47	Box/Tube	550	550	24	24
	ExCore01–16	Box/Tube	1300	1300	60	60
	ExCore17–32	Box/Tube	1100	1100	45	45
	ExCore33–47	Box/Tube	1000	1000	40	40
Beam	W10X39	I/Wide Flange	251.97	202.95	13.46	8.00
	W10X26	I/Wide Flange	261.62	146.56	11.18	6.60
	W10X19	I/Wide Flange	259.08	102.11	10.03	6.35

a) D01–16: Diagonal bracing from floors 1–16 and for all outriggers and belt trusses.
   D17–32: Diagonal bracing from floors 17–32.
   D33–40: Diagonal bracing from floors 33–40.
   D41–47: Diagonal bracing from floors 41–47.
   Int01–16: Internal column from floors 1–16.
   Int17–32: Internal column from floors 17–32.
   Int33–40: Internal column from floors 33–40.
   Int41–47: Internal column from floors 41–47.
   Core01–10: Core column from floors 1–10.
   Core11–20: Core column from floors 11–20.
   Core21–30: Core column from floors 21–30.
   Core31–40: Core column from floors 31–40.
   Core41–47: Core column from floors 41–47.
   ExCore01–16: External Core Column from floors 1–16.
   ExCore17–32: External Core Column from floors 17–32.
   ExCore33–47: External Core Column from floors 33–47.
   W10X39: Beam from floors 1–20.
   W10X26: Beam from floors 21–35.
   W10X19: Beam from floors 36–47.

presented in this section indicated that the use of $n_{pit} < 10$ could result in the underestimation of some peak DCIs by over 10% or more, whereas for $n_{pit} \geq 10$ the largest underestimation was almost constant at 3%.

To assess the efficiency of the ESWL procedure, the ratio r between ESWL and DAD computational times required to calculate design DCIs with MRI = 1700 years was obtained as a function of (i) the number of points n_{pit} and (ii) the number of members being analyzed. The dependence of the ratio r upon n_{pit} was found to be almost negligible. For 60 members r was approximately 0.4. The relative efficiency of the ESWL procedure increases when larger numbers of structural members are selected. For 1000 members r was approximately 0.2. The computation times for the DAD calculations were found to be fully compatible with practical capabilities of structural design offices. The computational times can be reduced by using parallel computing.

The differences between DAD- and ESWL-based DCIs are sufficiently small in this case that the designs D_1 and D_2 obtained by the DAD procedure on the one hand and the ESWL procedure on the other are the same for all the members considered in Tables 18.1 and 18.3. As pointed out in Chapter 14, this may not be the case for wind climates where winds from an unfavorable wind direction are dominant. As was also pointed out in Chapter 14, the ESWL procedure may not be practicable for buildings with irregular shapes.

For the number of members considered in the case study presented in this section, the ESWL procedure computation time on a personal computer was in the order of hours.[2] The computational time would have increased had the number of distinct members and the number of storm events been larger.

The amount of steel required for design D_1 was approximately 50% greater than for design D_0, that is, the capacities of the members in the preliminary design D_0 were too low. The iteration that followed the design D_1 resulted in a design D_2 for which the amount of steel was approximately 20% lower than for design D_1. The evolution of the successive designs can be followed by considering Table 18.2.

References

1 Park, S., Yeo, D., and Simiu, E., "Database-assisted design and equivalent static wind loads for mid- and high-rise structures: concepts, software, and user's manual," NIST Technical Note 2000, National Institute of Standards and Technology, Gaithersburg, MD, 2018. https://www.nist.gov/wind.

2 Ho, T., Surry, D., and Morrish, D. *NIST/TTU Cooperative Agreement/ Windstorm Mitigation Initiative: Wind Tunnel Experiments on Generic Low Buildings.* BLWT-SS20–2003, Boundary Layer Wind Tunnel Laboratory, University of Western Ontario, London, Canada, 2003.

3 Tamura, Y. "Aerodynamic Database of Low-Rise Buildings." Global Center of Excellence Program, Tokyo Polytechnic University, Tokyo, Japan, 2012.

---

2 The system specifications were as follows: Intel® Xeon® E5 CPU and 16 GB of RAM.

4 Yeo, D. (2014). Generation of large directional wind speed data sets for estimation of wind effects with long return periods. *Journal of Structural Engineering* 140: 04014073. https://www.nist.gov/wind.

5 ANSI/AISC, "Specification for Structural Steel Buildings," in *ANSI/AISC 360–10*, Chicago, Illinois: American Institute of Steel Construction, 2010.

6 TPU. TPU high-rise building aerodynamic database, Tokyo Polytechnic University (TPU). Available: http://wind.arch.t-kougei.ac.jp/system/eng/contents/code/tpu.

Part III

Aeroelastic Effects

Fundamentals and Applications

Certain types of civil engineering structures can experience aerodynamic forces generated by structural motions. These motions, called self-excited, are in turn affected by the aerodynamic forces they generate. The structural behavior associated with self-excited motions is called *aeroelastic*. The purpose of Part III is to provide an introduction to aeroelastic phenomena occurring in flexible civil engineering structures. Chapters 19, 20, and 21 consider, respectively, fundamental aspects of aeroelasticity phenomena associated with vortex lock-in, galloping and torsional divergence, and flutter. Presented here are applications are to chimneys with circular cross-sections and other slender structures including tall buildings (Chapter 22), and to suspended-span bridges (Chapter 23).

Iconic examples of aeroelastic instability are the flutter of the Brighton Chain Pier Bridge (termed "undulation" in the 1800s) (Figure III.1) and, more than one century later, the flutter of the original Tacoma-Narrows Bridge (Figure III.2).

To describe the interaction between aerodynamic forces and structural motions it is in principle necessary to solve the full equations of motion describing the flow, with time-dependent boundary conditions imposed by the moving structure. Even though progress is being made in the numerical solution of some aeroelastic problems, for bluff bodies immersed in shear, turbulent flow, the description of aeroelastic effects still relies largely on laboratory testing and empirical modeling. Owing to the violation of the Reynolds number similarity criterion, the applicability to the prototype of laboratory test results and of associated empirical models needs to be assessed as thoroughly as possible. However, for carefully modeled structures, aeroelastic test results are generally assumed to yield reasonably realistic results.

For additional fundamental and applied material on aeroelasticity in civil engineering, see [3]. The rich experience of the Japanese school of suspended-span bridge aeroelasticity is reflected in the abundant material contributed by Miyata in [4]. Ovalling oscillations, which can occur, for example, in certain types of silos, are considered in [5] and, using a Computational Wind Engineering (CWE) approach, in [6]. Aeroelastic motions of textile structures are considered in Chapter 26 (see, e.g., [7]).

Wind Effects on Structures: Modern Structural Design for Wind, Fourth Edition. Emil Simiu and DongHun Yeo.
© 2019 John Wiley & Sons Ltd. Published 2019 by John Wiley & Sons Ltd.

SKETCH showing the manner in which the 3rd span of the CHAIN PIER at BRIGHTON undulated
just before it gave way in a storm on the 20th of November 1836.

255 feet

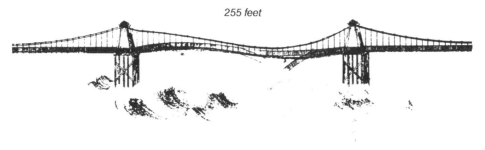

SKETCH showing the appearance of the 3rd span after it gave way.

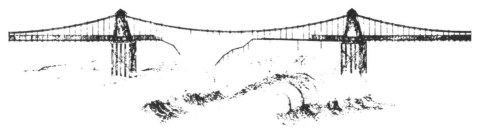

Figure III.1 Brighton chain pier failure, 1836. Source: From [1].

Figure III.2 Flutter of Tacoma Narrows suspension bridge, 1940. Source: From [2].

A number of empirical models allow design decisions to be based on results of relatively simple wind tunnel test results. For example, the designer of suspended-span bridges can account for the possibility of flutter by using empirical data, known as flutter derivatives, that can be measured in the laboratory. A more thorough approach can make use of detailed observations of flow patterns associated with the aeroelastic behavior of typically simple shapes. Fundamental studies of this type are considered in [4]; an example is reported in detail, with exemplary rigor, in [8].

References

1 Russel, J.D. On the vibration of suspension bridges and other structures, and the means of preventing injury from this cause. *Transactions of the Royal Society of Arts* 1841 (reproduced in [2]).

2 Farquharson, F.B. (ed.) (1949–1954). *Aerodynamic Stability of Suspension Bridges*, Part 1, Bulletin 116. Seattle, WA: University of Washington Engineering Experimental Station.

3 Scanlan, R.H. and Simiu, E. (2015). Aeroelasticity in civil engineering. In: *A modern course in aeroelasticity*, 5th ed. (ed. E.H. Dowell). Springer.

4 Simiu, E. and Miyata, T. (2006). *Design of Buildings and Bridges for Wind: A Practical Guide for ASCE-7 Standard Users and Designers of Special Structures*, 1st ed. Hoboken, NJ: John Wiley & Sons, Inc.

5 Paidoussis, M.P., Price, S.J., and de Langre, E. (2012). *Fluid-Structure Interactions*. Cambridge University Press.

6 Hillaewaere, J., Degroote, J., Lombaert, G. et al. (2015). Wind-structure interaction simulations of ovalling vibrations in silo groups. *Journal of Fluids and Structures* 59: 328–350. doi: 10.1016/j.jfluidstructs.2015.09.013.

7 Michalski, A., Kermel, P.D., Haug, E. et al. (2011). Validation of the computational fluid-structure interaction simulation at real-scale tests of a flexible 29 m umbrella in natural wind flow. *Journal of Wind Engineering and Industrial Aerodynamics* 99 (4): 400–413.

8 Hémon, P. and Santi, F. (2002). On the aeroelastic behaviour of rectangular cylinders in cross-flow. *Journal of Fluids and Structures* 16 (7): 855–889. doi: 10.1006/jfls.452.

19

Vortex-Induced Vibrations

19.1 Lock-In as an Aeroelastic Phenomenon

The shedding of vortices in the wake of a body gives rise to fluctuating *lift forces*. If the body is flexible, or if it has elastic supports, it will experience motions due to aerodynamic forces and, in particular, to the fluctuating lift force. As long as the motions are sufficiently small they do not affect the vortex-shedding frequency N_s, which remains proportional to the wind speed, in accordance with the relation

$$N_s = \frac{U\,St}{D} \tag{19.1}$$

(Section 4.4), where the Strouhal number, St, depends upon body geometry and the Reynolds number, D is a characteristic body dimension, and U is the mean velocity of the uniform flow, or a representative mean velocity in shear flow.

If the vortex-induced transverse deformations are sufficiently large, within an interval $N_s D/St - \Delta U < U < N_s D/St + \Delta U$, where $\Delta U/U$ is in the order of a few percent, the vortex shedding frequency no longer satisfies Eq. (19.1). Rather, because the body deformations influence the flow, the vortex shedding frequency will be constant for all wind speeds within that interval (Figure 19.1). This is an aeroelastic effect: while the flow affects the body motion, the body motion in turn affects the flow insofar as it produces *lock-in*; that is, a synchronization of the vortex-shedding frequency with the frequency of vibration of the body.

19.2 Vortex-Induced Oscillations of Circular Cylinders

A variety of vortex-induced oscillation models are available in which the aeroelastic forces depend upon adjustable parameters fitted to match experimental results. By construction, those models provide a reasonable description of the observed aeroelastic motions. However, the empirical models may not be valid as a motion predictor for conditions other than those of the experiments.

Consider a rigid circular cylinder in uniform, smooth flow. The across-wind force acting on the cylinder is approximately

$$F(t) = \frac{1}{2}\rho U^2 DC_{LS} \sin \omega_s t \tag{19.2}$$

Wind Effects on Structures: Modern Structural Design for Wind, Fourth Edition. Emil Simiu and DongHun Yeo.
© 2019 John Wiley & Sons Ltd. Published 2019 by John Wiley & Sons Ltd.

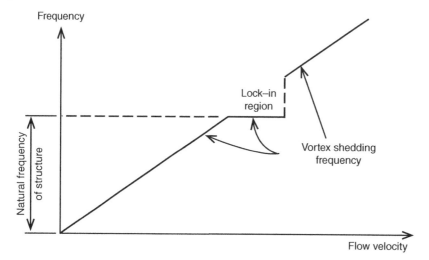

Frequency

Natural frequency of structure

Lock–in region

Vortex shedding frequency

Flow velocity

Figure 19.1 Frequency of vortex shedding in the wake of an elastic structure as a function of wind velocity.

where $\omega_s = 2\pi N_s$, N_s satisfies the Strouhal relation (Eq. [19.1]), and C_{LS} is a lift coefficient. For a circular cylinder in uniform smooth flow and Reynolds number $40 < Re < 3 \times 10^5$, $C_{LS} \approx 0.6$ [1, p. 7]).

For a cylinder allowed to oscillate, Eq. (19.2) is inadequate for two reasons. First, the across-wind force increases with oscillation amplitude until a limiting amplitude is reached. Second, the spanwise correlation of the across-wind force also increases, as indicated in Figure 19.2.

Let y denote the across-wind displacement of a cylinder of unit length for which the imperfect spanwise force correlation is not explicitly accounted for. The equation of motion of the cylinder can be written as

$$m\ddot{y} + c\dot{y} + ky = F(y, \dot{y}, \ddot{y}, t) \tag{19.3}$$

where m is the cylinder mass, c is the mechanical damping constant, k is the spring stiffness, and F is the fluid-induced force per unit span, which may be dependent on the displacement y and its first and second derivatives, as well as on time. Most empirical models recognize the near-sinusoidal response of the cylinder at the Strouhal frequency and the natural frequency of vibration of the structure. Unless the velocity is at the lock-in values the response gives rise to a beating oscillation. Figure 19.3 parts a–c show the displacements y and their spectral densities for an elastically supported cylinder before, at, and after lock-in, respectively.

Scanlan [4] proposed the following simple model:

$$m[\ddot{y} + 2\zeta\omega_1\dot{y} + \omega_1^2 y] = \frac{1}{2}\rho U^2 D \left[Y_1(K) \left(1 - \varepsilon\frac{y^2}{D^2} \right) \frac{\dot{y}}{U} + Y_2(K)\frac{y}{D} \right.$$
$$\left. + C_L(K)\sin(\omega t + \phi) \right] \tag{19.4}$$

where m is the body mass per unit length, ζ is the damping ratio, ω_1 is the frequency of vibration of the body, D is the cylinder's diameter, U is the flow velocity, ρ is the density of the fluid, $K = \omega D/U$ is the reduced frequency, and the vortex-shedding

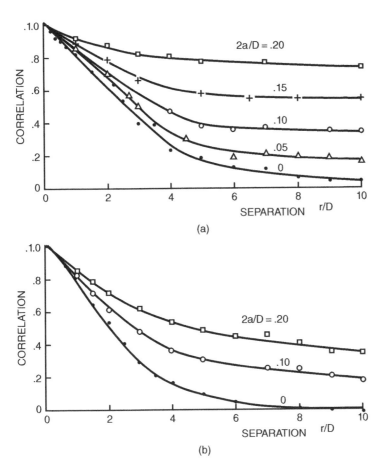

Figure 19.2 The effect of increasing the oscillation amplitude $a/2$ of a circular cylinder of diameter D on the correlation between pressures at points separated by distance r along a generator: (a) smooth flow; (b) flow with 11% turbulence intensity. Reynolds number: 2×10^4. Source: Reprinted from [2] with permission of Cambridge University Press.

frequency $n = \omega/2\pi$ satisfies the Strouhal relation $n = U St/D$; Y_1, Y_2, ε, and C_L are adjustable parameters that must be fitted to experimental results. As is the case for the van der Pol oscillator, the amplitude y is self-limiting. The first term within the brackets in the right-hand side of Eq. (19.4) is proportional to $\dot{y}$ and may therefore be viewed as a damping term of aerodynamic origin. For low amplitudes y that term is positive, meaning that the sum of the mechanical and aerodynamic damping forces can be negative, in agreement with the physical fact that the flow promotes the cylinder's motion by transferring energy to the body. The reverse is true for high amplitudes, where the body loses energy by transferring it to the flow.

At lock-in, $\omega \approx \omega_1$, and the last two terms in the right-hand side of Eq. (19.4) are relatively small and can be neglected. Then Y_1 and ε remain to be determined by experiment. At steady amplitudes the average energy dissipation per cycle is zero, so that

$$\int_0^T \left[4m\zeta\omega - \rho UDY_1 \left(1 - \varepsilon\frac{y^2}{D^2} \right) \right] \dot{y}^2 dt = 0 \tag{19.5}$$

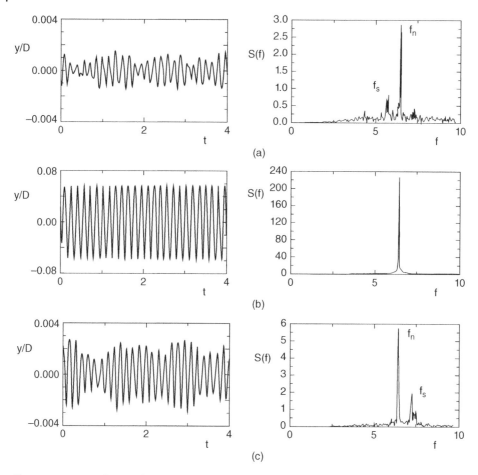

Figure 19.3 Across-flow oscillations y/D of elastically supported circular cylinder: (a) before lock-in; (b) at lock-in; (c) after lock-in. The Strouhal frequency and the natural frequency of vibration of the body, f_s and f_n, respectively, are shown in the spectral density plots $S(f)$ [3], with permission from the American Society of Civil Engineers (ASCE).

where $T = 2\pi/\omega$. Assuming that the oscillation $y(t)$ is practically harmonic,

$$y = y_0 \cos \omega t \tag{19.6}$$

leads to the results

$$\int_0^T \dot{y}^2 dt = \omega y_0^2 \pi \tag{19.7}$$

$$\int_0^T y^2 \dot{y}^2 dt = \omega y_0^4 \frac{\pi}{4} \tag{19.8}$$

Then Eq. (19.5) yields the steady amplitude solution

$$\frac{y_0}{D} = 2 \left[\frac{Y_1 - 8\pi S_{sc} St}{\varepsilon Y_1} \right]^{1/2} \tag{19.9}$$

where St is the Strouhal number and

$$S_{sc} = \frac{\zeta m}{\rho D^2} \tag{19.10}$$

is the Scruton number.

If, at lock-in velocity, the mechanical model is displaced to an initial amplitude A_0 and then released, it will undergo a decaying response until it reaches a steady state with amplitude y_0 given by Eq. (19.9) (Figure 19.4). A time-dependent expression for the decaying oscillation amplitude, derived in [5], yields the maximum amplitudes shown in Figure 19.5, which are close to those yielded by an empirical formula obtained in [5] and plotted in Figure 19.5:

$$\frac{y_0}{D} = \frac{1.29}{[1 + 0.43(8\pi^2 St^2 S_{sc})]^{3.35}} \tag{19.11}$$

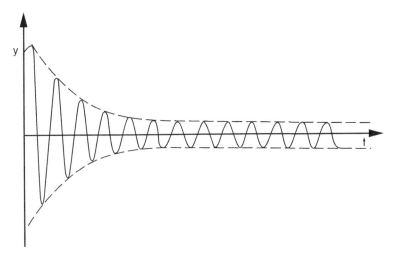

Figure 19.4 Decaying oscillation to steady state of bluff, elastically sprung model under vortex lock-in excitation.

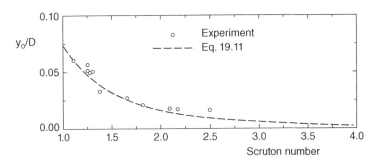

Figure 19.5 Maximum amplitudes versus Scruton number. Experiment: o; Eq. (19.11): - - - -. Source: Reprinted from [6], with permission from Elsevier.

19.3 Across-Wind Response of Chimneys and Towers with Circular Cross Section

A model similar to Eq. (19.4) was developed in [7] for application to the design of chimneys and towers with circular cross-section. Differences between this model and Eq. (19.4) are as follows. It is noted in [7] that $\rho U^2 Y_2(K) \ll m\omega_1^2$. The term $Y_2(K)y/D$ is therefore neglected, and since the actual motion of the chimneys or towers is random rather than periodic, the term $\varepsilon y^2/D^2$ of Eq. (19.4) is replaced by the ratio $\overline{y^2}/(\lambda D)^2$ where λ is a coefficient whose physical significance is discussed subsequently. The term

$$\frac{1}{2}\rho U^2 D Y_1(K)\left(1 - \varepsilon\frac{y^2}{D^2}\right)\frac{\dot{y}}{U} \tag{19.12}$$

of Eq. (19.4) is written in the form

$$2\omega_1 \rho D^2 K_{a0}\left(\frac{U}{U_{cr}}\right)\left(1 - \frac{\overline{y^2}}{(\lambda D)^2}\right)\dot{y} \tag{19.13}$$

where $K_{a0}(U/U_{cr})$ is an aerodynamic coefficient and $U_{cr} = \omega_1 D/(2\pi\, St)$ is the velocity that produces vortex shedding with frequency n_1. This term is equated to the product $-2m\zeta_a\omega_1$, where ζ_a is defined as the aerodynamic ratio, which may thus be written as

$$\zeta_a = -\frac{\rho D^2}{m}K_{a0}\left(\frac{U}{U_{cr}}\right)\left[1 - \frac{\overline{y^2}}{(\lambda D)^2}\right] \tag{19.14}$$

For $\overline{y^2}^{1/2} = \lambda D$ the aerodynamic damping vanishes, so the structure no longer experiences aeroelastic effects causing the response to increase. The coefficient λ may thus be interpreted as the ratio between the limiting r.m.s. value of the aeroelastic response and the diameter D. The total damping ratio of the system is then

$$\zeta_t = \zeta + \zeta_a \tag{19.15}$$

where ζ is the mechanical damping ratio. The aeroelastic effects are thus introduced by substituting into the equation of motion the total damping ζ_t for the mechanical damping ratio ζ.

This simple approach was validated in [7] against experimental results shown in Figure 19.6, which represents the dependence of the measured response $\eta_{rms} = \overline{y^2}^{1/2}/D$ upon the reduced wind speed $2\pi U/\omega_1 D$ for various damping ratios ζ. Figure 19.7 shows calculated versus measured ratios $\overline{y^2_{max}}^{1/2}/D$ for various values of the damping parameter $K_s = m\zeta/(\rho D^2)$, where $\overline{y^2_{max}}^{1/2}$ is the r.m.s. response corresponding to the most unfavorable reduced wind speed. In Figure 19.7, (i) the forced vibration regime corresponds to vibrations induced quasi-statically by the vorticity in the wake of the cylinder, and (ii) the lock-in regime corresponds to vibrations due to aeroelastic effects. A transition regime is observed between (i) and (ii). Turbulence in the oncoming flow decreases the coherence of the vorticity shed in the wake of the body, and reduces the magnitude of the across-wind response. Vibrations typical of these regimes are shown in Figure 19.8. The ratios of the peak to r.m.s. response are about 4.0 in the forced vibration regime and about $\sqrt{2}$ in the lock-in regime.

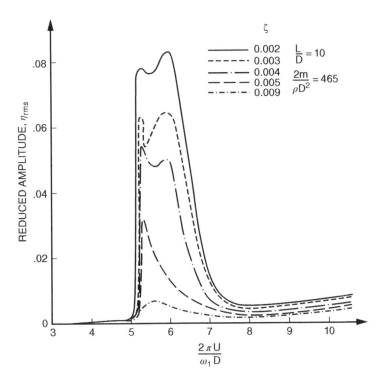

Figure 19.6 Response of a model stack of circular cross section and length L for different values of the mechanical damping (*Re* subcritical). Source: From [8]. Courtesy of National Physical Laboratory, U.K.

Based on inferences from experimental data available in the literature, [7] developed curves representing (i) the dependence upon Reynolds number of the largest value of $K_{a0}(U/U_{cr})$ in smooth flow, $K_{a0\max}$, and (ii) the dependence of the ratio $K_{a0}(U/U_{cr})/K_{a0\max}$ upon U/U_{cr} for smooth flow and flows with various turbulence intensities $\overline{u^2}^{1/2}/U$ (Figure 19.9).

For a vertical structure experiencing random motions described by the relation

$$\overline{y^2(z)} = \sum_i \overline{\xi_i^2} y_i^2(z) \tag{19.16}$$

where $\overline{y^2}$ is the r.m.s. response, $\overline{\xi_i}$ and y_i are the r.m.s. modal coefficient and the modal shape, respectively, for mode i [9], the following expression is proposed for the total damping in the ith mode:

$$\zeta_{ti} = \zeta_i + \zeta_{ai} \tag{19.17}$$

$$\zeta_{ai} = -\frac{\rho D_0^2}{m_{ei}} \left[2K_{1i} - K_{2i}\frac{\overline{\xi_i^2}}{D_0^2} \right] \tag{19.18}$$

$$K_{1i} = \frac{\int_0^h K_{a0}(z)\left[\frac{D(z)}{D_0}\right]^2 y_i^2(z)\,dz}{\int_0^h y_i^2(z)\,dz} \tag{19.19}$$

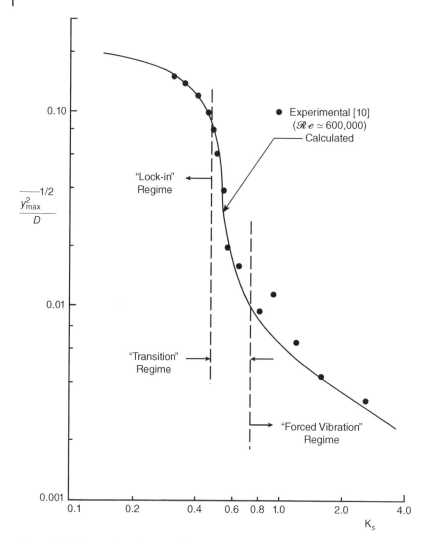

Figure 19.7 Measured and estimated response in smooth flow. Source: Reprinted from [7], with permission from Elsevier.

$$K_{2i} = \frac{\int_0^h K_{a0}(z) y_i^4(z) dz}{\lambda^2 \int_0^h y_i^2(z) dz} \tag{19.20}$$

where ζ_i and ζ_{ai} are the mechanical and the aerodynamic damping in the ith mode of vibration, respectively, D_0 is the diameter at elevation $z = 0$, $D(z)$ is the diameter at elevation z, h is the height of the structure, m_{e_i} is the equivalent mass per unit length in the i-th mode of vibration, defined as

$$m_{ei} = \frac{M_t}{\int_0^h y_i^2(z) dz} \tag{19.21}$$

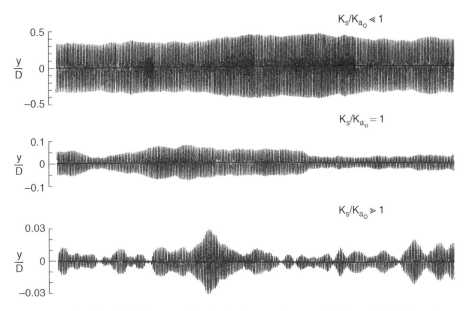

Figure 19.8 Simulated displacement histories for low, moderate, and high mechanical damping. Source: Reprinted from [7], with permission from Elsevier.

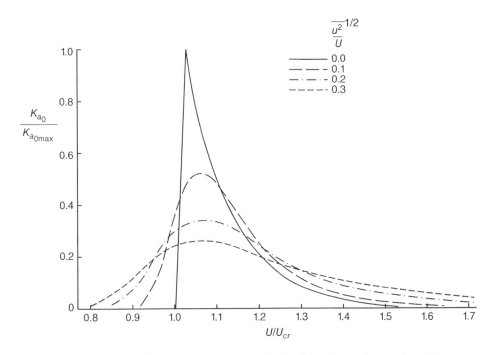

Figure 19.9 Dependence of ratio K_{a0}/K_{a0max} upon ratio U/U_{cr} for various turbulence intensities. Source: Reprinted from [7], with permission from Elsevier.

and M_i is the generalized mass in the ith mode. Equations 19.17–19.20 are based on the assumption that aeroelastic effects occurring at various elevations are linearly superposable.

For the relatively small values of the response that are acceptable for chimneys and stacks, the estimated response depends weakly upon the assumed value of λ. It is suggested in [9] that the value $\lambda = 0.4$ is reasonable for use in estimates of the response of concrete chimneys.

References

1 Bishop, R.E.D. and Hassan, A.Y. (1964). The lift and drag forces on a circular cylinder oscillating in a flowing fluid. *Proceedings of the Royal Society of London. Series A. Mathematical and Physical Sciences* 277: 51–75.

2 Novak, M. and Tanaka, H. (1976). Pressure correlations on vertical cylinders. In: *Fourth International Conference on Wind Effects on Structures* (ed. K.J. Eaton), 227–332. Heathrow, UK.

3 Goswami, I., Scanlan, R.H., and Jones, N.P. (1993). Vortex-induced vibration of circular cylinders. I: experimental data. *Journal of Engineering Mechanics* 119: 2270–2287.

4 Simiu, E. and Scanlan, R.H. (1996). *Wind Effects on Structures*, 3rd ed. Hoboken: John Wiley & Sons.

5 Ehsan, F. and Scanlan, R.H. (1990). Vortex-induced vibrations of flexible bridges. *Journal of Engineering Mechanics* 116: 1392–1411.

6 Griffin, O.M., Skop, R.A., and Ramberg, S.E., "The resonant, vortex-excited vibrations of structures and cable systems," presented at the Offshore Technology Conference, Houston, Texas, 1975.

7 Vickery, B.J. and Basu, R.I. (1983). Across-wind vibrations of structures of circular cross-section. Part I. Development of a mathematical model for two-dimensional conditions. *Journal of Wind Engineering and Industrial Aerodynamics* 12: 49–73.

8 Wooton, L. R. and Scruton, C., "Aerodynamic stability," in Modern Design of Wind-Sensitive Structures. London, UK: Construction Industry Research and Information Association, 1971, pp. 65–81

9 Basu, R.I. and Vickery, B.J. (1983). Across-wind vibrations of structure of circular cross-section. Part II. Development of a mathematical model for full-scale application. *Journal of Wind Engineering and Industrial Aerodynamics* 12: 75–97.

10 Wooton, L. R. (1969). Oscillations of Large Circular Stacks in Wind. *Proceedings of the Institution of Civil Engineers*, 43, pp. 573–598.

20

Galloping and Torsional Divergence

20.1 Galloping Motions

Galloping is a large-amplitude aeroelastic oscillation (one to ten or more cross-sectional dimensions of the body) that can be experienced by elastically restrained cylindrical bodies with certain types of cross-section (e.g., square section, D-section, ice laden power cables). For material on wake galloping of power transmission lines grouped in bundles, see, for example, Ref. [1].

20.1.1 Glauert–Den Hartog Necessary Condition for Galloping Motion

Consider first a fixed cylinder immersed in a flow with velocity U_r. Assume the angle of attack is α (Figure 20.1). The positive y-coordinate in Figure 20.1 is *downward*. The mean drag and lift are, respectively,

$$D(\alpha) = \frac{1}{2}\rho U_r^2 BC_D(\alpha), \tag{20.1}$$

$$L(\alpha) = \frac{1}{2}\rho U_r^2 BC_L(\alpha) \tag{20.2}$$

The sum of the projections of these components on the direction y is

$$F_y(\alpha) = -D(\alpha)\sin\alpha - L(\alpha)\cos\alpha \tag{20.3}$$

If $F_y(\alpha)$ is written in the alternative form

$$F_y(\alpha) = \frac{1}{2}\rho U^2 BC_{F_y}(\alpha) \tag{20.4}$$

where $U = U_r \cos\alpha$, it is easily verified that there follows from Eqs. (20.1)–(20.4)

$$C_{F_y}(\alpha) = \frac{-[C_L(\alpha) + C_D(\alpha)\tan\alpha]}{\cos\alpha} \tag{20.5}$$

Consider now the case in which, in a flow with velocity U, the body *oscillates* in the across-flow direction y (Figure 20.2). The magnitude of the relative velocity of the flow with respect to the moving body is denoted by U_r and can be written as

$$U_r = (U^2 + \dot{y}^2)^{1/2} \tag{20.6}$$

The angle of attack, denoted by α, is

$$\alpha = \arctan\frac{\dot{y}}{U} \tag{20.7}$$

Wind Effects on Structures: Modern Structural Design for Wind, Fourth Edition. Emil Simiu and DongHun Yeo.
© 2019 John Wiley & Sons Ltd. Published 2019 by John Wiley & Sons Ltd.

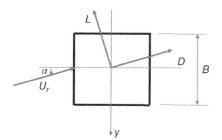

Figure 20.1 Lift and drag on a fixed bluff object.

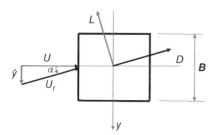

Figure 20.2 Effective angle of attack on an oscillating bluff object.

The equation of motion of the body in the y direction is

$$m[\ddot{y} + 2\zeta\omega_1\dot{y} + \omega_1^2 y] = F_y \tag{20.8}$$

where m is the mass per unit length, ζ is the damping ratio, and ω_1 is the natural circular frequency. F_y denotes the aerodynamic force acting on the body in the direction normal to the mean flow. It is assumed that the mean aerodynamic lift and drag coefficients $C_L(\alpha)$ and $C_D(\alpha)$ for the oscillating body and for the fixed body are the same, so that $F_y(\alpha)$ is given by Eq. (20.4) and $C_{F_y}(\alpha)$ is given by Eq. (20.5).

Consider now the case of incipient (small) motion, that is, the condition in the vicinity of $\dot{y} = 0$, wherein

$$\alpha \approx \frac{\dot{y}}{U} \approx 0 \tag{20.9}$$

For this condition

$$F_y \approx \left. \frac{dF_y}{d\alpha} \right|_{\alpha=0} \alpha \tag{20.10}$$

Differentiation of Eq. (20.5) yields

$$\left. \frac{dC_{F_y}}{d\alpha} \right|_{\alpha=0} = -\left(\frac{dC_L}{d\alpha} + C_D \right)_{\alpha=0} \tag{20.11}$$

The equation of motion thus takes the form

$$m[\ddot{y} + 2\zeta\omega_1\dot{y} + \omega_1^2 y] = -\frac{1}{2}\rho U^2 B \left(\frac{dC_L}{d\alpha} + C_D \right)_{\alpha=0} \frac{\dot{y}}{U} \tag{20.12}$$

Considering the aerodynamic (right-hand) side of Eq. (20.12) as a contribution to the overall system damping, the net damping coefficient of the system is

$$2m\omega_1\zeta + \frac{1}{2}\rho UB \left(\frac{dC_L}{d\alpha} + C_D \right)_{\alpha=0} = d \tag{20.13}$$

The condition for the occurrence of instability is that $d < 0$. Since $\zeta > 0$, for this condition to be satisfied it is necessary that

$$\left(\frac{dC_L}{d\alpha} + C_D \right)_{\alpha=0} < 0 \qquad (20.14)$$

Equation (20.14) is the Glauert–Den Hartog necessary condition for incipient galloping motion (a sufficient condition being $d < 0$) [1]. It follows from Eq. (20.14) that circular cylinders, for which $\frac{dC_L}{d\alpha} = 0$, cannot gallop.

The physical interpretation of Eq. (20.14) is the following. Let the body experience a small perturbation from its position of equilibrium that causes it to acquire a velocity $\dot{y}$. The perturbation causes an asymmetry in the aerodynamic forces that act on the body. If the body's aerodynamic properties are such that this asymmetry causes the initial velocity to increase, galloping motion will occur. Otherwise the body will be restored to its position of equilibrium.

To summarize: the susceptibility of a slender prismatic body to galloping instability can be assessed by evaluating its mean lift and drag coefficients and determining whether the left-hand side of Eq. (20.14) is negative. For example, plots of the drag and lift coefficients show that according to the Glauert–Den Hartog criterion the octagonal cylinder of Figure 20.3 is susceptible to galloping for angles $-5° < \alpha < 5°$ [2].

For a simple demonstration of the galloping motion of a square cylinder, see https://www.nist.gov/wind.

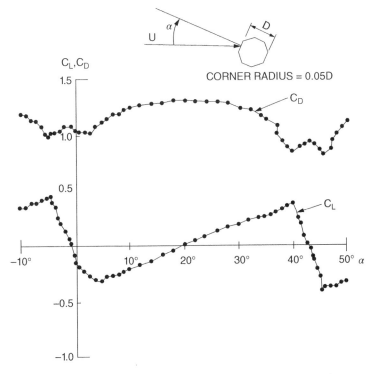

Figure 20.3 Force coefficients for an octagonal cylinder ($Re = 1.2 \times 10^6$). Source: Courtesy of Dr. R. H. Scanlan.

Tests have shown that the derivatives $\frac{dC_{F_y}}{d\alpha}$ are not dependent upon the frequency of the body motion and can be obtained from aerodynamic force measurements on the *fixed* body. The quantities $\frac{dC_{F_y}}{d\alpha}$ are called *steady-state aerodynamic lift coefficient derivatives* or, for short, steady-state aerodynamic derivatives. In the case of flutter, the aeroelastic behavior is characterized by quantities of a similar nature, called flutter aerodynamic derivatives that, unlike the steady-state derivatives that characterize galloping motion, depend upon the oscillation frequency. This difference is commented upon in Chapter 23.

20.1.2 Modeling of Galloping Motion

Galloping motion was described in [3] by developing the lift coefficient C_{F_y} in powers of $\frac{\dot{y}}{U}$:

$$C_{F_y}(\alpha) = A_1\left(\frac{\dot{y}}{U}\right) - A_2\left(\frac{\dot{y}}{U}\right)^2\frac{\dot{y}}{|\dot{y}|} - A_3\left(\frac{\dot{y}}{U}\right)^3 + A_5\left(\frac{\dot{y}}{U}\right)^5 - A_7\left(\frac{\dot{y}}{U}\right)^7 \quad (20.15)$$

If the dependence of C_D and C_L upon α is known, the coefficients A_1 through A_7 can be evaluated as follows. First, the coefficient C_{F_y} is plotted against $\tan \alpha = \frac{\dot{y}}{U}$ using Eq. (20.5). The coefficients in Eq. (20.15) can then be estimated on the basis of this plot, for example by using a least squares technique. Reference [3] applies the methods of Kryloff and Bogoliuboff [4] to the resulting nonlinear equation, postulating as a first approximation the solution

$$y = a \, \cos(\omega_1 t + \phi) \quad (20.16)$$

where a and ϕ are considered to be slowly varying functions of time. Depending upon whether the coefficient A_1 is less than, equal to, or larger than zero, three basic types of curves C_{F_y} are identified as functions of α, with the corresponding galloping response amplitudes as functions of the reduced velocity $U/D\omega_1$ – see Figure 20.4. The only possible oscillatory motions are those with amplitudes a traced in full lines. If the speed increases from U_0 to U_2 (Figure 20.4a), the amplitude of the response is likely to jump from the lower to the upper branch of the solid curve. If the speed decreases from U_2 to U_0 the jump occurs from the upper to the lower curve.

An elegant mathematical investigation into the nonlinear modeling of galloping motions is reported in [5].

20.1.3 Galloping of Two Elastically Coupled Square Cylinders

Reference [6] describes an experiment conducted in a water tunnel on the behavior of a system of two elastically restrained and coupled aluminum square bars with sides $h_1 = h_2 = 6.35$ mm and length 0.215 m. The spring constants were $k_1 = 56$, $k_2 = 78$, and $k_{12} = 145$ N m^{-1} (Figure 20.5). To prevent displacements due to drag, the bar ends were attached to fixed points by thin wires with lengths $r = 400$ mm. The bars were observed to gallop in phase, but except for relatively low flow speeds U, this oscillatory form alternated in unpredictable, chaotic fashion with a second oscillatory form wherein the two bars galloped with higher frequency in opposite phases (Figures 20.6a, b). The mean

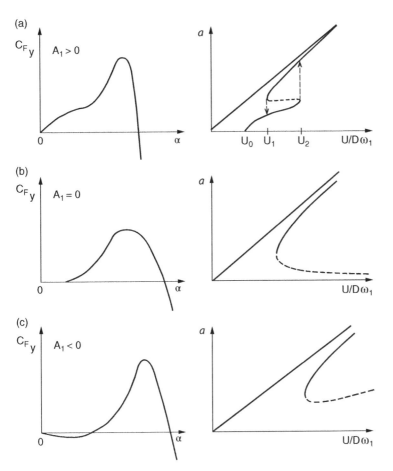

Figure 20.4 Three basic types of across-wind force coefficients and the corresponding galloping response amplitudes *a*. Source: From [3]. With permission from the American Society of Civil Engineers (ASCE).

Figure 20.5 Schematic of double galloping oscillator.

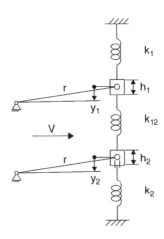

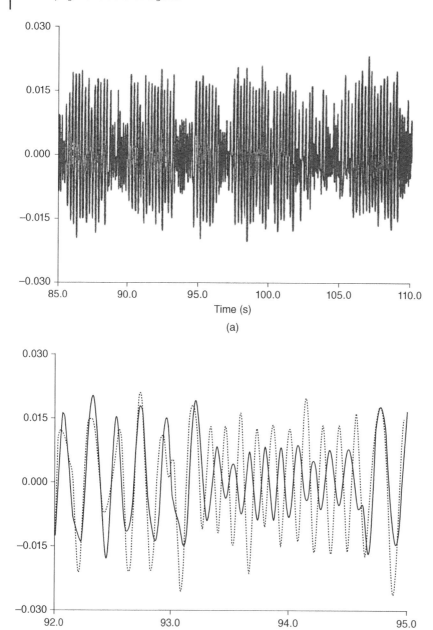

Figure 20.6 (a) Observed time history of displacement y_1; (b) observed time history of displacements y_1 (solid line) and y_2 (interrupted line). Source: From [6].

exit time of the system from the region of phase space corresponding to the in-phase oscillations decreased as the flow velocity increased.

20.2 Torsional Divergence

Torsional divergence, also called lateral buckling, can occur on airfoils or bridge decks. Like galloping, it can be modeled by using aerodynamic properties measured on the body at rest.

The parameters of the torsional divergence problem are shown in Figure 20.7, in which U is the horizontal wind velocity, α is the angle of rotation of the bridge deck about the elastic axis, and k_α is the torsional stiffness.

The aerodynamic moment per unit span is:

$$M(\alpha) = \frac{1}{2}\rho U^2 B^2 C_M(\alpha) \tag{20.17}$$

where B is the bridge deck width and $C_M(\alpha)$ is the aerodynamic moment coefficient about the elastic axis. For small α

$$M(\alpha) \approx \frac{1}{2}\rho U^2 B^2 \left[C_M(0) + \frac{dC_M}{d\alpha}\bigg|_{\alpha=0} \alpha \right] \tag{20.18}$$

Let $\lambda = \frac{1}{2}\rho U^2 B^2 > 0$. Equating $M(\alpha)$ to the internal torsional moment $k_\alpha \, \alpha$ yields

$$\alpha = \frac{\lambda C_M(0)}{k_\alpha - \lambda \frac{dC_M}{d\alpha}\big|_{\alpha=0}} \tag{20.19}$$

Divergence occurs when α goes to infinity for vanishing values of the denominator in Eq. (20.19). The critical divergence velocity is

$$U_{cr} = \sqrt{\frac{2k_\alpha}{\rho B^2 \frac{dC_M}{d\alpha}\big|_{\alpha=0}}} \tag{20.20}$$

In most cases of interest in civil engineering applications the critical divergence velocities are well beyond the range of velocities normally considered in design.

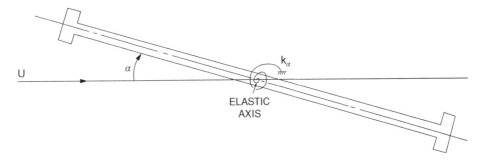

Figure 20.7 Parameters for the torsional divergence problem.

References

1 Den Hartog, J.P. (1956). *Mechanical Vibrations*, 4th ed. New York: McGraw-Hill.
2 Scanlan, R. H. and Wardlaw, R. L., "Reduction of flow-induced structural vibrations," in *Isolation of mechanical vibration, impact, and noise: A colloquium presented at the ASME Design Engineering Technical Conference*, Cincinnati, OH, 1973.
3 Novak, M. (1972). Galloping oscillations of prismatic structures. *Journal of the Engineering Mechanics Division* 98: 27–46.
4 Kryloff, N. and Bogoliuboff, N. (1947). *Introduction to Non-linear Mechanics*. Princeton: Princeton University Press.
5 Parkinson, G.V. and Smith, J.D. (1964). The square prism as an aeroelastic non-linear oscillator. *The Quarterly Journal of Mechanics and Applied Mathematics* 17: 225–239.
6 Simiu, E. and Cook, G.R. (1992). Empirical fluid-elastic models and chaotic galloping: A case study. *Journal of Sound and Vibration* 154: 45–66.

21

Flutter

Flutter is an aeroelastic phenomenon that occurs in flexible bodies with relatively flat shapes, such as airplane wings and bridge decks. It involves oscillations with amplitudes that grow in time and can result in catastrophic structural failure. Like other aeroelastic phenomena, flutter entails the solution of equations of motion involving inertial, mechanical damping, elastic restraint, and aerodynamic forces (including forces induced by self-excited motions) that depend upon the ambient flow and the shape and motion of the body.

Assume that the mechanical damping is negligible. The motion of the body is aeroe-lastically *stable* if, following a small perturbation away from its position of equilibrium, the body will revert to that position owing to stabilizing self-excited forces associated with the perturbation. As the flow velocity increases, the aerodynamic forces acting on the body change, and for certain elongated body shapes, at a critical value of the flow velocity the self-excited forces may cause the body to be neutrally stable. For velocities larger than the critical velocity the oscillations initiated by a small perturbation from the position of equilibrium will grow in time. The self-excited forces that cause these growing oscillations can be viewed as producing a *negative aerodynamic damping* effect.

The main difficulty in solving the flutter problem for bridges is the development of expressions for the self-excited forces. For thin airfoil flutter in incompressible flow, expressions for the self-excited forces due to small oscillations have been derived by Theodorsen [1]. However, the airfoil solutions are in general not applicable to bridge sections.

Although it is accompanied at all times by vortex shedding with frequency equal to the flutter frequency, *flutter is a phenomenon distinct from vortex-induced oscillation*. The latter entails aeroelastic flow-structure interactions only for flow velocities at which the frequency of the vortex shedding is equal or close to the structure's natural fre-quency; for velocities higher or lower than those at which lock-in occurs the oscillations are much weaker than at lock-in. In contrast, for velocities higher than those at which flutter sets in, the strength of the oscillations increases monotonically with velocity.

To date, one of the most influential contributions to solving the flutter problem for bridges is Scanlan's simple conceptual framework wherein the self-excited forces due to small bridge deck oscillations can be characterized by fundamental functions called flut-ter aerodynamic derivatives [2]. As shown earlier, in the galloping case the self-excited forces depend on the steady-state derivatives $dC_{F_y}/d\alpha$ that are not significantly affected by vorticity and may therefore be obtained from measurements on the *fixed* body. In contrast, owing to the elongated shapes of bodies susceptible to flutter, the aerodynamic

Wind Effects on Structures: Modern Structural Design for Wind, Fourth Edition. Emil Simiu and DongHun Yeo.
© 2019 John Wiley & Sons Ltd. Published 2019 by John Wiley & Sons Ltd.

derivatives of a body susceptible to flutter must be obtained from measurements on the *oscillating* body. This is the case because the aerodynamic pressures on the body are significantly affected by vortices induced by, and occurring at the frequency of, the torsional oscillations of the bridge.

In its detail, flutter in practically all cases involves nonlinear aerodynamics. It has been possible in a number of instances, however, to treat the problem successfully by linear analytical approaches. This is the case for two main reasons: First, the supporting structure is usually treatable as linearly elastic and its actions dominate the form of the response, which is usually an exponentially modified sinusoidal oscillation. Second, it is the incipient or starting condition, which may be treated as having only small amplitude, that separates the stable and unstable regimes. These two main features enable a flutter analysis to be based on the standard stability considerations of linear elastic systems.

It is characteristic of flutter as a typical self-excited oscillation that, by means of its deflections and their time derivatives, a structural system taps off energy from the wind flow. If the system is given an initial disturbance, its motion will either decay or grow according to whether the energy of motion extracted from the flow is less than or exceeds the energy dissipated by the system through mechanical damping. The theoretical dividing line between the decaying and the sustained sinusoidal oscillation due to an initial disturbance, is then recognized as the critical flutter condition.

Section 21.1 considers two-dimensional (2-D) bridge deck behavior in smooth flow. Section 21.2 briefly reviews the expression for the aerodynamic lift and moment acting on airfoils. Section 21.3 introduces the aerodynamic lift, drag and moment acting on bridge decks. Section 21.4 concerns the solution of the flutter equations for bridges. Section 21.5 discusses the bridge response to turbulent wind in the presence of aeroelastic effects.

21.1 Formulation of the Two-Dimensional Bridge Flutter Problem in Smooth Flow

In the 2-D case the bridge deformations are the same throughout the bridge span. Bridge decks are typically symmetrical, that is, their elastic and mass centers coincide. The dependence of flutter derivatives upon the oscillation frequency n of the fluttering body can be expressed in terms of the non-dimensional reduced frequency

$$K = 2\pi Bn/U, \tag{21.1}$$

where B is the width of the deck, and U is the mean wind flow velocity. If the horizontal displacement p of the deck is also taken into account, the equations of motion of a two-dimensional section of a symmetrical bridge deck with linear mechanical damping and elastic restoring forces in smooth flow can be written as

$$m\ddot{h} + c_h\dot{h} + k_h h = L_{ae} \tag{21.2a}$$

$$I\ddot{\alpha} + c_\alpha\dot{\alpha} + k_\alpha\alpha = M_{ae} \tag{21.2b}$$

$$m\ddot{p} + c_p\dot{p} + k_p p = D_{ae} \tag{21.2c}$$

where h, α, and p are the vertical displacement, torsional angle, and horizontal displacement of the bridge deck, respectively (see Figure 21.1 for notations pertaining to h and α;

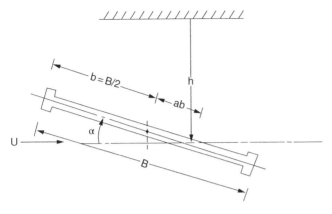

Figure 21.1 Notations.

similar notations are applicable to airfoils). A unit span is acted upon by the aeroelastic lift L_{ae}, moment M_{ae}, and drag D_{ae}, and has mass m, mass moment of inertia I, vertical, torsional and horizontal restoring forces with stiffness k_h, k_α, and k_p, respectively, and mechanical damping coefficients c_h, c_α, and c_p.

21.2   Aeroelastic Lift and Moment Acting on Airfoils

It is instructive at this point to briefly consider the modeling of the aeroelastic lift L_h and moment M_α acting on airfoils as shown in Figure 21.1. (For airfoils, p displacements are negligible.) Using basic principles of potential flow theory and an elegant mathematical technique involving conformal mapping, Theodorsen showed that, for small airfoil motions in incompressible flow, the expressions for L_h and M_α are linear in h and α and their first and second derivatives [1]. The coefficients in these expressions, called *aerodynamic coefficients*, are defined in terms of the complex function $C(K) = F(K) + iG(K)$, known as *Theodorsen's circulation function* (Figure 21.2), in which $K = b\,\omega/U$ is the reduced frequency, b is the half-chord of the airfoil, U is the flow velocity, and ω is the circular frequency of oscillation.

Theodorsen's theory yields the following expressions for the harmonically oscillating lift and moment:

$$L_{ae} = -\pi\rho b^2(U\dot\alpha + \ddot h - ab\ddot\alpha) - 2\pi\rho UC(K)\left[U\alpha + \dot h + b\left(\frac{1}{2} - a\right)\dot\alpha\right] \qquad (21.3a)$$

$$M_{ae} = -\pi\rho b^2\left\{\left(\frac{1}{2} - a\right)Ub\dot\alpha + b^2\left(\frac{1}{8} + a^2\right)\ddot\alpha - ab\ddot h\right\}$$
$$+ 2\pi\rho U\left(\frac{1}{2} + a\right)b^2C(K)\left[U\alpha + \dot h + b\left(\frac{1}{2} - a\right)\dot\alpha\right] \qquad (21.3b)$$

where a is the constant defining the distance ab from the mid-chord to the rotation point, ρ is the air density.

It was shown in Section 20.1.1 that a galloping body experiences a single-degree-of-freedom motion, and that, for small displacements, the aeroelastic force acting on the body is linear with respect to the time rate of change of the across-wind displacement y, the proportionality factor being a function of aerodynamic origin. Airfoil flutter entails

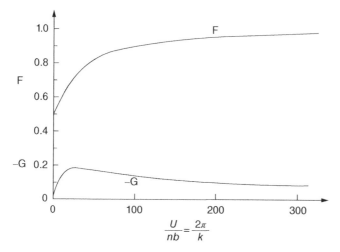

Figure 21.2 Functions $F(K)$ and $G(K)$.

motions with two degrees of freedom (h and α), and the expressions for the aeroelastic forces acting on the body are therefore more elaborate than in the galloping case, although conceptually they are related. Indeed, for small displacements the aeroelastic forces can be written as sums of terms that, like their galloping counterpart, are linear with respect to the rates of change of h and α, the factors of proportionality being also functions of aerodynamic origin. However, unlike in the case of galloping, terms proportional to α come into play as well, and the factors of proportionality depend upon the reduced frequency.

21.3 Aeroelastic Lift, Drag And Moment Acting on Bridge Decks

By analogy with Theodorsen's results, empirical expressions were proposed for the aeroelastic forces acting on bridge decks of the type [2–6]:

$$L_{ae} = \frac{1}{2}\rho U^2 B \left[KH_1^*(K)\frac{\dot{h}}{U} + KH_2^*(K)\frac{B\dot{\alpha}}{U} + K^2 H_3^*(K)\alpha \right.$$
$$\left. +K^2 H_4^*(K)\frac{h}{B} + KH_5^*(K)\frac{\dot{p}}{U} + K^2 H_6^*(K)\frac{p}{B} \right] \tag{21.4a}$$

$$M_{ae} = \frac{1}{2}\rho U^2 B^2 \left[KA_1^*(K)\frac{\dot{h}}{U} + KA_2^*(K)\frac{B\dot{\alpha}}{U} + K^2 A_3^*(K)\alpha \right.$$
$$\left. +K^2 A_4^*(K)\frac{h}{B} + KA_5^*(K)\frac{\dot{p}}{U} + K^2 A_6^*(K)\frac{p}{B} \right] \tag{21.4b}$$

$$D_{ae} = \frac{1}{2}\rho U^2 B \left[KP_1^*(K)\frac{\dot{p}}{U} + KP_2^*(K)\frac{B\dot{\alpha}}{U} + K^2 P_3^*(K)\alpha \right.$$
$$\left. +K^2 P_4^*(K)\frac{P}{B} + KP_5^*(K)\frac{\dot{h}}{U} + K^2 P_6^*(K)\frac{h}{B} \right] \tag{21.4c}$$

where $K = 2\pi nB/U$ and n is the oscillation frequency. For bridges the elastic and mass centers coincide; that is, $a = 0$. Terms proportional to $\ddot{h}$, $\ddot{\alpha}$, and $\ddot{p}$ (i.e., so-called added mass terms, reflecting the forces due to the body motion that result in fluid accelerations around the body) are negligible in bridge engineering applications and do not appear in Eqs. (21.4a–c). The role of the terms in h and p is to account for changes in the frequency of vibration of the body due to aeroelastic effects, while the terms in α reflect the role of the angle of attack. The quantities $\dot{h}/U$ and $B\dot{\alpha}/U$ are effective angles of attack (e.g., the ratio $\dot{h}/U$ has the same significance as in the case of galloping, i.e. it represents the angle of attack of the relative velocity of the flow with respect to the moving body). In Eqs. (21.4) the terms containing first derivatives of the displacements are measures of *aerodynamic damping*. If, among these terms, only those associated with the coefficients

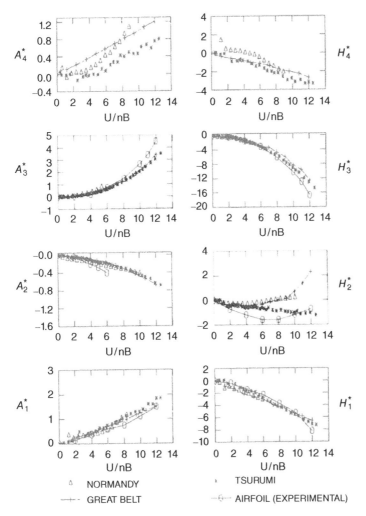

Figure 21.3 Aerodynamic coefficients H_i^* and A_i^* for bodies shown in Figure 21.4. Source: After [8] with permission of Professor Partha Sarkar.

H_1^*, A_2^*, and P_1^* are significant, the total (mechanical plus aerodynamic) damping can be written as

$$c_h - \tfrac{1}{2}\rho UBKH_1^*, \quad c_\alpha - \tfrac{1}{2}\rho UB^3 KA_2^*, \quad c_p - \tfrac{1}{2}\rho UBKP_1^* \qquad (21.5\text{a,b,c})$$

for the vertical, torsional, and horizontal degree of freedom, respectively.

The non-dimensional coefficients H_i^*, A_i^*, and P_i^* are known as *flutter derivatives*.[1] Unlike in the galloping case where, owing to the absence of significant vortex-induced pressures on the body, the derivatives can be obtained experimentally from static tests (that is, tests in which the body is at rest), for the flutter case the coefficients of the displacements and their time rate of change must be obtained experimentally from

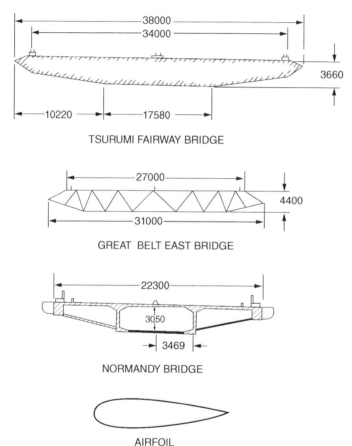

Figure 21.4 Box decks for three bridges (dimensions in millimeters) and airfoil. Source: After [8] with permission of Professor Partha Sarkar.

---

1 Equations (21.4a–c) are formulated in terms of real variables, viewed by some practitioners to be best suited for structural engineering purposes. An alternative wherein the aeroelastic forces and the displacements they induce in the bridge are expressed in terms of complex variables is preferred by some practitioners, insofar as it may offer insights into phase relationships among various aeroelastic forces and displacements – see [7].

measurements on the oscillating deck, which, owing to its elongated shape, is affected by vortex-induced pressures. For this reason, those coefficients are called *motional* aerodynamic derivatives, as opposed to the *steady-state* aerodynamic derivatives that characterize the galloping phenomenon.

Figure 21.3 shows aerodynamic coefficients H_i^* and A_i^* for a thin airfoil and three streamlined box decks depicted in Figure 21.4.

The original Tacoma Narrows bridge (Figure III.2) had negligible H_1^* values for all K, meaning that the total damping (Eqs. [21.5a,b,c]) for motion in the h direction was positive, thus precluding flutter in the vertical degree of freedom. The effect of horizontal deck motions $p(t)$ was not significant. However, A_2^* was positive for values of K such that for mean velocities greater than about $20\,\mathrm{m\,s^{-1}}$ the total damping given by Eqs. (21.5a,b,c) was negative, resulting in flutter motions involving only the torsional degree of freedom. The bridge's susceptibility to flutter was due to the use of an "H" section (the horizontal line in the "H" representing the deck, and the vertical lines representing the girders supporting it). Owing to their inherent instability "H" bridge sections are no longer used.

21.4 Solution of the Flutter Equations for Bridges

The solution of the flutter equations can be obtained if plots of the flutter derivatives H_i^*, A_i^*, and P_i^* are available from measurements as functions of K. It is assumed that the expressions for h, α, and p are proportional to $e^{i\omega t}$. These expressions are inserted into Eqs. (21.4), and the determinant of the amplitudes of h, α, and p is set to zero as the basic stability solution. For each value of K a complex equation in $\omega = \omega_1 + i\omega_2$ is obtained. The flutter velocity is the velocity for U_c for which $\omega_2 \approx 0$, that is

$$U_c = \frac{B\omega_1}{K_c} \tag{21.6}$$

where K_c is the value of K for which $\omega \approx \omega_1$.

A time-domain approach to the study of suspension bridge aeroelastic behavior is presented in [9] and [10]. For a simplified approach to determining the critical flutter velocity, see [11].

21.5 Two-Dimensional Bridge Deck Response to Turbulent Wind in the Presence of Aeroelastic Effects

The expressions for the aeroelastic forces in the turbulent flow have the same form as for the smooth flow case (Eqs. [21.4]). However, the aerodynamic coefficients H_i^*, A_i^*, P_i^* should be obtained from measurements in turbulent flow, since turbulence may affect the aerodynamics of the bridge deck by changing the configuration of the separation layers and the position of reattachment points. Through complex aerodynamic mechanisms, turbulence can affect the flutter derivatives and, therefore, the flutter velocity – in some instances favorably but possibly also unfavorably [12].

The buffeting forces per unit span may be written as follows:

$$L_b = -\frac{1}{2}\rho U^2 B \left[2C_L \frac{u(x,t)}{U} + \left(\frac{dC_L}{d\alpha} + C_D \right) \frac{w(x,t)}{U} \right] \tag{21.7a}$$

$$M_b = \frac{1}{2}\rho U^2 B^2 \left[2C_M \frac{u(x,t)}{U} + \left(\frac{dC_M}{d\alpha} \right) \frac{w(x,t)}{U} \right] \tag{21.7b}$$

$$D_b = \frac{1}{2}\rho U^2 B \left[2C_D \frac{u(x,t)}{U} \right] \tag{21.7c}$$

where B is the deck width, and $U + u(x,t)$ and $w(t)$ are the wind speed components in the x (along-wind) and vertical directions, respectively. For example, Eq. (21.7c) is derived from the expression for the total (mean plus fluctuating) drag force D:

$$D = \overline{D} + D_b = \frac{1}{2}\rho C_D B[U + u(t)]^2 \approx \frac{1}{2}\rho C_D B[U^2 + 2Uu(t)] \tag{21.8}$$

where U is the mean flow velocity, $u(t)$ is the along-wind (longitudinal) component of the turbulent velocity fluctuation at time t, and the mean drag force is defined as

$$\overline{D} = \frac{1}{2}\rho U^2 B C_D \tag{21.9}$$

the drag coefficient C_D is measured in turbulent flow, and the square of the ratio $u(x,t)/U$ is neglected. For the two-dimensional case, the solution of the buffeting problem in the presence of aeroelastic effects is obtained from Eqs. (21.2), in the right-hand sides of which the sums $L_{ae} + L_b, M_{ae} + M_b, D_{ae} + D_b$ are substituted, respectively, for L_{ae}, M_{ae}, and D_{ae} as defined by Eqs. (21.4) [12].

The two-dimensional case can provide useful insights into the behavior of a bridge. However, to be useful in applications to actual bridges, it is necessary to obtain the solution of the three-dimensional case, in which the bridge displacement and the aerodynamic forces are functions of position along the span. This solution is considered in Chapter 23.

References

1 Theodorsen, T., "General theory of aerodynamic instability and the mechanism of flutter," NACA-TR-496 National Advisory Committee for Aeronautics, Washington, DC, pp. 21–22, 1949.

2 Scanlan, R.H. and Tomko, J.J. (1971). Airfoil and bridge deck flutter derivatives. *Journal of the Engineering Mechanics Division* 97: 1717–1737.

3 Singh, L., Jones, N.P., Scanlan, R.H., and Lorendeaux, O. (1996). Identification of lateral flutter derivatives of bridge decks. *Journal of Wind Engineering and Industrial Aerodynamics* 60: 81–89.

4 Scanlan, R.H. and Simiu, E. (2015). Aeroelasticity in civil engineering. In: *A Modern Course in Aeroelasticity*, 5the (ed. E.H. Dowell), 285–347. Switzerland: Springer.

5 Simiu, E. and Scanlan, R.H. (1996). *Wind Effects on Structures*, 3rd ed. Hoboken, NJ: Wiley.

6 Gan Chowdhury, A. and Sarkar, P.P. (2003). A new technique for identification of eighteen flutter derivatives using three-degree-of-freedom section model. *Engineering Structures* 25: 1763–1772.

7 Simiu, E. and Miyata, T. (2006). *Design of Buildings and Bridges for Wind: A Practical Guide for ASCE-7 Standard Users and Designers of Special Structures*, 1st ed. Hoboken, NJ: Wiley.

8 Sarkar, P. P., "New identification methods applied to response of flexible bridges to wind," Doctoral dissertation, Civil Engineering, Johns Hopkins University, Baltimore, MD, 1992.

9 Caracoglia, L. and Jones, N.P. (2003). Time-domain vs. frequency domain characterization of aeroelastic forces for bridge deck sections. *Journal of Wind Engineering and Industrial Aerodynamics* 91: 371–402.

10 Cao, B. and Sarkar, P.P. (2013). Time-domain aeroelastic loads and response of flexible bridges in gusty wind: Prediction and experimental validation. *ASCE Journal of Engineering Mechanics* 139: 359–366.

11 Bartoli, G. and Mannini, C. (2008). A simplified approach to bridge deck flutter. *Journal of Wind Engineering and Industrial Aerodynamics* 96: 229–256.

12 Simiu, E., "Buffeting and aerodynamic stability of suspension bridges in turbulent wind," Doctoral dissertation, Civil Engineering, Princeton University, Princeton, NJ, 1971.

22

Slender Chimneys and Towers

This chapter presents material that complements Chapter 19 on the response of towers and chimneys with circular cross section and allows the practical calculation of that response (Section 22.1); it briefly discusses issues related to the aeroelastic response of slender structures with square or rectangular cross section (Section 22.2); and describes methods of alleviating wind-induced oscillations of slender structures (Section 22.3).

22.1 Slender Chimneys with Circular Cross Section

22.1.1 Slender Chimneys Assumed to be Rigid

In turbulent flow the nominal across-wind response σ_y^{nom} of a chimney is due to a super-position of two across-wind loads. The first across-wind load, due to vortex shedding in the tower's wake, has the expression

$$L_1(z, t) = \frac{1}{2}\rho C_L(z, t)D(z)U^2(z) \tag{22.1}$$

(the notations from Chapter 19 are used in this section). The spectral density of the lift force $L_1(z, t)$ is

$$S_{L_1}(z, n) = \left[\frac{1}{2}\rho D(z)U^2(z)\right]^2 S_{C_L}(z, n) \tag{22.2}$$

According to [1], measurements indicate that the spectral density $S_{C_L}(z, n)$ can be represented by the bell-shaped function

$$\frac{nS_{C_L}(z, n)}{C_L^2} = \frac{1}{\sqrt{\pi}Bn_s}\exp\left\{-\left[\frac{1-(n/n_s)}{B}\right]^2\right\} \tag{22.3}$$

where n denotes frequency, n_s is the vortex-shedding frequency given by the relation

$$n_s = \frac{St\,U(z)}{D(z)} \tag{22.4}$$

St is the Strouhal number, and B is an empirical parameter that determines the spread (bandwidth) of the spectral curve. This model is consistent with results of full-scale measurements (Figure 22.1).

The cross-spectral density of the load $L_1(z, t)$ can be expressed as [3]:

$$S_{L_1}(z_1, z_2, n) = S_{L_1}^{1/2}(z_1, n)S_{L_1}^{1/2}(z_2, n)R_0(z_1, z_2, n) \tag{22.5}$$

Wind Effects on Structures: Modern Structural Design for Wind, Fourth Edition. Emil Simiu and DongHun Yeo.
© 2019 John Wiley & Sons Ltd. Published 2019 by John Wiley & Sons Ltd.

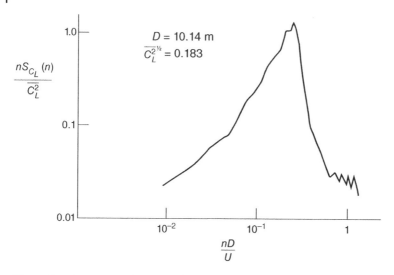

Figure 22.1 Power spectral density of lift force coefficients C_L measured on the Hamburg television tower. Source: Reprinted from [2], with permission from Elsevier.

where the coherence function is assumed to be

$$R_0(z_1, z_2, n) = \cos(2ar)\, \exp[-(ar)^2] \tag{22.6}$$

$$r = \frac{2\,|z_1 - z_2|}{D(z_1) + D(z_2)} \tag{22.7}$$

The parameter a is a measure of the decay of the cross-spectral function $S_{L_1}(z_1, z_2, n)$ with the distance $|z_1 - z_2|$. Associated with the parameter a is a correlation length l, a measure of the spatial separation beyond which the force fluctuations are no longer correlated.

The second lift force, denoted by $L_2(t)$, is the projection on the across-wind direction of the drag force induced by the resultant of the mean velocity $U(z)$ and the lateral turbulent velocity $v(z,t)$. In large-scale turbulence, this force has an angle of attack with respect to the along-wind direction equal to v/U, and its projection on that direction is

$$L_2(t) = \frac{1}{2}\rho C_D U^2(z) \frac{v(z, t)}{U(z)} \tag{22.8}$$

The aerodynamic parameters depend upon: the Reynolds number

$$Re(z) = 67{,}000\, U(z)\, D(z) \tag{22.9}$$

where $U(z)$ is the wind speed at elevation z in m s^{-1} and $D(z)$ is the outside diameter in meters; the turbulence in the oncoming flow; the aspect ratio $h/D(h)$, where h is the height of the structure; and the relative surface roughness k/D of the structure, where k is the height of the roughness elements. For steel stacks and reinforced concrete chimneys and towers $10^{-5} \lesssim k/D \lesssim 10^{-3}$ [4].

The dependence of C_D upon Reynolds number and surface roughness is represented in Figure 4.22 for cylinders with aspect ratios $h/D(h) \gtrsim 20$. For cylinders with aspect ratios $10 < h/D(h) < 20$ it may be assumed that up to the elevation $h - D(h)$ the drag coefficient has the value

$$C_D = C_D^s \left[1 - 0.015 \left(20 - \frac{h}{D(h)} \right) \right] \tag{22.10}$$

where C_D^s is the value of the drag coefficient taken from Figure 4.22. From elevation $h - D(h)$ to the top of the structure the drag coefficient may be assumed to have the value $C_D = 1.4 C_D^s$ regardless of aspect ratio [5]. The main effect of turbulence is to decrease the Reynolds number corresponding to the onset of the critical region defined in Figure 4.22.

The following values of the Strouhal number are suggested in [5]:

$$St = 0.20 \qquad\qquad Re \lesssim 2 \times 10^5 \tag{22.11a}$$

$$0.22 \lesssim St \lesssim 0.45 \qquad\qquad 2 \times 10^5 \lesssim Re \lesssim 2 \times 10^6 \tag{22.11b}$$

$$St = c \left\{ 0.23 - 0.007 \left[\log_{10} \left(\frac{k}{d} \right) + 5 \right] \right\} \qquad Re \gtrsim 2 \times 10^6 \tag{22.11c}$$

For $2 \times 10^5 \lesssim Re \lesssim 2 \times 10^6$ the vortex shedding is random, and the Strouhal number given by Eq. (22.11b) corresponds to the predominant frequencies of the flow fluctuations in the wake. In Eq. (22.11c) the coefficient c depends upon the aspect ratio as follows:

$$c = \begin{cases} 1.00 & \frac{h}{D(h)} \geq 30 \\ 0.736 + 0.012 \left[\frac{h}{D(h)} - 8.0 \right] & 8 < \frac{h}{D(h)} < 30 \end{cases} \tag{22.12a,b}$$

Note that the values given by Eq. (22.11b) differ from those, obtained in a more recent study, shown in Figure 4.15.

The following values of the r.m.s. lift coefficient are suggested for design purposes [5]:

$$\overline{C_L^2}^{1/2} \approx \begin{cases} 0.45 & Re \lesssim 2 \times 10^5 \\ 0.14 & 2 \times 10^5 \lesssim Re \lesssim 2 \times 10^6 \\ d \left\{ 0.15 + 0.035 \left[5 + \log_{10} \left(\frac{k}{D} \right) \right]^2 \right\} & Re \geq 2 \times 10^6 \end{cases} \tag{22.13a,b,c}$$

where

$$d \approx \begin{cases} 1.00 & \frac{h}{D(h)} \gtrsim 12 \\ 0.8 + 0.05 \left[\frac{h}{D(h)} - 8.0 \right] & 8 < \frac{h}{D(h)} < 12 \end{cases} \tag{22.14a,b}$$

No information appears to be available on the dependence of the lift coefficient upon turbulence intensity. It is suggested in [3, 5] that

$$B^2 \approx 0.08^2 + \frac{2\overline{u^2}}{U^2} \tag{22.15}$$

where $\overline{u^2}$ is the mean square value of the longitudinal velocity fluctuations and U is the mean speed. According to [6] it may be assumed $B = 0.18$ for all flows.

For $Re > 2 \times 10^5$ it is suggested in [3, 5] that $a = 1/3$ (see Eq. [22.6]), to which there corresponds a correlation length $l \approx D$. For $Re < 2 \times 10^5$, $l \approx 2.5D$ [7].

22.1.2 Flexible Slender Chimneys

The mechanical damping ratios ζ_i in the ith mode of vibration depend upon the type of structure. Suggested values are as follows [4]:

Unlined steel stacks and similar structures: 0.002–0.010
Lined steel stacks: 0.004–0.016
Reinforced concrete chimneys and towers: 0.004–0.020.

The following approximate expressions are suggested in [3, 5] for the aeroelastic parameter K_{a0}:

$$
K_{a0}\left(\frac{U}{U_{cr}}\right) = \begin{cases}
0 & \frac{U}{U_{cr}} < 0.85 \\
a_i\left(3.5\frac{U}{U_{cr}} - 2.95\right) & 0.85 \leq \frac{U}{U_{cr}} < 1.0 \\
0.55a_i & 1.0 \leq \frac{U}{U_{cr}} < 1.1 \\
a_i\left(2.75 - 2\frac{U}{U_{cr}}\right) & 1.1 \leq \frac{U}{U_{cr}} < 1.3 \\
a_i\left(0.46 - 0.25\frac{U}{U_{cr}}\right) & 1.3 \leq \frac{U}{U_{cr}} < 1.84 \\
0 & 1.84 \leq \frac{U}{U_{cr}}
\end{cases}
$$

$$(22.16a,b,c,d,e,f)$$

where

$$a_i = a_1\, a_2\, a_3\, a_4 \tag{22.17}$$

$$
a_1 = \begin{cases}
1.0 & Re < 10^4 \\
1.8 & 10^4 \leq Re < 10^5 \\
1.0 & Re \geq 10^5
\end{cases}
\tag{22.18a,b,c}
$$

$$
a_2 = \begin{cases}
2.0 & U(10\,\text{m}) \lesssim 12\,\text{ms}^{-1} \\
1.0 & U(10\,\text{m}) \gtrsim 12\,\text{ms}^{-1}
\end{cases}
\tag{22.19a,b}
$$

$$a_3 = 0.9 + 0.2\,[\log_{10}(k/D) + 5] \tag{22.20}$$

$$
a_4 = \begin{cases}
1.0 & \frac{h}{D(h)} \geq 12.5 \\
1.0 - 0.04\left(12.5 - \frac{h}{D(h)}\right) & \frac{h}{D(h)} < 12.5
\end{cases}
\tag{22.21a,b}
$$

where $U_{cr} = nD/St$ (see Section 19.3 and Figure 19.9).

22.1.3 Approximate Expressions for the Across-Wind Response

The across-wind response in the ith mode of vibration may be estimated as

$$\sigma_{yi}(z) = \overline{\xi_i^2}^{1/2}\, y_i(z) \tag{22.22}$$

$$Y_i(z) = g_{yi}\sigma_{yi}(z) \tag{22.23}$$

$$g_{yi} = [2\ln(3600\,n_i)]^{\frac{1}{2}} + \frac{0.577}{[2\ln(3600\,n_i)]^{\frac{1}{2}}} \tag{22.24}$$

$$\overline{\xi_i^2}^{1/2} = \overline{\xi_{nom,i}^2}^{1/2} \left| \frac{\zeta_i}{(\zeta_i + \zeta_{ai})} \right|^{1/2} \tag{22.25}$$

$$S_i(z) = (2\pi n_i)^2 \int_z^h m(z_1) Y_i(z_1) dz_1 \tag{22.26}$$

$$M_i(z) = (2\pi n_i)^2 \int_z^h m(z_1) Y_i(z_1)(z_1 - z) dz_1 \tag{22.27}$$

where $\sigma_{yi}(z)$ is the rms of the deflection at elevation z in the ith mode of vibration, $\overline{\xi_i^2}^{1/2}$ is the r.m.s. of the corresponding generalized coordinate, $y_i(z)$ is the ith modal shape, $Y_i(z)$ is the peak deflection in the ith mode of vibration, g_{yi} is the peak factor, n_i is the natural frequency in the ith mode in Hz, $\overline{\xi_{nom,i}^2}^{1/2}$ is the rms nominal generalized coordinate in the ith mode (which corresponds to the response estimated by assuming that no aeroelastic effects occur and that the motion is affected only by mechanical damping in the ith mode), ζ_i is the structural damping in the ith mode, ζ_{ai} is the aerodynamic damping in the ith mode. $S_i(z)$ and $M_i(z)$ are the shear force and the bending moment at elevation z due to the across-wind response in the ith mode, and $m(z)$ is the mass of the structure per unit length. Note that, for the ith mode, the ratio of peak acceleration to peak deflection is approximately $(2\pi n_i)^2$ (see Eq. [B.16b]).

To estimate the across-wind response, expressions are needed for the rms of the nominal generalized coordinate in the ith mode, $\overline{\xi_{nom,i}^2}^{1/2}$, and the aerodynamic damping in the ith mode, ζ_{ai}. These expressions are given next for (i) structures with constant cross section and (ii) tapered structures. In both cases, the expressions are valid only for relatively small ratios $\frac{\sigma_{yi}(h)}{D(h)}$, for example 3% or less, to which there would correspond negligible values of the second term within the bracket of Eq. (19.18). In practice, the design of the structure is acceptable only if the ratio $\frac{\sigma_{yi}(h)}{D(h)}$ is small.

Structures with Constant Cross Section. The following approximate expressions based on the approach described in Section 22.1.1 were proposed in [6]:

$$\overline{\xi_{nom,i}^2}^{1/2} = \frac{0.035 \overline{C_L^2}^{1/2} (l/D)^{1/2}}{\zeta_i^{1/2} St^2} \frac{\rho D^3}{M_i} \left[D \int_0^h y_i^2(z) dz \right]^{1/2} \tag{22.28}$$

$$\zeta_{ai} \approx -\frac{\rho D^2}{M_i} K_{a0}(1) \int_0^h y_i^2(z) dz \tag{22.29}$$

where $\rho \simeq 1.25 \, \mathrm{kg \, m^{-3}}$ is the air density, M_i is the generalized mass in the ith mode, l is the correlation length (see Section 22.1.1) and D is the outside diameter. The critical wind speed corresponding to the ith mode of vibration has the expression

$$U_{cr,i} = \frac{n_i D}{St} \tag{22.30}$$

Information on the mechanical damping ratios ζ_i is given in Section 22.1.2. Information on the parameters St, $\overline{C_L^2}$, and K_{a0} is given in Sections 22.1.1 and 22.1.2. In Eqs. (22.19a,b) the speed $U(10 \, \mathrm{m})$ corresponding to the ith mode is

$$U(10 \, \mathrm{m}) = \frac{\ln\left(\frac{10}{z_0}\right)}{\ln\left[\left(\frac{5}{6}\right)\frac{h}{z_0}\right]} U_{cr,i} \tag{22.31}$$

where h is the height of the structure in meters and z_0 is the roughness length in meters for the terrain that determines the wind profile over the upper half of the chimney (Table 2.1).

Example 22.1 Consider a chimney with $h = 193.6$ m, $D = 17.63$ m, $n_1 = 0.364$ Hz, $y_1(z/h) = (z/h)^{1.67}$, $m(z) = 58,000$ kg m^{-1} for $z \le h/2$, $m(z) = 41,000$ kg m^{-1} for $z > h/2$, $M_1 = 1.87 \times 10^6$ kg. It is assumed $\zeta_1 = 0.02$, $k/D = 10^{-5}$, and $z_0 = 0.05$ m. We seek the response in the first mode.

Assuming tentatively $St = 0.22$, the critical wind speed at elevation $5\,h/6 = 161.3$ m is $U_{cr,1} = 29.2$ m s^{-1} (Eq. [22.30]), to which there corresponds a Reynolds number $Re = 3.4 \times 10^7 > 2 \times 10^6$. The aspect ratio is $h/D \simeq 11$. It can be verified that

$St = 0.178$ (Eqs. [22.11c] and [22.12b])
$l = D$ (since $Re > 2 \times 10^5$)
$\overline{C_L^2}^{1/2} = 0.143$ (Eqs. [22.13a,b,c] and [22.14a,b])
$\int_0^h y_1^2(z)dz = 44.7$ m
$\overline{\xi_{nom,1}^2}^{1/2} = 0.115$ m (Eq. [22.28])
$U(10) = 19.1$ m s$^{-1} > 12$ m s^{-1} (Eqs. [22.30] and [22.31])
$K_{a0}(1) = 0.465$ (Eqs. [22.16–22.21])
$\zeta_{a1} = -0.0043$ (Eq. [22.29])
$\overline{\xi_1^2}^{1/2} = 0.130$ m (Eq. [22.25])
$g_{y1} = 3.94$ (Eq. [22.24])
$\sigma_{y1}(z) = 0.130\left(\frac{z}{193.6}\right)^{1.67}$ m (Eq. [22.22])
$Y_1(z) = 0.51\left(\frac{z}{193.6}\right)^{1.67}$ m (Eq. [22.23])
$M_1(0) = 1150 \times 10^6$ Nm (Eq. [22.27])

The results of the calculations depend strongly upon, in particular, the assumed value of the structural damping ratio ζ_1. Had the value $\zeta_1 = 0.01$ been appropriate, the results obtained would have been larger than those obtained in this example by a factor of $([0.02 - 0.0043]/[0.01 - 0.0043])^{1/2} \simeq 1.66$ (Eq. [22.25]).

Tapered Structures. The following approximate expressions based on the approach described in Section 22.1.1 were proposed in [6]:

$$\overline{\xi_{nom,i}^2}^{1/2}(z_{e_i}) \simeq \frac{0.016\overline{C_L^2}^{1/2}\left(\frac{l}{D}\right)^{1/2}\rho D^4(z_{e_i})y_i(z_{e_i})}{\zeta_i^{1/2}St^2 M_i \beta^{1/2}(z_{e_i})} \tag{22.32}$$

$$\beta(z_{e_i}) \simeq \frac{0.1D(z_{e_i})}{z_{e_i}} - \left.\frac{dD(z)}{dz}\right|_{z=z_{e_i}} \tag{22.33}$$

$$\zeta_{ai}(z_{e_i}) = -\frac{\rho D_0^2}{M_i}\int_0^h K_{a0}\left(\frac{U(z)}{U_{cr}(z_{e_i})}\right)\left[\frac{D(z)}{D_0}\right]^2 y_i^2(z)dz \tag{22.34}$$

where the notations of Eq. (22.28) are used, D_0 is the outside diameter at the base, z_{e_i} is the elevation corresponding to the critical velocity

$$U_{cr}(z_{e_i}) = \frac{n_i D(z_{e_i})}{St} \tag{22.35}$$

$$\frac{U(z)}{U_{cr}(z_{e_i})} = \frac{\ln(z/z_0)}{\ln(z_{e_i}/z_0)} \tag{22.36}$$

and z_0 is the terrain roughness that determines the wind profile over the upper half of the structure.

Since, as in Eq. (22.25),

$$\overline{\xi_i^2(z_{e_i})}^{1/2} = \overline{\xi_{nom,i}^2(z_{e_i})}^{1/2} \left(\frac{\zeta_i}{\zeta_i + \zeta_{ai}(z_{e_i})} \right)^{1/2} \tag{22.37}$$

it follows that the maximum response in the ith mode corresponds to the maximum value taken on by the function

$$F_i(z_{e_i}) = \frac{D^4(z_{e_i}) y_i(z_{e_i})}{\{\beta(z_{e_i})[\zeta_i + \zeta_{ai}(z_{e_i})]\}^{1/2}} \tag{22.38}$$

To determine that value, it is in practice necessary to calculate $F_i(z_{e_i})$, and in particular $\zeta_{ai}(z_{e_i})$, for a sufficient number of elevations $0 < z_{e_i} < h$.

As pointed out in [8], if the structure is very lightly tapered (i.e., if $\frac{dD(z)}{dz}\Big|_{z=z_{e_i}}$, and therefore $\beta(z_{e_i})$ is small – see Eq. [22.33]), the chimney is assumed to behave as if it had a constant outside diameter D equal to the average diameter of its top third [6], and Eqs. (22.28)–(22.30) are applied with the same values of the parameters St, $\overline{C_L^2}^{1/2}$ and correlation length D (or, for $Re < 10^5$, 2.5 D) as those used in Eq. (22.32). In practice, it is therefore necessary to calculate both the value of the response yielded by Eqs. (22.32) and (22.34) and the value yielded by Eqs. (22.28) and (22.29). The response to be assumed for structural design purposes is the smaller of these two values [1].

22.2 Aeroelastic Response of Slender Structures with Square and Rectangular Cross Section

Along-Wind Aeroelastic Response. Aerodynamic damping in tall buildings results from the interaction between the fluctuating aerodynamic forces acting on the building and the fluctuating building motions they induce. Since the aerodynamic damping is due to the building motion, it affects in the most general case the along-wind, across-wind, and torsional motions. In this section, attention is restricted to the aerodynamic damping affecting the along-wind motion of an isolated building with a rectangular shape in plan.

The aerodynamic along-wind force depends upon the relative wind speed with respect to the moving structure. If the structure is sufficiently rigid, it experiences no significant motion, and the relative wind speed with respect to the structure is equal, in practice, to the oncoming wind speed. However, if the structure is flexible, its motions can be significant, and the relative wind speed with respect to the structure is equal to the time-dependent difference between the oncoming fluctuating speed and the speed of the moving structure. The procedure for estimating the aerodynamic damping presented in this section was developed in [9].

The displacement at elevation z is written as

$$x(z, t) = \sum_{i=1}^{N} \phi_i(z)\xi_i(t) \tag{22.39}$$

where N = number of normal modes being considered and $\xi_i(t)$ and $\phi_i(z)$ = generalized coordinate and modal shapes corresponding to the ith normal mode of vibration, respectively. In the ith modal equation of motion the generalized force is

$$Q_i(t) = \sum_{l=1}^{L} \phi_i(z_l)F_l(z_l, t) \tag{22.40}$$

where L = total number of taps on the windward and leeward faces, $m(z)$ = mass distribution, and $F_l(z_l, t)$ = excitation force associated with tap l at elevation z_l.

The force $F_l(z_l, t)$ can be written as

$$F_l(z_l, t) = \frac{1}{2}\rho[U(z_l) + u(z_l, t) - \dot{x}(z_l, t)]^2 C_l(z_l)A_l \tag{22.41}$$

where $U(z_l)$ and $u(z_l)$ are the mean and the fluctuating wind speed at elevation z_l, $\dot{x}(z_l, t)$ is the time-dependent along-wind displacement of the building at elevation z_l, $C_l(z_l)$ is the mean pressure coefficient at z_l, and A_l is the tributary area of tap l. Equation (22.41) may be interpreted as follows. The aerodynamic damping depends upon the degree to which the fluctuating excitation of the structure is in phase or out of phase with the wind-induced velocity $\dot{x}$. If the excitation and the velocity are in phase, the relative fluctuating velocity $u(z_l, t) - \dot{x}(z_l, t)$ is lower than the fluctuating velocity $u(z_l, t)$, meaning that the fluctuating response of the structure will decrease; in other terms, the aerodynamic damping will be positive. The opposite is true if the excitation and the building velocity are in opposite phases.

In applying Eqs. (22.39)–(22.41), an iterative procedure is used. The force F_l is calculated first by neglecting the speeds $\dot{x}(z_l, t)$. The resulting equation of motion is used to calculate a first approximation to those speeds. This approximation is then used in Eq. (22.41) and the corresponding equation of motion to obtain a second approximation to $\dot{x}(z_l, t)$. This process continues until the nth and the $(n-1)$th approximations differ insignificantly. The aerodynamic damping value was obtained by a trial-and-error procedure where successive total damping ratios were input in the database-assisted design software described in Chapter 18, until the resulting r.m.s. displacements were approximately equal to the displacements calculated by the iterative procedure just described.

For a 60-story building with dimensions 45.7×30.5 m in plan and height $H = 185$ m, and mean wind speed normal to the building's wider face, the procedure described in [9] and summarized in this section yielded values of the aerodynamic damping that were positive, larger as the reduced wind increased, weakly dependent upon the modal shapes, and negligible for practical purposes even for mean wind speeds at the top of the building as high as 70 m s^{-1}. The results obtained for this case were comparable for practical purposes to those obtained in wind tunnel tests (Figure 22.2a) [10].

In view of the uncertainties associated with the estimation of the along-wind aerodynamic damping, it is prudent to neglect its favorable effect on the along-wind response.

Across-Wind Aeroelastic Response. Unlike for along-wind response, no practical analytical approach is available for the estimation of across-wind response. Based on wind tunnel test results, Figure 22.2b shows that for sufficiently high reduced velocities the aerodynamic damping can be negative (i.e., destabilizing), although this is not the case for ratios D/B sufficiently larger than unity. For details see [10], which notes that the wind velocities for which the across-wind aerodynamic damping becomes negative are much lower than the wind velocities at which galloping oscillations can occur.

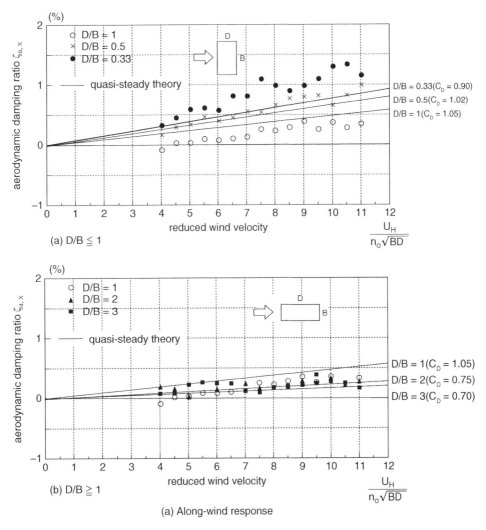

(a) Along-wind response

Figure 22.2 Aerodynamic damping as a function of reduced wind velocity and side ratio D/B (1% mechanical aerodynamic damping). Source: Reprinted from [10], with permission from Elsevier.

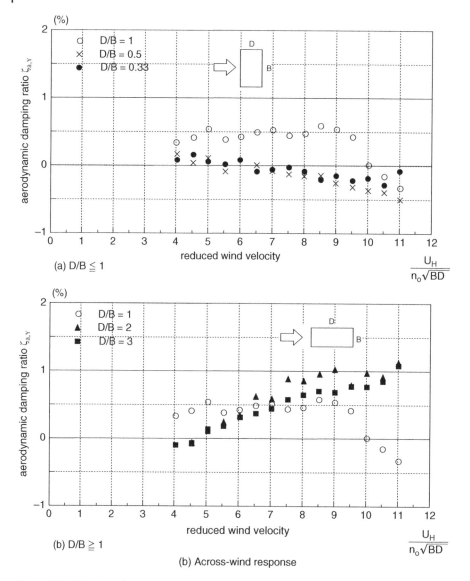

(a) D/B $\leqq$ 1

(b) D/B $\geqq$ 1

(b) Across-wind response

Figure 22.2 (*Continued*)

A schematic of the simple experimental set-up used to obtain the results reported in [10] is shown in Figure 22.3.

For the flexible structures with square cross section tested in [10] a sufficient condition assuring adequacy of the design from an aeroelastic point of view is that the wind speeds that may be expected during the life of the structure be lower than the lowest speed, denoted by U_l, which induces across-wind resonant oscillations. This statement is consistent with the test results of Figure 22.2b, which show that negative aerodynamic damping occurs at wind speeds higher than U_l. The necessary condition

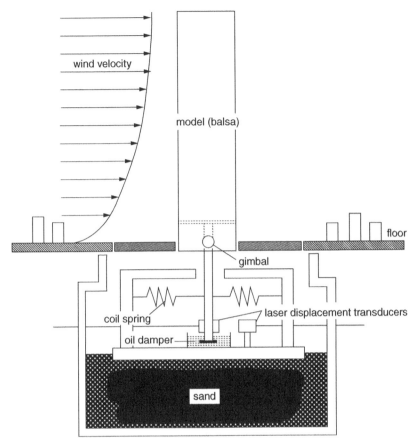

Figure 22.3 Experimental set-up for tests reported in [10]. Source: Reprinted from [10], with permission from Elsevier.

for galloping (Eq. [20.14], modified to account for shear, turbulent flow) should also be considered. A similar approach may be employed for structures with rectangular shapes in plan.

For additional material on aerodynamic damping see, for example, [11].

22.3 Alleviation of Vortex-Induced Oscillations

A common method of alleviating vortex-induced oscillations is the provision of "spoiler" devices that destroy or reduce the coherence of shed vortices [12, 13].

The helical strake system, first proposed by Scruton [14], consists of three rectangular strakes with a pitch of one revolution in five diameters and a strake (radial) height of 0.10 m diameter (to 0.13 m diameter for very lightly damped structures) applied over the top 33–40% of the stack height. The effectiveness of the system is not impaired by a gap of 0.005D between the strake and the cylinder surface [15]. Reference [16] reports

Figure 22.4 Steel chimney with helical strakes. Source: Reprinted from [16], with permission from Elsevier.

the remarkable results obtained by using this system (with 5-mm thick strakes, 0.6-m strake height, and 30-m pitch) in the case of a 145-m all and 6-m diameter steel stack (Figure 22.4).

For Reynolds numbers $Re < 2 \times 10^5$ or so, in flow with about 15% turbulence intensity, helical strakes were found to reduce the peak of the across-wind resonant oscillations by a factor of about two, as opposed to a factor of about 100 in smooth flow [17]. It appears that the performance of strakes can be unsatisfactory in the case of stacks grouped in a row [18]. Also, wind tunnel tests indicate that for large vibration amplitudes (e.g., 3–5% of the diameter), the vortex street re-establishes itself, and the aerodynamic devices become ineffective [19]. It is noted that strakes increase drag, as shown in Figure 22.5.

Shrouds were also found to be effective in reducing the coherence of shed vortices. A schematic view of a shroud fitted to a stack is shown in Figure 22.6. Results of wind tunnel experiments reported in [15, 20] showed that oscillations were substantially reduced with only the 25% of the model height shrouded. The most effective shrouds were found to be those with a gap width $w = 0.12D$ and an open-area ratio between 20 and 36% (with length of square $s = 0.052D$ to $0.070D$)

Improvements in the behavior of the structure under wind loads can be achieved by using tuned mass dampers (see Chapter 16) and similar devices, and/or by increasing the mechanical damping, and affecting the aerodynamic response of the structure by designing buildings with chamfered corners (see [21]), tapered shapes, and/or discontinuous changes of shape. An aerodynamic device used in the design of the New York City 85-floor, 425 m tall 432 Park Avenue building consists of leaving the mechanical floors open to allow air to pass through the building, thus disrupting the vorticity shed in the building's wake.

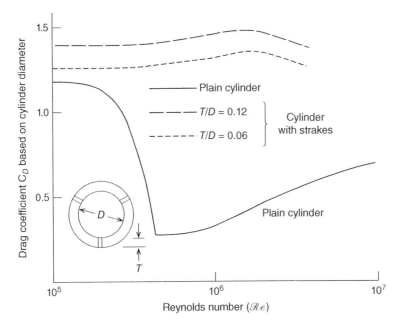

Figure 22.5 Effect of strakes on drag coefficient. Source: From [20]. Courtesy of National Physical Laboratory, U.K.

Figure 22.6 View of a shroud fitted to a stack. Source: After [15].

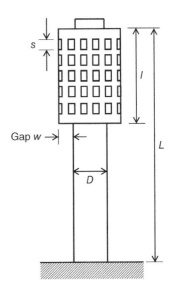

References

1 Vickery, B.J. and Clark, A.W. (1972). Lift or across-wind response of tapered stacks. *Journal of the Structural Division* 98: 1–20.
2 Ruscheweyh, H. (1976). Wind loading on the television tower, Hamburg, Germany. *Journal of Wind Engineering and Industrial Aerodynamics* 1: 315–333.

3 Vickery, B.J. and Basu, R.I. (1983). Across-wind vibrations of structures of circular cross-section. Part I. Development of a mathematical model for two-dimensional conditions. *Journal of Wind Engineering and Industrial Aerodynamics* 12: 49–73.

4 Basu, R. I. and Vickery, B. J., "A comparison of model and full-scale behavior in wind of towers and chimneys," in *Wind Tunnel Modeling for Civil Engineering Applications (Proceedings of the International Workshop on Wind Tunnel Modeling Criteria and Techniques in Civil Engineering Applications, Gaithersburg, MD, USA, April 1982),* T. A. Reinhold, ed., 1st ed., Cambridge, UK: Cambridge University Press, 1982.

5 Basu, R. I., "Across-Wind Response of Slender Structures of Circular Cross Section to Atmospheric Turbulence," Research Report BLWT-23983, University of Western Ontario, London, Ontario, Canada, 1983.

6 Vickery, B.J. and Basu, R. (1983). Simplified approaches to the evaluation of the across-wind response of chimneys. *Journal of Wind Engineering and Industrial Aerodynamics* 14: 153–166.

7 Davenport, A.G. and Novak, M. (2002). Vibrations of structures induced by Wind (Chapter 29, Part II). In: *Harris' Shock and Vibration Handbook*, 5th ed. (ed. C.M. Harris and A.G. Piersol), 29.21–29.46. New York: McGraw-Hill.

8 Vickery, B. J., "The aeroelastic modeling of chimneys and towers," in *Wind Tunnel Modeling for Civil Engineering Applications (Proceedings of the International Workshop on Wind Tunnel Modeling Criteria and Techniques in Civil Engineering Applications, Gaithersburg, MD, USA, April* 1982), T. A. Reinhold, ed., 1st ed. Cambridge, UK: Cambridge University Press, 1982.

9 Gabbai, R. and Simiu, E. (2010). Aerodynamic damping in the along-wind response of tall buildings. *Journal of Structural Engineering* 136: 117–119.

10 Marukawa, H., Kato, N., Fujii, K., and Tamura, Y. (1996). Experimental evaluation of aerodynamic damping of tall buildings. *Journal of Wind Engineering and Industrial Aerodynamics* 59: 177–190.

11 Kareem, A. and Gurley, K. (1996). Damping in structures: its evaluation and treatment of uncertainty. *Journal of Wind Engineering and Industrial Aerodynamics* 59: 131–157.

12 Zdravkovich, M.M. (1981). Review and classification of various aerodynamic and hydrodynamic means for suppressing vortex shedding. *Journal of Wind Engineering and Industrial Aerodynamics* 7: 145–189.

13 Zdravkovich, M.M. (1984). Reduction of effectiveness of means for suppressing wind-induced oscillation. *Engineering Structures* 6: 344–349.

14 Scruton, C., "Note on a device for the suppression of the vortex-induced oscillations of flexible structures of circular or near circular section, with special reference to its applications to tall stacks," NPL Aero Report No. 1012, National Physical Laboratory, Teddington, UK, 1963.

15 Walshe, D.E. and Wooton, L.R. (1970). Preventing wind-induced oscillations of structures of circular section. *Proceedings of the Institution of Civil Engineers* 47: 1–24.

16 Hirsch, G. and Ruscheweyh, H. (1975). Full-scale measurements on steel chimney stacks. *Journal of Wind Engineering and Industrial Aerodynamics* 1: 341–347.

17 Gartshore, I. S., Khanna, J., and Laccinole, S., "The Effectiveness of Vortex Spoilers on a Circular Cylinder In Smooth and Turbulent Flow," in *Wind Engineering (Proceedings of the Fifth International Conference, Fort Collins, Colorado, USA, July 1979)*, J. E. Cermak, ed., Pergamon, 1980, pp. 1371–1379.

18 Ruscheweyh, H. (1981). Straked in-line steel stacks with low mass-damping parameter. *Journal of Wind Engineering and Industrial Aerodynamics* 8: 203–210.

19 Ruscheweyh, H. (1994). Vortex excited vibrations. In: *Wind-Excited Vibrations of Structures* (ed. H. Sockel), 51–84. Wein/New York: Springer-Verlag.

20 Wooton, L.R. and Scruton, C. (1971). Aerodynamic stability. In: *Modern Design of Wind-Sensitive Structures*, 65–81. London, UK: Construction Industry Research and Information Association.

21 Simiu, E. and Miyata, T. (2006). *Design of Buildings and Bridges for Wind: A Practical Guide for ASCE-7 Standard Users and Designers of Special Structures*, 1st ed. Hoboken, NJ: Wiley.

23

Suspended-Span Bridges

23.1 Introduction

Suspended-span (i.e., suspension and cable-stayed) bridges must withstand drag forces induced by the mean wind. In addition, they may experience aeroelastic effects, which may include vortex-induced oscillations (Chapter 19), flutter, and buffeting in the presence of self-excited forces (Chapter 21). The study of these effects is possible only on the basis of information provided by wind tunnel tests. Various types of such tests are described in Section 23.2. Vortex-induced vibrations of bridge decks are considered in Section 23.3. Section 23.4 is concerned with bridge buffeting in the presence of aeroelastic effects. Vibrations occurring in cables of cable-stayed bridges are discussed in Section 23.5.

The action of wind must be taken into account not only for the completed bridge, but also for the bridge in the construction stage. In general, the same methods of testing and analysis apply in both cases. To decrease the vulnerability of the partially completed bridge to wind, temporary ties and damping devices are used [1, 2]. Also, to minimize the risk of strong wind loading, construction usually takes place in seasons with low probabilities of occurrence of severe storms.

In addition to the deck and stay cables, aeroelastic phenomena may affect the bridge's tower and hangers, on which detailed material is available in [1].

23.2 Wind Tunnel Testing

The following three types of wind tunnel tests are commonly used to obtain information on the aerodynamic behavior of suspended-span bridges:

1) *Tests on models of the full bridge.* In addition to being geometrically similar to the full bridge, such models must satisfy similarity requirements pertaining to mass distribution, reduced frequency, mechanical damping, and shapes of vibration modes (see Chapter 5). The construction of full bridge models is therefore elaborate and their cost is high. Their usual scale is in the order of 1/300. A view of a full-scale bridge model in a large, specially built wind tunnel, is shown in Figure 23.1.
2) *Tests on three-dimensional partial-bridge models.* In such models the main span (or half of the main span) is reproduced in the laboratory. Typically, a support structure consisting of taut wires or a catenary supports the simulated deck. The model is typically immersed in a simulated boundary-layer flow.

Wind Effects on Structures: Modern Structural Design for Wind, Fourth Edition. Emil Simiu and DongHun Yeo.
© 2019 John Wiley & Sons Ltd. Published 2019 by John Wiley & Sons Ltd.

Figure 23.1 Model of Akashi Strait suspension bridge. Source: Courtesy of T. Miyata of Yokohama University and M. Kitagawa, Honshu-Shikoku Bridge Authority, Tokyo.

3) *Tests on section models.* Section models consist of representative spanwise sections of the deck constructed to scale, with spring supports at the ends to allow both vertical and torsional motion. The model is provided with end plates or other devices that reduce aerodynamic end effects (Figure 23.2). Section models are relatively inexpensive and are built to scales in the order of $1:50$–$1:25$. They are useful for performing initial assessments of a bridge deck's aeroelastic stability, and allow the measurement of the fundamental aerodynamic characteristics of the bridge deck on the basis of which comprehensive analytical studies can be carried out. These characteristics include:

a) The steady-state drag, lift, and moment coefficient, defined as

$$C_D = \frac{\overline{D}}{\frac{1}{2}\rho U^2 B} \tag{23.1a}$$

$$C_L = \frac{\overline{L}}{\frac{1}{2}\rho U^2 B} \tag{23.1b}$$

$$C_M = \frac{\overline{M}}{\frac{1}{2}\rho U^2 B^2} \tag{23.1c}$$

where $\overline{D}, \overline{L}$, and $\overline{M}$ are the mean drag, lift and moment per unit span, respectively, ρ is the air density, B is the deck width, and U is the mean wind speed in the

Figure 23.2 Section model of the Halifax Narrows Bridge. Source: Courtesy of Boundary-Layer Wind Tunnel Laboratory, University of Western Ontario.

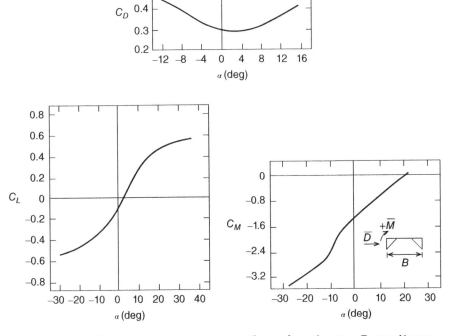

Figure 23.3 Drag, lift and aerodynamic moment coefficients for replacement Tacoma Narrows Bridge [3].

oncoming flow at the deck elevation. The aerodynamic coefficients are usually plotted as functions of the angle α between the horizontal plane and the plane of the bridge deck. Coefficients C_D, C_L, and C_M are shown in Figure 23.3 for the open-truss bridge deck of the replacement Tacoma Narrows Bridge [3] and in Figure 23.4 for a proposed streamlined box section of the New Burrard Inlet Crossing [4].

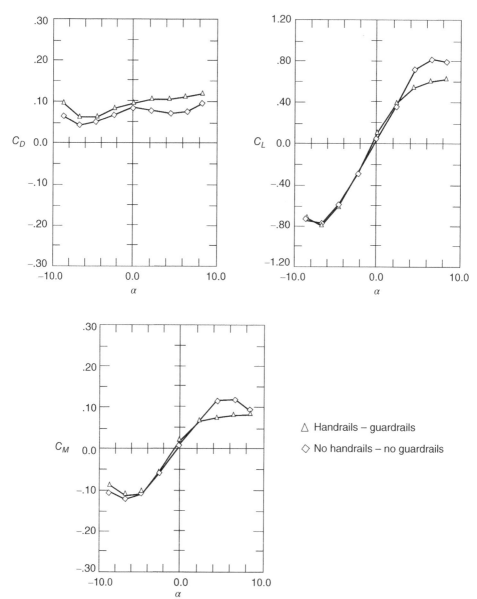

Figure 23.4 Drag, lift, and aerodynamic moment coefficients for proposed deck of New Burrard Inlet Crossing [4]. Source: Courtesy of National Aeronautical Establishment, National Research Council of Canada.

b) The motional aerodynamic coefficients. These coefficients characterize the self-excited forces acting on the oscillating bridge and are discussed in Section 21.3. Examples of motional aerodynamic coefficients, first introduced for bridge decks by Scanlan and Tomko in 1971 [5], are given in Figure 21.3.

c) The Strouhal number, *St.*

For details on a type of test that allows along-wind motion of the model, see [6].

23.3 Response to Vortex Shedding

Open truss sections generally "shred" the oncoming flow to such an extent that large, coherent vortices cannot occur, and vortex-induced oscillations of the deck are weak. However, severe vortex-induced oscillations of bluff deck sections of the box type are known to have occurred. A soffit plate and fairings with various dimensions were added to the original section, with the results shown in Figure 23.5 [7]. The water surface was in this case close to the underside of the bridge and was modeled in the wind tunnel tests. Additional shapes of streamlined bridge deck forms are shown in Figure 23.6. For additional material on remedial aerodynamic measures, see [1].

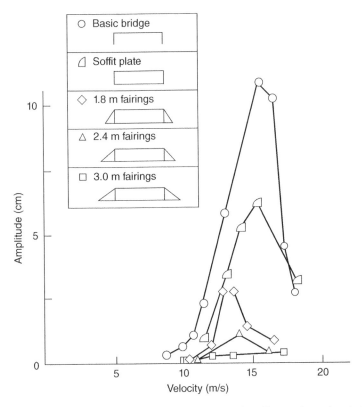

Figure 23.5 Vertical amplitudes of vortex-induced oscillations for various bridge deck sections of the proposed Long Creek's Bridge. Source: Courtesy of National Research Council, Canada.

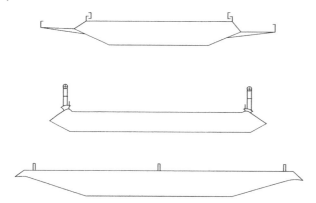

Figure 23.6 Streamlined bridge deck forms.

We now present an approach to the estimation of the bridge vortex-induced response [8]. Let the equation of motion of the section with unit spanwise length have the form, similar to the simplified version of Eq. (19.4):

$$m(\ddot{h} + 2\zeta_h\omega_h\dot{h} + \omega_h^2 h) = \frac{1}{2}\rho U^2 BKH_1^* \left(1 - \varepsilon\frac{h^2}{B^2}\right)\frac{\dot{h}}{U} \tag{23.2}$$

where $K = \omega B/U$, and H_1^*, ε are aerodynamic parameters obtainable experimentally. Derivations similar to those of Section 19.2 yield the non-dimensionalized steady-state amplitude h_0 of the bridge deck section model:

$$\frac{h_0}{B} = 2\left(\frac{H_1^* - 4S_{cr}}{\varepsilon H_1^*}\right)^{1/2} \tag{23.3}$$

where $S_{cr} = \zeta_h m/(\rho B^2)$ is the Scruton number. The coefficient H_1^* may be viewed as the value obtained at low oscillation amplitudes by any one of the several identification schemes employed to obtain flutter derivatives. If the steady-state vortex-induced amplitude h_0 is measured in a section model test, then

$$\varepsilon = 4\frac{H_1^* - 4S_{cr}}{(h_0/B)^2 H_1^*} \tag{23.4}$$

However, if H_1^* is not obtained from a low-amplitude model test, but instead the model is allowed to oscillate down from an initial larger amplitude A_0 to a steady, locked-in state of measured amplitude h_0, the value of H_1^* may be determined from

$$KH_1^* = \frac{m}{2\rho B^2}\left(\alpha\frac{h_0^2}{B^2} + 16\pi\zeta_h St\right) \tag{23.5}$$

where $K = 2\pi St$, St is the Strouhal number,

$$\alpha = -\frac{4StB^2}{nh_0^2}\ln\left(\frac{A_0^2 - R_n^2 h_0^2}{A_0^2 - h_0^2}\right) \tag{23.6}$$

and R_n is the ratio of the response amplitudes of the first to the nth cycle of amplitude decay [8].

The information given in Eq. (23.3) is applicable to the section model only. To extrapolate it to the full bridge, it is necessary to consider the oscillatory structural mode involved (usually a simple, low-frequency mode) as well as the spanwise correlation of the lock-in forces. In Eq. (23.2) it is therefore further assumed that

$$h(x,t) = \phi(x)B\xi(t) \tag{23.7}$$

where $\phi(x)$ is the dimensionless modal shape associated with the frequency ω_h of the deck excited by the locked-in vorticity. The corresponding generalized coordinate $\xi(t)$ is assumed to undergo purely sinusoidal oscillations

$$\xi(t) = \xi_0 \cos \omega t \tag{23.8}$$

at the Strouhal frequency, that is, where

$$\omega = \frac{2\pi StU}{B} = \omega_h \tag{23.9}$$

If h from Eq. (23.7) is inserted into Eq. (23.2) and the result is multiplied by $B\,\phi(x)$, the action of the section dx of the structure associated with the spanwise point with coordinate x is described by the equation

$$m(x)B^2\phi^2(x)[\ddot{\xi}(t) + 2\zeta_h\omega_h\dot{\xi}(t) + \omega_h^2\xi(t)]dx$$
$$= \frac{1}{2}\rho UB^3 KH_1^*[1 - \varepsilon\phi^2(x)\xi^2(t)]\phi^2(x)\dot{\xi}(t)f(x)dx \tag{23.10}$$

in which $f(x)$ is a function inserted to account for spanwise loss of coherence in the vortex-related forces.

If integration of Eq. (23.10) is performed over the full bridge span, there results

$$I[\ddot{\xi}(t) + 2\zeta_h\omega_h\dot{\xi}(t) + \omega_h^2\xi(t)] = \frac{1}{2}\rho UB^3 LKH_1^*[C_2 - \varepsilon C_3\xi^2(t)]\dot{\xi}(t) \tag{23.11}$$

where I is the generalized full-bridge inertia of the mode in question and

$$C_2 = \int_{\text{span}} \frac{\phi^2(x)f(x)dx}{L} \tag{23.12}$$

$$C_4 = \int_{\text{span}} \frac{\phi^4(x)f(x)dx}{L} \tag{23.13}$$

The strength of the vortex-induced forces is dependent upon the local oscillation amplitude of the structure. There is also a loss of coherence with spanwise separation. For example, Figure 19.2 shows the correlations between local lateral pressures separated spanwise along cylinders displaced sinusoidally in the vertical direction with various amplitudes. It is suggested that the correlation loss can be approximated by selecting $f(x)$ to be the mode shape $\varphi(x)$ normalized to unit value at its highest point. For example, with a mode representing a half-sinusoid over a span $L, f(x)$ may be estimated as

$$f(x) = \sin(\pi x/L) \tag{23.14}$$

At steady-state amplitude, as noted earlier, the damping energy balance per cycle of oscillation will be zero, a condition that defines the vortex-induced amplitude

$$\xi_0 = 2\left(\frac{C_2 H_1^* - 4S_{cr}}{\varepsilon C_4 H_1^*}\right)^{1/2} \tag{23.15}$$

where the Scruton number is defined as

$$S_{cr} = \frac{\zeta_h I}{\rho B^4 L} \tag{23.16}$$

For the case of a sinusoidal mode the values of C_2 and C_4, respectively, are

$$C_2 = \int_0^L \sin^3 \frac{\pi x}{L} \frac{dx}{L} = 0.4244 \tag{23.17a}$$

$$C_4 = \int_0^L \sin^5 \frac{\pi x}{L} \frac{dx}{L} = 0.3395 \tag{23.17b}$$

For a study of conditions for the occurrence of vortex shedding on a large cable stayed bridge based on full-scale data obtained by a monitoring system, see [9].

23.4 Flutter and Buffeting of the Full-Span Bridge

23.4.1 Theory

The flutter phenomenon was studied in some detail in Chapter 21 under the assumption that two-dimensional geometrical conditions hold. For a full-span bridge, the deformations of the deck are functions of position along the span so that this assumption is no longer valid. This section presents a generalization of the results of Chapter 21 to the case of a full-span bridge. An example is included.

Let $h(x, t), p(x, t)$, and $\alpha(x, t)$ represent, respectively, the vertical, sway, and twist deflections of a reference spanwise point x of the deck of a full bridge:

$$h(x, t) = \sum_{i=1}^N h_i(x) B \xi_i(t) \tag{23.18a}$$

$$\alpha(x, t) = \sum_{i=1}^N \alpha_i(x) \xi_i(t) \tag{23.18b}$$

$$p(x, t) = \sum_{i=1}^N p_i(x) \xi_i(t) \tag{23.18c}$$

where $h_i(x)$, $\alpha_i(x)$, $p_i(x)$ are, respectively, the values of the ith modal shape at point x of the deck and $\xi_i(t)$ is the generalized coordinate of the ith mode.

If I_i is the generalized inertia of the full bridge in mode i, the equation of motion for that mode is

$$I_i(\ddot{\xi}_i + 2\zeta_i \omega_i \dot{\xi}_i + \omega_i^2 \xi_i) = Q_i \tag{23.19}$$

where ζ_i, ω_i are the mechanical damping ratio and the circular natural frequency (in radians) of the ith mode, respectively, and

$$Q_i = \int_{deck} [(L_{ae} + L_b) h_i B + (D_{ae} + D_b) p_i B + (M_{ae} + M_b) \alpha_i] dx \tag{23.20}$$

is the generalized force in the ith mode of vibration. The subscripts "ae" and "b" signify "aeroelastic" and "buffeting," respectively. It is assumed that the following definitions of forces per unit span at section x hold:

Aeroelastic (self-excited) forces under sinusoidal motion:

$$L_{ae} = \frac{1}{2}\rho U^2 B \left[KH_1^*(K)\frac{\dot{h}}{U} + KH_2^*(K)\frac{B\dot{\alpha}}{U} + K^2 H_3^*(K)\alpha + K^2 H_4^*(K)\frac{h}{B} \right] \qquad (23.21a)$$

$$M_{ae} = \frac{1}{2}\rho U^2 B^2 \left[KA_1^*(K)\frac{\dot{h}}{U} + KA_2^*(K)\frac{B\dot{\alpha}}{U} + K^2 A_3^*(K)\alpha + K^2 A_4^*(K)\frac{h}{B} \right] \qquad (23.21b)$$

$$D_{ae} = \frac{1}{2}\rho U^2 B \left[KP_1^*(K)\frac{\dot{p}}{U} + KP_2^*(K)\frac{B\dot{\alpha}}{U} + K^2 P_3^*(K)\alpha + K^2 P_4^*(K)\frac{p}{B} \right] \qquad (23.21c)$$

Buffeting forces:

$$L_b = \frac{1}{2}\rho U^2 B \left[2C_L \frac{u(x,t)}{U} + \left(\frac{dC_L}{d\alpha} + C_D \right)\frac{w(x,t)}{U} \right] \qquad (23.22a)$$

$$M_b = \frac{1}{2}\rho U^2 B^2 \left[2C_M \frac{u(x,t)}{U} + \left(\frac{dC_M}{d\alpha} \right)\frac{w(x,t)}{U} \right] \qquad (23.22b)$$

$$D_b = \frac{1}{2}\rho U^2 B \left[2C_D \frac{u(x,t)}{U} \right] \qquad (23.22c)$$

In Eqs. (23.21) and (23.22) it is assumed that there is no interaction between the aeroelastic and the buffeting forces. However, the interaction is implicit in Eq. (23.22) if the aeroelastic forces are measured in turbulent flow (see, e.g., [10, 12]).

In what follows only a single-mode approximation to the total response will be postulated. This is justifiable by the observation that typically just one prominent mode will become unstable and dominate the flutter response of a three-dimensional bridge model in the wind tunnel. In this single-mode form of analysis, any mode i may be considered in Eqs. (23.18)–(23.22). When all but the flutter derivatives shown in Eq. (23.21) are considered as being of lesser importance, the expression for the generalized force is

$$Q_i = \frac{1}{2}\rho U^2 B^2 l \left\{ \frac{KB}{U}[H_1^* G_{h_i h_i} + P_1^* G_{p_i p_i} + A_2^* G_{\alpha_i \alpha_i}]\dot{\xi} + K^2 A_3^* G_{\alpha_i \alpha_i}\xi_i \right\}$$
$$+ \int_{\text{deck}} [L_b h_i B + D_b p_i B + M_b \alpha_i] dx \qquad (23.23)$$

in which

$$G_{q_i q_i} = \int_{\text{deck}} q_i^2(x)\frac{dx}{l} \qquad [q_i = h_i, p_i \text{ or } \alpha_i] \qquad (23.24)$$

and l is the span length. Because of the linearity of the resulting equation of motion the conditions of system stability are independent of the buffeting forces.

Equation (23.19) may be rewritten with a new frequency ω_{i0}, a new damping ratio γ_i, and a buffeting force Q_{ib}, as follows:

$$\ddot{\xi}_i + 2\gamma_i \omega_{i0}\dot{\xi}_i + \omega_{i0}^2 \xi_i = \frac{Q_{ib}(t)}{I_i} \qquad (23.25)$$

where

$$\omega_{i0}^2 = \omega_i^2 - \frac{\rho B^4 l}{2I_i}\omega^2 A_3^* G_{\alpha_i \alpha_i} \qquad (23.26)$$

$$2\gamma\omega_{i0} = 2\zeta_i\omega_i - \frac{\rho B^4 l}{2I_i}\omega[H_1^* G_{h_i h_i} + P_1^* G_{p_i p_i} + A_2^* G_{\alpha_i \alpha_i}] \qquad (23.27)$$

$$Q_{ib}(t) = \frac{1}{2}\rho U^2 B^2 l \int_{\text{deck}} [L_b h_i B + D_b p_i B + M_b \alpha_i] dx \tag{23.28}$$

Flutter. For instability to occur it is necessary that the damping ratio $\gamma_i \leq 0$. This leads to the single-mode flutter instability criterion

$$H_1^* G_{h_i h_i} + P_1^* G_{p_i p_i} + A_2^* G_{\alpha_i \alpha_i} \geq \frac{4\zeta_i I_i}{\rho B^4 l}\left[1 + \frac{\rho B^4 l}{2I_i}A_3^* G_{\alpha_i \alpha_i}\right]^{1/2} \tag{23.29}$$

in which the only significant flutter derivatives $H_1^*, P_1^*, A_2^*, A_3^*$ have been retained. An assumption inherent in this criterion is that the flutter derivatives retain full coherence throughout the deck span. The effect of the reduced coherence can be seen in a reduction of the values of the quantities $G_{q_i q_i}$.

In practice, the flutter derivatives H_1^* and P_1^* are typically negative, while A_2^* may take on positive values for sufficiently large values of the reduced velocity $U/(nB)$. The effect of the flutter derivative A_3^* – an "aerodynamic stiffness" effect – is in many practical cases negligible since the structural stiffness is typically considerably larger than the aerodynamic stiffness.

Buffeting. The generalized force may be written as

$$\frac{Q_{ib}(t)}{I_i} = \frac{\rho U^2 B^2 l}{2I_i} \int_{\text{deck}} [L h_i + D p_i + M \alpha_i] \frac{dx}{l} \tag{23.30}$$

where L, M, and D are, respectively, the quantities between brackets in Eqs. (23.22a, b and c). Defining the functions

$$\varphi(x) = 2[C_L h_i(x) + C_D p_i(x) + C_M \alpha_i(x)] \tag{23.31a}$$

$$\psi(x) = \left(\frac{dC_L}{d\alpha} + C_D\right) h_i(x) + \frac{dC_M}{d\alpha}\alpha_i(x) \tag{23.31b}$$

the integrand of Eq. (23.30) becomes

$$L h_i + D p_i + M \alpha_i = \varphi(x)\frac{u(x,t)}{U} + \psi(x)\frac{w(x,t)}{U} \tag{23.32}$$

Information on the turbulent flow fluctuations u and w is available in the form of spectral densities $S_u(n)$ and $S_w(n)$, respectively (see Chapter 2). This motivates the adoption of a frequency domain approach to the solution of Eq. (23.25). It is shown in [11] that the frequency domain counterpart of Eq. (23.25) yields the result

$$S_{\xi_i \xi_i}(\omega) = \frac{[\rho U^2 B^2 l/(2I_i)]^2}{\omega_{i0}^4[(1 - (\omega/\omega_{i0})^2)^2 + (2\gamma_i \omega/\omega_{i0})^2]} \iint_{\text{deck}} \frac{1}{U^2}\left[\varphi(x_a)\varphi(x_b)S_{uu}(x_a, x_b, \omega)\right.$$
$$\left. + \psi(x_a)\psi(x_b)S_{ww}(x_a, x_b, \omega)\right]\frac{dx_a}{l}\frac{dx_b}{l} \tag{23.33}$$

In Eq. (23.33) the effect of the cross-spectra of the fluctuations u and w has been neglected. The distributed cross power spectral densities are assumed to take the real forms (neglecting their imaginary components).

$$S_{uu}(x_a, x_b, \omega) = S_u(\omega)\exp\left(-\frac{C_u |x_a - x_b|}{l}\right) \tag{23.34a}$$

$$S_{ww}(x_a, x_b, \omega) = S_w(\omega)\exp\left(-\frac{C_w |x_a - x_b|}{l}\right) \tag{23.34b}$$

Expressions for the spectra $S_u(\omega)$ and $S_w(\omega)$ and values of C_u and C_w are suggested in Chapter 2. The standard deviation of ξ_i is

$$\sigma_{\xi_i} = \left[\int_0^\infty S_{\xi_i \xi_i}(n)dn \right]^{1/2} \tag{23.36}$$

where $n = \omega/2\pi$. From Eqs. (23.18) it follows that

$$\sigma_{h_i}(x) = h_i(x)B\sigma_{\xi_i} \tag{23.37a}$$

$$\sigma_{p_i}(x) = p_i(x)B\sigma_{\xi_i} \tag{23.37b}$$

$$\sigma_{\alpha_i}(x) = \alpha_i(x)\sigma_{\xi_i} \tag{23.37c}$$

23.4.2 Example: Critical Flutter Velocity and Buffeting Response of Golden Gate Bridge

This section presents a set of calculations developed by Scanlan on the basis of tests performed by Ragget that illustrate the approaches developed in Section 23.4.1 [11]. A $1:50$ scale model section was used to obtain flutter derivatives H_i^* and A_i^* ($i = 1, \dots,$ 4). A set of those derivatives for zero-degree angle of attack in smooth flow is shown in Figures 23.7 and 23.8. This example presents calculations that illustrate the use of the approach described in this section.

The vibration modes and frequencies of the bridge, together with their modal integrals $G_{q_i q_i}$, were obtained for the first eight modes with the results given in Table 23.1. Modal forms are suggested by the notations S (symmetric), AS (antisymmetric), L (lateral), V (vertical), and T (torsion). Values of the modal integrals $G_{q_i q_i}$ suggest the importance of the mode, in Table 23.1 the largest in each category (i.e., vertical, lateral, torsion) is underlined. The most pronounced modes are mode 6 (vertical), mode 1 (lateral), and mode 7 (antisymmetric torsion).

Flutter. The torsional aerodynamic damping coefficient A_2^* exhibits a pronounced change of sign with increasing velocity, indicating the possibility of single-degree of

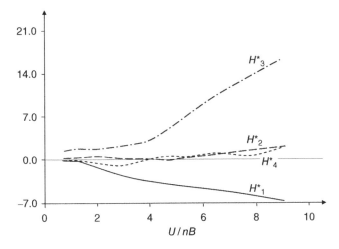

Figure 23.7 Flutter derivatives H_i^* ($i = 1, 2, 3, 4$). Golden Gate Bridge. Source: Courtesy of Dr. J. D. Raggett, West Wind Laboratory, Carmel, CA.

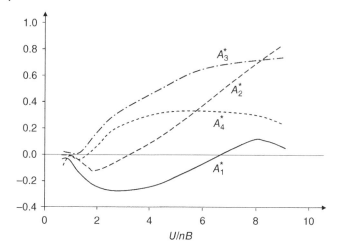

Figure 23.8 Flutter derivatives A_i^* ($i = 1, 2, 3, 4$). Golden Gate Bridge. Source: Courtesy of Dr. J. D. Raggett, West Wind Laboratory, Carmel, CA.

Table 23.1 Frequencies, types of modal forms, and modal integrals for Golden Gate Bridge.

	Frequency	Type	$G_{h_i h_i}$	$G_{p_i p_i}$	$G_{\alpha_i \alpha_i}$
1	0.049	L	2.62E-16	3.33E-01	8.03E-05
2	0.087	ASV_1	3.25E-01	7.39E-15	1.77E-15
3	0.112	L	1.72E-14	3.09E-01	1.24E-02
4	0.129	SV_1	1.90E-01	7.82E-14	1.16E-14
5	0.140	V	1.91E-01	5.58E-14	2.43E-14
6	0.164	V	3.44E-01	3.87E-13	1.25E-14
7	0.192	AST_1	6.67E-12	3.32E-02	1.29E + 00
8	0.197	ST_1	2.50E-12	2.47E-01	2.55E-01

freedom torsional flutter (Figure 23.8). Mode 7 is the torsional mode with both the lowest frequency and the greatest $G_{q_i q_i}$ value, and was selected as the most vulnerable to flutter instability (Figure 23.9). In the case of the original Tacoma Narrows Bridge the lowest antisymmetric mode was also the most-flutter prone. In the Golden Gate Bridge case this mode is practically a complete sine wave along the main span, with a node at the center, and practically zero amplitude on the two side spans.

The pertinent parameters are

$\zeta_7 = 0.005$ (assumed)
$I_7 = 8.5 \times 10^9$ lb ft s^2
$\rho = 2.38 \times 10^{-6}$ kip ft^{-4} s$^2 = 0.00238$ lb ft^{-4} s$^2 = 0.002378$ slugs/ft^3
$B = 90$ ft
$L = 6451$ ft
$G_{\alpha_7 \alpha_7} = 1.29$

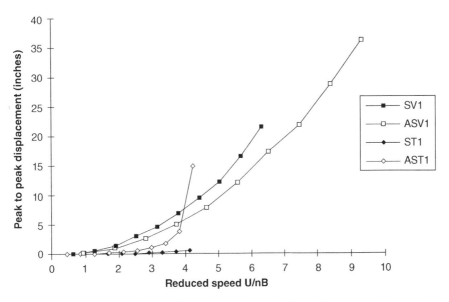

Figure 23.9 Calculated peak-to-peak displacements induced by buffeting in four selected modes.

The flutter criterion in this case reduces to

$$A_2^* \geq \frac{4\zeta_7 I_7}{\rho B^4 G_{\alpha_7 \alpha_7}} = 0.131$$

From the graph for A_2^* (Figure 23.8) the corresponding reduced velocity value (with $n = n_7 = 0.192\,\text{Hz}$) is

$$\frac{U}{nB} = 4.32$$

which corresponds to a critical laminar flow flutter velocity

$$U_{cr} = (4, 32)(0.192)(90) = 74.65\,\text{ft s}^{-1}\,(22.75\,\text{m s}^{-1})$$

Buffeting. The four modes listed in Table 23.2 are mainly active over the main span.

The following data were used: $l_{MS} = 4144\,\text{ft.}$ (main span length); $z_0 = 0.02\,\text{ft}$; $z = 220\,\text{ft}$ (deck height); $u_* = U/[2.5\,\ln(z/z_0)]$; $C_D = 0.34$; $C_L = 0.215$; $dC_L/d\alpha = 3.15$; $C_M = 0$; $dC_M/d\alpha = -0.111$. The modal shapes were assumed to have the shape of simple sinusoids: $h_{SV1} = h_0\,\sin(\pi x/l_{MS})$; $h_{ASV1} = h_0\,\sin(2\pi x/l_{MS})$; $\alpha_{SV1} = \alpha_0\,\sin(\pi x/l_{MS})$; $\alpha_{ASV1} = \alpha_0\,\sin(2\pi x/l_{MS})$. Equations (23.33)–(23.37) were then used to obtain the results of Figure 23.9.

Table 23.2 Generalized inertia of full bridge for four modes.

	Mode i	Frequency (Hz)	$10^{-9}\,I_i$ (lb ft s^2)
ASV_1	2	0.0870	15.71
SV_1	4	0.1285	6.15
AST_1	7	0.1916	8.50
ST_1	8	0.1972	8.59

23.5   Stay Cable Vibrations

23.5.1   Cable Vibration Characteristics

Stayed-bridge cables have low damping, small mass, and low bending stiffness. They can experience two types of vibration: (i) low-amplitude, high-frequency vortex-induced vibrations, and (ii) large-amplitude, low-frequency vibrations under skewed winds that include rain-wind-induced vibrations and dry galloping. This section considers only vibrations of the latter type.

According to full-scale field observations,

- Wind speeds at the onset of the vibrations can vary from 6 to 40 m s^{-1} [13, 14].
- Stays are vulnerable to excitation by skewed winds with directions making either a negative or a positive angle with the direction normal to the plane containing the cable [15–17].
- Vibrations were observed mostly in rainy weather [18, 19], but also occurred in the absence of rain [13, 17].
- Vibrations occurred in winds with both low and high turbulence intensity [19, 20].
- Vibrations occurred in low modes of vibration, mostly with frequencies of 1~3 Hz [15], but also in simultaneous multiple vibration modes [13, 21].
- Stay cables susceptible to excitation (e.g., polyethylene tube-lapped cables) had smooth surfaces [15, 22].
- Reynolds numbers ranged from 6×10^4 to 2×10^5 (sub-critical regime) [22, 23].
- The maximum acceleration of the cables varied between $4g$ and $10g$, where g is the gravitational acceleration (≈ 9.81 m s^{-2}) [16, 19, 24].
- Peak-to-peak amplitudes could several times a cable diameter [13, 24].

The wide range of the observed data suggests that no single mechanism can explain the cable vibration phenomenon. The proposed mechanisms can be roughly divided into two main categories: "high speed vortex-induced vibration" and "galloping instability."

According to [25], for vibrations occurring in rainy weather, rain water flowing downward owing to gravity forms a rivulet on the lower surface of the cable. As the wind becomes stronger, another rivulet forms on the upper surface, in which the forces due to wind, gravity and water surface tension are balanced. Cable oscillations cause the rivulets to oscillate around their mean positions, thus changing the points of separation of the wind flow, and affecting the pressure distribution, around the cable. This results in forces that cause the cable to vibrate. However, to date the fundamental mechanism of these vibrations remains uncertain, and could not be clarified by wind tunnel testing. For references on wind-rain induced vibrations see [1, 25, 26].

23.5.2   Mitigation Approaches

Common mechanical approaches to mitigating low-frequency, large-amplitude stay cable vibrations include increasing the damping by installing dampers and using cross-ties. The damping ratios of stay cables are typically in the range of 0.1–0.5% [26]. Most types of wind-induced vibrations can be reduced to acceptable levels by increasing the Scruton number $S_c = m\zeta/(\rho D^2)$ (Eq. [19.10]) by increasing the cable mass and damping. For rain-wind-induced vibrations it is recommended that $S_c > 10$ [26].

Owing to the geometrical constraints of bridge decks, dampers are typically attached to stay cables near cable anchorage and are designed to mitigate the cable vibrations in the fundamental modes. Transverse restrainers (e.g., cross-ties) between stay cables are commonly used to effectively mitigate the in-plane global mode cable vibrations [27], which give rise to local modes of vibration of the interconnected stays. Their excessive use may affect the aesthetics of the bridges.

Aerodynamic countermeasures include the modification of cable cross sections (by using, e.g., helical strakes and pattern-indented surfaces), with a view to disturbing the formation of water rivulets on the stay cables, which could cause the rain-wind induced vibrations [1]. According to [26], a Scruton number $S_c > 5$ is recommended if both mechanical and aerodynamic countermeasures are used in a cable system. It was reported in [28] that the effectiveness of aerodynamic countermeasures can be weaken if the Scruton number is less than 8.

References

1 Simiu, E. and Miyata, T. (2006). *Design of Buildings and Bridges for Wind: A Practical Guide for ASCE-7 Standard Users and Designers of Special Structures*, 1st ed. Hoboken, NJ: Wiley.

2 Davenport, A. G., Isyumov, N., Fader, D. J., and Bowen, C. F. P., "A Study of Wind Action on a Suspension Bridge during Erection and Completion," Report No. BWLT-3-69 with Appendix BLWT-4-70, Faculty of Engineering Science, University of Western Ontario, London, Canada, 1969–1970.

3 Farquharson, F.B. (ed.) (1949–1954). *Aerodynamic Stability of Suspension Bridges*, Bulletin No. 116. Seattle, WA: University of Washington Engineering Experiment Station.

4 Wardlaw, R. L., "Static Force Measurements of Six Deck Sections for the Proposed New Burrard Inlet Crossing," Report No. LTR-LA-53, National Aeronautical Establishment, National Research Council, Ottawa, Canada, 1970.

5 Scanlan, R.H. and Tomko, J.J. (1971). Airfoil and bridge deck flutter derivatives. *Journal of the Engineering Mechanics Division* 97: 1717–1737.

6 Gan Chowdhury, A. and Sarkar, P.P. (2003). A new technique for identification of eighteen flutter derivatives using three-degree-of-freedom section model. *Engineering Structures* 25: 1763–1772.

7 Wardlaw, R. L. and Goettler, L. L., "A Wind Tunnel Study of Modifications to Improve the Aerodynamic Stability of the Long Creek's Bridge," Report LTR-LA-8, National Aeronautical Establishment, National Research Council, Ottawa, Canada, 1968.

8 Ehsan, F. and Scanlan, R.H. (1990). Vortex-induced vibrations of flexible bridges. *Journal of Engineering Mechanics* 116: 1392–1411.

9 Flamand, O., De Oliveira, F., Stathopoulos-Vlamis, A., and Papanikolas, P. (2014). Conditions for occurrence of vortex shedding on a large cable stayed bridge: Full scale data from monitoring system. *Journal of Wind Engineering and Industrial Aerodynamics* 135: 163–169.

10 Sarkar, P. P., "New identification methods applied to response of flexible bridges to wind," Doctoral dissertation, Civil Engineering, Johns Hopkins University, Baltimore, MD, 1992.

11 Simiu, E. and Scanlan, R.H. (1996). *Wind Effects on Structures*, 3rd ed. Hoboken, NJ: John Wiley & Sons.

12 Scanlan, R.H. and Lin, W.-H. (1978). Effects of turbulence on bridge flutter derivatives. *Journal of the Engineering Mechanics Division* 104: 719–733.

13 Matsumoto, M., Daito, Y., Kanamura, T. et al. (1998). Wind-induced vibration of cables of cable-stayed bridges. *Journal of Wind Engineering and Industrial Aerodynamics* 74–76: 1015–1027.

14 Matsumoto, M., Saitoh, T., Kitazawa, M., Shirato, H., and Nishizaki, T., "Response characteristics of rain-wind induced vibration of stay-cables of cable-stayed bridges," *Journal of Wind Engineering and Industrial Aerodynamics*, 57, 323–333, 1995.

15 Hikami, Y. and Shiraishi, N. (1988). Rain-wind induced vibrations of cables stayed bridges. *Journal of Wind Engineering and Industrial Aerodynamics* 29: 409–418.

16 Phelan, R.S., Sarkar, P.P., and Mehta, K.C. (2006). Full-scale measurements to investigate rain-wind induced cable-stay vibration and its mitigation. *Journal of Bridge Engineering* 11: 293–304.

17 Zuo, D. and Jones, N. P., "Understanding wind- and rain-wind-induced stay cable vibrations from field observations and wind tunnel tests," in *4th U.S.-Japan Workshop on Wind Engineering*, Tsukuba, Japan, 2006.

18 Main, J. A. and Jones, N. P., "Full-scale measurements of stay cable vibration," in *10th International Conference on Wind Engineering*, Copenhagen, Denmark, 1999, pp. 963–970.

19 Ni, Y.Q., Wang, X.Y., Chen, Z.Q., and Ko, J.M. (2007). Field observations of rain-wind-induced cable vibration in cable-stayed Dongting Lake Bridge. *Journal of Wind Engineering and Industrial Aerodynamics* 95: 303–328.

20 Matsumoto, M., Shiraishi, N., and Shirato, H. (1992). Rain-wind induced vibration of cables of cable-stayed bridges. *Journal of Wind Engineering and Industrial Aerodynamics* 43: 2011–2022.

21 Zuo, D., Jones, N.P., and Main, J.A. (2008). Field observation of vortex- and rain-wind-induced stay-cable vibrations in a three-dimensional environment. *Journal of Wind Engineering and Industrial Aerodynamics* 96: 1124–1133.

22 Matsumoto, M. (1998). Observed behavior of prototype cable vibration and its generation mechanism. In: *Advances in Bridge Aerodynamics* (ed. A. Larsen), 189–211. Rotterdam, The Netherlands: Balkema.

23 Zuo, D., "Understanding wind- and rain-wind induced stay cable vibrations," Doctoral dissertation, Civil Engineering, Johns Hopkins University, Baltimore, 2005.

24 Main, J. A., Jones, N. P., and Yamaguchi, H., "Characterization of rain-wind induced stay-cable vibrations from full-scale measurements," in *4th International Symposium on Cable Dynamics*, Montreal, Canada, 2001, pp. 235–242.

25 Caetano, E., *Cable vibrations in cable-stayed bridges* 9: IABSE (International Association for Bridge and Structural Engineering), 2007.

26 Kumarasena, S., Jones, N. P., Irwin, P. A., and Taylor, P., "Wind-induced vibration of stay cables," FHWA-RD-05-083, Federal Highway Administration, McLean, VA, 2007.

27 Yamaguchi, H. and Nagahawatta, H.D. (1995). Damping effects of cable cross ties in cable-stayed bridges. *Journal of Wind Engineering and Industrial Aerodynamics* 54-55: 35–43.

28 Ruscheweyh, H. (1994). Vortex excited vibrations. In: *Wind-Excited Vibrations of Structures* (ed. H. Sockel), 51–84. Wien/New York: Springer-Verlag.

Part IV

Other Structures and Special Topics

24

Trussed Frameworks and Plate Girders

This chapter reviews the aerodynamic behavior of trussed frameworks and plate girders, including single trusses and girders, systems consisting of two or more parallel trusses or girders, and square and triangular towers. Test results are often presented from several sources with a view to allowing an assessment of the errors that may be expected in typical wind tunnel measurements. Throughout this chapter, the aerodynamic coefficients are referred to and should be used in conjunction with the effective area of the framework, A_f.

For any given wind speed, the principal factors that determine the wind load acting on a trussed framework are:

- The aspect ratio, λ; that is, the ratio of the length of the framework to its width. If end plates or abutments are provided, the flow around the framework is essentially two-dimensional.
- The solidity ratio, φ; that is, the ratio of the effective to the gross area of the framework.[1] For any solidity ratio, φ, the wind load is for practical purposes independent of the truss configuration, that is, of whether a diagonal truss, a K-truss, and so forth, is involved.
- The shielding of portions of the framework by other portions located upwind. The degree to which shielding occurs depends upon the configuration of the spatial framework. If the framework consists of parallel trusses (or girders), the shielding depends on the number and spacing of the trusses (or girders).
- The shape of the members; that is, whether the members are rounded or have sharp edges. Forces on rounded members depend on Reynolds number Re and on the roughness of the member surface (see Figure 4.22). For trusses with sharp edges, the effect of the Reynolds number and of the shape and surface roughness of the member is, in practice, negligible.
- The turbulence in the oncoming flow. The effect of turbulence on the drag force acting on frameworks with sharp-edged members is relatively small in most cases of practical interest [1–6]. A similar conclusion appears to be valid for frameworks composed of members with circular cross section in flows with subcritical Reynolds numbers.

1 The effective area of a plane truss is the area of the shadow projected by its members on a plane parallel to the truss, the projection being normal to that plane. The gross area of a plane truss is the area contained within the outside contour of that truss. The effective area and the gross area of a spatial framework are defined, respectively, as the effective area and the gross area of its upwind face.

Wind Effects on Structures: Modern Structural Design for Wind, Fourth Edition. Emil Simiu and DongHun Yeo.
© 2019 John Wiley & Sons Ltd. Published 2019 by John Wiley & Sons Ltd.

For this reason, and owing to scaling difficulties, in most cases wind tunnel tests for trussed frameworks are to this day conducted in smooth flow [3–6].

- The orientation of the framework with respect to mean wind direction.

Wind forces on ancillary parts (e.g., ladders, antenna dishes, solar panels) must be taken into account in design in addition to the wind forces on the trussed frameworks themselves. Drag and interference effects on microwave dish antennas and their supporting towers were studied in [7]. Drag coefficients for an unshrouded isolated microwave dish with depth-to-diameter ratio 0.24 were found to be largest for angles of 0–30° between wind direction and the horizontal projection of the normal to the dish surface, and are almost independent of the flow turbulence ($C_D \approx 1.4$). For a single dish, the ratio f_a between the incremental total drag on the tower due to the addition of a single dish and the drag for the isolated dish depends on the wind direction, and it is higher than unity (as high as 1.3) for the most unfavorable directions. This is due to the flow accelerations induced in the dish. As more dishes are added at the same level of a tower, interference factors are still greater than unity, but tend to decrease as the number of dishes increases.

Various petrochemical and other industrial facilities consist of complex assemblies of pipes, reservoirs, vessels, ladders, frames, trusses, beams, and so forth, for which the determination of overall wind loads is typically difficult. Estimates of wind loads for such facilities are discussed in some detail in [8].

24.1 Single Trusses and Girders

Figure 24.1 summarizes measurement of the drag coefficient $C_D^{(1)}$ for a single truss with infinite aspect ratio normal to the wind. The data of Figure 24.1 were obtained in the 1930s in Göttingen by Flachsbart for trusses with sharp-edged members [1, 2], and in the late 1970s at the National Maritime Institute, U.K. (NMI) for trusses with sharp-edged and trusses with members of circular cross section (all NMI measurements reported in this chapter were conducted at Reynolds numbers $10^4 < Re \lesssim 7 \times 10^4$). It is seen that the differences between the Göttingen and the NMI results are approximately 15% or less. For single trusses normal to the wind and composed of sharp-edged members, Figure 24.2 shows ratios $C_D^{(1)}(\lambda)/C_D^{(1)}(\lambda = \infty)$ of the drag coefficients corresponding to an aspect ratio λ and to an infinite aspect ratio.

Drag coefficients $C_D^{(1)}$ reported in [3] for trusses normal to the wind, composed of sharp-edged members, and having aspect ratios $1/6 < \lambda < 6$ are listed in the first line of Table 24.1. The second line of Table 24.1 lists values $C_D^{(1)}(\lambda = \infty)$ obtained from the drag coefficients of [3] through multiplication by the appropriate correction factor taken from Figure 24.2.

Figure 24.3 [7] summarizes results of tests on trusses with members of circular cross section ($\lambda = \infty$) conducted in the subsonic wind tunnel at Porz-Wahn, Germany [9, 10] and in the compressed air tunnel of the National Physical Laboratory, UK [11].[2] Note

---

2 Figures 24.3 and 24.16–24.19 are reproduced with permission of CIDECT (Comité International pour le Développenent et l'Etude de la de la Construction Tubulaire) from H.B. Walker, ed., *Wind Forces on Unclad Tubular Structures*. They are based in part on research work carried out by CIDECT and reported in [9, 10].

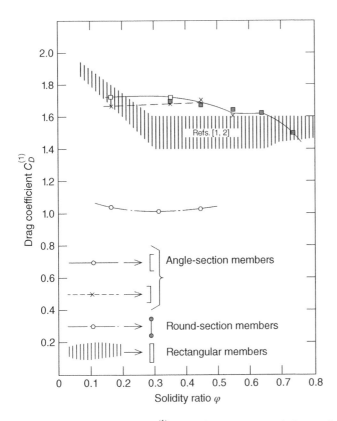

Figure 24.1 Drag coefficient $C_D^{(1)}$ for single truss, $\lambda = \infty$, wind normal to truss. Source: From [6].

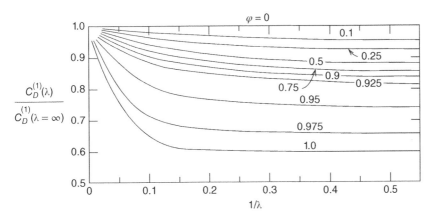

Figure 24.2 Ratios $C_D^{(1)}(\lambda)/C_D^{(1)}(\lambda = \infty)$, wind normal to truss [2].

Table 24.1 Drag coefficients for simple trusses.

φ	0.14	0.29	0.47	0.77	1.0
$C_D^{(1)}\left(\frac{1}{6}<\lambda<6\right)$ [3]	$1.40\pm5\%$	$1.54\pm5\%$	$1.27\pm5\%$	$1.18\pm5\%$	$1.28\pm5\%$
$C_D^{(1)}(\lambda=\infty)$	~1.45	~1.65	~1.45	~1.35	~2.10

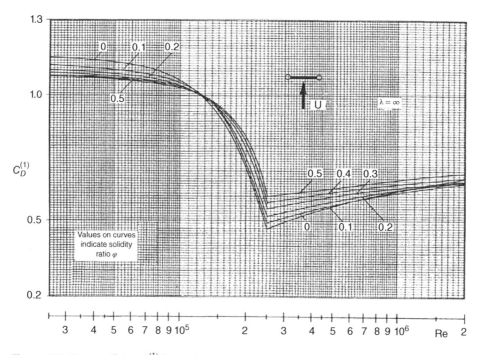

Figure 24.3 Drag coefficient $C_D^{(1)}$ for single truss with members of circular cross section, $\lambda=\infty$, wind normal to truss [7]. Source: Courtesy Comité International pour le Développement et l'Etude de la Construction Tubulaire and Constructional Steel Research and Development Organization.

that for $Re<10^5$ the drag coefficients of Figure 24.3 differ by about 5% or less from the corresponding results of Figure 24.1.

The aerodynamic force normal to a rectangular plate with aspect ratio $\lambda=5$–10 is larger when the yaw angle (i.e., the horizontal angle between the mean wind direction and the normal to the trusses) is $\alpha\approx40°$ than if the wind is normal to the plate; however, for trusses with solidity ratio $\varphi<4$ or so the maximum drag occurs when the wind is normal to the truss [1].

24.2 Pairs of Trusses and of Plate Girders

We consider a pair of identical, parallel trusses and denote the drag coefficient corresponding to the total aerodynamic force normal to the trusses by $C_D^{(2)}(\alpha)$, where α is the yaw angle. For brevity, $C_D^{(2)}(0)$ is denoted by $C_D^{(2)}$. The cases where the wind is normal to the truss ($\alpha = 0°$) and where $\alpha \neq 0°$ are considered in Sections 24.2.1 and 24.2.2, respectively.

24.2.1 Trusses Normal to Wind

Two parallel trusses normal to the wind affect each other aerodynamically, so that the drag on the upwind and on the downwind truss will have drag coefficients $\Psi_I C_D^{(1)}$ and $\Psi_{II} C_D^{(1)}$, respectively, where $C_D^{(1)}$ is the drag coefficient for a single truss normal to the wind and, in general, $\Psi_I \neq \Psi_{II} \neq 1$. It follows that

$$C_D^{(2)} = C_D^{(1)}(\Psi_I + \Psi_{II}) \tag{24.1}$$

Figure 24.4 shows values of Ψ_I and Ψ_{II} reported in [12], as functions of the solidity ratio φ, the ratio between the truss spacing in the along-wind direction, e, and the truss width, d. Values of Ψ_I and Ψ_{II}, also reported in [12], for four types of truss with sharp-edged members and aspect ratio $\lambda = 9.5$ are shown in Figure 24.5. On the basis of the data in Figures 24.4 and 24.5, [12] proposed the use for design purposes of the conservative values $C_D^{(2)}/C_D^{(1)}$ given, for $e/d > 1.0$, in Figure 24.6.

Measurements conducted at NMI on trusses with infinite aspect ratios are summarized in Figure 24.7. The following approximate expressions based on the results of Figure 24.7 are suggested in [6]:

$$\frac{C_D^{(2)}}{C_D^{(1)}} = 2 - \varphi^{0.45}\left(\frac{e}{d}\right)^{\varphi-0.45} \qquad \text{for } 0 < \varphi < 0.5 \tag{24.2}$$

for trusses with sharp-edged members, and

$$\frac{C_D^{(2)}}{C_D^{(1)}} = 2 - \varphi_e^{0.45}\left(\frac{e}{d}\right)^{\varphi_e-0.45} \tag{24.3}$$

for trusses composed of members with circular cross section. The nominal solidity ratio φ_e in Eq. (24.3) is related to the actual solidity ratio as shown in Figure 24.8.

Table 24.2 lists ratios $C_D^{(2)}/C_D^{(1)}$ for trusses with sharp-edged members and aspect ratio $\lambda = 8$ [4].

Example 24.1 Consider a truss with sharp-edged members, solidity ratio $\varphi = 0.18$, spacing ratio $e/d = 1.0$, and aspect ratio $\lambda = \infty$. According to both the Flachsbart and the NMI tests, $C_D^{(1)} \approx 1.70$ (Figure 24.1), and $C_D^{(2)}/C_D^{(1)} = \Psi_I + \Psi_{II} \approx 1.5$ (Figures 24.4a and 24.7a), so $C_D^{(2)} = 1.70 \times 1.55 \approx 2.65$. According to the deliberately conservative Figure 24.6, $C_D^{(2)}/C_D^{(1)} \approx 1.83$, which exceeds by about 20% the value based on Figures 24.4a and 24.7a.

24.2.2 Trusses Skewed with Respect to Wind Direction

We now consider the case in which the yaw angle is $\alpha \neq 0°$. For certain values of α the effectiveness of the shielding decreases, and the drag coefficient $C_D^{(2)}(\alpha)$ characterizing the total force normal to the trusses is larger than the value $C_D^{(2)}$.

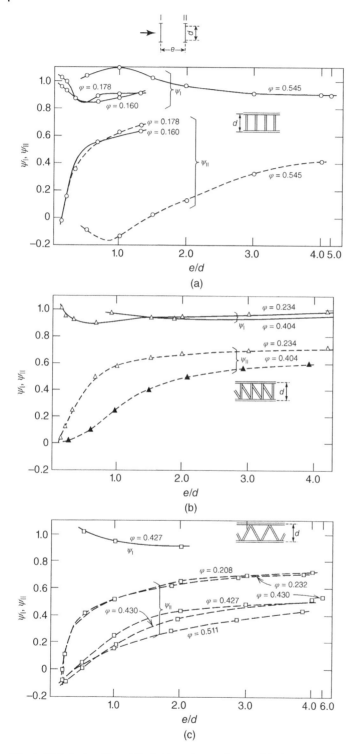

Figure 24.4 Factors Ψ_I and Ψ_{II} for three types of truss with sharp-edged members and infinite aspect ratio [12].

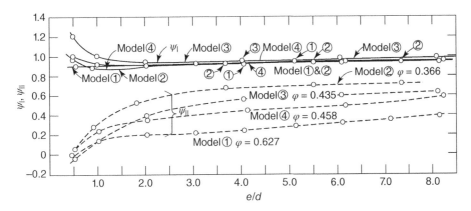

Figure 24.5 Factors Ψ_I and Ψ_{II} for four sets of two parallel trusses with sharp-edgd members, $\lambda = 9.5$, wind normal to trusses [12].

Figure 24.6 Approximate ratios $C_D^{(2)}/C_D^{(1)}$ proposed for design purposes in [12].

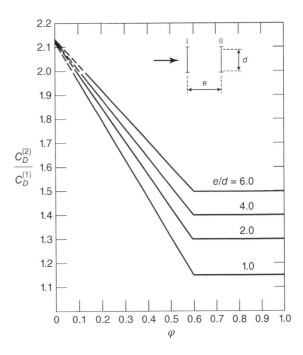

Ratios max $\{C_D^{(2)}(\alpha)\}/C_D^{(1)}$ reported in [3] for trusses with sharp-edged members and aspect ratio $\lambda = 8$ are shown in Table 24.3. For example, for $e/d = 1.0$, $\varphi = 0.286$, and $\lambda = 8$, the ratio $\{C_D^{(2)}(\alpha)\}/C_D^{(1)} \approx 1.77$ (Table 24.3) versus $C_D^{(2)}/C_D^{(1)} \approx 1.59$ (Table 24.2).

24.2.3 Pairs of Solid Plates and Girders

Figure 24.9 shows the dependence of the factors Ψ_I and Ψ_{II} (see Eq. [24.1]) upon the spacing ratio e/d for a solid disk and for three girders normal to the wind [12, 13]. For certain values of the horizontal angle α between the wind direction and the normal to the

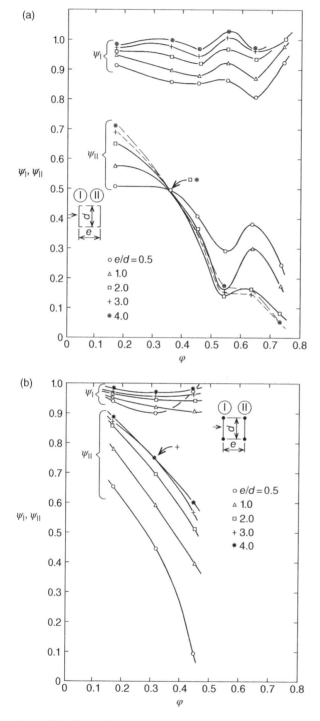

Figure 24.7 Factors Ψ_I and Ψ_{II} for two parallel trusses with (a) sharp-edged members and (b) members of circular cross section, infinite aspect ratio, wind normal to trusses. Source: From [6].

Figure 24.8 Equivalent solidity ratio φ_e for trusses with members of circular cross section and solidity ratio φ. Source: From [6].

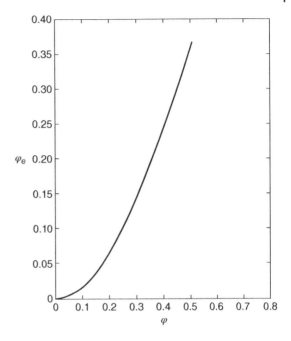

Table 24.2 Ratios $C_D^{(2)}/C_D^{(1)}$ for trusses with sharp-edged members and aspect ratio $\lambda = 8$, wind normal to trusses.

e/d φ	0.12	0.2	0.26	0.5	0.75	1.0	1.5	2.0
0.136	1.35	1.67	1.73	1.84	1.83	1.84	—	—
0.286	1.14	1.47	1.43	1.56	1.59	1.59	—	—
0.464	1.22	1.29	1.32	1.32	1.33	1.33	1.34	—
0.773	1.16	1.15	1.13	1.10	1.09	1.08	1.01	1.01
1.0	1.01	1.01	1.01	1.00	1.01	0.99	0.95	0.91

Source: After [5].

plates, the ratio $C_D^{(2)}(\alpha)/C_D^{(2)}$ may be larger than unity. For example, for a plate with aspect ratio $\lambda = 4$ and spacing ratio $e/d = 0.5$, if $40° \leq \alpha \leq 65°$, then $\{C_D^{(2)}(\alpha)\}/C_D^{(2)} \approx 1.20$ [5].

24.3 Multiple Frame Arrays

The first attempts to measure aerodynamic forces on multiple frame arrays were reported in [5, 6]. For frames normal to the wind, the drag coefficients for the first, second, ..., n-th frame may be written as $\Psi_1 C_D^{(1)}, \Psi_2 C_D^{(1)}, ..., \Psi_n C_D^{(1)}$, where $C_D^{(1)}$ is the

Table 24.3 Ratios max $\{C_D^{(2)}(\alpha)\}/C_D^{(1)}$ for trusses with sharp-edged members, $\lambda = 8$.

e/d	0.25	0.50	0.75	1.0	1.5	2.0
φ						
0.15	1.85	1.85	1.86	1.88	1.93	1.99
0.3	1.62	1.66	1.71	1.77	1.87	1.97
0.5	1.40	1.48	1.54	1.61	1.76	1.94
0.8	1.14	1.19	1.38	1.48	1.71	1.84
1.0	1.01	1.27	1.36	1.43	1.61	1.69

Source: After [5].

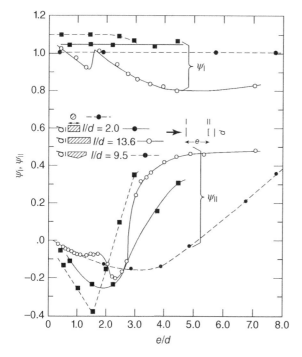

Figure 24.9 Factors $\Psi_\perp$ and $\Psi_\parallel$ for two parallel solid plates (girders) [12, 13].

drag coefficient for a single frame normal to the wind. The drag coefficient for the array of frames normal to the wind is then

$$C_D^{(n)} = C_D^{(1)}(\Psi_1 + \Psi_2 + \cdots + \Psi_n) \tag{24.4}$$

Factors Ψ_j ($j = 1, 2,..., n$) for arrays of three, four, and five parallel trusses with sharp-edged members and infinite aspect ratio are given in Figure 24.10 for spacings $e/d = 0.5$ and $e/d = 1$ [6]. Figure 24.11 show plots of drag coefficients $C_D^{(n)}$ for the same arrays with members of circular and angle cross section [6].

Figure 24.10 Factors Ψ_j ($j = 1, 2,..., n$) for arrays of n parallel trusses ($n = 3$, 4, and 5) with sharp-edged members, $\lambda = \infty$, wind normal to trusses. (a) Spacing ratio $e/d = 0.5$. (b) Spacing ratio $e/d = 1.0$. Source: From [6].

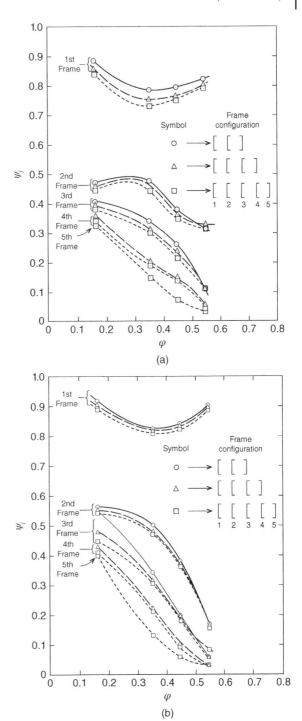

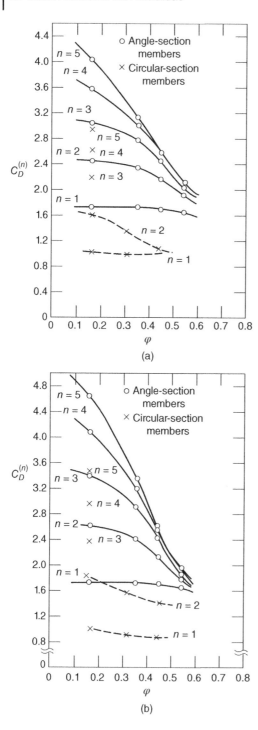

Figure 24.11 Drag coefficients $C_D^{(n)}$ for arrays of n parallel trusses, $\lambda = \infty$, wind normal to trusses. (a) Spacing ratio $e/d = 0.5$. (b) Spacing ratio $e/d = 1.0$. Source: From [6].

24.4 Square and Triangular Towers

The aerodynamic coefficients given in this chapter are in all cases referred to, and should be used in conjunction with, the effective area of the framework, A_f. For square and rectangular towers A_f is the effective area of one of the identical faces of the tower. The dynamic response of the towers can be determined conservatively as shown in Appendix D. The width of the structure used as input should be the actual width of the framework. This ensures that the lateral coherence of the load fluctuations is taken into account. The depth (along-wind dimension) of the framework should be assumed to be zero in order not to overestimate the favorable effect of the imperfect along-wind cross-correlations of the fluctuating loads. Finally, the area of the framework per unit height at any given elevation, used to estimate the mean and the fluctuating drag forces, should be equal to the effective area per unit height at that elevation.

24.4.1 Aerodynamic Data for Square and Triangular Towers

The results of wind force measurement on square towers can be expressed in terms of the aerodynamic coefficients $C_N(\alpha)$ and $C_T(\alpha)$ associated, respectively, with the wind force components N and T ($N \geq T$) normal to the faces of the tower (Figure 24.12) and in terms of the aerodynamic coefficient $C_F(\alpha)$ associated with the total wind force F acting at a yaw angle $\alpha = \tan^{-1}(T/N)$. Note that $C_F(\alpha) = [C_N^2(\alpha) + C_T^2(\alpha)]^{1/2}$, since, as indicated earlier, all aerodynamic coefficients are referenced to the effective area of one face of the framework, A_f.

For a triangular tower (which has in practice and is therefore assumed here to have equal sides in plan), the results of the measurements can be expressed in terms of the aerodynamic coefficients $C_F(\alpha)$ (Figure 24.13). The aerodynamic coefficients $C_F(0°)$ and $C_F(60°)$ correspond, respectively, to wind forces acting in a direction normal to a side and along the direction of a median (Figure 24.13).

Square Towers Composed of Sharp-Edged Members. Measurements of loads on a tapered square tower model with sharp-edges members, aspect ratio $\lambda \approx \infty$, and solidity ratio averaged over the height of the tower $\varphi \approx 0.19$ (ranging from $\varphi = 0.13$ at the base to $\varphi = 0.47$ at the top) were reported in the 1930s [14]. Until recently these measurements have been the principal source of data on square towers. The coefficients $C_N(\alpha)$, $C_T(\alpha)$, and $C_F(\alpha)$ obtained in [4] are listed for various angles α in Table 24.4.

For $\alpha = 45°$ the values of $C_N(\alpha)$ and $C_T(\alpha)$ should be equal; as pointed out in [14], the 4% difference between these values in Table 24.4 is due to measurement errors. Note that the value $C_N(0°) = 2.54$ is close to the values inferred from [3] and [6], which are, respectively, $C_N(0°) = C_D^{(2)} \approx 1.5 \times 1.73 = 2.60$ (as obtained by linear interpolation for $\varphi = 0.19$ and $e/d = 1.0$ from Tables 24.1 and 24.2), and $C_N(0°) = C_D^{(2)} \approx 1.7(0.93 + 0.58) = 2.57$ (Eq. [24.1], Figures 24.1 and 24.7a). Note also that while the largest tension (compression) in the tower columns is caused by

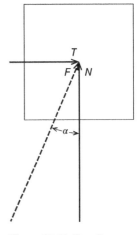

Figure 24.12 Notations.

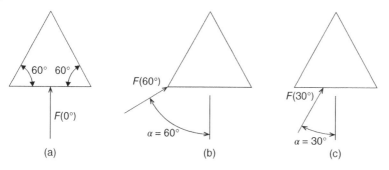

Figure 24.13 Notations.

Table 24.4 Aerodynamic coefficients: $C_N(\alpha)$, $C_T(\alpha)$, and $C_F(\alpha)$ for a square tower, $\varphi \approx 0.19$ and $\lambda \approx \infty$ [14].

α	0°	9°	18°	27°	36°	45°
$C_N(\alpha)$	2.54	2.75	2.97	3.01	2.84	2.60
$C_T(\alpha)$	—	0.19	0.70	1.36	2.05	2.49
$C_F(\alpha)$	2.54	2.76	3.05	3.30	3.50	3.60

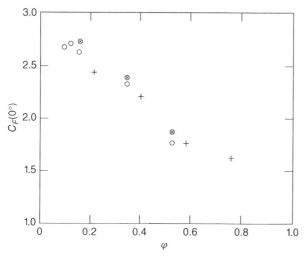

o Angle members—smooth flow.
⊗ Angle members—turbulent flow.
+ Square shped members—smooth flow.

Figure 24.14 Drag coefficients $C_F(0°)$ for square tower with sharp-edged members measured at NMI. Source: From [4].

winds acting in the direction $\alpha = 45°$, the largest stresses in the bracing members occur for $\alpha = 27°$.

Measurements of forces on square towers with sharp-edged members ($\lambda \approx \infty$) were more recently conducted at NMI [4]. Coefficients $C_F(0°)$ and ratios $C_F(\alpha)/C_F(0°)$ based on these measurements are shown in Figures 24.14 and 24.15, respectively. Note, for

Figure 24.15 Ratios $C_F(\alpha)/C_F(0°)$ for a square tower with sharp-edged members measured at NMI. Source: From [4].

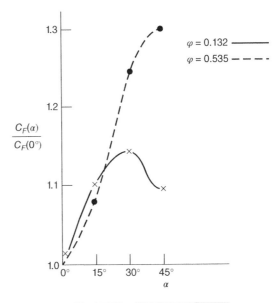

$\varphi = 0.132$ ———
$\varphi = 0.535$ — — —

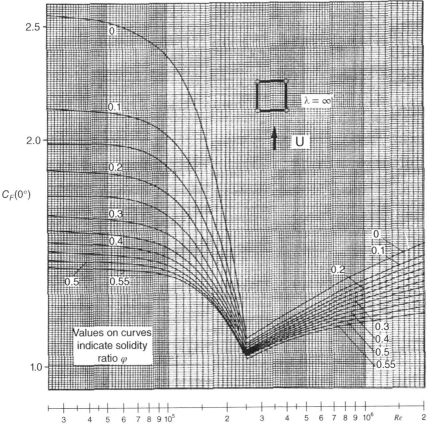

Figure 24.16 Drag coefficients $C_F(0°)$ for a square tower with members of circular cross section measured at the National Maritime Institute [7]. Source: Courtesy Comité International pour le Développenent et l'Etude de la de la Construction Tubulaire and Constructional Steel Research and Development Organization.

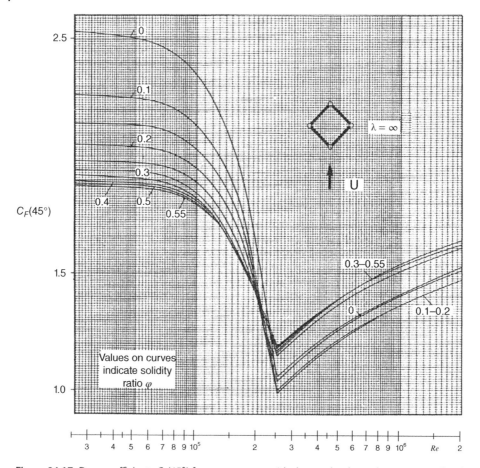

Figure 24.17 Drag coefficients $C_F(45°)$ for square tower with sharp-edged members measured at the National Maritime Institute [7]. Source: Courtesy Comité International pour le Développenent et l'Etude de la de la Construction Tubulaire and Constructional Steel Research and Development Organization.

example, that for $\varphi \approx 0.19$, $C_F(0°) \approx 2.60$ (Figure 24.14), versus $C_F(0°) = 2.54$, as obtained in [14] (Table 24.4). The agreement is less good for the ratio $C_F(45°)/C_F(0°)$, which is about 1.12 according to Figure 24.15, and about 1.40 according to data of Table 24.4.

Square Towers Composed of Members with Circular Cross Sections. Figures 24.16 and 24.17 [9] represent, respectively, aerodynamic coefficients $C_F(0°)$ and $C_F(45°)$ as functions of Reynolds number *Re* for towers with aspect ratio $\lambda = \infty$, based on wind tunnel test results reported in [9, 10]. The values $C_F(45°)$ of Figure 24.17 may be regarded as conservative envelopes that account for the loadings in the most unfavorable directions. Results of NMI tests in both smooth and turbulent flow at Reynolds numbers $Re \approx 2 \times 10^4$ for solidity ratios $\varphi = 0.17$, $\varphi = 0.23$, and $\varphi = 0.31$ ($\lambda = \infty$) match the curves of Figures 24.16 and 24.17 to within about 5% [4].

Triangular Towers Composed of Members with Circular Cross Sections. Figures 24.18 and 24.19 [9] represent proposed aerodynamic coefficients $C_F(0°) \approx C_F(60°)$ and

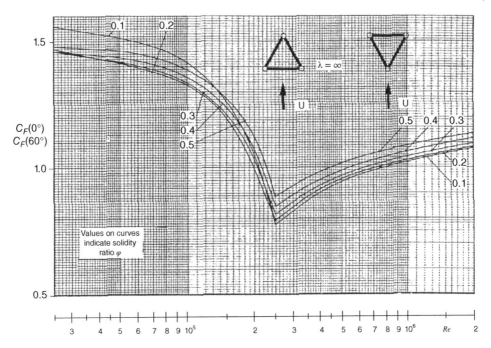

Figure 24.18 Drag coefficients $C_F(0°)$ and $C_F(60°)$ for triangular tower with members of circular cross section [7]. Source: Courtesy Comité International pour le Développenent et l'Etude de la de la Construction Tubulaire and Constructional Steel Research and Development Organization.

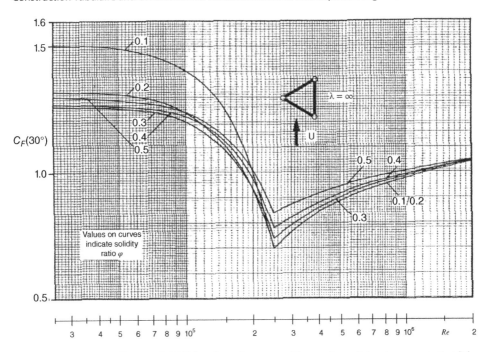

Figure 24.19 Drag coefficients $C_F(30°)$ for triangular tower with members of circular cross section [8]. Source: Courtesy Comité International pour le Développenent et l'Etude de la de la Construction Tubulaire and Constructional Steel Research and Development Organization.

C_F (30°) as functions of Reynolds number *Re* for towers with aspect ratio $\lambda = \infty$, based on measurements reported in [9–11].

References

1 Flachsbart, O. (1934). Modellversuche über die Belastung von Gitterfachweken durch Windkräfte: 1. Teil: Einzelne ebene Gitterträger. *Der Stahlbau* 7: 65–69.

2 Flachsbart, O. (1934). Modellversuche über die Belastung von Gitterfachweken durch Windkräfte: 1. Teil: Einzelne ebene Gitterträger. *Der Stahlbau* 7: 73–79.

3 Georgiou, P. N., Vickery, B. J., and Church, R., "Wind loading on open framed structures," presented at the *Third Canadian Workshop on Wind Engineering*, Vancouver, Canada, 1981.

4 Flint, A. R. and Smith, B. W., "The development of the British draft code of practice for the loading of lattice towers" in *Proceedings of the Fifth International Conference, Wind Engineering, Fort Collins, Colorado, July 1979.* vol. 2, J. E. Cermak, ed., NY: Pergamon, 1980, pp. 1293–1304.

5 Georgiou, P. N. and Vickery, B. J., "Wind loads on building frames," in *Proceedings of the Fifth International Conference, Wind Engineering, Fort Collins, Colorado, July 1979.* vol. 1, J. E. Cermak, ed., NY: Pergamon, 1980, pp. 421–433.

6 Whitbread, R. E., "The influence of shielding on the wind forces experienced by arrays of lattice frames," in *Proceedings of the Fifth International Conference, Wind Engineering, Fort Collins, Colorado, July 1979.* vol. 1, J. E. Cermak, ed., NY: Pergamon, 1980, pp. 405–420.

7 Walker, H.B. (ed.) (1975). *Wind Forces on Unclad Tubular Structures.* Croydon, UK: Constructional Steel Research and Development Organization.

8 ASCE (2011). *Wind Loads for Petrochemical and Other Industrial Facilities.* Task Committee on Wind-Induced Forces, Petrochemical Committee of the Energy Division, American Society of Civil Engineers: Reston, VA.

9 Schulz, G., "The Drag of Lattice Structures Constructed from Cylindrical Members (Tubes) and its Calculation (in German)," CIDECT Report No. 69/21, Düsseldorf, Germany, 1969.

10 Schulz, G., "International Comparison of Standards on the Wind Loading of Structures," CIDECT Report No. 69/29, Düsseldorf, Germany, 1969.

11 Gould, R. W. and Raymer, W. G., "Measurements over a Wide Range of Reynolds Numbers of the Wind Forces on Models and Lattice Frameworks," Sc. Rep. No. 5-72, National Physical Laboratory, Teddington, U.K., 1972.

12 Flachsbart, O. (1935). Modellversuche über die Belastung von Gitterfachwerken durch Windkräfte: 2. Teil: Räumliche Gitterfachwerke. *Der Stahlbau* 8: 73–79.

13 Eiffel, G. (1911). *La Résistance de l'Air et l'Aviation.* Paris, France: H. Dunod & E. Pinat.

14 Katzmayr, D. and Seitz, H. (1934). Winddruck auf Fachwerktürme von quadratischem Querschnitt. *Der Bauingenieur* 21/22: 218–251.

25

Offshore Structures

Wind loads affect offshore structures during construction, towing, and in service. They are a significant design factor, especially in the case of large compliant platforms such as guyed towers and tension leg platforms.

Wind also affects the flight of helicopters near offshore platform landing decks [1–3], as potentially dangerous conditions may be created by flow separation at the edges of the platform. Let the horizontal distance between the upstream edge of the platform and of the helideck be denoted by d, and the depth of the upstream surface producing the separated flow be denoted by t. On the basis of wind tunnel tests it has been suggested that the elevation h of the helideck with respect to the platform edge should vary from at least $h \approx 0.2\, t$ if $d \approx 0$ to at least $h \approx 0.5\, t$ if $d \approx t$ [2].

This chapter contains information on wind loads on offshore structures of various types (Section 25.1) and on the treatment of dynamic effects on compliant structures (Section 25.2).

25.1 Wind Loading on Offshore Structures

Methods for calculating wind loads on offshore platforms are recommended in [4–8]. However, laboratory and full-scale measurements indicate that these methods may, in some instances, have serious limitations, particularly insofar as they do not account for the presence of lift forces, and account insufficiently or not at all for shielding and mutual interference effects. For example, according to wind tunnel test results obtained for a jack-up (self-elevating) platform [9], the methods of [4] and [5] overestimate wind loads on jack-up units by at least 35%. Estimates based on full-scale data for an anchored semisubmersible platform [10] suggest that the method of [5] overpredicts wind loads by as much as 100%.

This section briefly reviews a number of wind tunnel tests conducted for semisubmersible units and for a large guyed tower platform. Wind tunnel test information on jack-up units, jacket structures in the towing mode, and on two types of concrete platform is available, for example, in [9] and [11–14].

Wind Effects on Structures: Modern Structural Design for Wind, Fourth Edition. Emil Simiu and DongHun Yeo.
© 2019 John Wiley & Sons Ltd. Published 2019 by John Wiley & Sons Ltd.

25.1.1   Wind Loads on Semisubmersible Units

A schematic view of a semisubmersible unit used for tests reported in [15] is shown in Figure 25.1.[1]

The side force and the heeling moment coefficients are defined as

$$CY = \frac{Y}{(1/2)\rho U^2 (50)A_s} \tag{25.1}$$

$$CK = \frac{K}{(1/2)\rho U^2 (50)A_s H_s} \tag{25.2}$$

where Y is the side force, K is the heeling moment, ρ is the air density, $U(50)$ is the mean wind speed at 50 m above sea level, A_s is the projected side area, and H_s is the elevation of the center of gravity of A_s. The coefficients CY and CK are obtained separately for the overwater and the underwater part of the unit. The overwater coefficients reflect the action of wind and should be obtained in flow simulating the atmospheric boundary layer. The overwater coefficients reflect account for hydrodynamic effects and should therefore be measured in uniform flow.

Figures 25.2 and 25.3 show values of CY and CK measured in [15] for the case of an upright draft[2] $T_{M0^\circ} \approx 10.85$ m (corresponding, for the unit being modeled, to a displacement[3] of 17,730 tons). As noted in [15], the purpose of the tests for the underwater part is to determine the elevation of the center of reaction (i.e., the point of application of the resultant of the underwater forces) or the free-floating unit.

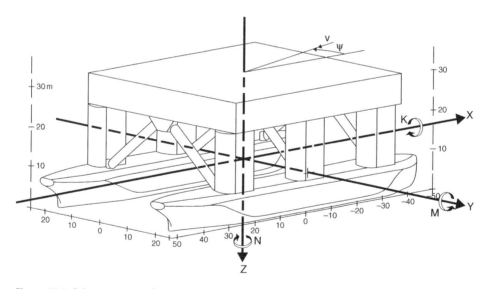

Figure 25.1 Schematic view of a semisubmersible unit model [15].

---

1 Figures 25.1–25.6 are excerpted from Bjeregaard, E. and Velschou, S. (May 1978). Wind Overturning Effect on a Semisubmersible. Paper OTC 3063, *Proceedings, Offshore Technology Conference*, Houston, TX. Copyright 1978 Offshore Technology Conference.
2 The upright draft T_{M0° is the depth of immersion of the unit in even heel condition (i.e., for an angle of heel $\phi = 0^\circ$).
3 The displacement is the volume of water displaced by the immersed part of the unit.

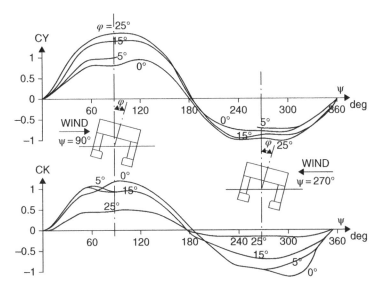

Figure 25.2 Values *CY* and *CK* as functions of wind direction Ψ at different angles of heel φ for the overwater part [15].

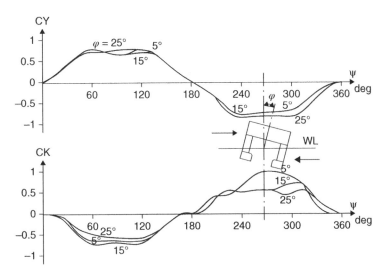

Figure 25.3 Values *CY* and *CK* as functions of wind direction Ψ at different angles of heel φ for the underwater part [15].

Figure 25.4 shows estimated values of the heeling forces induced by 100-knot beam winds (winds blowing along the *x*-axis) for various values of the upright draft $T_{M0°}$ and of the angle of heel φ. The elevations of the center of action of the overwater (wind) force and of the center of reaction on the underwater part are shown in Figure 25.5. It is seen that as the angle of heel increases the elevation of the center of action of the wind force decreases. This decrease is due to lift forces arising at nonzero angles of heel φ.

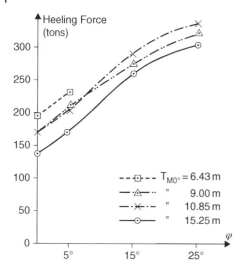

Figure 25.4 Wind heeling forces corresponding to 100-knot beam winds [15].

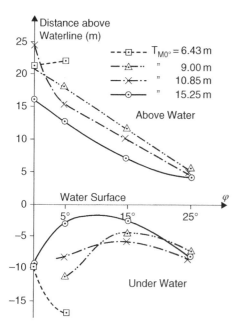

Figure 25.5 Elevation of center of action of wind forces and corresponding center of reaction on the underwater part [15].

The healing lever is defined as the ratio of the overturning moment to the displacement of the vessel. Values of the heeling lever for 100-knot beam winds, obtained from wind tests of [15] on the one hand and by using the American Bureau of Shipping method [4] on the other, are shown in Figure 25.6. (The displacements listed in [15] for the 6.43, 9.00, and 15.25 m drafts are 12,740, 16,963, and 19,495 tons, respectively.) It is seen that for large angles of heel the differences between the two sets of values are significant. This is largely due to the failure of [4] to account for the effects of lift. It is noted in [16] that the largest overturning moments are commonly induced by quartering winds.

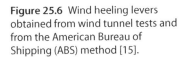

Figure 25.6 Wind heeling levers obtained from wind tunnel tests and from the American Bureau of Shipping (ABS) method [15].

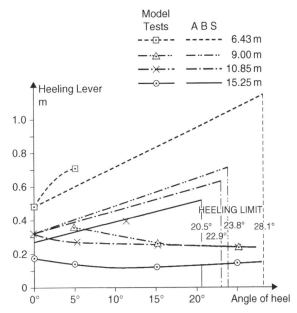

In the tests of [15] and [16] the water surface was modeled by the rigid horizontal surface of the wind tunnel flow. Following the method described in [17], texts reported in [18] were also conducted by placing the model in a tank filled with viscoelastic material up to the level of the wind tunnel flow. This facilitates the testing of models of partially submerged units. Reference [18] also includes results of tests conducted in the presence of rigid obstructions aimed at representing water waves. The results revealed that water waves could increase the overturning moments substantially. This suggests the need for improving the simulation of the sea surface in laboratory tests.

The aerodynamic tests of the Ocean Ranger semisubmersible[4] is reported in [19]. The problem of combining hydrodynamic and wind loads was addressed by conducting 1 : 100 scale aerodynamic model tests in turbulent flow over a floor with rigid waves, and using lightweight lines to apply the measured mean and fluctuating wind forces and moments to a 1 : 40 hydrodynamic model tested in conditions simulating those experienced during the storm. Additional wind tunnel tests of semisubmersible units are reported in [20–22].

25.1.2 Wind Loads on a Guyed Tower Platform

Reference [23] presents results of wind tunnel measurements on a 1 : 120 scale model of the overwater part of a structure similar to Exxon's Lena guyed tower platform. A schematic of the platform, installed in over 300 m of water in the Gulf of Mexico, is shown in Figure 25.7; see also Figure 25.8 and the expression for wind speeds averaged

---

4 The Ocean Ranger had capsized on February 15, 1982 in Hibernia Field, 315 km southeast of St. John's Newfoundland, in a storm with 17–20 m wave heights and 120–130 km h^{-1}. wind speeds. It was the world's largest submersible offshore drilling platform, 46 m high from keel to operations deck and with 120 m long pontoons. All of the 84 crew members were lost in the accident.

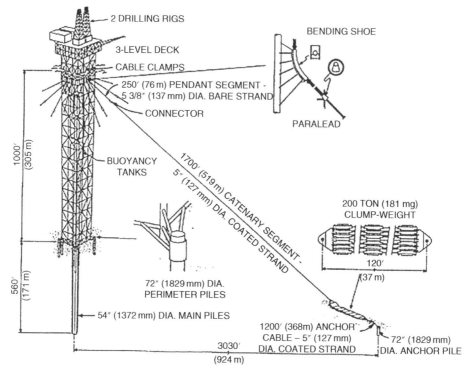

Figure 25.7 Schematic view of the Lena guyed tower platform [24]. Source: With permission from ASCE.

over at least one minute recommended by the U.S. Geological Survey [7] for use within the Gulf of Mexico:

$$U(z) = U(10)\left(\frac{z - z_d}{10 - z_d}\right)^{0.1128} \tag{25.3}$$

where z is the elevation above still water in meters, and $z_d = 2.2$ m. The air/water boundary was modeled by the rigid horizontal surface of the wind tunnel floor. Force and moment coefficients were defined by relations of the type

$$C_F = \frac{F}{(1/2)\rho U^2(16)A_R} \tag{25.4}$$

$$C_M = \frac{M}{(1/2)\rho U^2(16)A_R L_R} \tag{25.5}$$

where F and M are the mean force and moment of interest, ρ is the air density, $U(16)$ is the mean wind speed at 16 ft above the surface, and the reference area A_R and length L_R were chosen as 1 ft^2 and 1 ft, respectively. The force and moments obtained in [23] are represented in Figure 25.9, which also shows the notations for the respective aerodynamic coefficients. The moments characterized by the coefficients *CMD* and *CMT* were taken with respect to a distance of 6.2 in (62 ft full scale) below the still water level. The measured values of the aerodynamic coefficients are represented in Figure 25.10 for several platform configurations. The configuration for the base case was the same

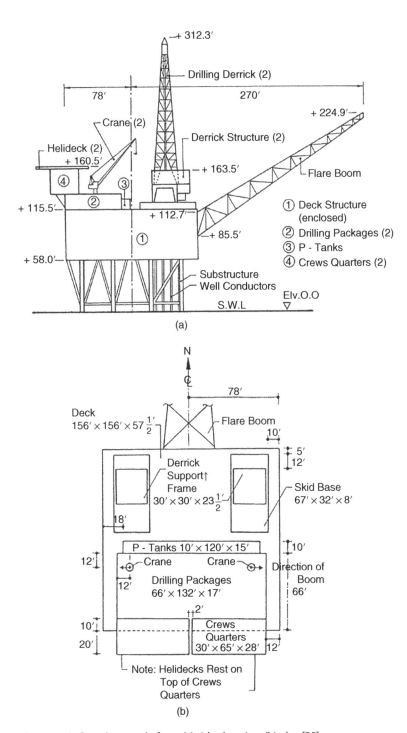

Figure 25.8 Guyed tower platform: (a) side elevation; (b) plan [25].

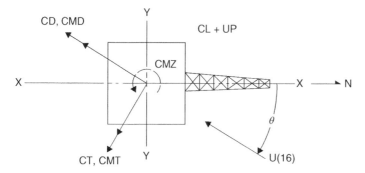

Figure 25.9 Notations. Source: From [23]. Copyright 1982 Offshore Technology Conference.

as in Figure 25.8, except that the deck structure was not enclosed. Additional results in [23] show that the effect of enclosing the deck is negligible, as is the effect of the well conductors. Removing the flares boom results in torsional moment reductions, but has negligible effects otherwise. It is shown in [23] that drag forces and drag moments based on wind tunnel measurements are smaller by about 30 and 20%, respectively, than calculated values based on [7]. To check the extent to which the results depend upon the laboratory facility being used, the same structure was subsequently tested independently in a different wind tunnel [25]. In most cases of significance from a design viewpoint the results obtained in [25] were larger than those of [23] by amounts that did not exceed 20–30%.

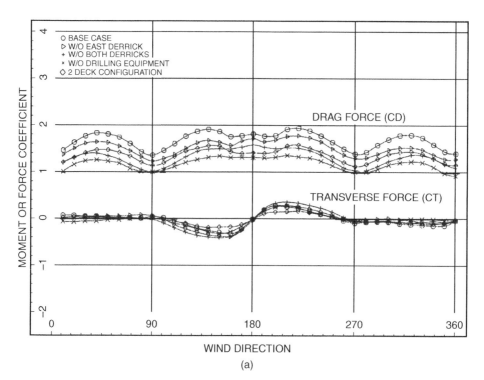

Figure 25.10 Wind tunnel test results. Source: From [23]. Copyright 1982 Offshore Technology Conference.

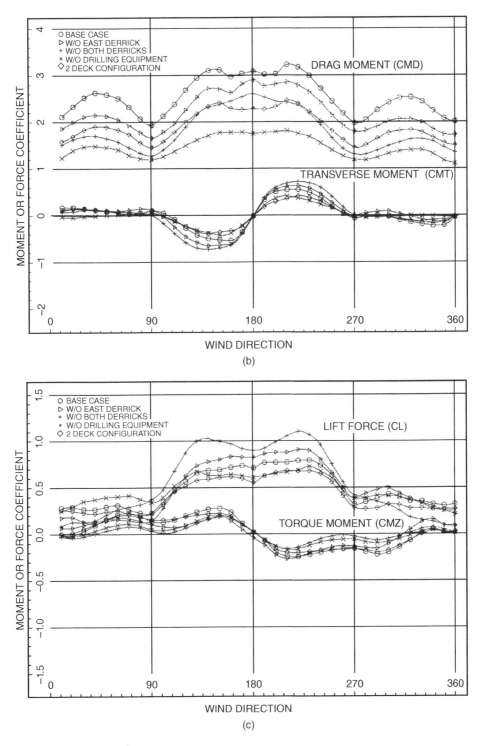

Figure 25.10 (*Continued*)

25.2   Dynamic Wind Effects on Compliant Offshore Structures

Compliant offshore platforms are designed to experience significant motions under load. An advantage of compliance is that the forces of inertia due to the motion of the platform tend to counteract the external loads. For large offshore platforms installed in deep water an additional advantage is that the natural frequencies of the platform motions in the surge, sway, and yaw[5] degrees of freedom are very low (e.g., from 1/30 to 1/150 Hz, depending upon type of platform and water depth). Wave motions have narrow spectra centered about relatively high frequencies (e.g., from 1/15 Hz for extreme events to about 1 Hz for service conditions). Thus, aside from possible second-order effects, compliant platforms do not exhibit dynamic amplifications of wave-induced response.

Unlike wave motions, wind speed fluctuations in the atmospheric boundary layer are characterized by broadband spectra. For this reason, it has been surmised that wind-induced dynamic amplification effects on compliant structures are significant [23, 26]. A more guarded assessment of the effects of wind gustiness was presented in [27] as a part of an evaluation of the response to environmental loads of the North Sea Hutton tension leg platform (Figure 25.11, see also [28]). According to [27]: "Wind gusts are typically broad-banded and may contain energy which could excite surge motions at the natural period. These would be controlled by surge damping. Theoretical and experimental research is required to clarify the importance of this matter."

Investigations into the behavior of tension leg platforms under wind loads reported in [29] and [30] were based on the assumption that the response to wind is described by a system with proportional damping, with damping ratio in the order of 5%. However, it was shown in [31] that for structures comparable to the Hutton platform the effective hydrodynamic damping is considerably stronger, and that the wind-induced dynamic amplification for low-frequency motions is for this reason negligible. Section 25.2.1 describes the approach used in [31] to estimate the response of a tension leg platform to wind in the presence of current and waves, and a simple method for estimating the order of magnitude of the damping inherent in the hydrodynamic loads.

25.2.1   Turbulent Wind Effects on Tension Leg Platform Surge

Under the assumption that the external loads are parallel to one of the sides of the platform shown in Figure 25.11, the equation of surge motion can be written as

$$M\ddot{x} = F_x(t) \tag{25.6}$$

where

$$F_x(t) \approx F_u(t) + F_h(t) + R(t) \tag{25.7}$$

and $F_u(t)$, $F_h(t)$, and $R(t)$ denote the wind force, the hydrodynamic force, and the restoring force, respectively. Not included in Eq. (25.7) is the damping force due to internal

---

5 Displacements in the longitudinal, transverse, and vertical direction are called *surge*, *sway*, and *heave*, respectively. Rotations in a transverse, longitudinal, and horizontal plane are called *roll*, *pitch*, and *yaw*, respectively.

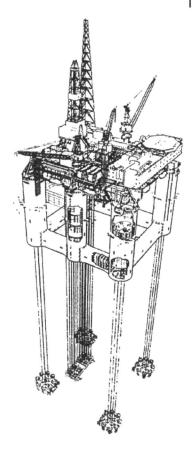

Figure 25.11 Schematic view of the Hutton tension leg platform. Source: From [28]. Copyright 1982 Offshore Technology Conference.

friction within the structure, which is negligible compared to the damping forces due to hydrodynamic effects.

Wind Loads. To estimate the wind-induced drag force it is assumed that the elemental drag force per unit of area projected on a plane P normal to the mean wind speed is

$$p(y, z, t) = \frac{1}{2}\rho_a C_p(y, z)[u(y, z, t) - \dot{x}(t)]^2 \tag{25.8}$$

where ρ_a is the air density, $C_p(y, z)$ is the pressure coefficient at elevation z and horizontal coordinate y in the plane P, t is the time, x is the surge displacement, the dot denotes differentiation with respect to time, and $u(z, y, t)$ is the wind speed upwind of the structure in the direction of the mean wind. The speed $u(z, y, t)$ can be expressed as a sum of the mean speed $U(z)$ and the fluctuating speed $u'(y, z, t)$:

$$u(z, y, t) = U(z) + u'(z, y, t) \tag{25.9}$$

The total wind-induced drag force is

$$F_u(t) = \int_{A_a} p(y, z, t) dy dz \tag{25.10}$$

where A_a is the projection of the above-water part of the platform on a plane normal to the mean wind speed.

The mean wind speeds and the turbulence spectrum and co-spectrum can be modeled as in Chapter 2. Neglecting second-order terms, it follows from Eqs. (25.8)–(25.10) that the mean wind-induced drag is

$$\overline{F_u(t)} = \frac{1}{2}\rho_a C_a A_a U^2(z_a) \tag{25.11}$$

where the overall aerodynamic drag coefficient is

$$C_a = \frac{1}{A_a U^2(z_a)} \int_{A_a} C_p(y,z) U^2(z) dy dz \tag{25.12}$$

and z_a is the elevation of the aerodynamic center of the above-water part of the platform. The fluctuating part of the wind-induced drag is

$$F'_{u,r}(t) = \rho_a \int_{A_a} C_p(y,z) U(z) u'(z,t) dy dz \tag{25.13}$$

where the subscript r refers to the fact that the platform is at rest. As shown in [31], the spectral density of the fluctuating part of the wind-induced drag is

$$S_{F_{u,r}}(n) = [\rho_a C_a A_a U(z_a)]^2 S_{u,eq}(n) \tag{25.14}$$

where, for typical drilling and production platform geometries, the equivalent wind fluctuation spectrum can be defined as

$$S_{u,eq}(n) \approx S_u(z_a, n) J(n) \tag{25.15}$$

$J(n)$ is a reduction factor that accounts for the imperfect spatial coherence if the fluctuations u' with the expression

$$J(n) \approx -\frac{2}{E}\left\{-\exp(-E) + \left(1 - \frac{1}{E}\right)[\exp(-E) - 1]\right\} \tag{25.16}$$

$$E = C_y b \frac{n}{U(z_a)} \tag{25.17}$$

where b is the width of main deck and C_y is the horizontal exponential decay coefficient in Eq. (2.94).

Hydrodynamic Loads. The total hydrodynamic load F_h can be written

$$F_h = F_v + F_e - A\ddot{x} - B\dot{x} \tag{25.18}$$

where F_v is the total hydrodynamic viscous force, F_e is the total wave-induced exciting force, A is the surge-added mass, and B is the surge wave-radiation damping coefficient. It was assumed for convenience in [31] that the wave motion is monochromatic, hence the absence of second-order drift forces in Eq. (25.18). It was also assumed that $B = 0$, since the radiation damping at low frequencies is negligible [32, 33].

The total wave-induced exciting force and the surge-added mass can be estimated numerically on the basis of potential theory. Alternatively, they may be assumed to be given by the inertia component of the Morison equation

$$A \approx \rho_w \sum_i \sum_j \forall_{ij}(C_{mij} - 1) \tag{25.19}$$

$$F_e \approx \rho_w \sum_i \sum_j \forall_{ij} C_{mij} \left\{ \frac{\partial v_{ij}}{\partial t} + [\bar{v}_i + v_{ij} - \dot{x}] \frac{\partial v_{ij}}{\partial X} \right\} \tag{25.20}$$

[34, p. 31], where ρ_w is the water density, $\forall_{ij}$ is the elemental volume of the submerged structure, C_{mij} is the surge inertia coefficient corresponding to $\forall_{ij}$, X is the horizontal distance from some arbitrary origin to the center of $\forall_{ij}$ along the direction parallel to surge motion, $\bar{v}_i$ and v_{ij} are the current velocity and horizontal particle velocity due to wave motion, respectively, at the center of $\forall_{ij}$. Equations (25.19) and (25.20) may be employed if for the component being considered the ratio of diameter to wave length, $D/L \leq 0.2$ [34, p. 283]. Since for $T_w \approx 15\,\text{s}$, $L = gT_w^2/2\pi$, it follows that, for members of typical tension leg platform structures for which $D < 20\,\text{m}$ or so, the use of Eqs. (25.19) and (25.20) is acceptable if three-dimensional flow effects are not taken into account. The wave motion can be described by deep water linear theory, so

$$v_{ij} = \frac{\pi H}{T_w} e^{-k_w z_i} \cos\left(k_w X_j - \frac{2\pi t}{T_w} \right) \tag{25.21}$$

where H is the wave height and k_w is the wave number given by

$$k_w = \frac{1}{g} \left(\frac{2\pi}{T_w} \right)^2 \tag{25.22}$$

[34, p. 157]. The total hydrodynamic viscous load may be described by the viscous component of Morison's equation

$$F_v = 0.5\rho_w \sum_i \sum_j C_{dij} A_{pij} \, | \bar{v}_i + v_{ij} - \dot{x} | \, [\bar{v}_i + v_{ij} - \dot{x}] \tag{25.23}$$

where A_{pij} is the area of elemental volume $\forall_{ij}$ projected on a plane normal to the direction of the current, and C_{dij} is the drag coefficient corresponding to A_{pij}.

If the relative motion of the body with respect to the fluid is harmonic, the drag and inertia coefficients in the Morison equation can be determined on the basis of experimental results as functions of local oscillatory Reynolds number, $Re = 2\pi\,D^2/(\nu\,T_f)$, Keulegan–Carpenter number, $K = VT_f/D$, and relative body roughness, where D is the diameter of the body, ν is the kinematic viscosity, and V and T_j are the amplitude and period of the relative fluid-body velocity. However, actual relative fluid-body motions are not harmonic. This introduces uncertainties in the determination of the drag and inertia coefficients even if experimental information for harmonic relative motions were available in terms of Re and K. Unfortunately, such information is not available for the small numbers K (in the order of 2) and the large Reynolds numbers (in the order of 10^6) of interest in tension leg platform design. For this reason calculations should be carried out for various sets of values C_d, C_m, and investigations should be conducted into the sensitivity of the results to changes in these values.

Restoring Force. The surge-restoring force in a tension leg platform is supplied by the horizontal projection of the total force in the tethers. Most of this force is the result of pretensioning, which is achieved by ballasting the floating platform, tying it by means of the tethers to the foundations at the sea floor, then deballasting it. The tension forces

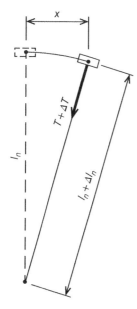

Figure 25.12 Notations.

in the tethers should exceed the compression forces induced by pitching and rolling moments due to extreme loads.

Under the assumption that the tethers are straight at all times, the restoring force can be written as

$$R(t) \simeq -(T + \Delta T)\frac{x}{l_n + \Delta l_n} \tag{25.24}$$

where T is the initial pretensioning force, ΔT is the incremental tension due to surge motion, l_n is the nominal length of the tethers at $x = 0$, Δl_n is the incremental length, and

$$\frac{T + \Delta T}{l_n + \Delta l_n} \simeq \frac{T}{l_n} + C_{NL}[1 - \sqrt{1 - (x/l_n)^2}] \tag{25.25}$$

where C_{NL} is the downdraw coefficient, equal to the weight of water displaced as the draft is increased by a unit length [32] (Figure 25.12).

In reality, hydrodynamic and inertia forces cause the tethers to deform transversely. The angle between the horizontal and the tangent to the tether axis at the platform heel can therefore differ significantly for the values corresponding to the case of a straight tether. Nevertheless, owing to the relatively small role of the restoring force in the dynamics of typical tension leg platforms, the effect of such differences on the motion of the platforms appears to be negligible for practical purposes [35–37].

Surge Response. The surge response is obtained by solving Eq. (25.6). This equation is nonlinear, the strongest contribution to the nonlinearity being due to the hydrodynamic viscous load F_v. Its solution is sought in the time domain.

The nominal natural period in surge is

$$T_n = 2\pi \left(\frac{M_{eff}}{k}\right)^{1/2} \tag{25.26}$$

where M_{eff} is the coefficient of the term in $\ddot{x}$ and k is the coefficient of the term in x in Eq. (25.6). From Eqs. (25.6), (25.18), and (25.24) it follows that

$$T_n = 2\pi \left[\frac{(M + A)l_n}{T}\right]^{1/2} \tag{25.27}$$

A calculated time history of the surge response is represented in Figure 25.13 as a function of time for a platform with the geometrical configuration of Figure 25.14, under the following assumptions: platform mass $M = 34.3 \times 10^6$ kg; total initial tension in legs $T = 1.56 \times 10^5$ kN (it follows from these assumptions and Eqs. (25.19) and (25.27) that for the platform of Figure 25.14 the nominal natural frequency is $T_n = 100$ s); Morison equation coefficients $C_{mij} = 1.8$, $C_{dij} = 0.6$; wave height and period $H = 25$ m and $T_w = 15$ s, respectively; current speed varying from 1.4 m s^{-1} at the mean water level to 0.15 m s^{-1} at 550 m depth; aerodynamic parameter $C_a A_a = 4320$ m^2; elevation of aerodynamic center $z_a = 50$ m; atmospheric boundary layer parameters $\kappa = 0.002$, $\beta = 6.0$, $L_u^x = 180$ m, $C_y = 16$ (see Chapter 2); and mean wind speed $U(z_a) = 45$ m s^{-1}. It is shown

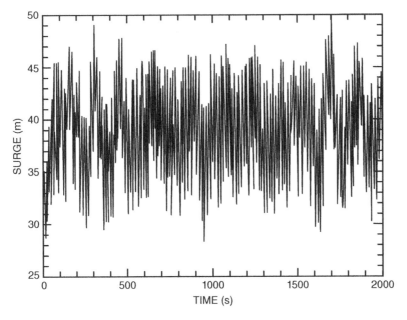

Figure 25.13 Calculated time history of a surge response [31].

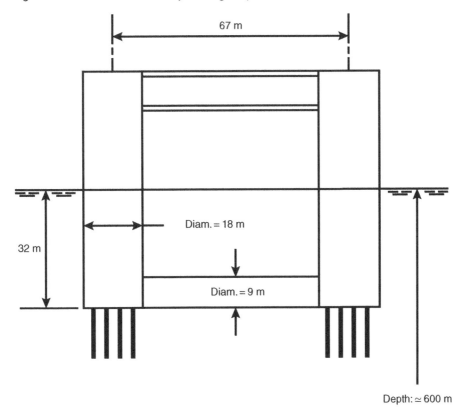

Figure 25.14 Geometry of a platform [31].

in [31] that the contributions of the mean wind speed and the wind speed fluctuations about the mean are about 40 and 12%, respectively. It can be verified that this conclusion is equivalent to stating that wind-induced resonant amplification effects are negligible in the cases investigated in [31].

Sensitivity studies showed that the results were affected insignificantly by uncertainties with respect to the actual values of the atmospheric boundary-layer parameters. It is shown in [31] that the damping ratio in a linear system equivalent to the nonlinear system studies in this section is in the order of $\zeta \approx 0.5$ and $\zeta \approx 0.2$ for the platforms with $l_n \approx 600$ m and $l_n \approx 150$ m, respectively. The coefficients $C_{dij} = 0.6$ and $C_{mij} = 1.8$ on which these results were based may not be realistic for members with large diameters, such as those depicted in Figure 25.14. The use of alternative values for those coefficients showed that the damping ratios were in all cases sufficiently large to prevent the occurrence of significant wind-induced dynamic amplification effects. However, for some values of C_{dij}, calculations in which the assumed currents would be lower than those of [31] could result in reduced nominal damping rations for certain wind climatological conditions. Because wind-wave tests violate Reynolds number and Keulegan–Carpenter number similarity, they cannot provide a reliable indication of the equivalent damping ratio for the prototype. This is a continuing cause of uncertainty in the assessment of dynamic effects induced by wind acting alone or, in the case of a nonlinear analysis, in conjunction with wave-induced slow drift.

References

1 Davies, M. E., Cole, L. R., and O'Neill, P. G. G., *The Nature of Air Flows Over Off-shore Platforms*, NMI R14 (OT-R-7726), National Maritime Institute, Feltham, UK, June 1977.

2 Davies, M. E., *Wind Tunnel Modeling of the Local Atmospheric Environment of Offshore Platforms*, NMI R58 (OT-R-7935), National Maritime Institute, Feltham, UK, May 1979.

3 Littleburg, K. H., "Wind tunnel testing techniques for offshore gas/oil production platforms," Paper OTC 4125, *Proceedings, Offshore Technology Conference*, Houston, TX, 1981

4 American Bureau of Shipping (1980). *Rules for Building and Classing Mobile Offshore Drilling Units*. New York: American Bureau of Shipping.

5 Det Norske Veritas (1981). *Rules for the Construction and Classification of Mobile Offshore Units*. Oslo: Det Norske Veritas.

6 Det Norske Veritas. *Rules for the Design, Construction, and Inspection of Offshore Structures*, Appendix B, Loads, Det Norske, Veritas, Oslo, 1977 (Reprint with corrections, 1979).

7 USGS. *Requirements for Verifying the Structural Integrity of OCS Platforms*, Appendices, United States Geological Survey, OCS Platform Verification Division, Reston, VA, 1979.

8 API. *API Recommended Practice for Planning, Designing, and Constructing Fixed Offshore Platforms*, API RP 2A, American Petroleum Institute, Washington, DC, 1981.

9 Norton, D. J. and Wolff, C. V., "Mobile offshore platform wind loads," Paper OTC 4126, *Proceedings, Offshore Technology Conference*, Houston, TX, 1981.

10 Boonstra, H. (1980). Analysis of full-scale wind forces on a semisubmersible platform using operators' data. *Journal of Petroleum Technology* 32: 771–776.

11 Ponsford, P. J., *Measurements of the Wind Forces and Measurements of an Oil Production Jacket Structure in Tow-Out Mode*, NMI R30 (OT-R-7801), National Maritime Institute, Feltham, UK, January 1978.

12 Cowdrey, C. F., *Time-Averaged Aerodynamic Forces and Moments on a Model of a Three-Legged Concrete Production Platform*, NMI R36 (OT-R-7808), National Maritime Institute, Feltham, UK, June 1982

13 Miller, B. L. and Davies, M. E., *Wind Loading on Offshore Structures: A Summary of Wind Tunnel Studies*, NMI R36 (OT-R-7808), National Maritime Institute, Feltham, UK, September 1982.

14 Davenport, A. G. and Hambly, E. C., "Turbulent wind loading and dynamic response of of jackup platform, OTC Paper 4824, *Proceedings, Offshore Technology Conference*, Houston, TX, May 1984.

15 Bjerregaard, E. and Velschou, S., "Wind overturning effects on a semisubmersible" OTC Paper 3063, *Proceedings, Offshore Technology Conference*, Houston, TX, May 1978.

16 Bjerregaard, E. and Sorensen, E., "Wind overturning effects obtained from wind tunnel tests with various submersible models," OTC Paper 4124, *Proceedings, Offshore Technology Conference*, Houston, TX, May 1981.

17 Ribbe, J. H. and Brusse, J. C., "Simulation of the air/water interface for wind tunnel testing of floating structures," *Proceedings: Fourth U.S. National Conference, Wind Engineering Research*, B. J. Hartz, Ed. Department of Civil Engineering, University of Washington, Seattle, July 1981

18 Macha, J. M. and Reid, D. F., "Semisubmersible wind loads and wind effects,: Paper no. 3, *Annual Meeting, New York, November 1984*. New York: The Society of Naval Architects and Marine Engineers, 1984.

19 Wardlaw, R. L., Laurich, P. H., and Mogridge, G. R., "Modeling of dynamic loads in wave basin tests of the semisubmersible drilling platform ocean ranger," *Proceedings, International Conference on Flow-Induced Vibrations, Bowness-on-Windermere*, UK, May 12–14, 1987.

20 Cowdrey, C. F. and Gould, R. F., *Time-Averaged Aerodynamic Forces and Moments on a National Model of a Submersible Offshore Rig*. NMI R25 (OT-R-7748), National Maritime Institute, Feltham, UK, September 1982.

21 Ponsword, P. J., *Wind Tunnel Measurements of Aerodynamic Forces and Moments on a Model of a Semisubmersible Offshore Rig*, NMI R34 (OT-R-7807), National Maritime Institute, Feltham, UK, June 1982.

22 Troesch, A.W., Van Gunst, R.W., and Lee, S. (1983). Wind loads on a 1:115 model of a semisubmersible. *Marine Technology* 20: 283–289.

23 Pike, P. J. and Vickery, B. J., "A wind tunnel investigation of loads and pressure on a typical guyed tower offshore platform," OTC Paper 4288, *Proceedings, Offshore Technology Conference*, Houston, TX, May 1982.

24 Glasscock, M.S. and Finn, L.D. (1984). Design of a guyed tower for 1000 ft of water in the Gulf of Mexico. *Journal of Structural Engineering* 110: 1083–1098.

25 Morreale, T. A., Gergely, P., and Grigoriu, M., *Wind Tunnel Study of Wind Loading on a Compliant Offshore Platform*, NBS-GCR-84-465, National Bureau of Standards, Washington, DC, December 1983.

26 Smith, J. R. and Taylor, R. S., "The development of articulated buoyant column systems as an aid to economic offshore production," *Proceedings, European Offshore Petroleum Conference Exhibition*, London, October 1980, pp. 545–557.

27 Mercier, J. A., Leverette, S. J., and Bliault, A. L., "Evaluation of Hutton TLP response to environmental loads, OTC Paper 4429, *Proceedings, Offshore Technology Conference*, Houston, TX, May 1982.

28 Ellis, N., Tetlow, J. H., Anderson, F., and Woodhead, A. L., "Hutton TLP vessel – Structural configuration and design features," OTC Paper 4427, *Proceedings, Offshore Technology Conference*, Houston, TX, May 1982.

29 Kareem, A. and Dalton, C., "Dynamic effects of wind on tension leg platforms," OTC Paper 4229, *Proceedings, Offshore Technology Conference*, Houston, TX, May 1982.

30 Vickery, B. J., "Wind loads on compliant offshore structures," *Proceedings, Ocean Structural Dynamics Symposium*, Department of Civil Engineering, Corvallis, OR, September 1982, pp. 632–648.

31 Simiu, E. and Leigh, S.D. (1984). Turbulent wind and tension leg platform surge. *Journal of Structural Engineering* 110: 785–802. https://www.nist.gov/wind.

32 Salvesen, N., von Kerczek, C. H., Vue, D. K. et al. "Computations of nonlinear surge motions of tension leg platforms," OTC Paper 4394, *Proceedings, Offshore Technology Conference*, Houston, TX, May 1982.

33 Pinkster, J. A. and Van Oortmerssen, G., "Computation of first- and second-order forces on oscillating bodies in regular waves," *Proceedings, Second International Conference on Ship Hydrodynamics*, University of California, Berkeley, 1977.

34 Sarpkaya, T. and Isaacson, M. (1981). *Mechanics of Wave Forces on Offshore Structures*. New York: Van Nostrand Reinhold.

35 Jefferys, E.R. and Patel, M.H. (1982). On the dynamics of taut mooring systems. *Engineering Structures* 4: 37–43.

36 Simiu, E., Carasso, A., and Smith, C.E. (1984). Tether deformation and tension leg platform surge. *Journal of Structural Engineering* 110: 1419–1422.

37 Simiu, E. and Carasso, A., "Interdependence between dynamic surge motions of platform and tethers for a deep water TLP," *Proceedings, Fourth International Conference on Behavior of Offshore Structures (BOSS)*, 1–5 July 1985, Delft, The Netherlands, pp. 557–562.

26

Tensile Membrane Structures

Tensile membrane structures owe their capacity to resist loads to tension stresses in membranes supported by cables, columns, other members such as beams or arches, and/or pressurized air [1, 2].

For a number of small structures with commonly used simple shapes (cones, ridge-and-valley shapes, hyperbolic paraboloids also known as saddle shapes, cantilevered canopies), external and internal constant pressure coefficients specified for well-defined zones on the membrane surfaces are available in the literature (e.g., [3, 4]) in formats similar to those used in codes and standards for ordinary structures. Tentative aerodynamic information is also provided in [3] for the preliminary design of certain types of open stadium roofs.

For tensile membrane structures with unusual shapes and/or with long-spans (e.g., exceeding 100 m, say) it is necessary to resort to wind tunnel testing. Commonly performed on rigid models, such testing can provide time histories of pressures at large numbers of points on the structures' surfaces. The deformations induced by the time- and space-dependent aerodynamic pressures can be calculated by accounting for geometric and material non-linearities and for dynamic effects. Because these deformations are typically large and can therefore significantly affect the structure's shape, the rigid model that reproduces the original surface needs to be modified accordingly. The modified rigid model is used to measure a new set of pressure time histories. The stresses and deformations induced by those pressures can then be determined with improved accuracy [5].

Deformations measured in aeroelastic tests are reported in [6], which notes that the prototype Froude number was reproduced in the laboratory. No other information on the aeroelastic testing technique is provided in [6].

Computational Wind Engineering (CWE) simulations are increasingly being performed with a view to modeling aerodynamic or aeroelastic response [7]. Their results have been validated in some cases (see, e.g., [8]). In the absence of appropriate validation CWE results are generally not accepted for design purposes.

The form of tensile membrane structures must be consistent with specified (i) geometric boundary conditions (support geometry, and cable or fixed edges) and (ii) cable and fabric prestress. Form finding is an intricate process that requires the use of specialized software (see, e.g., www.formfinder.at). Prestressing and anticlastic shapes (shapes with double curvature, i.e., saddle forms) are designed to prevent the occurrence of membrane flutter and of compression in the membrane and cables. For structures with common shapes, classified as small (i.e., with dimensions in the order of 10 m to less

Wind Effects on Structures: Modern Structural Design for Wind, Fourth Edition. Emil Simiu and DongHun Yeo.
© 2019 John Wiley & Sons Ltd. Published 2019 by John Wiley & Sons Ltd.

than 100 m), it is suggested in [3] that the sum of the ratios of prestress in the warp and weft directions (in kN m^{-1}) to the respective radii of curvature (in m) is a useful indicator of structural behavior: if the sum exceeds 0.3 kN m^{-2} the performance was typically found to be sound, whereas if it is less than 0.2 kN m^{-2} a detailed investigation of wind effects is in order.

Based on results of a carefully designed round robin exercise, it is noted in [9] that different form-finding procedures can yield significantly different forms. It is strongly emphasized in [9] that geometric and material nonlinearities render the structural analysis far more complex than is the case for typical structures. For this reason, and in the absence of a clear and consistent basis for ensuring structural safety by accounting for the various uncertainties inherent in the analysis, it was found in [9] that estimated design stress factors varied among the round robin participants between 2.8 and 7.1.

In addition to aerodynamic information applicable to the design of small membrane structures with simple shapes, [3] provides tentative information that may be used for the preliminary design of a few types of open stadium roofs. Measurements of pressures performed on rigid models by using pneumatic averaging are described extensively in [6]. Similar, though much more complete and accurate pressure measurements, can currently be performed by using the pressure scanner technique. Such measurements, performed iteratively following the approach described in [5], can be employed in nonlinear finite-element static and dynamic analyses to obtain the requisite design information. While analyses of this type can in principle follow the database-assisted design approaches described in Part II of this book, it is shown in [10] that they present formidable difficulties that can result in incorrect response predictions. This can be the case even if the use of follower wind forces (see, e.g., [11]) (i.e., wind forces that change direction by remaining normal to the moving membrane surface) is included in the analyses; see also [12]. However, according to [13], for a low-profile cable-reinforced air-supported structure, full-scale measurements in strong winds showed that wind-tunnel pressure measurements on a rigid model, used in conjunction with a straightforward linear model of the dynamic response, provided a reasonable representation of the structure's behavior under wind loads.

References

1 ASCE, "Tensile Membrane Structures," in *ASCE Standard ASCE/SEI 55-16*, Reston, VA: American Society of Civil Engineers, 2016.

2 Beccarelli, P. (2015). *Biaxial Testing for Fabrics and Foils*. Milan: Springer.

3 Forster, B. and Mollaert, M. (2015). *European Design Guide for Tensile Surface Structures*. Brussels: Tensinet Publications.

4 CEN, "Eurocode 1: Actions on structures – Part 1–4: General actions – Wind actions," in *EN 1991-1-4*: European Committee for Standardization (CEN), 2005.

5 Hincz, K. and Gamboa-Marrufo, M. (2016). Deformed shape wind analysis of tensile membrane structures. *Journal of Structural Engineering* 142: 04015153.

6 Vickery, B.J. and Majowiecki, M. (1992). Wind induced response of a cable supported stadium roof. *Journal of Wind Engineering and Industrial Aerodynamics* 41–44: 1447–1458.

7 Heil, M., Andrew, L.H., and Jonathan, B. (2008). Solvers for large-displacement fluid-structure interaction problems: segregated versus monolithic approaches. *Computational Mechanics* 43 (1): 91–101.

8 Michalski, A., Kermel, P.D., Haug, E. et al. (2011). Validation of the computational fluid-structure interaction simulation at real-scale tests of a flexible 29 m umbrella in natural wind flow. *Journal of Wind Engineering and Industrial Aerodynamics* 99 (4): 400–413.

9 Gosling, P.D., Bridgens, B.N., Albrecht, A. et al. (2013). Analysis and design of membrane structures: results of a round robin exercise. *Engineering Structures* 48: 313–328. doi: 10.1016/j.engstruct.2012.10.008.

10 Lazzari, M., Masowiecki, M., Vitaliani, R.V., and Saetta, A.V. (2009). Nonlinear FE analysis of Montreal Olympic Stadium roof under natural loading conditions. *Engineering Structures* 31: 16–31.

11 Lazzari, M., Vitaliani, R.V., Majowiecki, M., and Saetta, A.V. (2003). Dynamic behavior of a transgrity system subjected to follower wind loading. *Computers and Structures* 81: 2199–2217.

12 Gil Pérez, M., Kang, T.H.-K., Sin, I., and Kim, S.D. (2016). Nonlinear analysis and design of membrane fabric structures: modeling procedure and case studies. *Journal of Structural Engineering*, 142: 05016001, Nov.

13 Mataki, Y., Iwasa, Y., Fukao, Y., and Okada, A. (1988). Wind-induced response of low-profile, cable reinforced air-supported structures. *Journal of Wind Engineering and Industrial Aerodynamics* 29: 253–262.

27

Tornado Wind and Atmospheric Pressure Change Effects

27.1   Introduction

Tornadoes are storms containing the most powerful of all winds. Their probabilities of occurrence at any one location are low compared to those of other extreme winds. It has therefore been generally considered that the cost of designing structures to withstand tornado effects is significantly higher than the expected loss associated with the risk of a tornado strike, risk being defined as the product of the loss by its probability of occurrence. For this reason, tornado-resistant design requirements are not included in current building codes or standards. This is changing, however, as efforts are underway to develop standard requirements for the design of such facilities as fire stations, police stations, hospitals, and power plants, whose survival of a tornado strike is considered essential from a community resilience point of view. The consequences of failure would be especially grave for nuclear power plants. In the United States, construction permits or operating licenses for nuclear power plants are issued or continued only if their design is consistent with Regulatory Guides issued by the U.S. Nuclear Regulatory Commission or is otherwise acceptable to the Regulatory staff of that agency [1, 2].

Tornado effects may be divided into three groups:

1) Wind pressures caused by the direct aerodynamic action of the air flow on the structure.
2) Atmospheric pressure change effects.
3) Impactive forces caused by tornado-borne missiles.

This chapter and Chapter 28 present design criteria and procedures developed to ensure an adequate representation of tornado effects on nuclear power plants.

Reference [1] uses a model of the tornado wind flow characterized by the following parameters: (i) maximum rotational wind speed, (ii) translational wind speed of the tornado vortex V_{tr}, (iii) radius of maximum rotational wind speed R_m, (iv) pressure drop p_a, and (v) rate of pressure drop dp_a/dt. Values of these parameters specified by the U.S. Nuclear Regulatory Commission [1] as a design basis for nuclear power plants are listed in Section 3.4.3. The use of this model for the estimation of wind pressures on structures is discussed in Section 27.2. Section 27.3 is concerned with atmospheric pressure change loading. Recent experimental work on the modeling of tornadoes and of the pressures they induce on buildings is briefly reviewed in Section 27.4. Tornado-borne missile speeds are discussed in Chapter 28, which also discusses hurricane-borne missile speeds.

Wind Effects on Structures: Modern Structural Design for Wind, Fourth Edition. Emil Simiu and DongHun Yeo.
© 2019 John Wiley & Sons Ltd. Published 2019 by John Wiley & Sons Ltd.

27.2 Wind Pressures

A procedure for calculating wind pressures, proposed in [3], assumes the following:

1) The wind velocities, and therefore the wind pressures, do not vary with height above ground.
2) The rotational velocity component (Figure 27.1) is given by the expressions:

$$V_R = \frac{r}{R_m} V_{R_m} \quad (0 \leq r \leq Rm) \tag{27.1}$$

$$V_R = \frac{R_m}{r} V_{R_m} \quad (r \geq Rm) \tag{27.2}$$

where V_{R_m} is the maximum rotational wind speed and R_m is the radius of maximum rotational wind speed.
3) The wind flow model described by Eqs. (27.1) and (27.2) moves horizontally with a translation velocity V_{tr}. The corresponding maximum wind speed is

$$V_{\max} = V_{R_m} + V_{tr} \tag{27.3}$$

The flow described by Eqs. (27.1)–(27.3) is called the combined Rankine vortex (Figure 27.1).

The wind pressure p_w, used in designing structures or parts thereof, may be written as

$$p_w = q_F C_p + q_M C_{pi} \tag{27.4}$$

where C_p is the external pressure coefficient, C_{pi} is the internal pressure coefficient, q_F is the basic external pressure, and q_M is the basic internal pressure. The quantities q_F and q_M may be calculated as follows:

$$q_F = C_s^F p_{\max} \tag{27.5}$$
$$q_M = C_s^M p_{\max} \tag{27.6}$$
$$p_{\max} = \frac{1}{2}\rho V_{\max}^2 \tag{27.7}$$

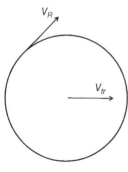

Figure 27.1 Rankine combined vortex model. Source: From [1].

where ρ is the air density and $V_{\max}$ is the maximum horizontal wind speed (Table 3.2). If $V_{\max}$ is expressed in mph and $p_{\max}$ in lb ft^{-2}, $(1/2)\rho = 0.00256$ lb ft^{-2} mph^{-2}. The quantities C_s^F and C_s^M are reduction (or size) coefficients that account for the non-uniformity in space of the tornado wind field. C_s^F can be determined from Figure 27.2 as a function of the ratio L/R_m, where L is the horizontal dimension, normal to the wind direction, of the tributary area of the structural element concerned (if the wind load is distributed among several structural elements, e.g., by a horizontal diaphragm, L is the horizontal dimension, normal to the wind direction, of the total area tributary to those elements). If the size and distribution of the openings are relatively uniform around the periphery of the structure, C_s^M is determined in the same way as C_s^F using a value of L equal to the horizontal dimension of

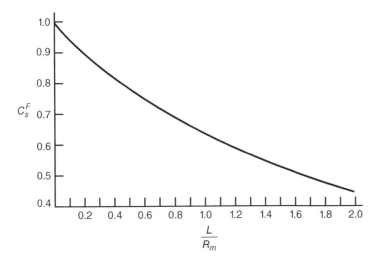

Figure 27.2 Size coefficient C_s^F [3].

the structure normal to the wind direction. If the sizes and distribution of the openings are not uniform, the following weighted averaging procedure is used:

1) Determine the quantity r_1/R_m such that

$$\frac{r_1}{R_m} = \frac{R_m}{r_1 + L} \tag{27.8}$$

2) Locate the plan of the structure drawn at appropriate scale within the non-dimensional pressure profile of Figure 27.3, with the left end of the structure at the coordinate r_1/R_m.
3) Determine C_q from Figure 27.3 for each exposed opening.
4) Determine C_s^M from Eq. (27.9).

$$C_s^M = \frac{\sum_i^N A_{0i} C_{qi}}{\sum_i^N A_{0i}} \tag{27.9}$$

where A_{0i} is the area of the opening at location i, C_{qi} is the factor C_q at location i, and N is the number of openings. The coefficient C_q in Figure 27.3 represents non-dimensionalized wind pressures and was calculated using Eqs. (27.1)–(27.3) and (27.7). To obtain Figure 27.2, the non-dimensionalized pressures of Figure 27.3 were integrated between the limits r_1 and $r_1 + L$, where r_1 is given by Eq. (27.8), and the results of the integration were normalized; the coefficient C_s^F is thus an approximate measure of the average pressure coefficient over the interval L [3].

Numerical Example. The sizes and distribution of the openings (not represented in Figure 27.4) are assumed to be uniform around the periphery of the structure. The ratio between area of openings and the total wall area is $A_0/A_w = 0.25$. It is assumed that $V_{max} = 200$ mph (89.4 m s^{-1}), $R_m = 150$ ft (46 m). The pressures on the 100 ft (30.5 m) side walls induced by wind blowing in the direction shown in Figure 27.4 are calculated as follows:

$$p_{max} = 0.00256 \times 200^2 = 102.4 \text{ lb ft}^{-2} \quad (4900 \text{ N m}^{-2}) \quad (\text{Eq.}[27.7])$$

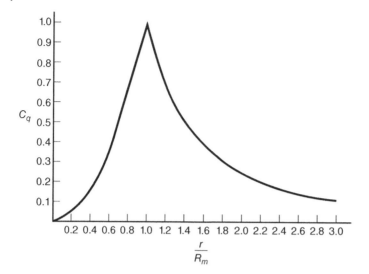

Figure 27.3 Coefficient C_q [3].

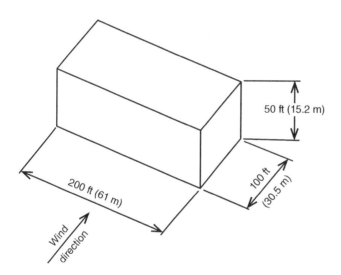

50 ft (15.2 m)

100 ft (30.5 m)

200 ft (61 m)

Wind direction

Figure 27.4 Schematic view of building.

Basic external pressures:

$L = 200\,\text{ft (61 m)}$

$\dfrac{L}{R_m} = 1.33$

$C_s^F = 0.57$ (Figure 27.2)

$q_F = 0.57 \times 102.4 = 58\,\text{lb ft}^{-2}$ (2800 N m^{-2}) (Eq. [27.5])

Basic internal pressures:

$C_s^F = 0.57$ (Figure 27.2)

$q_M = 0.57 \times 102.4 = 58\,\text{lb ft}^{-2}$ (2800 N m^{-2}) (Eq. [27.6])

The pressure coefficients are assumed in this example to have the following values:

$C_p = -0.7$

$C_{pi} = \pm 0.3$

Wind pressure:

$p_w = -0.7 \times 58 - 0.3 \times 58 = -58 \, \text{lb ft}^{-2} \, (-2800 \, \text{N m}^{-2}) \, (\text{Eq. [27.4]})$

27.3 Atmospheric Pressure Change Loading

Consider the cyclostrophic equation (Eq. [1.4], in which the term affected by the Coriolis acceleration may be neglected), written as

$$\frac{dp_a}{dr} = -\rho \frac{V_R^2}{r} \tag{27.10}$$

where dp_a/dr is the atmospheric pressure gradient at radius r from the center of the tornado vortex. To obtain the pressure drop p_a, Eq. (27.10) is integrated from infinity to r. Using the expression for V_t given by Eqs. (27.1) and (27.2),

$$p_a(r) = \rho \frac{V_{R_m}^2}{2} \left(2 - \frac{r^2}{R_m^2} \right) \qquad (0 \le r \le R_m) \tag{27.11a}$$

$$p_a(r) = \rho \frac{V_{R_m}^2}{2} \frac{R_m^2}{r^2} \qquad (R_m \le r < \infty) \tag{27.11b}$$

In structures with no openings (i.e., unvented structures), the internal pressure remains equal to the atmospheric pressure before the passage of the tornado. Therefore, during the passage of the tornado the difference between the internal pressure and the atmospheric pressure is equal to p_a. It follows from Eqs. (27.11) that the maximum value of p_a, which occurs at $r = 0$, is

$$p_a^{\max} = \rho V_{R_m}^2 \tag{27.12}$$

If the structures are completely open, the internal and external pressures are equalized, for practical purposes, instantaneously, so the loading due to atmospheric pressure changes approaches zero. In structure with openings (i.e., vented structures), the internal pressures change during the tornado passage by an amount $p_i(t)$. Denoting the atmospheric pressure change by $p_a(t)$, the atmospheric differential pressure that acts on the external walls is $p_a(t) - p_i(t)$.

A useful model for $p_a(t)$ can be obtained by assuming in Eqs. (27.11) that $r = V_{tr} \, t$, where V_{tr} is the tornado translation speed and t is the time. A simpler model in which the variation of $p_a(t)$ with time is given by the graph of Figure 27.5 may also be used. The time varying internal pressures $p_i(t)$ may be estimated by iteration as follows. Assume that the building consists of a number n of compartments. The air mass in compartment N, where $N \le n$, at time t_{j+1} is denoted by $W_N(t_{j+1})$ and may be written as

$$W_N(t_{j+1}) = W_N(t_j) + [G_{N(\text{in})}(t_j) - G_{N(\text{out})}(t_j)]\Delta t \tag{27.13}$$

where $G_{N(\text{in})}$ and $G_{N(\text{out})}$ denote the mass of air flowing into and out of compartment N per unit of time, respectively, and Δt is the time increment. The air mass flow rates G_N can be calculated as functions of the pressures outside and within the compartment N,

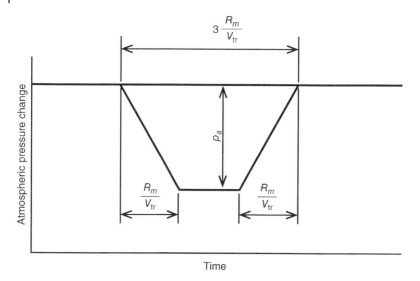

Figure 27.5 Idealized atmospheric pressure change versus time function [3].

and of relevant geometrical parameters, including opening sizes, as shown subsequently. The internal pressure in compartment N at time t_{j+1}, $p_{iN}(t_{j+1})$, is then written as

$$p_{iN}(t_{j+1}) = \left[\frac{W_N(t_{j+1})}{W_N(t_j)}\right]^k p_{iN}(t_j) \tag{27.14}$$

where $k = 1.4$ is the ratio of specific heat of air at constant pressure to specific heat of air at constant volume. The air mass flow rate can be modeled as follows:

$$G = 0.6 C_c A_2 [2\gamma_1(p_1 - p_2)]^{1/2} \tag{27.15}$$

where the non-dimensional compressibility coefficient C_c has the expression

$$C_c = \left\{ \left(\frac{p_2}{p_1}\right)^{2/k} \frac{k}{k-1} \left[\frac{1 - (p_2/p_1)^{(k-1)/k}}{1 - p_2/p_1}\right] \left[\frac{1 - (A_2/A_1)^2}{1 - (A_2/A_1)^2(p_2/p_1)^{2/k}}\right] \right\}^{1/2} \tag{27.16}$$

and A_1 is the area (on the side of compartment 1) of the wall between compartments 1 and 2, A_2 is the area connecting compartments 1 and 2, $k = 1.4$, p_1 is the pressure in compartment 1, p_2 is the pressure in compartment 2 ($p_2 < p_1$), and γ_1 is the mass per unit volume of air in compartment 1. If, in compartments provided with a blowout panel, the differential pressure exceeds the design pressure for a panel, the blowout area is transformed into a wall opening. To account for three-dimensional effects disregarded in Eq. (27.15), the atmospheric differential pressures on external walls obtained by the procedure just described are multiplied by a factor of 1.2 [3].

Figure 27.6 is an illustration of the pressure distribution and of the flow pattern in a building during depressurization. An illustration of a structure depressurization model with values of geometric parameters required as input in the calculations, and an example of a calculated corresponding differential pressure–time history, are shown in Figures 27.7 and 27.8, respectively.

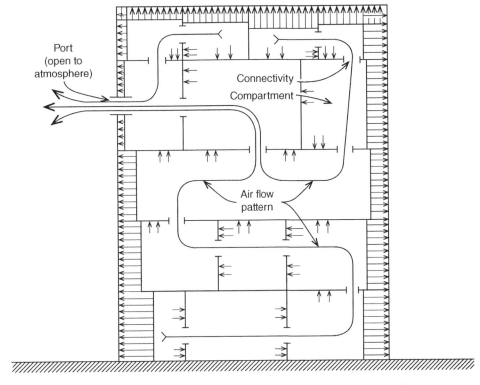

Figure 27.6 Pressure distribution and flow pattern during building depressurization [3].

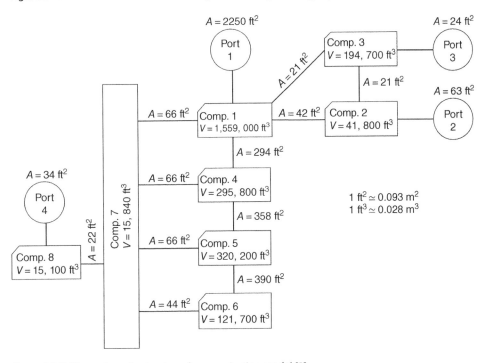

Figure 27.7 Illustration of a structure depressurization model [3].

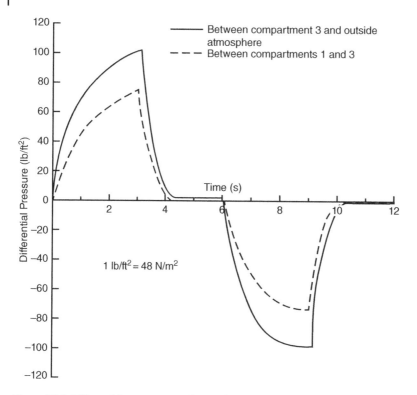

Figure 27.8 Differential pressure–time history for compartments 1 and 3. Input time history based on Figure 27.5 using $3R_m/V_{tr} = 9$ s and $p_a = 432$ lb ft^{-2} [3].

27.4 Experimental Modeling of Tornado-Like Wind Flows

Facilities aimed to simulate tornado flows have been developed since the early in the early 1970s (e.g., [4–7]; see Figures 5.12 and 5.13 for two examples). Their objective is to produce vortex flows with a strong rotation combined with a radial sink flow [8].

From the non-dimensional Navier–Stokes equations in cylindrical coordinates, Lewellen [9] established that such flows depend upon three non-dimensional parameters: the aspect ratio $a = h/r_0$, where h and r_0 are the axial inflow height and the updraft radius, respectively; the swirl ratio S; and a radial Reynolds number $Re_r = Q/(2\pi v)$, where v is the kinematic viscosity and Q is the volumetric flow rate per unit axial length. A commonly used expression for the swirl ratio is

$$S = \frac{r_0 \Gamma}{2Qh} \tag{27.17}$$

where the circulation Γ

$$\Gamma = 2\pi r_0 V_{t_{max}} \tag{27.18}$$

and $V_{t,max}$ is the maximum tangential velocity. Experimental and numerical results showed that flows with, approximately, swirl ratios $S < 0.5$ and $S > 1.0$ produced one-cell and multiple-cell vortices, respectively [10, 11]. In addition, it is established that the wind flows depend upon terrain roughness (e.g., [12]). Descriptions of flow fields associated with various values of S and Re_r are presented in [8] and, with the added benefit of modern measurement and flow visualization techniques, in [7]. A transition, from laminar axisymmetric core to a turbulent core with greatly expanded radius, termed vortex breakdown, is noted in [13]; the transition is due to development of an adverse pressure gradient as the laminar core spreads out in radius with increasing downstream distance [8, 13, 14].

Measurements of pressures induced by tornadoes on structures are reported in [6, 11, 15–17]. It appears that aerodynamic pressures induced by tornadoes on lateral walls of low-rise buildings may differ in some cases than those induced by straight winds; this is especially the case for suctions on roofs, owing to suctions induced by atmospheric pressure defects on fully unvented structures.

In situ observations reported in [18] are a first, promising attempt to document the structure of tornado wind flows near the ground.

References

1 U.S. Nuclear Regulatory Commission, *Regulatory Guide 1.76, Design-Basis Tornado and Tornado Missiles for Nuclear Power Plants,* Revision 1, 2007.

2 U.S. Nuclear Regulatory Commission, NUREG-0800, *Standard Review Plan, 3.3.2 Tornado Loads,* p. 3.3.2-6, Revision 3, March 2007.

3 Rotz, J. V., Yeh, G. C. K., and Bertwell, W., *Tornado and Extreme Wind Criteria for Nuclear Power Plants,* Topical Report No. BC-TOP-3A Revision 3, Bechtel Power Corporation, San Francisco, 1974.

4 Ward, N.B. (1962). The exploration of certain features of tornado dynamics using a laboratory model. *Journal of the Atmospheric Sciences* 29: 1194–1204.

5 Haan, F.L., Sarkar, P.P., and Gallus, W.A. (2008). Design, construction and performance of a large tornado simulator for wind engineering applications. *Engineering Structures* 30: 1146–1159.

6 Tang, Z., Feng, C., Wu, L. et al. (2017). Characteristics of tornado-like vortices simulated in a large-scale ward-type simulator. *Boundary-Layer Meteorology* doi: 10.1007/s10546-017-0305-7.

7 Refan, M. and Hangan, H.M. (2014). Characterization of tornado-like flow fields in a new model scale wind testing chamber. *Journal of Wind Engineering and Industrial Aerodynamics* 151: 107–121.

8 Church, C.R., Snow, J.T., Baker, G.L., and Agee, E.M. (1979). Characteristics of tornado-like vortices as a function of swirl ratio: a laboratory investigation. *Journal of the Atmospheric Sciences* 36: 1755–1776.

9 Lewellen, W.S. (1962). A solution for three-dimensional vortex flows with strong circulation. *Journal of Fluid Mechanics* 14: 420–432.

10 Davies-Jones, R.P. (1973). The dependence of core radius on swirl ratio in a tornado simulator. *Journal of the Atmospheric Sciences* 30: 1427–1430.

11 Haan, F.L. Jr., Kumar Balaramudu, V., and Sarkar, P.P. (2010). Tornado-induced wind loads on a low-rise building. *Journal of Structural Engineering* 136: 106–116.

12 Natarajan, D. and Hangan, H. (2012). Large eddy simulations of translation and surface roughness effects on tornado-like vortices. *Journal of Wind Engineering and Industrial Aerodynamics* 104–106,: 577–584.

13 Hall, M.G. (1972). Vortex breakdown. *Annual Review of Fluid Mechanics* 4: 195–218.

14 Tari, P.H., Gurka, R., and Hangan, H. (2010). Experimental investigation of tornado-like vortex dynamics with swirl ratio: the mean and turbulent flow fields. *Journal of Wind Engineering and Industrial Aerodynamics* 98: 936–944.

15 Kikitsu, H., Sarkar, P. P., and Haan, F. L. Jr., "Experimental study on tornado-induced loads of low-rise buildings using a large tornado simulator," *Proceedings of the 13th International Conference on Wind Engineering, Amsterdam, Netherlands*, July 10–15, 2011.

16 Thampi, H., Dayal, V., and Sarkar, P.P. (2011). Finite element analysis of interaction of tornadoes with a low-rise timber building. *Journal of Wind Engineering and Industrial Aerodynamics* 99: 369–377.

17 Mishra, A.R., James, D.L., and Letchford, C.W. (2008). Physical simulation of a single-celled tornado-like vortex: flow field characterization. *Journal of Wind Engineering and Industrial Aerodynamics* 96: 1243–1257. doi: 10.1016/j.jweia.2008.02.063.

18 Wurman, J., Kosiba, K., and Robinson, P. (2013). In situ, Doppler radar, and video observations of the interior structure of a tornado and the wind-damage relationship. *Bulletin of the American Meteorological Society* 94: 835–846.

28

Tornado- and Hurricane-Borne Missile Speeds

28.1   Introduction

Debris produced by wind-induced damage to structures, and various other objects that may be carried by strong winds, can acquire sufficiently high speeds to cause serious damage to the structures or building components they impact during their flight. Damage that may be produced by certain types of objects, for example roof gravel and light fences, can be avoided by appropriately regulating their use in high wind zones; objects such as roof pavers can be prevented from becoming wind-borne by adequately attaching them to their supporting structure; and openings can be protected from damage through the use of shutters.

However, for the design of nuclear power plants or other facilities whose failure to perform adequately could be catastrophic, specific allowance must be made in design for the impacts produced by wind-borne missiles in tornadoes or hurricanes. The purpose of this chapter is to review approaches to determining tornado- and hurricane-borne missile speeds for structural design purposes. Sections 28.2 and 28.3 concern tornado-borne and hurricane-borne missiles, respectively. For additional information on wind-borne debris hazards see [17].

28.2   Tornado-Borne Missile Speeds

To estimate speeds attained by an object under the action of aerodynamic forces induced by tornado winds, a set of assumptions is needed concerning:

- The aerodynamic characteristics of the object.
- The detailed features of the wind flow field.
- The initial position of the object with respect to the ground and to the tornado center and the translation velocity vector.

For the design of nuclear power plants, objects commonly considered as potential missiles include bluff bodies such as planks, steel rods, steel pipes, utility poles, and automobiles.[1] This section reviews approaches to the tornado-borne missile problem based on (i) deterministic and (ii) probabilistic modeling.

---

1 Information on the behavior of automobiles in strong winds is presented in [1, 2].

Wind Effects on Structures: Modern Structural Design for Wind, Fourth Edition. Emil Simiu and DongHun Yeo.
© 2019 John Wiley & Sons Ltd. Published 2019 by John Wiley & Sons Ltd.

28.2.1 Deterministic Modeling of Design-Basis Missile Speeds

Equations of Motion and Aerodynamic Modeling. The motion of an object can be described by solving a system of three equations of balance of momenta and three equations of balance of moments of momenta. For bluff bodies in motion, a major difficulty in writing these six equations is that the aerodynamic forcing functions are not known.

In the absence of a satisfactory model for the aerodynamic description of the missile as a rigid body, it is customary to resort to the alternative of describing the missile as a material point acted upon by a drag force

$$\mathbf{D} = \tfrac{1}{2}\rho \, C_D A |\mathbf{V}_w - \mathbf{V}_M|(\mathbf{V}_w - \mathbf{V}_M) \tag{28.1}$$

where ρ is the air density, $\mathbf{V}_w$ is the wind velocity, $\mathbf{V}_M$ is the missile velocity, A is a suitably chosen area, and C_D is the corresponding drag coefficient. This model is reasonable if, during its motion, the missile either maintains a constant or almost constant attitude with respect to the relative velocity vector $\mathbf{V}_w - \mathbf{V}_M$, or has a tumbling motion such that, with no significant errors, a mean value of the quantity $C_D A$ can be used in the expression for the drag D. The assumption of a constant body attitude with respect to the flow would be credible if the aerodynamic force were applied at all times exactly at the center of mass of the body – which is highly unlikely – or if the body rotation induced by a non-zero aerodynamic moment with respect to the center of mass were prevented by aerodynamic forces intrinsic in the body-fluid system. There is no evidence to this effect, so the assumption that wind-borne missiles will tumble during their flight is reasonable.

Assuming then that Eq. (28.1) is valid and that the average lift force vanishes under tumbling conditions, the motion of the missile viewed as a three-degree-of freedom system is governed by the relation

$$\frac{d\mathbf{V}_M}{dt} = \frac{1}{2}\rho\frac{C_D A}{m}|\mathbf{V}_w - \mathbf{V}_M|(\mathbf{V}_w - \mathbf{V}_M) - g\mathbf{k} \tag{28.2}$$

where g is the acceleration of gravity, $\mathbf{k}$ is the unit vector along the vertical axis, and m is the mass of the missile. It follows form Eq. (28.2) that for a given flow field and given initial conditions the motion depends only upon the value of the parameter $a = C_D A/m$. For a tumbling body this value can, in principle, be determined experimentally. Unfortunately, little information on this topic appears to be available. Information on tumbling motions under flow conditions corresponding to Mach numbers 0.5–3.5 is available in [3]. Those data were extrapolated in [4] to lower subsonic speeds; according to this extrapolation, for a randomly tumbling cube the quantity $C_D A/m$ equals, approximately, the average of the projected areas corresponding to "all positions statistically possible" times the respective static drag coefficients [4, pp. 13–17 and 14–16]. In the absence of more experimental information, it appears reasonable to assume that the effective product $C_D A$ is given by the expression

$$C_D A = c(C_{D_1} + C_{D_2} + C_{D_3}) \tag{28.3}$$

where $C_{D_i} A_i (i = 1, \ 2, \ 3)$ are products of the projected areas corresponding to the cases in which the principal axes of the body are parallel to the vector $\mathbf{V}_w - \mathbf{V}_M$ times the respective static drag coefficients, and c is a coefficient assumed to be 0.50 for planks, rods, pipes and poles, and 0.33 for automobiles. In the case of circular cylindrical bodies (rods, pipes, poles) the assumption $c = 0.5$ is conservative.

Computation of Missile Speeds. A computer program for calculating and plotting trajectories and velocities of tornado-borne missiles is listed in [5]. The program includes specialized subroutines incorporating the assumed model for the tornado wind field and the assumed drag coefficients (which may vary as functions of Reynolds number). Input statements include values of relevant parameters and the initial conditions of the missile motion.

In Eq. (28.2) both $\mathbf{V}_w$ and $\mathbf{V}_M$ are referenced with respect to an absolute frame. The velocity $\mathbf{V}_w$ is usually specified as a sum of two parts. The first part represents the wind velocity of a stationary tornado vortex and is referenced with respect to a cylindrical coordinate system. The second part represents a translation velocity of the tornado with respect to an absolute frame of reference. Transformations required to represent $\mathbf{V}_w$ in an absolute frame are derived in [5] and are incorporated in the computer program.

Maximum calculated horizontal missile speed V_{Mh}^{max} are reported in [5] as functions of the parameter $C_D A/m$ under the following assumptions:

- The rotational velocity of the tornado vortex V_R is described by Eq. (28.1).
- The radial velocity component V_r and the vertical velocity component V_z are given by the expressions suggested in [6]:

$$V_r = 0.50 V_R \tag{28.4}$$
$$V_z = 0.67 V_R \tag{28.5}$$

The radial component is directed toward the center of the vortex; the vertical component is directed upward.
- The translation velocity of the tornado vortex V_{tr} is directed along the x-axis.
- The initial conditions (at time $t = 0$) are $x(0) = R_m$, $y(0) = 0$, $z(0) = 40$ m, $V_{M_x} = 0$, $V_{M_y} = 0$, $V_{M_z} = 0$, where x, y, z are the coordinates of the center of mass of the missile and V_{M_x}, V_{M_y}, V_{M_z} are the missile velocity components along the x-, y-, and z-axes. Also, at $t = 0$, the center of the tornado vortex coincides with the origin O of the coordinate axes.

Similar calculations were performed independently by the U.S. Nuclear Regulatory Commission for a set of potential missiles listed in Table 28.1, assuming the validity of the tornado model with the characteristics listed in Table 3.2 for Regions I, II, and III (corresponding to Regions 1, 2, and 3 in Figure 3.5). For details see [1].

The ANSI/ANS-2.3-2011; R2016 Standard [7] contains a number of differences in the specification of missile speeds with respect to the values of [1].

A critique of various models of the wind field in tornadoes was recently presented in [8], and a novel, improved modeling of tornado-borne missile flight was proposed in [9].

28.2.2 Probabilistic Modeling of Design-Basis Missile Speeds

Reference [10] proposed a procedure for estimating speeds with 10^7-year mean recurrence intervals of postulated missiles that strike a given set of targets within a nuclear

Table 28.1 Design-basis tornado missile spectrum and maximum horizontal speeds V_{Mh}^{max}.

Missile Type		Schedule 40 Pipe	Automobile	Solid Steel Sphere
Dimensions		0.168 m dia. × 4.58 m long	Regions I and II 5 m × 2 m × 1.3 m Region III (1 in. dia.) 4.5 m × 1.7 m × 1.5 m	2.54 cm dia.
Mass		130 kg	Regions I and II 1810 kg Region III 1178 kg	0.0669 kg
$C_D A/m$		0.0043 m² kg⁻¹	Regions I and II 0.0070 m² kg⁻¹ Region III 0.0095 m² kg⁻¹	0.0034 m² kg⁻¹
V_{Mh}^{max}	Region I	41 m s⁻¹	41 m s⁻¹	8 m s⁻¹
	Region II	34 m s⁻¹	34 m s⁻¹	7 m s⁻¹
	Region III	24 m s⁻¹	24 m s⁻¹	6 m s⁻¹

power plant or similar installation. The procedure is based on assumptions concerning the number and location of potential missiles, the magnitude of the force opposing missile takeoff, the direction of the tornado axis of translation, and the size of the target area. The results of the calculations depend upon the parameter $C_D A/m$ and the ratio k between the minimum aerodynamic force required to cause missile takeoff and the weight of the missile. A listing of the computer program used in the procedure is available in [11].

A more elaborate approach to the development of a risk-informed approach is proposed in [12], which defines a missile impact probability (MIP) as the number of hits per missile per unit of target area. In this approach the hit frequency given a target structure is proportional to the tornado frequency, the number of missiles, the target area, and the MIP, and can be used for probabilistic risk assessments of core damage and radioactive release. In [12] the MIP was computed using data from [13]. The MIP depends on tornado characteristics, height of target, shielding inherent in the configuration of buildings in a plant, and area of spread of the missiles' initial location, and is independent of tornado frequency.

An innovative approach that does not require the use of Monte Carlo simulations is described in [14], which uses a three-degree-of-freedom model of the missile motion rather than a six-degree-of freedom model. The translating tornado wind velocity field can be described either by using the Rankine vortex or the Fujita model. Also included in this approach is a model for the lifting of potential missiles initially located on the ground in the tornado path.

28.3 Hurricane-Borne Missile Speeds

Calculated hurricane-borne missile speeds for the design of nuclear power plants are listed in [15] for the missiles considered in [1] and, in addition, for a plate-like and a plank-like missile that arise from metallic siding dislodged during a tornado event.

The assumptions on the basis of which the calculations were performed and the properties of the missiles being considered are considered in Section 28.3.1. A sample of results of the numerical calculations is presented in Section 28.3.2. Closed form as opposed to numerical solutions can be obtained for the case of wind speeds independent of height above ground, and are presented in Section 28.3.3. The closed form equations provide useful insights into the missiles dynamic behavior as a function of the various parameters of the motion (initial conditions, hurricane wind speeds, parameters defining missile properties). A summary of the numerical results of interest for regulatory purposes is presented in [16].

28.3.1 Basic Assumptions

This section considers the assumptions on the basis of which the calculations were performed.

1) Unlike for tornadoes, for hurricanes winds updraft speeds may be neglected. It follows that forces tending to increase the elevation of the missile with respect to the ground level may be assumed to be negligible as well. In particular, no updraft forces are available to lift automobiles.

2) The missiles start their motion with zero initial velocity from an elevation h above ground. As was the case for the tornado missile analyses performed for Regulatory Guide 1.76, it was assumed $h = 40$ m. In addition, the assumptions $h = 30, 20,$ and 10 m were used.

These assumptions imply that the change in the hurricane wind field through which the missile travel during its flight time is small. Indeed, for $h = 40$ m the flight time t_{max}, that is, the time it takes the missile to reach the ground from its initial elevation is

$$t_{max} = \left(2 \times \frac{40}{g} \right)^{1/2} \approx 2.86 \text{ seconds}$$

where $g = 9.81 \text{ m s}^{-2}$ is the acceleration of gravity. Therefore, for all the elevations h assumed in the calculations, $t_{max} < 3$ s. Let the hurricane speed be 100 m s^{-1}, say, and the radius of maximum wind speed be 1.5 km (the vast majority of hurricanes have radii of maximum wind speeds one order of magnitude larger). Assume conservatively that the horizontal distance traveled by the missile is in the order of $100 \text{ m s}^{-1} \times 3 \text{ s} = 300$ m, and that the missile's horizontal trajectory is tangent to the circle with radius 1.5 km, assumed conservatively to represent the hurricane's radius of maximum wind speeds. At the end of the trajectory the distance from the center of the circle to the missile will then be

$$r = \frac{1500}{\cos \left[\tan^{-1} (300/1500) \right]} = 1530 \text{ m}. \tag{28.6}$$

For practical purposes, the wind flows at 1500 and 1530 m from the center can be assumed to be the same. The differences between wind fields at the beginning and end of the missile trajectory (i.e., over a time interval in the order of 3 s) may similarly be assumed to be small.

3) Suburban terrain exposure and open terrain exposure represent, respectively, Exposure B and C as defined in the ASCE 7 Standard. For open terrain exposure, the wind speed v_h considered in the calculations represents the peak 3-second gust speed, and varies with height above ground z in accordance with the power law

$$\frac{v_h^{open}(z)}{v_h^{open}(10)} = \left(\frac{z}{10}\right)^{1/9.5} \tag{28.7a}$$

where (10) is the peak 3-second gust speed at 10 m above ground in open terrain. A simplified model of the wind field adopted in the ASCE 7-05 Standard (2006) is based on the assumption that the retardation of the wind flow by friction at the ground surface becomes negligible at an elevation, referred to conventionally as the gradient height, $z = 274$ m. At the gradient height the wind speed is, in accordance with Eq. (28.7a), (274 m) = 1.42 (10 m). In that simplified model it is further assumed that for suburban terrain exposure the retardation of the wind flow by friction at the ground surface becomes negligible at a gradient height $z = 366$ m. (The retardation of the wind flow by surface friction is effective up to higher elevations than over open exposure because the friction is stronger over suburban than over open terrain.)

For suburban terrain exposure the wind speed considered in the calculations represents the peak 3-second gust speed, and varies with height above ground z in accordance with the power law

$$\frac{v_h^{sub}(z)}{v_h^{sub}(366\ m)} = \left(\frac{z}{366\ m}\right)^{1/7} \tag{28.7b}$$

(z in meters). Since (366 m) = 1.42 (10 m), Eq. (28.7b) can be written as

$$\frac{v_h^{sub}(z)}{v_h^{open}(10)} = 1.42\left(\frac{z}{366}\right)^{1/7} \tag{28.7c}$$

For example, if v_h^{open} (10 m) = 40 and 150 m s^{-1} ($\alpha = 1/9.5$), then v_h^{sub} (10 m) = 34 and 127.5 m s^{-1} ($\alpha = 1/7$), respectively. The equations of motion of the missiles used in conjunction with Eqs. (28.7a) and (28.7b) can only be solved numerically. Results of numerical calculations are presented in Section 28.3.2. For simplified representations of the hurricane flow field it is possible to solve the equations of motion in closed form. Such closed form solutions are presented in Section 28.3.3.

4) As in the case of tornado-borne missiles, the aerodynamic force acting on a missile at any point of its trajectory was assumed to be proportional to the square of the velocity at that point times the parameter

$$a = \frac{1}{2}\rho C_D \frac{A}{m} \tag{28.8}$$

where ρ is the air density ($\approx 1.2\,\mathrm{kg\,m^{-3}}$), C_D is the drag coefficient characterizing the average aerodynamic pressure acting on the missile, A is the effective area of the missile, that is, the area by which pressures must be multiplied to yield the aerodynamic force, and m is the mass of the missile. For a plank with length and width $3.05\,\mathrm{m} \times 0.305\,\mathrm{m}$, $A = 0.93\,\mathrm{m^2}$, mass $m = 3.8\,\mathrm{kg}$ (steel board batten siding coated in PVC); for a slab with length and width $3.05\,\mathrm{m} \times 1.53\,\mathrm{m}$, $A = 4.67\,\mathrm{m^2}$, mass $m = 38\,\mathrm{kg}$. The assumptions concerning the areas A are conservative. For these two missiles it is assumed $C_D = 1.2$. Therefore, $a = 0.176$ and $a = 0.0885\,\mathrm{m^{-1}}$, respectively. For the other missiles being considered the parameters a have the same values as in Table 28.1.

Software for the calculation of hurricane-borne missile speeds based on the assumptions listed in this section is available at https://www.nist.gov/wind.

28.3.2 Numerical Solutions

Reference [15] lists:

- Terminal horizontal missile speeds (i.e., horizontal speeds at the time the missile reaches the ground).
- Terminal total missile speeds (i.e., resultants of the horizontal and vertical missile speeds at the time the missile reaches the ground).
- Maximum horizontal wind speeds (i.e., largest horizontal wind speeds reached during the missile flight).
- Maximum total missile speeds.

for the following conditions:

- Wind flows corresponding to 3-second wind speeds $(10\,\mathrm{m}) = 40\text{–}150\,\mathrm{m\,s^{-1}}$ in increments of $10\,\mathrm{m\,s^{-1}}$ at $10\,\mathrm{m}$ above terrain with open exposure, (i) over open terrain, and (ii) over suburban terrain.
- Missiles starting from rest from elevations 40, 30, 20, and $10\,\mathrm{m}$.

For values of the parameter $a < 0.006$ (in particular, for the four missiles covered by Regulatory Guide 1.76) the differences between the maximum missile speeds and the speeds at the time the missiles reach the ground level are not significant. However, for values of the parameter $a > 0.006\,\mathrm{m^{-1}}$ those differences can be large. The explanation for the decrease of the missile speeds from their maximum values is the following. After reaching those maximum speeds, the difference $v_h - v_{mh}$ between the hurricane wind speed and the horizontal missile speed can become negative as the missile moves at lower elevations where, owing to friction at the ground level, hurricane speeds are low. The missile motion is then decelerated.

Figure 28.1 shows an example of results obtained by numerical calculations. For example, for hurricanes and tornadoes with 230 mph ($103\,\mathrm{m\,s^{-1}}$) maximum 3-second wind speeds at $10\,\mathrm{m}$ above terrain with open exposure, calculated maximum horizontal speeds of missiles listed in Table 28.1 are shown in Table 28.2.

Results obtained in [15] were used to develop Regulatory Guide 1.221 [16].

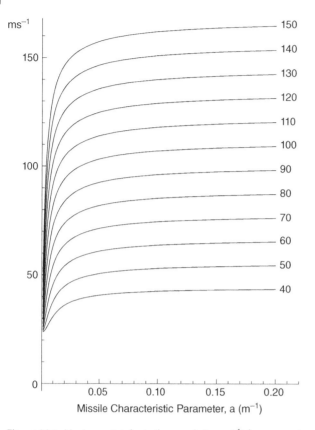

Figure 28.1 Maximum total missile speeds (in m s^{-1}) for parameters 0.005 m^{-1} < a < 0.200 m^{-1} and wind speeds over terrain with open exposure (10 m) = 40, 50,…,150 m s^{-1}. Missiles start at 40 m above ground level.

Table 28.2 Calculated maximum horizontal missile speeds in hurricanes and tornadoes, in m s^{-1}.

	Hurricanes	Tornadoes (Region I)
Solid steel sphere	48	8
Schedule 40 pipe	54	41
5 m automobile	68	41

28.3.3 Simplified Flow Field: Closed Form Solutions

It is now shown that a closed form solution can be obtained under the assumption that the wind speed v_h does not depend on height above ground. To check the validity of the algorithm by which they were obtained, numerical solutions corresponding to that assumption, were compared to their closed form counterparts. It was assumed that the vertical drag force is negligible and that the parameter a is given by Eq. (28.8).

The equation of horizontal motion of the missile can be written as:

$$\frac{dv_{mh}}{dt} = a(v_h - v_{mh})^2 \tag{28.9}$$

where v_{mh} is the horizontal missile velocity.
Equation (28.9) can be written as follows:

$$-\frac{d(v_h - v_{mh})}{dt} = a(v_h - v_{mh})^2 \tag{28.10}$$

Let $v_h - v_{mh} = y$. Eq. (28.10) becomes

$$-\frac{dy}{dt} = ay^2 \tag{28.11}$$

It follows that

$$-\frac{dy}{y^2} = adt \tag{28.12}$$

$$\frac{1}{y} = at + C \tag{28.13}$$

$$v_h - v_{mh} = \frac{1}{at + C} \tag{28.14}$$

$$v_{mh} = v_h - \frac{1}{at + C} \tag{28.15}$$

For $t = 0$, $v_{mh} = 0$, so $C = 1/v_h$. Therefore

$$v_{mh} = v_h - \frac{v_h}{av_h t + 1}. \tag{28.16}$$

For example, for $v_h = 100\,\text{m s}^{-1}$, $a = 0.0042\,\text{m}^{-1}$, a 40 m initial elevation of the missile and, therefore, it takes the missile a time $t = (2 \times 40/9.81)^{1/2} = 2.86\,\text{s}$ to reach the ground level under the action of gravity, and the horizontal missile speed at that time is $v_{mh} = 100 - 100/(1.20 + 1) = 54.55\,\text{m s}^{-1}$.

The horizontal distance traveled by the missile in 2.86 s is a small fraction of the hurricane's radius of maximum wind speeds, assumed conservatively to be 1.5 km. Denoting the horizontal position of the missile by x_{mh}, with the change of variable

$$t + \frac{1}{(av_h)} = \tau \tag{28.17}$$

integration of Eq. (28.18), in which $v_{mh} = dx_{mh}/dx$, yields

$$x_{mh} = v_h \tau - (1/a)\log\frac{\tau}{\tau_0} + B \tag{28.18}$$

where the integration constant C was written in the form $C = B + (1/a)\ln \tau_0$, and τ_0 is the value taken on by τ for $t = 0$. After some algebra, since for $t = 0$, $x_{mh} = 0$,

$$x = v_h t - \frac{1}{a}\log(1 + av_h t). \tag{28.19}$$

For $v_h = 100\,\text{m s}^{-1}$, $a = 0.0042\,\text{m}^{-1}$, $t = 2.86$ seconds, $x_{mh} = 286 - (1/0.0042)\log(1 + 2.86 \times 0.0042 \times 100) = 98\,\text{m}$.

It is shown in [15] that this result differs negligibly from its counterparts obtained numerically, thus verifying the numerical procedure being used. A similar verification was performed for tornado-borne missile speeds.

References

1 Paulikas, M.J., Schmidlin, T.W., and Marshall, T.P. (2016). The stability of passenger vehicles at tornado wind intensities of the (enhanced) Fujita scale. *Weather, Climate and Society* 8: 85–91.

2 Haan, F.L., Sarkar, P.P., Kop, G.A., and Stedman, D.A. (2017). Critical wind speeds or tornado-induced vehicle movements. *Journal of Wind Engineering and Industrial Aerodynamics* 168: 1–8.

3 Hansche, E. and Rinehart, J.S. (1952). Air drag on cubes at Mach numbers 0.5 to 3.5. *Journal of the Atmospheric Sciences* 19: 83–84.

4 Hoerner, S. F., Fluid-Dynamic Drag (published by the author, 1958).

5 Simiu, E. and Cordes, M.R., *Tornado-Borne Missile Speeds*, NBSIR 76-1050, National Bureau of Standards, Washington, DC, 1976. https://www.nist.gov/wind.

6 McDonald, J.R., Mehta, K.C., and Minor, J.E. (1974). Tornado-resistant design of nuclear power-plant structures. *Nuclear Safety* 15: 432–439.

7 American Nuclear Society, ANSI/ANS-2.3-2011. *Estimating tornado, hurricane, and extreme straight wind characteristics at nuclear facility sites.* La Grange Park, Illinois, reaffirmed Jun 29, 2016.

8 Gillmeier, S., Sterling, M., Hemida, H., and Baker, C.J. (2018). A reflection on analytical tornado-like vortex flow field models. *Journal of Wind Engineering and Industrial Aerodynamics* 174: 10–27.

9 Baker, C.J. and Sterling, M. (2017). Modelling wind fields and debris flight in tornadoes. *Journal of Wind Engineering and Industrial Aerodynamics* 168: 312–321.

10 Simiu, E. and Cordes, M.R. (1983). Tornado-borne missile speed probabilities. *Journal of Structural Engineering* 109: 154–168. Online publication date: January 1, 1983. https://www.nist.gov/wind.

11 Cordes, M. R. and Simiu, E., *Probabilistic Assessment of Tornado-Borne Missile Speeds,* NBSIR 80-2117, National Bureau of Standards, Washington, DC, 1980. https://www.nist.gov/wind.

12 Pensado, O. *Analysis of Missile Impact Probability for Generic Tornado Hazard Assessments.* Prepared for U.S. Nuclear Regulatory Commission Division of Risk Assessment of the Office of Nuclear Reactor Regulation, Southwest Research Institute Center for Nuclear Waste Regulatory Analyses, 2016.

13 EPRI. "Tornado Missile Risk Analysis." Report NP-768. Electric Power Research Institute: Washington DC, 1978.

14 Eguchi, Y., Murakami, T., Hirakuchi, H. et al. (2017). An evaluation method for Tornado missile strike probability with stochastic correlation. *Nuclear Engineering and Technology* 49: 395–403.

15 Simiu, E. and Potra, F., *Technical Basis for Regulatory Guidance on Design-Basis Hurricane- Hurricane-Borne Missile Speeds for Nuclear Power Plants,* NUREG/CR-7004, S. Sancaktar, NRC Project Manager, National Institute of Standards and Technology, Gaithersburg, MD, NRC Code 6726, Nov. 2011. https://www.nist.gov/wind.

16 U.S. Nuclear Regulatory Commission (NRC). *Design-Basis Hurricane and Hurricane Missiles for Nuclear Power Plants.* Regulatory Guide 1.221, 2011.

17 ASCE. *Wind-Borne Debris Hazards.* N.B. Kaye, ed., Environmental Wind Engineering Committee, Wind Engineering Division, American Society of Civil Engineers, 2018.

Appendices

Appendix A

Elements of Probability and Statistics

A.1   Introduction

A.1.1   Definition and Purpose of Probability Theory

Following Cramér [1], probability theory will be defined as a mathematical model for the description and interpretation of phenomena showing *statistical regularity*. Examples are phenomena such as the wind intensity at a given location, the turbulent wind speed fluctuations at a point, the pressure fluctuations on the surface of a building, or the fluctuating response of a structure to wind loads. Probabilistic models arising in connection with the wind loading of structures are discussed in Sections A.1–A.6.

Consider an experiment that can be repeated an indefinite number of times and whose outcome can be the occurrence or non-occurrence of an event A. If, for large values of trials n, the ratio m/n, called the *relative frequency* of the event A, differs little from some unique limiting value $P(A)$, the number $P(A)$ is defined as the *probability* of occurrence of event A. For example, if a coin is tossed, the ratio of the number of heads observed in a very large recorded sequence of H's (heads) and T's (tails) should be close to $\frac{1}{2}$ so that, in any one toss, the probability of occurrence of a head would be $\frac{1}{2}$. Consider, however, the recorded sequence

$$\text{H T H T H T H T H T H T H T}$$

consisting of alternating H's and T's. If, in this sequence, the observed outcome of a toss is a head, the probability of a head in the next toss will not be $\frac{1}{2}$ [2].

Indeed, for the definition of probability just advanced to be meaningful, it is required that the sequence (S) previously referred to satisfy the condition of *randomness*. This condition states that the relative frequency of event A must have the same limiting value in the sequence (S) as in any partial sequence that might be selected from it in any arbitrary way, the number of terms in any sequence being sufficiently large, and the selection being made in the absence of any information on the outcomes of the experiment [3]. The hypothesis that limiting values of the relative frequencies exist is confirmed for a wide variety of random phenomena by a large body of empirical evidence.

A.1.2   Statistical Estimation

Data obtained from observations must be fitted to mathematical models provided by probability theory by using statistical methods. Such methods fall into two broad

Wind Effects on Structures: Modern Structural Design for Wind, Fourth Edition. Emil Simiu and DongHun Yeo.
© 2019 John Wiley & Sons Ltd. Published 2019 by John Wiley & Sons Ltd.

categories: parametric and non-parametric. Parametric models aim to estimate parameters of the probabilistic models. Non-parametric (parameter-free) models are typically applied to large samples of rank-ordered data, obtained in some applications by numerical simulation.

Like probability theory, statistics is a vast field. Basic statistical notions and methods used in applications connected with the wind loading of structures are discussed in Chapter 3 and Appendices C and E.

References that complement the material covered in this Appendix include [4–13].

A.2 Fundamental Relations

A.2.1 Addition of Probabilities

Consider two events, A_1 and A_2, associated with an experiment. Assume that these events are mutually exclusive (i.e., cannot occur at the same time). The event that either A_1 or A_2 will occur is denoted by $A_1 \cup A_2$. The probability of this event is

$$P(A_1 \cup A_2) = P(A_1) + P(A_2) \tag{A.1}$$

The empirical basis of the addition rule (Eq. [A.1]) is that, if the relative frequency of event A_1 is m_1/n and that of event A_2 is m_2/n, the frequency of either A_1 or A_2 is $(m_1 + m_2)/n$. Equation (A.1) then follows from the relation between frequencies and probabilities, and can obviously be extended to any number of mutually exclusive events $A_1, A_2, ..., A_n$.

Example A.1 For a fair die the probability of throwing a "five" is 1/6 and the probability of throwing a "six" is 1/6. The probability of throwing either a "five" or a "six" is then $1/6 + 1/6 = 1/3$.

Let the non-occurrence of event A be denoted by $\overline{A}$. Events A and $\overline{A}$ are mutually exclusive. Also, the event that A either occurs or does not occur is certain; that is, its probability is unity:

$$P(A \cup \overline{A}) = 1 \tag{A.2a}$$

Equation (A.2a) follows immediately from the addition rule (Eq. [A.1]) applied to the events A and $\overline{A}$, the probabilities of which are the limiting values of the relative frequencies m/n and $(n - m)/n$, respectively. The probability that A does not occur can be written as

$$P(\overline{A}) = 1 - P(A) \tag{A.2b}$$

Two events for which Eq. (A.2b) holds are said to be *complementary*.

A.2.2 Compound and Conditional Probabilities: The Multiplication Rule

Consider events A and B that may occur at the same time. The probability of the event that A and B will occur simultaneously is called the *compound probability* of events A and B, and is denoted by $P(A_1 \cap A_2)$. The probability of event A given that event B has already occurred is denoted by $P(A|B)$ and is known as the *conditional probability* of event A under the condition that B has already occurred. Formally, $P(A|B)$ is defined as follows:

$$P(A \mid B) = \frac{P(A \cap B)}{P(B)} \tag{A.3a}$$

In Eq. (A.3a) it is assumed that $P(B) \neq 0$. Similarly, if $P(A) \neq 0$,

$$P(B \mid A) = \frac{P(A \cap B)}{P(A)} \tag{A.3b}$$

Example A.2   In a certain region, records show that in an average year 60 days are windy, 200 days are cold, and 50 days are both windy and cold. Let the probability that a day will be windy and the probability that a day will be cold be denoted by $P(W)$ and $P(C)$, respectively. If it is known that condition C (i.e., cold weather) prevails, the probability that a day is windy, $P(W|C)$, is

$$\frac{P(W \cap C)}{P(C)} = \frac{(50/365)}{(200/365)} = \frac{50}{200}.$$

From Eqs. (A.3a) and (A.3b) it follows that

$$P(A \cap B) = P(B)P(A \mid B)$$
$$= P(A)P(B \mid A) \tag{A.4}$$

Equation (A.4) is referred to as the *multiplication rule* of probability theory.

A.2.3   Total Probabilities

If the events $B_1, B_2, \ldots, B_n$ are mutually exclusive and $P(B_1) + P(B_2) + \cdots + P(B_n) = 1$, the probability of event A is

$$P(A) = P(A \mid B_1)P(B_1) + P(A \mid B_2)P(B_2) + \cdots + P(A \mid B_n)P(B_n) \tag{A.5}$$

Equation (A.5) is referred to as the theorem of *total probability*.

Example A.3   With reference to the previous example, we denote the probability of occurrence of winds as $P(W)$, the probability of occurrence of winds given that a day is cold as $P(W|C)$, the probability that a day is not cold as $P(W \mid \overline{C})$, the probability that a day is cold as $P(C)$, and the probability that a day is not cold as $P(\overline{C})$. From Eq. (A.5) it follows that

$$P(W) = P(W \mid C)P(C) + P(W \mid \overline{C})P(\overline{C})$$
$$= \left(\frac{50}{200}\right)\left(\frac{200}{365}\right) + \left(\frac{10}{165}\right)\left(\frac{165}{365}\right) = \left(\frac{60}{365}\right)$$

A.2.4   Bayes' Rule

If $B_1, B_2, \ldots, B_n$ are n simultaneously exclusive events, the conditional probability of occurrence of B_i given that the event A has occurred is

$$P(B_i \mid A) = \frac{P(A \mid B_i)P(B_i)}{P(A \mid B_1)P(B_1) + \cdots + P(A \mid B_n)P(B_n)} \tag{A.6}$$

Equation (A.6) follows immediately from Eqs. (A.3b) and (A.4) (in which B is replaced by B_i) and Eq. (A.5). Equation (A.6) allows the calculation of the *posterior probabilities* $P(B_i|A)$ in terms of the *prior probabilities* $P(B_1)$, $P(B_2)$, ..., $P(B_n)$ and the conditional probabilities $P(A|B_1)$, $P(A|B_2)$, ..., $P(A|B_n)$.

Example A.4 On the basis of experience with destructive effects of previous tornadoes, it was estimated subjectively that the maximum wind speeds in a tornado were 50–70 m s^{-1}. It was further estimated, also subjectively, that the likelihood of the speeds being about 50, 60, and 70 m s^{-1} is $P(50) = 0.3$, $P(60) = 0.5$, and $P(70) = 0.2$. These values are prior probabilities. According to a subsequent failure investigation the speed was 50 m s^{-1}. However, associated with the investigation were uncertainties that were estimated subjectively in terms of conditional probabilities $P(\widehat{50}|V_{\text{true}})$, that is, of probabilities that the speed estimated on the basis of the investigation is 50 m s^{-1} given that the actual speed of the tornado was V_{true}. The estimated values of $P(\widehat{50}|V_{\text{true}})$ were

$$P(\widehat{50}|50) = 0.6$$

$$P(\widehat{50}|60) = 0.3$$

$$P(\widehat{50}|70) = 0.1$$

It follows from Eq. (A.6) that the posterior probabilities, that is, the probabilities calculated by taking into account the information due to the failure investigation, are

$$P(50 \mid \widehat{50}) = \frac{P(\widehat{50} \mid 50)P(50)}{P(\widehat{50} \mid 50)P(50) + P(\widehat{50} \mid 60)(P(60) + P(\widehat{50} \mid 70)P(70)}$$

$$= 0.51$$

$$P(60|\widehat{50}) = 0.43$$

$$P(70|\widehat{50}) = 0.06$$

Whereas the prior probabilities favored the assumption that the speed was 60 m s^{-1}, according to the calculated posterior probabilities it is more likely that the speed was only 50 m s^{-1}. This result is, of course, useful only to the extent that the various subjective estimates assumed in the calculations are reasonably correct.

A.2.5 Independence

In the example following Eq. (A.3b), the occurrence of winds and the occurrence of low temperatures were not independent events. Indeed, in the region in question, if the weather is cold, the probability of windiness increases.

Assume now that event A consists of the occurrence of a rainy day in Pensacola, Florida, and event B consists of an increase in the world market price of gold. It is reasonable to state that the probability of rain in Pensacola is in no way dependent upon whether such an increase has occurred or not. In this case it is then natural to state that

$$P(A \mid B) = P(A) \tag{A.7}$$

Two events A and B for which Eq. (A.7) holds are called stochastically[1] *independent*. By virtue of Eqs. (A.1) and (A.7), an alternative definition of independence is

$$P(A \cap B) = P(A)P(B) \tag{A.8}$$

Example A.5 The probability that one part of a mechanism will be defective is 0.01; for another part, independent of the first, this probability is 0.02. The probability that both parts will be defective is $0.01 \times 0.02 = 0.0002$.

Three events A, B, and C are (stochastically) independent only if, in addition to Eq. (A.8), the following relations hold:

$$P(A \cap C) = P(A)P(C)$$
$$P(B \cap C) = P(B)P(C)$$
$$P(A \cap B \cap C) = P(A)P(B)P(C) \tag{A.9}$$

In general, n events are said to be independent if relations similar to Eq. (A.9) hold for all combinations of two or more events.

A.3 Random Variables and Probability Distributions

A.3.1 Random Variables: Definition

Let a numerical value be assigned to each of the events that may occur as a result of an experiment. The resulting set of possible numbers is defined as a random variable.

Example A.6

(1) A coin is tossed. The numbers zero and one are assigned to the outcome heads and the outcome tails, respectively. The set of numbers zero and one constitutes a random variable.
(2) To each measurement of a quantity, a number is assigned equal to the result of that measurement. The set of all possible results of the measurements constitutes a random variable.

Random variables are called *discrete* or *continuous* according to whether they may take on values restricted to a set of integers (as in Example A.6 (1)), or any value on a segment of the real axis (as in Example A.6 (2)). It is customary to denote random variables by capital letters (e.g., X, Y, Z). Specific values that may be taken on by these random numbers are then denoted by the corresponding lower case letters (x, y, or z).

A.3.2 Histograms, Probability Density Functions, Cumulative Distribution Functions

Let the range of the continuous random variable X associated with an experiment be divided into equal intervals ΔX. Assume that, if the experiment is carried out n times,

---

[1] The word stochastic means "connected with random experiments and reliability," and is derived from the Greek $\sigma\tau o\chi\alpha\zeta o\mu\alpha\iota$, meaning "to aim at, seek after, guess, surmise."

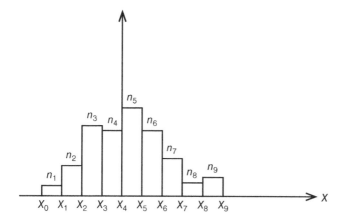

Figure A.1 Histogram.

the number of times that X has taken on values in the given intervals $X_1 - X_0, X_2 - X_1,$ $\ldots, X_i - X_{i-1}, \ldots$ is $n_1, n_2, \ldots, n_i, \ldots$, respectively. A graph in which the numbers n_i are plotted as in Figure A.1 is called a *histogram* (similar graphs may be plotted for discrete variables.)

Let the ordinates of the histogram in Figure A.1 be divided by $n\Delta X$. The resulting diagram is called the *frequency density distribution.* The relative frequency of the event $X_{i-1} < X \le X_i$ is then equal to the product of the ordinate of the frequency distribution, $n_i/(n\Delta X)$ by the interval ΔX. Since the area under the histogram is $(n_1 + n_2 + \cdots + n_i + \cdots)\Delta X = n\Delta X$, the total area under the frequency density diagram is unity.

As ΔX becomes very small so that $\Delta X = dx$ and as n becomes very large, the ordinates of the frequency density distribution approach in the limit values denoted by $f(x)$, where x denotes a value that may be taken on by the random variable X. The function $f(x)$ is known as the *probability density function* (PDF) of the random variable X (Figure A.2a). It follows from this definition that the probability of the event $x < X \le x + dx$ is equal to $f(x)dx$, and that

$$\int_{-\infty}^{\infty} f(x)dx = 1. \tag{A.10a}$$

In the experiment reflected in Figure A.1 the number of times that X has assumed values smaller than X_i is equal to the sum $n_1 + n_2 + \cdots + n_i$. Similarly, the probability that $X \le x$, called the *cumulative distribution function* (CDF) of the random variable X and denoted by $F(x)$, can be written as

$$F(x) = \int_{-\infty}^{x} f(x)dx \tag{A.10b}$$

that is, the ordinate at X in Figure A.2b is equal to the shaded area of Figure A.2a.

It follows from Eq. (A.10b) that

$$f(x) = \frac{dF(x)}{dx} \tag{A.11}$$

Figure A.2 (a) Probability density function.
(b) Cumulative distribution function.

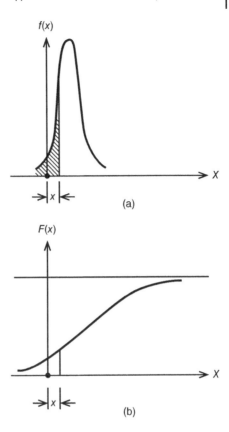

(a)

(b)

A.3.3 Changes of Variable

We consider here only the change of variable $y = (x - a)/b$, where a and b are constants. We assume the CDF $F_X(x)$ is known, and we seek the CDF $F_Y(y)$ and the PDF $f_Y(y)$. We can write

$$F_X(x) = P(X \leq x)$$
$$= P\left(\frac{X - a}{b} \leq \frac{x - a}{b}\right)$$
$$= F_Y(y) \qquad\qquad\qquad\qquad\qquad\text{(A.12a,b,c)}$$

Since Eq. (A.12c) implies that $dF_X(x) = dF_Y(y)$, or $f_X(x)dx = f_Y(y)dy$, it follows that

$$f_X(x) = \frac{1}{b} f_Y(y) \qquad\qquad\qquad\qquad\text{(A.13)}$$

A.3.4 Joint Probability Distributions

Let X and Y be two continuous random variables, and let $f(x, y)dxdy$ be the probability that $x < X \leq x + dx$ and $y < Y \leq y + dy$. The quantity $f(x, y)$ is called the *joint PDF* of the random variables X and Y (Figure A.3). The probability that $X \leq x$ and $Y \leq y$ is called the *joint cumulative probability distribution* of X and Y and is denoted by $F(x, y)$.

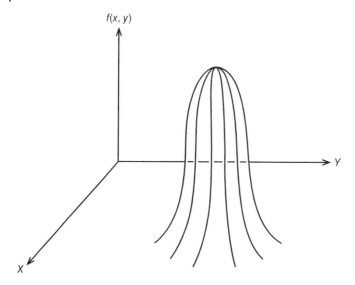

Figure A.3 Probability density function $f(x, y)$.

From the definition of $f(x, y)dxdy$ it follows that

$$F(x, y) = \int_{-\infty}^{x} \int_{-\infty}^{y} f(x, y)dx\, dy \tag{A.14a}$$

and

$$\int_{-\infty}^{\infty} \int_{-\infty}^{\infty} f(x, y)dx\, dy = 1 \tag{A.14b}$$

It follows from Eq. (A.14a) that

$$f(x, y) = \frac{\partial^2 F(x, y)}{\partial x \partial y} \tag{A.15}$$

If $f(x, y)$ is known, the probability that $x < X \le x + dx$, denoted by $f_X(x)dx$, is obtained by applying the addition rule to the probabilities $f(x, y)\, dx\, dy$ over the entire Y domain:

$$f_X(x) = \int_{-\infty}^{\infty} f(x, y)\, dy \tag{A.16}$$

The function $f_X(x)$ is called the *marginal PDF* of X.

Finally, the probability that $y < Y \le y + dy$ under the condition that $x < X \le x + dx$ is denoted by $f(y|x)dy$. The function $f(y|x)$ is known as the *conditional probability function* of Y given that $X = x$. If Eq. (A.3a) is used, it follows that

$$f(y \mid x) = \frac{f(x, y)}{f_X(x)} \tag{A.17}$$

If X and Y are independent, $f(y|x) = f_Y(y)$ and

$$f(x, y) = f_X(x)f_Y(y) \tag{A.18}$$

Similar definitions hold for any number of discrete or continuous random variables.

A.4 Descriptors of Random Variable Behavior

A.4.1 Mean Value, Median, Mode, Standard Deviation, Coefficient of Variation, and Correlation Coefficient

The complete description of the behavior of a random variable is provided by its probability distribution (in the case of several variables, by their joint probability distribution). Useful if less detailed information is provided by such descriptors as the mean value, the median, the mode, the standard deviation. and, in the case of two variables, their correlation coefficient.

The *mean value*, also known as the *expected value* or the *expectation*, of the discrete random variable, X, is defined as

$$E(X) = \sum_{i=1}^{m} x_i f_i \tag{A.19}$$

where m is the number of values taken on by x. The counterpart of Eq. (A.19) in terms of relative frequencies of the quantity $E(X)$ is

$$E(X) = \sum_{i=1}^{m} x_i \frac{n_i}{n} \tag{A.20}$$

If the random variable X is continuous, the expected value of X is written in complete analogy with Eq. (A.18) as

$$E(X) = \int_{-\infty}^{\infty} x f(x)\, dx \tag{A.21}$$

The *median* of a continuous random variable X is the value x that corresponds to the value $\frac{1}{2}$ of the CDF. The *mode* of X corresponds to the maximum value of the PDF. Since Prob $(x < X \leq x + dx) = f(x)\, dx$, the mode may be interpreted as the value of the variable that has the largest probability of occurrence in any given trial. The mean value, the median, and the mode are measures of *location*.

The expected value of the quantity $[x - E(X)]^2$ is the *variance* of the variable X. By virtue of the definition of the expected value (Eq. [A.21]), the variance can be written as

$$\mathrm{Var}(x) = E\{[X - E(X)]^2\}$$
$$= \int_{-\infty}^{\infty} [x - E(X)]^2 f(x) dx \tag{A.22}$$

The quantity $SD(X) = [\mathrm{Var}(X)]^{1/2}$ is the *standard deviation* of the random variable X. The ratio $SD(X)/E(X)$ is the *coefficient of variation* (CoV) of X. The variance, the standard deviation, and the CoV are useful measures of the scatter (or dispersion) of the random variable about its mean.

The *correlation coefficient* of two continuous random variables X and Y is defined as

$$\mathrm{Corr}(X, Y) = \frac{\int_{-\infty}^{\infty} \int_{-\infty}^{\infty} [x - E(X)]\, [y - E(Y)] f(x, y)\, dx\, dy}{SD(X)SD(Y)} \tag{A.23}$$

The correlation coefficient is similarly defined if the variables are discrete. It can be shown that

$$-1 \leq \mathrm{Corr}(X, Y) \leq 1 \tag{A.24}$$

It follows from Eq. (A.23) that if two random variables are linearly related:

$$Y = a + bX \tag{A.25}$$

then

$$Corr(X, Y) = \pm 1 \tag{A.26}$$

The sign in the right-hand side of Eq. (A.26) is the same as that of the coefficient b in Eq. (A.25). It can be proved that, conversely, Eq. (A.26) implies Eq. (A.25). The correlation coefficient may thus be viewed as an index of the extent to which two variables are linearly related.

If X and Y are independent, then $Corr(X, Y) = 0$. This follows from Eqs. (A.23), (A.18), and (A.21). However, the relation $Corr(X, Y) = 0$ does not necessarily imply the independence of X and Y [4].

A.5 Geometric, Poisson, Normal, and Lognormal Distributions

A.5.1 The Geometric Distribution

Consider an experiment of the type known as *Bernoulli trials*, in which (i) the only possible outcomes are the occurrence and the non-occurrence of an event A, (ii) the probability s of the event A is the same for all trials, and (iii) the outcomes of the trials are independent of each other.

Let the random variable N be equal to the number of the trial in which the event A occurs for the first time. The probability $p(n)$ that event A will occur on the nth trial is equal to the probability that event A will not occur on each of the first $n - 1$ trials and *will* occur on the nth trial. Since the probability of non-occurrence of event A in one trial is $1 - s$ (Eq. [A.2]), and since the n trials are independent, it follows from the multiplication rule (Eq. [A.8])

$$p(n) = (1 - s)^{n-1} s \qquad (n = 1, 2, 3, \ldots) \tag{A.27}$$

This probability distribution is known as the *geometric distribution* with parameter s.

The probability $P(n)$ that event A will occur at least once in n trials can be found as follows. The probability that event A will *not* occur in n trials is $(1 - s)^n$. The probability that it will occur at least once is therefore

$$P(n) = 1 - (1 - s)^n \tag{A.28}$$

The expected value of N is, by virtue of Eqs. (A.19) and (A.27),

$$\overline{N} = \sum_{n=1}^{\infty} n(1 - s)^{n-1} s \tag{A.29}$$

The sum of this series can be shown to be

$$\overline{N} = \frac{1}{s} \tag{A.30}$$

The quantity $\overline{N}$ is called the *mean return period*, or the *mean recurrence interval* (MRI).

Example A.7 For a die, the probability that a "four" occurs in a trial is $s = 1/6$. If the total number of trials is large, it may be expected that in the long run a "four" will appear on average once in $\overline{N} = 1/(1/6) = 6$ trials. The extension of the Mean Recurrence Intervals (MRI) concept to extreme wind speeds is discussed in Section 3.1.1.

A.5.2 The Poisson Distribution

Consider a class of events, each of which occurs independently of the other and with equal likelihood at any time $0 \le t \le T$. A random variable is defined, consisting of the number N of events that will occur during an arbitrary time interval $\tau = t_2 - t_1 (t_1 \ge 0$, $t_1 < t_2 \le T)$. Let $p(n, \tau)$ denote the probability that n events will occur during the interval τ. If it is assumed that $p(n, \tau)$ is not influenced by the occurrence of any number of events at times outside this interval, it can be shown that

$$p(n, \tau) = \frac{(\lambda\tau)^n}{n!}e^{-\lambda\tau} \qquad (n = 0, 1, 2, 3, \ldots) \tag{A.31}$$

If Eqs. (A.21) and (A.22) are used, it is found that the expected value and the variance of n are both equal to $\lambda\tau$. Since $\lambda\tau$ is the expected number of events occurring during time τ, the parameter λ is called the *average rate of arrival* of the process and represents the expected number of events per unit of time.

The applicability of Poisson's distribution may be illustrated in connection with the incidence of telephone calls in a telephone exchange. Consider an interval of, say, 15 minutes, during which the average rate of arrival of calls is constant. During any subinterval of those 15 minutes the incidence of a number n of calls is as likely as during any other equal subinterval. In addition, it may be assumed that individual calls are independent of each other. Therefore, Eq. (A.31) applies to any time subinterval τ lying within the 15-minute interval.

Example A.8 The estimated mean annual rate of arrival of hurricanes in Miami is $\lambda = 0.56/\text{year}$. Consider a period $\tau = 3$ years. Therefore, $\lambda\tau = 1.68$. What is the probability that there will be two hurricane occurrences in Miami during a period $\tau = 3$ years? From Eq. (A.31), $p(n = 2, \tau = 3) = 0.263$.

A.5.3 Normal and Lognormal Distributions

Consider a random variable X that consists of a sum of small, independent contributions $X_1, X_2, \ldots, X_n$. It can be proved that, under very general conditions, if n is large, the PDF of X is

$$f(x) = \frac{1}{\sqrt{2\pi}\sigma_x} \exp\left[-\frac{(x - \mu_x)^2}{2\sigma_x^2}\right] \tag{A.32}$$

where $\mu_x = E(X)$ and $\sigma_x^2 = \text{Var}(X)$ are the mean value and the variance of X, respectively. This statement is known as the *central limit theorem*. The distribution represented by Eq. (A.32) is called *normal* or *Gaussian*. It can be shown that the distribution of a linear function of a normally distributed variable is also normal, as is the sum of independent normally distributed variables.

If the distribution of the variable $Z = \ln X$ is normal, the distribution of X is called *lognormal*. Lognormal distributions are *heavy-tailed*, meaning that the ordinates of its

PDF are still significant for values X for which the ordinates of the Gaussian PDF are negligibly small.

A.6 Extreme Value Distributions

A.6.1 Extreme Value Distribution Types

Let the variable X be the largest of n *independent* random variables Y_1, Y_2, ..., Y_n. The inequality $X \leq x$ implies $Y_1 \leq x$, $Y_2 \leq x$, ..., $Y_n \leq x$. Therefore

$$F(X \leq x) = \text{Prob}(Y_1 \leq x, Y_2 \leq x, \ldots, Y_n \leq x)$$
$$= F_{Y_1}(x)F_{Y_2}(x) \ldots F_{Y_n}(x) \tag{A.33a,b}$$

where, to obtain Eq. (A.33b) from Eq. (A.33a), the generalized form of Eq. (A.8) is used. In the particular case in which the variables Y_i are *identically distributed* (i.e., have the same distribution $F_Y(x)F_Y(x)$), Eq. (A.33b) becomes

$$F_X(x) = [F_Y(x)]^n \tag{A.34}$$

The distribution $F_Y(y)$ is called the underlying or the *initial* distribution of the variable Y, which constitutes the *parent population* from which the largest values X have been extracted. It has been shown that, depending upon the properties of the initial distribution, there exist three types of extreme value distributions: the Fisher–Tippett Type I, Type II, and Type III distributions of the largest values, also known as the Gumbel, Fréchet, and reverse Weibull distributions. In extreme wind climatology the initial distributions can be tentatively determined only for a few types of storm, that do not include, for example, tropical storms. For this reason, in practice, the choice among the three distributions can only be made on an empirical basis (see Section A.7).

A.6.1.1 Extreme Value Type I Distribution

$$F_I(x) = \exp\left[-\exp\left(-\frac{x-\mu}{\sigma}\right)\right] \qquad (-\infty < x < \infty; -\infty < \mu < \infty; 0 < \sigma < \infty) \tag{A.35}$$

where μ and σ are the location and scale parameter, respectively. Equations (A.35), (A.21), and (A.22) yield the mean value and the standard deviation of the variate X:

$$E(X) = \mu + 0.5722\sigma \tag{A.36a,b}$$
$$SD(X) = \frac{\pi}{\sqrt{6}}\sigma$$

The *percentage point function*, defined as the inverse of the CDF, is

$$x(F_I) = \mu - \sigma \ln(-\ln F_I) \tag{A.37}$$

The estimated extreme value with MRI $= \overline{N}$ years can be determined from Eqs. (A.35)–(A.37),

$$v_I(\overline{N}) = E(X) + 0.78SD(X)(\ln \overline{N} - 0.577) \tag{A.38}$$

where $\overline{N} = 1/[1 - F(x)]$.

A.6.1.2   Extreme Value Type II Distribution

$$F_{II}(x) = \exp\left[-\left(\frac{x-\mu}{\sigma}\right)^{-\gamma}\right] \qquad (\mu < x < \infty; -\infty < \mu < \infty; 0 < \sigma < \infty; \gamma > 0)$$

(A.39)

where μ, σ and γ are the location, scale and shape (or tail length) parameters. For $\gamma > 2$ both the mean value and the standard deviation of the variate X diverge.

A.6.1.3   Extreme Value Type III Distribution

$$F_{III}(x) = \exp\left[-\left(-\frac{x-\mu}{\sigma}\right)^{\gamma}\right] \qquad (x < \mu)$$

(A.40)

$$x(F_{III}) = \mu - \sigma[-\ln(F_{III})]^{1/\gamma}$$

(A.41)

The mean value and the standard deviation of the variate X are related to the parameters μ, σ and γ as follows:

$$SD(X) = \sigma\left\{\Gamma\left(1+\frac{2}{\gamma}\right) - \left[\Gamma\left(1+\frac{1}{\gamma}\right)\right]^2\right\}^{1/2}$$

(A.42)

$$E(X) = \mu - \sigma\Gamma\left(1+\frac{1}{\gamma}\right)$$

(A.43)

where Γ is the gamma function.

In wind engineering practice it is typically assumed that the Extreme Value Type I (Gumbel) distribution is an appropriate distributional model. The rationale for this assumption is discussed in Section 3.3.2.

A.6.2   Generalized Extreme Value (GEV) Distribution

The GEV distribution is applied to independent extreme data (e.g., extreme wind speeds, peak wind effects) that exceed an optimal threshold. Its CDF is

$$F_{GEV}(x; \mu, \sigma, k) = \exp\left\{-\left[1 + k\left(\frac{x-\mu}{\sigma}\right)\right]^{-1/k}\right\}$$

(A.44)

where $1 + k(x-\mu)/\sigma > 0$, $-\infty < \mu < \infty$, and $0 < \sigma < \infty$.

For the shape parameter $k > 0$ and $k < 0$ Eq. (A.44) corresponds to the EV II and EV III distribution, respectively. In the limit $k = 0$, the GEV CDF is

$$F_{GEV}(x; \mu, \sigma, 0) = \exp\left[-\exp\left(-\frac{x-\mu}{\sigma}\right)\right]$$

(A.45)

and corresponds to the EV I distribution. Equation (A.45) is the conditional CDF of the variate X, given that $X > u$, where u is a sufficiently large, optimal threshold. The GEV is used with a different notation in Section C.2.

A.6.3   Generalized Pareto Distribution (GPD)

The GPD is applied to *differences* between independent extreme data and an optimal threshold. Its expression is

$$\text{for } c \neq 0 \qquad F_{GPD}(y; a, c) = 1 - \left(1 + c\frac{y}{a}\right)^{-1/c}$$

(A.46)

$$\text{for } c = 0 \qquad F_{GPD}(y; a, 0) = 1 - \exp\left(-\frac{y}{a}\right) \qquad (A.47)$$

where $a > 0$; $y \geq 0$ when $c \geq 0$, and $0 \leq y \leq -a/c$ when $c < 0$.

Equation (A.46) is the conditional CDF of the excess of the variate X over the optimal threshold u, $Y = X - u$, given $X > u$ for u sufficiently large. The tail length parameters $c > 0$, $c = 0$, and $c < 0$ correspond, respectively, to EV II, EV I, and EV III distribution tails. For $c = 0$ (Eq. [A.47]) the expression between braces is understood in a limiting sense as the exponential $\exp(-y/a)$.

The relations between the parameters a and c and the mean value $E(Y)$ and standard deviation $SD(Y)$ of the variate Y are [14]

$$a = \frac{1}{2}E(Y)\left[1 + \left(\frac{E(Y)}{SD(Y)}\right)^2\right] \qquad (A.48a)$$

$$c = \frac{1}{2}\left[1 - \left(\frac{E(Y)}{SD(Y)}\right)^2\right] \qquad (A.48b)$$

A.6.4 Mean Recurrence Intervals (MRIs) for Epochal and Peaks-over-Threshold (POT) Approaches

Epochal Approach. Consider the largest value of the variate X within each of number of *fixed* epochs, each assumed to be one year. Given the CDF $F(x)$ of the variate X, the probability of exceedance of x is $1 - F(x)$, and the MRI in years is $\overline{N} = 1/[1 - F(x)]$.

POT Approach. We first consider the *GEV distribution*. Let λ denote the average number per unit time (i.e., the mean rate of arrival) of exceedances of the threshold u by the variate X, and let the unit of time be 1 year. The average number of exceedances in $\overline{N}$ years is then $\lambda \overline{N}$. An *average* epoch – the average length of time between successive exceedances – is then equal to $1/\lambda$ years. For example, if $\lambda = 2$ exceedances/year, the average epoch is ½ years; if $\lambda = 0.5$ exceedances/year the average epoch is 2 years. The MRI, in terms of the number of average epochs between exceedances of the value x, is $1/F(X > x) = \lambda\overline{N}$. Therefore, the MRI of the event $X > x$ in years is

$$\overline{N} = \frac{1}{\lambda[1 - F(X < x)]} \qquad (A.49)$$

$$F(X < x) = 1 - \frac{1}{\lambda\overline{N}} \qquad (A.50)$$

A similar equation, in which $Y = X - u$ and $y = x - u$ are substituted in Eq. (A.50) for X and x, applies to the *Generalized Pareto Distribution*, that is,

$$F(Y < y) = 1 - \frac{1}{\lambda\overline{N}} \qquad (A.51)$$

$$1 - \left(1 + c\frac{y}{a}\right)^{-1/c} = 1 - \frac{1}{\lambda\overline{N}} \qquad (A.52)$$

Therefore

$$y = \frac{-a[1 - (\lambda\overline{N})^c]}{c} \qquad (A.53)$$

and the value being sought is

$$x(\overline{N}) = y + u \tag{A.54}$$

where $\overline{N}$ is the MRI of x in years.

A.7 Statistical Estimates

A.7.1 Goodness of Fit, Confidence Intervals, Estimator Efficiency

Data obtained from observations may be viewed as observed values of random variables. The behavior of the data may then be assumed to be described by models governing the behavior of random variables, that is, by mathematical models used in probability theory.

In practical applications, from the nature of the phenomenon being investigated and on the basis of observations, one must infer the probability distribution that will adequately describe the behavior of the data and, unless a non-parametric approach is used, the parameters of that distribution; or at least some characteristics of that distribution, for example the mean and the standard deviation.

In practice, given a set of observed data, or a *data sample,* it is hypothesized in the parametric approach that its behavior can be modeled by means of some probability distribution believed to be appropriate. This hypothesis must then be tested. Techniques are available that incorporate some measure of the degree of agreement, or *goodness of fit,* between the model (including hypothesized values of its parameters) and the data or, conversely, of the degree to which the data deviate from the model. Techniques that allow the selection of the most appropriate distributional model and the estimation of its best fitting parameters include, among others, the method of moments, least squares, the probability plot correlation coefficient and DATAPLOT, and maximum likelihood. For details on such techniques see also the publicly available *NIST SEMATECH e-Handbook of Statistical Methods* [13] and *R: A Language and Environment for Statistical Computing* [12]. For details on W-statistics, see Appendix C.

An *estimator* is defined as a function $\hat{\alpha}(X_1, X_2, \ldots, X_n)$ of the sample data such that $\hat{\alpha}$ is a reasonable approximation of the unknown value α of the distribution parameter or characteristic being sought. As a function of random variables X_i $(i = 1, 2, \ldots, n)$, $\hat{\alpha}$ is itself a random variable. This is illustrated by the following example.

Consider the observed sequence of 14 outcomes of an experiment consisting of the tossing of a coin:

$$\text{H T T T H T H H T H H H T H} \tag{A.55a}$$

The random numbers associated with this experiment are the numbers zero and one, which are assigned to the outcome heads and the outcome tails, respectively. The data sample corresponding to the observed outcome is then

$$0, 1, 1, 1, 0, 1, 0, 0, 1, 0, 0, 0, 1, 0 \tag{A.55b}$$

This sample is assumed to be extracted from an infinite population that, in the case of an ideally fair coin, will have a mean value, denoted in this case by α, equal to $\frac{1}{2}$. A reasonable estimator for the mean α is the sample mean:

$$\hat{a} = \frac{1}{n} \sum_{i=1}^{n} X_i \tag{A.56}$$

where n is the sample size and X_i are the observed data. For the sample of size 14 in Eq. (A.55b), $\hat{\alpha} = 3/7$. If the samples consisting of the first seven and the last seven observations in Eq. (A.55b) are used, $\hat{\alpha} = 4/7$ and $\hat{\alpha} = 2/7$, respectively.

As a random variable an estimator $\hat{\alpha}$ will have a certain probability distribution with non-zero dispersion about the true value α. Thus, given a sample of statistical data, it is not possible to calculate the true value α being sought. Rather, *confidence intervals* can be estimated of which it can be stated, with a specified confidence level q, that they contain the unknown value α. Typically, a nominal 95% confidence interval is considered, which corresponds for the Gaussian distribution to $\widehat{E(X)} \pm 2\widehat{SD(X)}$ where $\widehat{E(X)}$ and $\widehat{SD(X)}$ denote the estimated mean value and standard deviation of the variate X.

In order for the confidence interval corresponding to a given confidence level q to be as narrow as possible it is desirable that the estimator being used be *efficient*. Of two different estimators $\hat{\alpha}_1$ and $\hat{\alpha}_2$ of the same quantity being estimated, the estimator $\hat{\alpha}_1$ is said to be more efficient if $E[(\hat{\alpha}_1 - \alpha)^2] < E[(\hat{\alpha}_2 - \alpha)^2]$.

A.7.2 Parameter Estimation for Extreme Wind Speed Distributions

Among the numerous methods for estimation of extreme wind distribution parameters by the *epochal approach* we mention the method of moments as applied to the EV I distribution, and the Lieblein method, which was developed specifically for the EV I distribution. Both are covered in Section 3.3.3.

For the POT approach, wind speed data separated by intervals of five days or more may be regarded as independent, although more rigorous methods for declustering data are available (see Appendix C, in which Poisson processes are applied to the estimation of extremes). Let the wind speed data be denoted by x_i.

Generalized Pareto Distribution (GPD). The analysis is performed on data $x_i - u$, where u denotes the threshold. If the threshold u is too large, the size of the data sample will be small and the estimated values will be affected by large sampling errors. If the threshold is too low, the estimates biased by the presence in the sample of non-extreme wind data. The analysis is carried out for a sufficiently large set of thresholds u. For a subset of those thresholds the analysis will yield approximately the same estimated values of the parameters being sought. A threshold within that subset, referred to as optimal, yields the estimates being sought. The determination of the subset is performed visually, and is subjective and slow. An objective approach is presented in Appendix C. Two methods for the estimation of the GPD are now presented. In the *method of moments* the estimated GPD parameters are obtained by applying Eq. (A.48) to the sample mean value and standard deviation of the data y_i. From Eq. (A.54) it follows that the estimated wind speed with an $\overline{N}$-year MRI is

$$\hat{x}(\overline{N}) = \hat{y}(\overline{N}) + u \tag{A.57}$$

where u is an optimal threshold. In the *de Haan method* [15], the number of data *above* the threshold is denoted by k, so that the threshold u represents the $(k+1)$th highest data point. We have $\lambda = k/n_{years}$, where n_{years} is the length of the record in years. The highest, second highest, ..., kth highest, $(k+1)$th highest data points are denoted by $X_{n,n}$, $X_{n-1,n}$, ..., $X_{n-(k+1),n}$, $X_{n-k,n} \equiv u$, respectively. Compute the quantities

$$M_n^{(r)} = \frac{1}{k} \sum_{i=0}^{k-1} [\ln(X_{n-1,n}) - \ln(X_{n-k,n})]^r, \qquad r = 1, 2 \qquad (A.58)$$

The estimators of c and a are

$$\hat{c} = M_n^{(1)} + 1 - \frac{1}{2[1 - (M_n^{(1)})^2/M_n^{(2)}]}, \qquad \hat{a} = \frac{u M_n^{(1)}}{\rho_1} \qquad (A.59a,b)$$

$$\rho_1 = 1, \ \hat{c} \geq 0; \quad \rho_1 = 1/(1 - \hat{c}), \ \hat{c} \leq 0. \qquad (A.60a,b)$$

Figure 3.3 is a POT plot of the estimated wind speeds obtained by Eqs. (A.59) and (A.60) as functions of threshold u (in mph), and of sample size corresponding to the threshold u.

Generalized Extreme Value Distribution (GEV). The GEV distribution is applied to data that exceed a threshold u. Unlike in the GPD, the statistical analysis is performed on the data themselves, rather than on the differences between the data and the threshold, see Appendix C.

A.8 Monte Carlo Methods

Monte Carlo methods are a branch of mathematics pertaining to experiments on random numbers. The simulation of the statistics of interest is achieved by appropriate transformations of sequences of random numbers. The new sequences thus obtained may be viewed as data, the sample statistics of which are representative of the statistical properties of interest.

The following example illustrates the application of Monte Carlo techniques. We consider a sequence of uniformly distributed random numbers $0 < y_i < 1$ ($i = 1, 2, ..., n$). The numbers y_i are viewed as values of the CDF $F_I(x_i)$ of a variate X with EV I distribution, that is, $y_i = F_I(x_i)$. From Eq. (A.37) it the follows that

$$x(y_i) = \mu - \sigma \ln(-\ln y_i) \qquad (A.61)$$

From the sample $x(y_i)$ ($i = 1, 2, ..., n$) of the variate X, it is possible to obtain estimates of μ, σ, and percentage points $x(F_I)$ for any specified F_I. The procedure is repeated a large number m of times. A number m of sets of values $\hat{\mu}$, $\hat{\sigma}$, and $\hat{x}(F_I)$, and corresponding histograms, can then be obtained. From the m sets, statistics of those estimates can be produced. For example, large directional wind speed datasets of synoptic windstorms can be generated from relatively short measured wind datasets by using Monte Carlo simulations [16].

A.9 Non-Parametric Statistical Estimates

A.9.1 Single Hazards

Consider a data sample of size n at a location where the mean arrival rate of the variate of interest λ/year. If the rate were $\lambda = 1$/year, the estimated probability that the highest value of the variate in the set would be exceeded is $1/(n+1)$, and the corresponding estimated MRI would be $\overline{N} = n + 1$ years (on average $n + 1$ "trials" would be required for a storm to exceed that highest valued, Section 3.1.1.2, Example 3.2). The estimated probability that the qth highest value of the variate in the set is exceeded is $q/(n+1)$, the corresponding estimated MRI in years is $\overline{N} = (n+1)/q$, and the rank of the variate with MRI $\overline{N}$ is $q = (n+1)/\overline{N}$.

In general $\lambda \neq 1$, and the estimated MRI is therefore $\overline{N} = (n+1)/(q\lambda)$ years. For example, if $n = 999$ hurricane wind speed data, and $\lambda = 0.5$/year, the estimated MRI of the event that the highest wind speed in the sample will occur is $\overline{N} = (n+1)/q\lambda = 1000/0.5 = 2000$ years, the estimated MRI of the second highest speed is 1000 years, and so forth. The rank of the speed with a specified MRI $\overline{N}$ is $q = (n+1)/(\overline{N}\lambda)$.[2]

Example A.9 *Non-parametric MRI estimates for hurricane wind speeds from a specified directional sector at a specified coastal location.* The use of non-parametric estimates of MRIs is illustrated for quantities forming a vector v_k ($k = 1, 2, \ldots, n$, where n is the number of trials). The methodology is the same regardless of the nature of the variate, which can represent wind effects or, as in this example, hurricane wind speeds. We consider speeds blowing from the 22.5° sector centered on the SW (i.e., 225°) direction at milestone 2250 (near New York City), where $\lambda = 0.305$/year. The data being used were obtained from the site https://www.nist.gov/wind, as indicated in Section 3.1. They are rank-ordered in Table A.1. It is sufficient to consider the first 55 rank-ordered data, since higher-rank data are small.

The q-th largest speed in the set of 999 speeds corresponds to a MRI $\overline{N} = (n+1)/(q\lambda) = 1000/(0.305q)$. For the first highest and second highest speeds listed in Table A.1 $\overline{N} = 1000/0.305 = 3279$ years and $\overline{N} = 1000/(0.305 \times 2) = 1639$ years, respectively. The peak 3-second gust speed with a 100-year MRI has rank $q = 1000/(0.305 \times 100) = 32.78$, that is, 33, and is seen from Table A.1 to be 17 m s^{-1}. Note that the precision of the estimates is poorer for higher-ranking speeds, owing to the relatively large differences between successive higher-ranking speeds in Table A.1 (e.g., 54 vs. 39 m s^{-1} for the highest vs. the second highest speed). For this reason, it is appropriate to develop datasets covering periods longer by a factor of 3, say, than the specified design MRI.

A.9.2 Multiple Hazards

We now consider the case of multiple hazards, for example synoptic wind speeds and thunderstorm wind speeds, or hurricanes and earthquakes.

---

2 A formula that takes into account the possibility that two or more hurricanes may occur at a site in any one year, and is more exact for short MRIs (e.g., 5 years), is: $\overline{N} = 1/\{1 - \exp[-\lambda q/(n+1)]\}$. For example, for $n = 999$, $\lambda = 0.5$, and $q = 2$, $\overline{N} = 1000.5$ years.

Table A.1 Rank-ordered peak 3-second gust speeds (in m s^{-1}) from SW direction at 10 m above open terrain for 22.5° sector at milepost 2550 (1-minute speed in knots $= 0.625 \times$ 3-second speed in m s^{-1}).

Rank q	SW 225°	Rank q	SW 225°	Rank q	SW 225°
1	54	19	19	39	14
2	39	20	19	40	14
3	33	21	18	41	14
4	30	22	18	42	13
5	27	23	18	43	13
6	26	24	17	44	13
7	26	25	17	45	13
8	23	26	17	46	13
9	23	27	17	47	13
10	22	28	17	48	12
11	22	29	17	49	12
12	21	30	17	50	12
13	20	31	17	51	11
14	20	32	17	52	10
15	20	33	17	53	10
16	19	34	16	54	9
17	19	35	16	55	2
18	19	36	16	...	...

Example A.10 Assume that the mean annual rates of synoptic storm and thunderstorm arrival at the location of interest are $\lambda_s = 4/\text{year}$ and $\lambda_t = 3.5/\text{year}$. The rank-ordered DCIs induced in a structural member by 10 000 synthetic synoptic storms and 10 000 thunderstorms are listed in Table A.2.

The MRI of DCIs > 1.00 induced by synoptic storms is $\overline{N}_s = (n_s + 1)/(q_s \lambda_s) = 10001/(5 \times 4) = 500$ years, so the probability that the DCI induced by synoptic winds is greater than 1.00 is 1/500 in any one year. Similarly, the probability that the DCI induced by thunderstorm winds is greater than 1.00 is $1/[10\,001/(9 \times 3.5)] = 1/317$ in any one year. The probability that the DCI induced by synoptic winds or by thunderstorms is greater than 1.00 is $1/500 + 1/317$ in any one year (hint: see Section 3.1.2, Eq. [3.3]). This corresponds to an MRI of the occurrence of the event DCI > 1.00 equal to $\overline{N} \approx 194$ years.

A similar approach can be used for regions, such as South Carolina and Hawaii, subjected to both hurricane and earthquake hazards, see [17]. For the approach to be applicable in this case it is necessary to provide – in addition to a probabilistic model of the extreme wind speeds at the location of interest and a procedure for determining the DCI (demand-to-capacity) indexes induced in the structure by those speeds, – a probabilistic model of the strength of the seismic events at that location and a procedure for determining the DCIs induced in the structure by those events.

Table A.2 Rank-ordered DCIs Induced by synoptic storm and thunderstorm winds.

DCIs Induced by Synoptic Storms		DCIs Induced by Thunderstorms	
Rank	DCI	Rank	DCI
		1	1.34
		2	1.30
		3	1.26
		4	1.23
1	1.22		
		5	1.21
2	1.16		
		6	1.18
3	1.10	7	1.10
4	1.04		
		8	1.02
5	1.01	9	1.01
6	0.99		
		10	0.98

References

1 Cramér, H. (1955). *The Elements of Probability Theory*. New York: Wiley.
2 Mihram, A.G. (1972). *Simulation*. New York: Academic Press.
3 von Mises, R. (1957). *Probability, Statistics and Truth*. London: G. Allen & Unwin.
4 Benjamin, J.R. and Cornell, C.A. (1970). *Probability, Statistics and Decision for Civil Engineers*. New York: McGraw-Hill.
5 Montgomery, D.C. and Runger, G.C. (2013). *Applied Statistics and Probability for Engineers*, 6th ed. Hoboken: Wiley.
6 Montgomery, D.C., Runger, G.C., and Hobele, N.F. (2011). *Student Solutions Manual Engineering Statistics*, 5the. Hoboken: Wiley.
7 Kay, S. (2006). *Intuitive Probability and Random Processes Using MATLAB*. New York: Springer.
8 Gumbel, E.J. (1958). *Statistics of Extremes*. New York: Columbia University Press.
9 Coles, S. (2001). *An Introduction to Statistical Modeling of Extreme Values*. London: Springer.
10 Castillo, E., Hadi, A.S., Balakrishnan, N., and Sarabia, J.M. (2004). *Extreme Value and Related Models with Applications in Engineering and Science*, 1st ed. Hoboken, New Jersey: Wiley.
11 Beirlant, J., Goegebeur, Y., Segers, J., and Teugels, J. (2004). *Statistics of Extremes. Theory and Applications*. Chichester: Wiley.

12 R_Development_Core_Team. *R: A Language and Environment for Statistical Computing, R Foundation for Statistical Computing*. Available: http://www.R-project.org, 2011.

13 NIST/SEMATECH. *e-Handbook of Statistical Methods*. Available: https://www.itl.nist .gov/div898/handbook, 2012.

14 Hosking, J.R.M. and Wallis, J.R. (1987). Parameter and quantile estimation for the generalized Pareto distribution. *Technometrics* 29: 339–349.

15 de Haan, L. (1994). Extreme value statistics. In: *Extreme Value Theory and Applications*, vol. 1 (ed. J. Galambos, J. Lechner and E. Simiu), 93–122. Boston, MA: Kluwer Academic Publishers.

16 Yeo, D. (2014). Generation of large directional wind speed data sets for estimation of wind effects with long return periods. *Journal of Structural Engineering* 140: 04014073. https://www.nist.gov/wind.

17 Duthinh, D. and Simiu, E. (2010). Safety of structures in strong winds and earthquakes: multihazard considerations. *Journal of Structural Engineering* 136: 330–333. https://www.nist.gov/wind.

Appendix B

Random Processes

Consider a process the possible outcomes of which form a collection (or an ensemble) of functions of time $\{y(t)\}$. A member of the ensemble is called a *sample function* or a *random signal*. The process is called a *random process* if the values of the sample functions at any particular time constitute a random variable.

Let a numerical value be assigned to each of the events that may occur as a result of an experiment. The resulting set of possible numbers is defined as a *random variable*. Examples: (i) If a coin is tossed, the numbers zero and one assigned to the outcome heads and to the outcome tails constitute a discrete random variable. (ii) To each measurement of a quantity a number is assigned to the result of that measurement. The set of all possible results of the measurements constitutes a continuous random variable.

A time-dependent random process is *stationary* if its statistical properties (e.g., the mean and the mean square value) do not depend upon the choice of the time origin and do not vary with time. A stationary random signal is thus assumed to extend over the entire time domain. The *ensemble average*, or *expectation*, of a random process is the average of the values of the member functions at any particular time. A stationary random process is *ergodic* if its time averages equal its ensemble averages. Ergodicity requires that every sample function be typical of the entire ensemble.

A stationary random signal may be viewed as a superposition of harmonic oscillations over a continuous range of frequencies. Some basic results of harmonic analysis are reviewed in Sections B.1 and B.2. The spectral density function (Section B.3), the autocovariance function (Section B.4), the cross-covariance function, the co-spectrum, the quadrature spectrum, and the coherence function (Section B.5) are defined next. Mean upcrossing and outcrossing rates are introduced in Section B.6. The estimation of peaks of Gaussian random signals is considered in Section B.7.

B.1 Fourier Series and Fourier Integrals

Consider a *periodic* function $x(t)$ with zero mean and period T. It can be easily shown that

$$x(t) = C_0 + \sum_{k=1}^{\infty} C_k \cos(2\pi k n_1 t - \phi_k) \tag{B.1}$$

Wind Effects on Structures: Modern Structural Design for Wind, Fourth Edition. Emil Simiu and DongHun Yeo.
© 2019 John Wiley & Sons Ltd. Published 2019 by John Wiley & Sons Ltd.

where $n_1 = 1/T$ is the *fundamental frequency* and

$$C_0 = \frac{1}{T} \int_{-T/2}^{T/2} x(t)dt \tag{B.1a}$$

$$C_k = (A_k^2 + B_k^2)^{1/2} \tag{B.1b}$$

$$\phi_k = \tan^{-1} \frac{B_k}{A_k} \tag{B.1c}$$

$$A_k = \frac{2}{T} \int_{-T/2}^{T/2} x(t) \cos(2\pi k n_1 t)dt \tag{B.1d}$$

$$B_k = \frac{2}{T} \int_{-T/2}^{T/2} x(t) \sin(2\pi k n_1 t)dt \tag{B.1e}$$

Equation (B.1) is the *Fourier series expansion* of the periodic function $x(t)$.

If a function $y(t)$ is *nonperiodic*, it is still possible to regard it as periodic with infinite period. It can be shown that if $y(t)$ is piecewise differentiable in every finite interval, and if the integral

$$\int_{-\infty}^{\infty} |y(t)|dt \tag{B.2}$$

exists, the following relation holds:

$$y(t) = \int_{-\infty}^{\infty} C(n) \cos[2\pi nt - \phi(n)]dn \tag{B.3}$$

In Eq. (B.3), called the *Fourier integral* of $y(t)$ in real form, n is a continuously varying frequency, and

$$C(n) = (A^2(n) + B^2(n))^{1/2} \tag{B.3a}$$

$$\phi(n) = \tan^{-1} \frac{B(n)}{A(n)} \tag{B.3b}$$

$$A(n) = \int_{-\infty}^{\infty} y(t) \cos(2\pi nt)dt \tag{B.3c}$$

$$B(n) = \int_{-\infty}^{\infty} y(t) \sin(2\pi nt)dt \tag{B.3d}$$

From Eqs. (B.3a) through (B.3d) and the identities

$$\sin \phi = \frac{\tan \phi}{(1 + \tan^2 \phi)^{1/2}} \tag{B.4a}$$

$$\cos \phi = \frac{1}{(1 + \tan^2 \phi)^{1/2}} \tag{B.4b}$$

it follows that

$$\int_{-\infty}^{\infty} y(t) \cos[2\pi nt - \phi(n)]dt = C(n) \tag{B.5}$$

The functions $y(t)$ and $C(n)$, which satisfy the symmetrical relations Eqs. (B.3) and (B.5), form a *Fourier transform pair*.

Successive differentiation of Eq. (B.3) yields

$$\dot{y}(t) = -\int_{-\infty}^{\infty} 2\pi n C(n) \sin[2\pi nt - \phi(n)]dn \tag{B.6a}$$

$$\ddot{y}(t) = -\int_{-\infty}^{\infty} 4\pi^2 n^2 C(n) \cos[2\pi nt - \phi(n)]dn \tag{B.6b}$$

B.2 Parseval's Equality

The mean square value of the periodic function $x(t)$ with period T (Eq. (B.1)) is

$$\sigma_x^2 = \frac{1}{T}\int_{-T/2}^{T/2} x^2(t)dt \tag{B.7}$$

Substitution of Eq. (B.1) into Eq. (B.7) yields

$$\sigma_x^2 = \sum_{k=0}^{\infty} S_k \tag{B.8}$$

where $S_0 = C_0^2$ and $S_k = \frac{1}{2} C_k^2$ $(k = 1, 2, \ldots)$. The quantity S_k is the contribution to the mean square value of $x(t)$ of the harmonic component with frequency kn_1. Equation (B.8) is a form of Parseval's equality.

For a nonperiodic function for which an integral Fourier expression exists, Eqs. (B.3) and (B.5) yield

$$\int_{-\infty}^{\infty} y^2(t)dt = \int_{-\infty}^{\infty} y(t) \int_{-\infty}^{\infty} C(n) \cos[2\pi nt - \phi(n)]dn \, dt$$

$$= \int_{-\infty}^{\infty} C(n) \int_{-\infty}^{\infty} y(t) \cos[2\pi nt - \phi(n)]dt \, dn$$

$$= \int_{-\infty}^{\infty} C^2(n)dn$$

$$= 2\int_{0}^{\infty} C^2(n)dn. \tag{B.9}$$

Equation (B.9) is the form taken by Parseval's equality in the case of a nonperiodic function.

B.3 Spectral Density Function of a Random Stationary Signal

A relation similar to Eq. (B.8) is now sought for functions generated by stationary processes. The spectral density of such functions is defined as the counterpart of the quantities S_k.

Let $z(t)$ be a stationary random signal with zero mean. Because it does not satisfy the condition (B.2), $z(t)$ does not have a Fourier transform. An auxiliary function $y(t)$ is therefore defined as follows (Figure B.1):

$$y(t) = z(t) \qquad \left(-\frac{T}{2} < t < \frac{T}{2}\right) \tag{B.10a}$$

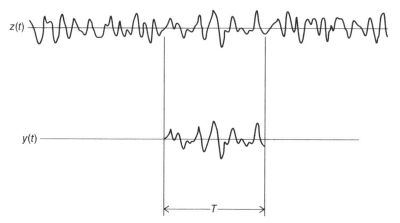

Figure B.1 Definition of function $y(t)$.

$$y(t) = 0 \qquad \text{elsewhere} \tag{B.10b}$$

The function $y(t)$ so defined is nonperiodic, satisfies condition (B.2), and thus has a Fourier integral. From the definition of $y(t)$ it follows that

$$\lim_{T\to\infty} y(t) = z(t) \tag{B.11}$$

By virtue of Eqs. (B.9) and (B.10), the mean square value of $y(t)$ is

$$\sigma_y^2 = \lim_{T\to\infty} \frac{1}{T} \int_{-T/2}^{T/2} y^2(t)dt$$

$$= \frac{1}{T} \int_{-\infty}^{\infty} y^2(t)dt$$

$$= \frac{2}{T} \int_0^{\infty} C^2(n)dn \tag{B.12}$$

The mean square of the function $z(t)$ is then

$$\sigma_z^2 = \lim_{T\to\infty} \sigma_y^2$$

$$= \lim_{T\to\infty} \frac{2}{T} \int_0^{\infty} C^2(n)dn \tag{B.13}$$

With the notation

$$S_z(n) = \lim_{T\to\infty} \frac{2}{T} C^2(n) \tag{B.14}$$

Equation (B.13) becomes

$$\sigma_z^2 = \int_0^{\infty} S_z(n)dn. \tag{B.15}$$

The function $S_z(n)$ is defined as *the spectral density function of $z(t)$*. To each frequency n $(0 < n < \infty)$ there corresponds an elemental contribution $S(n)\,dn$ to the mean square value σ_z^2; σ_z^2 is equal to the area under the spectral density curve $S_z(n)$. Because in

Eq. (B.15) the spectrum is defined for $0 < n < \infty$ only, $S_z(n)$ is called the *one-sided* spectral density function of $z(t)$. This definition of the spectrum is used throughout this text. A different convention may be used where the spectrum is defined for $-\infty < n < \infty$, and the integration limits in Eq. (B.15) are $-\infty$ to ∞. This convention yields the *two-sided* spectral density function of $z(t)$.

From Eqs. (B.6a,b), following the same steps that led from Eq. (B.3) to Eq. (B.14), there result the expressions for the spectral density of the first and second derivative of a random process:

$$S_{\dot{z}}(n) = 4\pi^2 n^2 S_z(n) \tag{B.16a}$$

$$S_{\ddot{z}}(n) = 16\pi^4 n^4 S_z(n) \tag{B.16b}$$

B.4 Autocorrelation Function of a Random Stationary Signal

From Eqs. (B.3a), (B.3c) and (B.3d), it follows that

$$
\begin{aligned}
\frac{2}{T}C^2(n) &= \frac{2}{T}[A^2(n) + B^2(n)] \\
&= \frac{2}{T}[A(n)A(n) + B(n)B(n)] \\
&= \frac{2}{T}\left[\int_{-\infty}^{\infty} y(t_1)\cos(2\pi n t_1)dt_1 \int_{-\infty}^{\infty} y(t_2)\cos(2\pi n t_2)dt_2 \right. \\
&\quad \left. + \int_{-\infty}^{\infty} y(t_1)\sin(2\pi n t_1)dt_1 \int_{-\infty}^{\infty} y(t_2)\sin(2\pi n t_2)dt_2\right] \\
&= \frac{2}{T}\int_{-\infty}^{\infty}\int_{-\infty}^{\infty} y(t_1)y(t_2)\cos[2\pi n(t_2 - t_1)]dt_1 dt_2.
\end{aligned}
\tag{B.17}
$$

Using the notations $\tau = t_2 - t_1$ and

$$\widetilde{R}(\tau) = \frac{1}{T}\int_{-\infty}^{\infty} y(t_1)y(t_1 + \tau)dt_1, \tag{B.18}$$

Equation (B.17) can be written as

$$\frac{2}{T}C^2(n) = 2\int_{-\infty}^{\infty} \widetilde{R}(\tau)\cos(2\pi n\tau)d\tau \tag{B.19}$$

Equations (B.19), (B.11), and (B.14) thus yield

$$S_z(n) = \int_{-\infty}^{\infty} 2R_z(\tau)\cos(2\pi n\tau)d\tau \tag{B.20}$$

where

$$R_z(\tau) = \lim_{T\to\infty} \frac{1}{T}\int_{-T/2}^{T/2} z(t)z(t + \tau)dt. \tag{B.21}$$

The function $R_z(\tau)$ is defined as the *autocovariance function* of $z(t)$ and provides a measure of the interdependence of the variable z at times t and $t + \tau$. From the stationarity of $z(t)$, it follows that

$$R_z(\tau) = R_z(-\tau). \tag{B.22}$$

Since $R_z(\tau)$ is an even function of τ,

$$\int_{-\infty}^{\infty} 2R_z(\tau) \sin(2\pi n\tau) d\tau = 0. \tag{B.23}$$

A comparison of Eqs. (B.5) and (B.20) shows that $S_z(n)$ and $2R_z(\tau)$ form a Fourier transform pair. Therefore,

$$R_z(\tau) = \frac{1}{2} \int_{-\infty}^{\infty} S_z(n) \cos(2\pi n\tau) dn. \tag{B.24a}$$

Since, as follows from Eq. (B.20), $S_z(n)$ is an even function of n, Eq. (B.24a) may be written as

$$R_z(\tau) = \int_{0}^{\infty} S_z(n) \cos(2\pi n\tau) dn. \tag{B.24b}$$

Similarly, by virtue of Eqs. (B.20) and (B.22),

$$S_z(n) = 4 \int_{0}^{\infty} R_z(\tau) \cos(2\pi n\tau) d\tau \tag{B.25}$$

The definition of the autocovariance function (Eq. (B.21)) yields

$$R_z(0) = \sigma_z^2 \tag{B.26}$$

For $\tau > 0$ the products $z(t)z(t+\tau)$ are not always positive as is the case for $\tau = 0$, so

$$R_z(\tau) < \sigma_z^2 \tag{B.27}$$

For large values of τ, the values $z(t)$ and $z(t+\tau)$ bear no relationship to each other, so

$$\lim_{\tau \to \infty} R_z(\tau) = 0 \tag{B.28}$$

The non-dimensional quantity $R_z(\tau)/\sigma_z^2$, called the *autocorrelation function* of the function $z(t)$, is equal to unity for $\tau = 0$ and vanishes for $\tau = \infty$.

B.5 Cross-Covariance Function, Co-Spectrum, Quadrature Spectrum, Coherence

Consider two stationary signals $z_1(t)$ and $z_2(t)$ with zero means. The function

$$R_{z_1 z_2}(\tau) = \lim_{T \to \infty} \frac{1}{T} \int_{-T/2}^{T/2} z_1(t)z_2(t+\tau) \, dt \tag{B.29}$$

is defined as the *cross-covariance function* of the signals $z_1(t)$ and $z_2(t)$. From this definition and the stationarity of the signals, it follows that

$$R_{z1z2}(\tau) = R_{z2z1}(-\tau). \tag{B.30}$$

However, in general, $R_{z_1 z_2}(\tau) \neq R_{z_1 z_2}(-\tau)$. For example, if $z_2(t) \equiv z_1(t - \tau_0)$, it can immediately be seen from Figure B.2 that

$$R_{z1z2}(\tau_0) = R_{z1}(0) \tag{B.31}$$

$$R_{z1z2}(-\tau_0) = R_{z1}(2\tau_0) \tag{B.32}$$

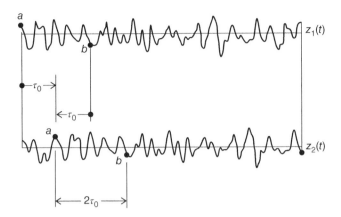

Figure B.2 Functions $z_1(t)$ and $z_2(t) = z_1(t-\tau_0)$.

The *co-spectrum* and the *quadrature spectrum* of the signals $z_1(t)$ and $z_2(t)$ are defined, respectively, as

$$S_{z1z2}^C(n) = \int_{-\infty}^{\infty} 2R_{z1z2}(\tau)\cos(2\pi n\tau)d\tau \qquad\qquad (B.33)$$

$$S_{z1z2}^Q(n) = \int_{-\infty}^{\infty} 2R_{z1z2}(\tau)\sin(2\pi n\tau)d\tau. \qquad\qquad (B.34)$$

It follows from Eq. (B.30) that

$$S_{z1z2}^C(n) = S_{z2z1}^C(n) \qquad\qquad (B.35a)$$

$$S_{z1z2}^Q(n) = -S_{z2z1}^Q(n) \qquad\qquad (B.35b)$$

The *coherence function* is a measure of the correlation between components with frequency n of two signals $z_1(t)$ and $z_2(t)$, and is defined as

$$\text{Coh}_{z1z2}(n) = \left\{ \frac{[S_{z1z2}^C(n)]^2 + [S_{z1z2}^Q(n)]^2}{S_{z1}(n)S_{z2}(n)} \right\}^{1/2} \qquad\qquad (B.36)$$

Example B.1 The animation in Figure 4.27 shows pressures on the exterior surface of a building, induced by wind blowing in the direction shown by the arrow. If pressures at any two points were perfectly coherent spatially, at any given time the shades representing their intensity would be the same regardless of the distance between the points.

B.6 Mean Upcrossing and Outcrossing Rate for a Gaussian Process

Let $z(t)$ be a stationary differentiable process with mean zero. The process crosses a level k at least once in a time interval $(t, t+\Delta t)$ if $z(t) < k$ and $z(t+\Delta t) > k$. If $z(t)$ has smooth samples and Δt is sufficiently small, $z(t)$ will have a single k-crossing with positive slope, (i.e. a single k-upcrossing). The probability of occurrence of the event

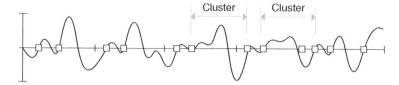

Figure B.3 Upcrossings of a random process (indicated by rectangles). Clusters are groups of two or more local peaks within an interval defined by two successive upcrossings.

$\{z(t) < k, \ z(t + \Delta t) > k\}$ can be approximated by the probability of the event $\{z(t) < k < z(t) + \dot{z}(t) \Delta t\}$. The mean rate of k-upcrossings of $z(t)$ is

$$v(k) = \int_0^\infty \dot{z} f_{z,\dot{z}}(b, \dot{z}) d\dot{z} \qquad (B.37a)$$

where $f_{z,\dot{z}}$ denotes the joint probability density function of $z(t), \dot{z}(t)$. For a stationary process the variables z and $\dot{z}$ are independent,[1] so

$$v(k) = E[\dot{z}(t)_+ \mid z(t) = k] f_z(k) \qquad (B.37b)$$
$$= E[\dot{z}(t)_+] f_z(k) \qquad (B.37c)$$

where f_z denotes the probability density function of $z(t)$, and $E[\dot{z}(t)_+ \mid z(t) = k]$ denotes the expectation of the positive part of $\dot{z}(t)$ conditional on $z(t) = k$. A plot showing zero upcrossings of a random process is shown in Figure B.3.

If $z(t)$ is a stationary Gaussian process with mean zero,

$$f_{\dot{z},z}(z, \dot{z}) = \frac{1}{2\pi \sigma_{\dot{z}} \sigma_z} \exp\left[-\frac{1}{2} \left(\frac{z^2}{\sigma_z^2} + \frac{\dot{z}^2}{\sigma_{\dot{z}}^2} \right) \right] \qquad (B.38)$$

and the mean k-upcrossing rate is

$$v(k) = E[\dot{z}(t)_+] f(k)$$
$$= \frac{\sigma_{\dot{z}}}{\sqrt{2\pi}} \frac{1}{\sqrt{2\pi} \sigma_z} \exp\left(-\frac{k^2}{2\sigma_z^2} \right) \qquad (B.39)$$

where σ_z and $\sigma_{\dot{z}}$ denote the standard deviations of $z(t)$ and $\dot{z}(t)$.

Equation (B.37a) can be extended to the case in which the random process is a vector **x**. Let v_D denote the mean rate at which the random process (i.e., the tip of the vector with specified origin O) crosses in an outward direction the boundary F_D of a region containing the point O. The rate v_D has the expression

$$v_D = \int_{F_D} d\mathbf{x} \int_0^\infty \dot{x}_n f_{\mathbf{x}, \dot{x}_n}(\mathbf{x}, \dot{x}_n) d\dot{x}_n \qquad (B.40)$$

---

1 For a stationary process $E[z^2(t)] = \text{const.}$, so $dE[z^2(t)]/dt = 2E[z(t)dz(t)/dt] = 0$ for a fixed arbitrary time t, meaning that $z(t)$ and $dz(t)/dt$ are uncorrelated. If $z(t)$ is Gaussian, so is $dz(t)/dt$. It then follows from the expression for the joint Gaussian distribution of two correlated variables that if their correlation vanishes the two variates are independent.

where $\dot{x}_n$ is the projection of the vector $\dot{\mathbf{x}}$ on the normal to F_D, and $f_{\mathbf{x},\dot{x}_n}(\mathbf{x}, \dot{x}_n)$ is the joint probability distribution of $\mathbf{x}$ and $\dot{x}_n$. Eq. (B.40) can be written as

$$v_D = \int_{F_D} \left\{ \int_0^\infty \dot{x}_n f_{\dot{x}_n}[\dot{x}_n | \mathbf{X} = \mathbf{x}] d\dot{x}_n \right\} f_{\mathbf{X}}(\mathbf{x}) d\mathbf{x}$$

$$= \int_{F_D} E_0^\infty[\dot{X}_n \mid \mathbf{X} = \mathbf{x}] f_{\mathbf{X}}(\mathbf{x}) d\mathbf{x} \tag{B.41}$$

where $f_{\mathbf{X}}(\mathbf{X})$ = probability density of the vector $\mathbf{X}$, and $E_0^\infty[\dot{X}_n | \mathbf{X} = \mathbf{x}]$ is the average of the positive values of $\dot{X}_n$ given that $\mathbf{X} = \mathbf{x}$. If $\dot{X}_n$ and $\mathbf{X}$ are independent, $E_0^\infty[\dot{X}_n | \mathbf{X} = \mathbf{x}] = E_0^\infty[\dot{X}_n]$.

Equation (B.41) has been used in an attempt to estimate mean recurrence intervals of directional wind effects that exceed (outcross) a limit state defined by a boundary F_D. Objections to this approach include: the perception by structural engineers that it lacks transparency (see Appendix F); the fact that the vector $\mathbf{x}$, which represents a structural response to wind (e.g., a demand-to-capacity index) may be non-Gaussian; the fact that the limit state boundary cannot be defined unless the structural design is finalized, which is in practice not the case at the time the outcrossing calculations are performed; and the fact that, if the size of the available directional wind speeds data is small, rather than creating a larger data set by Monte Carlo simulation some practitioners make use of what are purported to be parent population data; that is, non-extreme wind speeds that may include morning breezes and other types of wind that differ from a meteorological point of view from the extremes, and cannot therefore constitute a reliable basis for estimating extreme values.

B.7 Probability Distribution of the Peak Value of a Random Signal with Gaussian Marginal Distribution

The probability distribution of the set of values $z(t)$ of the random process is called the *marginal distribution* of that process. Since

$$\sigma_z^2 = \int_0^\infty S_z(n) \, dn \tag{B.42}$$

$$\sigma_{\dot{z}}^2 = 4\pi^2 \int_0^\infty n^2 S_z(n) \, dn \tag{B.43}$$

(Eq. [B.16a]), denoting

$$v = (1/2\pi)(\sigma_{\dot{x}}/\sigma_x) \tag{B.44}$$

$$\kappa = k/\sigma_x, \tag{B.45}$$

it follows from Eq. (B.39) that the upcrossing rate of the level κ (in units of standards deviations of the process) is

$$E(\kappa) = v \exp\left(-\frac{\kappa^2}{2}\right) \tag{B.46}$$

where

$$v = \left[\frac{\int_0^\infty n^2 S_z(n)dn}{\int_0^\infty S_z(n)dn} \right]^{1/2} \tag{B.47}$$

is the mean zero upcrossing rate, that is,

$$v = E(0). \tag{B.48}$$

Peaks greater than $k\sigma_z$ may be regarded as rare events. Their probability distribution may therefore be assumed to be of the Poisson type. The probability that in the time interval T there will be no peaks equal to or larger than $k\sigma_z$ can therefore be written as

$$p(0, T) = \exp[-E(k)T] \tag{B.49}$$

The probability $p(0, T)$ can be viewed as the probability that, given the interval T, the ratio K of the largest peak to the r.m.s. value of $z(t)$ is less than κ, that is,

$$P(K < \kappa | T) = \exp[-E(\kappa)T] \tag{B.50}$$

The probability density function of K, that is, the probability $p_K(\kappa | T)$ that $\kappa < K < \kappa + d\kappa$, is obtained from Eq. (B.50) by differentiation:

$$P(\kappa | T) = \kappa T E(\kappa) \exp[-E(\kappa)T] \tag{B.51}$$

The expected value of the largest peak occurring in the interval T may then be calculated as

$$\overline{K} = \int_0^\infty \kappa p_K(\kappa | T)d\kappa \tag{B.52}$$

The integral of Eq. (B.52) is, approximately,

$$\overline{K} = (2 \ln vT)^{1/2} + \frac{0.577}{(2 \ln vT)^{1/2}} \tag{B.53}$$

[1], where v is given by Eq. (B.47).

The estimation of statistics of peaks of random signals with arbitrary marginal probability distributions is discussed in detail in Appendix C.

Reference

1 Davenport, A.G. (1964). Note on the distribution of the largest value of a random function with application to gust loading. *Journal of the Institution of Civil Engineers* 24: 187–196.

Appendix C

Peaks-Over-Threshold Poisson-Process Procedure for Estimating Peaks*

C.1   Introduction

The estimation of the distribution of the peak of a random process $y(t)$ with specified duration T from a single finite time series of length $T_1 \leq T$, and of the corresponding uncertainties, has applications in:

- Extreme wind climatology, where the time series consists of a record of extreme wind speeds over a time interval $T_1 = N_1$ years, and the statistics of the largest wind speed during a longer time interval $T = N$ years are of interest.
- Aerodynamics and structural engineering, where a time series of length T_1 of wind effects (e.g., measured pressure coefficients, or calculated internal forces, demand-to-capacity indexes, inter-story drift, accelerations) is available, and the statistics of the peak wind effect for a time series with length $T \geq T_1$ are of interest.

For the particular case in which the marginal distribution of a process $y(t)$ is Gaussian, a closed-form expression for the distribution of the peak is available (see Section B.7). If the distribution is not Gaussian, a nonlinear mapping procedure, referred to as "translation," has been developed by which those statistics can be obtained [1]. The translation procedure depends heavily on the user's ability to choose an appropriate marginal probability distribution. In practice, because of the difficulty of this task, the performance of the translation method can be unsatisfactory.

A simple procedure in which the time history of length T_1 is divided into n equal segments (epochs) was proposed in [2]. A data sample is created consisting of the peak of each of those segments, and a Gumbel Cumulative Distribution Function (CDF) is fitted to that sample. The length T_1/n of the segments must be sufficient for the peaks of different segments to be mutually independent. To obtain the largest peak for a time history of length $T = rT_1/n$ ($r \geq n$) the Gumbel CDF describing the probabilistic behavior of the segment peaks is raised to the r-th power. Because that CDF is an exponential function, this operation results in an alternative Gumbel distribution that describes the probabilistic behavior of the peak of the time history of length T [3]. This procedure is most efficiently implemented by using the BLUE (*B*est *L*inear *U*nbiased *E*stimator) method to estimate the parameters of the Gumbel distribution of the segment peaks (see Section 3.3.3 and [4]; https://www.nist.gov/wind). However, as shown in Section C.3, a

* Dr. A. L. Pintar's leading role in the development and application of the procedure described in this Appendix is acknowledged with thanks.

Wind Effects on Structures: Modern Structural Design for Wind, Fourth Edition. Emil Simiu and DongHun Yeo.
© 2019 John Wiley & Sons Ltd. Published 2019 by John Wiley & Sons Ltd.

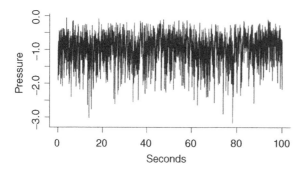

Figure C.1 Time history of pressure coefficients.

drawback of the BLUE method is that the estimates can depend significantly on n, with no criterion for an optimal choice of n being available in the literature.

The purposes of this Appendix are: (i) to describe a Peaks-over-Threshold (POT), Poisson-process procedure for the estimation of the distribution of the peak of a stationary random process of specified duration (Section A.2) and of the corresponding uncertainties; and (ii) to assess the performance of the procedure through comparison of its results with observed data and with results obtained by the BLUE method. A software implementation of the procedure applicable to time histories of pressures or pressure effects (e.g. internal forces in structural members), that leverages the R environment for statistical computing and graphics [5], is available in [6] (https://github.com/usnistgov/potMax), which also contains detailed instructions for installation and use. The procedure is described and illustrated in what follows with reference to the time history of pressure coefficients of Figure C.1. To allow the reader to replicate the calculations described herein we note that the data were obtained from the NIST-UWO Aerodynamic Database for Rigid Buildings [7], [https://www.nist.gov/wind], dataset jp1 Building 7, open terrain, tap 1715 at middle of eave, sampling rate 500 Hz, wind direction 270°. For similar software applied to the estimation of extreme wind speeds, see Section 3.3.5.

C.2   Peak Estimation by Peaks-Over-Threshold Poisson-Process Procedure

Description of Procedure. The POT approach is applied to observations $y(t)$ within a time series that exceed a threshold u. The POT approach is chosen over the epochal approach for two reasons. First, the POT approach generally allows the use of more observations than does the epochal approach, potentially leading to less uncertainty. Second, and more important, a procedure is available for an optimal selection of the threshold u [8].

The steps of the procedure are as follows:

1) *Reverse the signs of the time series, if necessary.* The procedure is developed for positive peaks. The peaks of interest in Figure C.1 being negative, the signs of this time series were reversed. If analysts are interested in both positive and negative peaks, the procedure is applied twice, first with the original signs, and second with reversed signs.

2) *Choose a model.* For the reasons indicated in Section 3.3.2, it is assumed for wind climatological purposes that the peaks of the variate $y(t)$ are described probabilistically by an Extreme Value Type I (EV I) distribution. For the same reasons, the restriction to the EV I distribution also holds for the peak of time series considered in aerodynamics and structural engineering applications. However, if interested in considering EV II or EV III distributional models, the analyst can choose to do so, as is indicated subsequently. Supposing that the variables (t, y) follow a Poisson process with intensity function $\lambda(t, y)$, the random number of peak values of y events that occur in a time interval $t_2 - t_1$ and have magnitude between y_1 and y_2 can be described by the Poisson distribution

$$p(n) = \frac{\left[\int_{t_1}^{t_2} \int_{y_1}^{y_2} \lambda(t, y) dt \, dy \right]^n}{n!} \exp\left[- \int_{t_1}^{t_2} \int_{y_1}^{y_2} \lambda(t, y) dt \, dy \right] \tag{C.1}$$

Let us consider the particular case in which the *intensity function* $\lambda(t, y) = $ const. and y_2 is the largest possible value of y (under the assumptions that the peaks y have an EV I or EV II distribution, y_2 is infinitely large; if y has an EV III distribution it has a finite upper bound). In that case the expected number of events is $(y_1 - y_2)(t_2 - t_1) \lambda$, where the constant intensity function λ is the rate of arrival of those events. However, Eq. (C.1) allows for more complex cases. In one such case the random process is not stationary. For example, if y represents wind speeds in either synoptic storms or thunderstorms, the process y should have two different constant intensity functions (rates of arrival), λ_{syn} and λ_{th}, applicable to the time intervals in which there occur synoptic storms and thunderstorms, respectively. In the case of a stationary process, for peak values y that cross a high threshold, asymptotic arguments lead to the expressions

$$\lambda(t, y) = \frac{1}{\sigma} \left[1 + \frac{k(y - \mu)}{\sigma} \right]_+^{-1-1/k} \tag{C.2}$$

$$\lambda(t, y) = \frac{1}{\sigma} \exp\left\{ \frac{-(y - \mu)}{\sigma} \right\} \tag{C.3}$$

[9]. In Eq. (C.2) the subscript "+" means that negative values of the quantity $1 + \frac{k(y-\mu)}{\sigma}$ are raised to zero. Depending upon whether $k > 0$ or $k < 0$, Eq. (C.2) is the POT equivalent of a Type II (Fréchet) or Type III (reverse Weibull) extreme value distribution, respectively. Equation (C.3) is the POT equivalent to the Type I (Gumbel) extreme value distribution; it is the limit as k approaches zero of Eq. (C.2). The POT Poisson-process procedure is designated as FpotMax if used with Eq. (C.2) and GpotMax if used with Eq. (C.3); the letters F and G stand for "full" and "Gumbel," respectively. The parameters μ and σ are, respectively, the location and scale parameters of the distribution of the peak value of $y(t)$. The volume $\lambda(t, y) dt \, dy$ is equal to the expected number of peaks per elemental area $dt \, dy$.

3) *Decluster.* Figure C.2a depicts the same raw time series as Figure C.1. Thresholded variants with the threshold $u = 1.8$ and $u = 2.0$ are depicted in Figure C.2b and Figure C.2c, respectively. In raw time series successive peaks can be separated by time intervals smaller than the time between an upcrossing of the mean and the subsequent downcrossing of the mean (see Appendix B, Figure B.3). Such successive

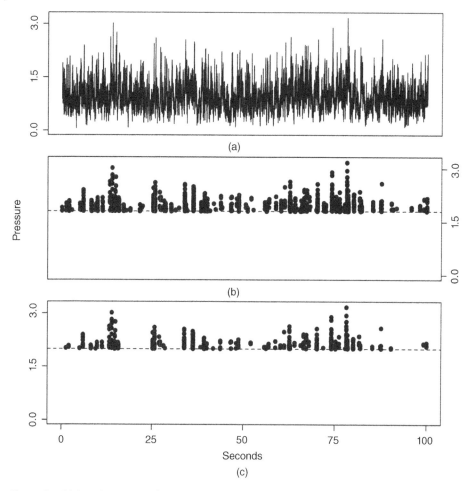

Figure C.2 (a) Raw time series, observations in raw time series, (b) above threshold $u = 1.8$, and (c) above threshold $u = 2.0$.

peaks are typically strongly correlated, as shown by Figure C.3, where it is seen that the autocorrelation function remains strong and positive for observations separated by more than 40 increments of time (in this case $40/[500\,\text{Hz}] = 0.08$ seconds). Poisson processes are not appropriate for highly autocorrelated data without further processing because of the independence assumption that underlies them.

Clusters are data blocks within time intervals defined by an upcrossing of the mean and the subsequent downcrossing of the mean (see Figure B.4). Declustering is an operation that is effective in removing the high autocorrelation from the data. It proceeds by discarding, in each cluster, all data other than the cluster maximum. Figure C.4 displays the counterparts of Figure C.2 after declustering. The estimated autocorrelation function of the data analysis of the time series in Figure C.4a shows that declustering is highly effective. After removing the autocorrelation in the series, or declustering, the use of Poisson processes as models for crossings of a high threshold is justified. They are used for such purposes in many papers, for example [8, 10–12].

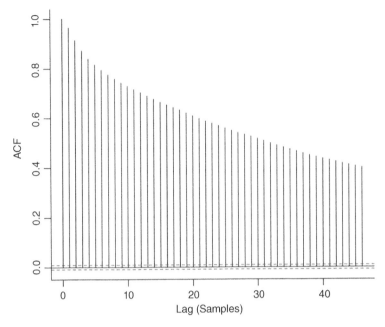

Figure C.3 Estimated autocorrelation function for the time series in Figure C.2a.

4) *Select optimal threshold.* Historically, a hurdle to the use of the POT models has been the appropriate choice of a threshold. Since the threshold dictates the data that are included in (or omitted from) the sample used to fit the model, its impact on the results can be large. The extreme value model becomes more appropriate as the threshold increases (as more non-extreme values of the variate are excluded from the sample being analyzed), but the threshold cannot be too high because too few observations will remain for fitting the model, since observations are taken over a finite period of time. Any approach to choosing a threshold must balance these competing aspects. A common and easy to implement approach – though not necessarily optimal – is to pick a high quantile of the series, e.g. 95% [13, p. 489]. The approach of [8] is superior insofar as it uses an *optimal* threshold based on the fit of the model to the data, as judged by the statistics, called W-statistics, defined in [11, Eq. (1.30)]. The W-statistic is unitless and defines a transformation of the data such that, if the Poisson-process model were perfectly correct, the transformed data would follow exactly an exponential distribution with mean one. Figure C.5 shows a plot of the ordered W-statistics versus quantiles of the standard exponential distribution using the optimal threshold for the series in Figure C.4a. If the data fitted perfectly to the model, the points would fall exactly on the diagonal line. The threshold is chosen by creating such a plot for a sequence of potential thresholds and selecting the threshold that minimizes the maximum absolute vertical distance to the diagonal line. This method for selecting the threshold is comparable to the method used in [14].

5) *Estimate model parameters.* The model parameters, $\eta = (\mu, \sigma)$ for the intensity function in Eq. (C.1) are estimated by maximum likelihood from the set of declustered

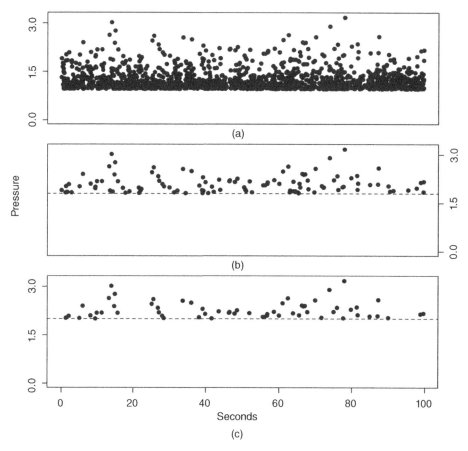

Figure C.4 (a) Declustered time series; resulting observations (b) above $u = 1.8$, and (c) above $u = 2.0$.

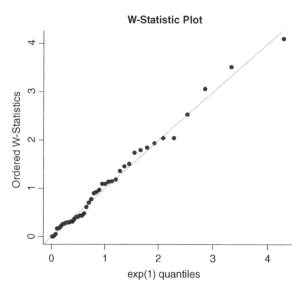

Figure C.5 Plot of the W-statistics versus their corresponding standard exponential quantiles for the declustered series depicted in Figure C.4a using the optimal threshold. Best fit of data using GpotMax with a 2.1 threshold (45 data points).

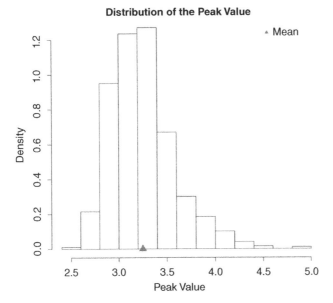

Figure C.6 Histogram of the estimated distribution of the peak value starting with the time series depicted in Figure C.4a. The triangle shows the mean of the distribution.

data corresponding to the chosen threshold. The likelihood is given in Equation (2) of [8], and maximum likelihood is discussed in, for example [15, Section 7.2.2].

6) *Empirically build the distribution of the peak by Monte Carlo simulation.* A series of desired length T, is generated from the fitted model, and the peak of the generated series is recorded. This is repeated n_{mc} times. The recorded peaks form an empirical approximation to the distribution of the peak. A histogram of the simulated peaks over 100 seconds, with $n_{mc} = 1000$ for the example data set, is shown in Figure C.6, in which the mean value is marked by the triangle.

7) *Quantify uncertainty.* The objective of the computations is to estimate the *distribution* of the peak of the time series under study. Thus, the uncertainty in the estimate of the entire distribution of the peak is being quantified, not just, for example, the uncertainty in the mean of that distribution. To accomplish this, a second layer of Monte Carlo sampling is performed. The input to step 6 was the maximum likelihood estimate of the vector η, denoted by $\hat{\eta}$. However, because only a finite sample is available, these estimates are uncertain. That uncertainty may be described using the multivariate Gaussian distribution. More specifically, one may sample values of η that are also consistent with observed time series, and repeat step 6 for those new parameter values a number n_{boot} of times. The result of step 7 is n_{boot} empirical approximations to the distribution of the peak. For clarity Figure C.7 shows only $n_{boot} = 50$ replicates of the distribution of the peak for the example data set. Typically, 1000 replicates, say, may be used. The bar shown in Figure C. 7 depicts an 80% confidence interval for the mean, which is calculated from 1000 replicates. This technique is an approximation to a bootstrap algorithm [16, 20].

Discussion of Results. The dashed line in Figure C.8 shows the peak estimated by GpotMax applied to the entire time series of duration 100 seconds. This estimate is

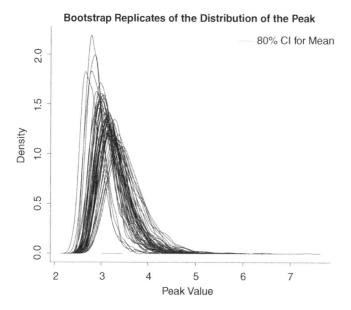

Figure C.7 Replicates of the distribution of the peak starting with the series shown in Figure C.4a. The short horizontal line shows an 80% confidence interval for the mean of the distribution of the peak value.

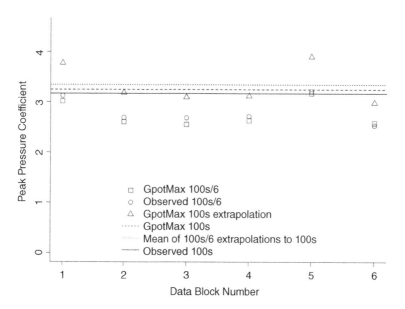

Figure C.8 Comparison of estimates based on six equal data blocks and on global analysis, using GpotMax.

close to the observed peak of the time series, shown by the solid line. The squares show the results of six analyses performed on six partitions of the same time series, each of length $100 \, s/6$. The GpotMax estimates closely track the observed peaks (i.e., the circles) for each of the partitions. For each partition, GpotMax may also be used to calculate the mean of the distribution of the peak for a duration of 100 seconds, shown by the triangles in Figure C.8. The six individual partitions can yield estimates that differ by as much as approximately 25% from the estimate based on the entire 100-second time series. However, the average of these six estimates, shown by the dashed line, is reasonably close to the global estimate and the observed 100-second peak.

As noted earlier, the full version of the algorithm based on Eq. (C.1), referred to as FpotMax, does not assume that the tail length parameter of the distribution of peaks is zero. It is shown in [17] that the estimates by GpotMax and FpotMax of the distributions of the peaks are similar for five representative pressure taps of the building model examined herein, and are close to the observed peaks. GpotMax, rather than FpotMax, may therefore be used in practice, unless there were one or two very large peaks relative to other threshold crossings.

C.3   Dependence of Peak Estimates by BLUE Upon Number of Partitions

Peaks were estimated using the epochal method for two probabilities of non-exceedance, $p = 0.78$ and $p = 0.5704$. The latter corresponds to the mean of the Gumbel distribution, while the former is commonly used by wind tunnel operators [18] and is close to the number 0.80 specified in the ISO 4354 [19]. For a number of partitions $10 \leq n \leq 24$ the estimated peaks for tap 708, wind direction $360°$, varied between 3.72 and 4.20 for $p = 0.78$, and between 3.48 and 3.82 for $p = 0.57$. For comparison, the single GpotMax and FpotMax estimates were 3.41 and 3.35, respectively, and the observed peak was 3.24.

C.4   Summary

Current procedures for estimating peaks of pressure time series have drawbacks that motivated the development of the new procedure, one advantage of which is that it typically results in an extreme value data set larger than is the case for epochal procedures. The translation procedure has the drawback that it depends upon the estimate of the marginal distribution of a non-Gaussian time series, which is typically difficult to perform reliably. The epochal procedure used in conjunction with the BLUE estimation of the Gumbel parameters depends, in some cases very significantly, upon the number of partitions being used.

The procedure described in this Appendix is based on a Poisson process model for quantities y that exceed a specified threshold u of the time series being considered. The estimate depends upon the choice of the threshold. A criterion is available that allows the analyst to make an optimal choice (according to a specified metric) of the threshold value.

Two versions of the proposed procedure are available. One version, denoted by Fpot-Max, includes estimation of a tail length parameter resulting in a tail of the Fréchet or the reverse Weibull distribution type. The second version, denoted by GpotMax, assumes that the tail length parameter vanishes, resulting in a tail of the Gumbel distribution type. Typically GpotMax results in fully satisfactory estimates and should in practice be used for structural design applications, which include the analysis of wind speed time series, of time series of pressure coefficients, or of wind effects such as internal forces, demand-to-capacity indexes, inter-story drift, and floor accelerations.

References

1 Sadek, F. and Simiu, E. (2002). Peak non-Gaussian wind effects for database-assisted low-rise building design. *Journal of Engineering Mechanics* 128 (5): 530–539. https://www.nist.gov/wind.

2 Eaton, K.J. and Mayne, J.R. (1975). The measurement of wind pressures on two-story houses at Aylesbury. *Journal of Industrial Aerodynamics* 1: 67–109.

3 Simiu, E., Pintar, A.L., Duthinh, D., and Yeo, D. (2017). Wind load factors for use in the wind tunnel procedure. *ASCE-ASME Journal of Risk and Uncertainty in Engineering Systems, Part A, Civil Engineering* 3 (4): 04017007. https://www.nist.gov/wind.

4 Lieblein, J., *Efficient Methods of Extreme Value Methodology*, NBSIR74-602, National Bureau of Standards, Washington, DC. 1974. https://www.nist.gov/wind.

5 Core Team, R. (2015). *R: A Language and Environment for Statistical Computing*. Vienna, Austria, Available: www.R-project.org: R Foundation for Statistical Computing.

6 Pintar, A., "potMax – Estimating the distribution of the maximum of a time series using peaks-over-threshold models." Available: https://github.com/usnistgov/potMax, (Aug. 16, 2017), 2016.

7 Ho, T.C.E., Surry, D., Morrish, D., and Kopp, G.A. (2005). The UWO contribution to the NIST aerodynamic database for wind loads on low buildings. Part 1: archiving format and basic aerodynamic data,. *Journal of Wind Engineering and Industrial Aerodynamics* 93 (1): 1–30. (For the aerodynamic data see https://www.nist.gov/wind).

8 Pintar, A. L., Simiu, E., Lombardo, F. T., and Levitan, M., *Maps of Non-Hurricane Non-Tornadic Winds Speeds With Specified Mean Recurrence Intervals for the Contiguous United States Using a Two-Dimensional Poisson Process Extreme Value Model and Local Regression*, NIST Special Publication 500–301, https://nvlpubs.nist.gov/nistpubs/SpecialPublications/NIST.SP.500-301.pdf, 2015.

9 Pickands, J. III, The two-dimensional Poisson process and extremal processes. *Journal of Applied Probability* 8: 745–756, 1971.

10 Smith, R.L. (1989). Extreme value analysis of environmental time series: an application to trend detection in ground-level Ozone. *Statistical Science* 4: 367–393.

11 Smith, R.L. (2004). Statistics of extremes, with applications in environment, insurance, and finance. In: *Extreme Values in Finance, Telecommunications, and the Environment* (ed. B. Finkenstädt and H. Rootzén), 1–78. Chapman & Hall/CRC, Ch. 1.

12 Coles, S. (2004). The use and misuse of extreme value models in practice. In: *Extreme Values in Finance, Telecommunications, and the Environment* (ed. B. Finkenstädt and H. Rootzén), 79–100. Chapman & Hall/CRC, Ch. 2.

13 Mannshardt-Shamseldin, E.C., Smith, R.L., Sain, S.R. et al. (2010). Downscaling extremes: a comparison of extreme value distributions in point-source and gridded precipitation data. *The Annals of Applied Statistics* 4: 484–502.

14 Pickands, J. III, "Bayes quantile estimation and threshold selection for the Generalized Pareto family," in *Proceedings of the Conference on Extreme Value Theory and Applications, Gaithersburg, MD, 1993*, J. Galambos, J. Lechner, and E. Simiu, ed., Boston, MA: Kluwer Academic Publishers, 1994.

15 Casella, G. and Berger, R.L. (2002). *Statistical Inference*, vol. 2. Pacific Grove, CA: Duxbury.

16 Yeo, D. (2014). Generation of large directional wind speed data sets for estimation of wind effects with long return periods. *Journal of Structural Engineering* 140: 04014073. https://www.nist.gov/wind.

17 Duthinh, D., Pintar, A.L., and Simiu, E. (2017). Estimating peaks of stationary random processes: a peaks-over-threshold approach. *ASCE – ASME Journal of Risk and Uncertainty in Engineering Systems, Part A, Civil Engineering* 3 (4): 04017028. https://www.nist.gov/wind.

18 Peng, X., Yang, L., Gurley, K., Prevatt, D., and Gavanski, E. "Prediction of peak wind loads on low-rise building," *Proceedings of the 12th Americas Conference on Wind Engineering, Seattle, WA*, 2013.

19 International Standard, ISO 4354 (2009-06-01), 2nd ed., "Wind actions on structures," Annex D (informative) "Aerodynamic pressure and force coefficients," Geneva, Switzerland, p. 22.

20 Efron, B. and Tibshirani, R.J. *An Introduction to the Bootstrap*. CRC Press, 1994.

Appendix D

Structural Dynamics

Frequency-Domain Approach

D.1   Introduction

The mathematical model for wind-induced dynamic response is Newton's second law, that is, an ordinary second-order differential equation. In Part II of the book, the solution to this equation was obtained by time-domain methods. This approach is currently feasible because (i) forcing functions can be obtained as functions of time from simultaneously measured aerodynamic time histories, and (ii) computer capabilities allow the ready solution of the differential equations of motion of the dynamical systems of interest. Neither of these two capabilities was available until relatively recently. For this reason, the differential equations were transformed via Fourier transformation into more tractable algebraic functions in the frequency domain, and forcing functions were thus defined via spectral and cross-spectral densities. Frequency-domain solutions of structural dynamics problems remain useful for certain applications and can provide helpful insights into wind-induced structural dynamics.

Section D.2 presents the building blocks of the frequency-domain approach for the single-degree-of-freedom system. Section D.3 presents basic results obtained for continuously distributed linear systems. Section D.4 is an interesting application of those results: the determination of the along-wind response of a tall building with rectangular shape in plan to wind normal to one of its faces.

D.2   The Single-Degree-of-Freedom Linear System

Consider the single-degree-of-freedom motion of a particle of mass M subjected to a time-dependent force $F(t)$. The particle is restrained by an elastic spring with stiffness k. Its motion is damped by a viscous damper with coefficient c. The particle's displacement $x(t)$ is opposed by (i) a restoring force $-kx$ and (ii) a damping force $-c\, dx/dt \equiv -c\dot{x}$, where the stiffness k and the damping coefficient c are assumed to be constant. Newton's second law states that the product of the particle's mass by its acceleration, $M\ddot{x}$, equals the total force applied to the particle. The equation of motion of the system is

$$M\ddot{x} = -c\dot{x} - kx + F(t) \tag{D.1}$$

With the notations $n_1 = \sqrt{k/M}/(2\pi)$ and $\zeta_1 = c/(2\sqrt{kM})$, where n_1 denotes the frequency of vibration of the oscillator[1], and ζ_1 is the damping ratio (i.e., the ratio of the

1 The quantity $2\pi n$ is called *circular frequency* and is commonly denoted by ω.

Wind Effects on Structures: Modern Structural Design for Wind, Fourth Edition. Emil Simiu and DongHun Yeo.
© 2019 John Wiley & Sons Ltd. Published 2019 by John Wiley & Sons Ltd.

damping c to the critical damping $\zeta_{cr} = 2\sqrt{kM}$ beyond which the system's motion would no longer be oscillatory), Eq. (D.1) becomes

$$\ddot{x} + 2\zeta(2\pi n_1)\dot{x} + (2\pi n_1)^2 x = \frac{F(t)}{M} \qquad (D.2)$$

For structures, ζ_1 is typically small (in the order of 1%).

D.2.1 Response to a Harmonic Load

In the particular case of a harmonic load $F(t) = F_0 \cos 2\pi n t$, it can be verified by substitution that the steady-state solution of Eq. (D.2) is

$$x(t) = H(n)F_0 \cos(2\pi n t - \theta) \qquad (D.3)$$

where

$$H(n) = \frac{1}{4\pi^2 n_1^2 M\{[1 - (n/n_1)^2]^2 + 4\zeta_1^2(n/n_1)^2\}^{1/2}} \qquad (D.4)$$

$$\theta = \tan^{-1}\frac{2\zeta_1(n/n_1)}{1 - (n/n_1)^2} \qquad (D.5)$$

The quantity θ is the *phase angle*, and $H(n)$ is the system's *mechanical admittance function* (or *mechanical amplification factor*). For $n = n_1$, that is, if the frequency of the harmonic forcing function coincides with the frequency of vibration of the oscillator, the amplitude of the response is largest, and is inversely proportional to the damping ratio ζ_1. In this case, the motion exhibits *resonance*. In the particular case $F(t) = F_0 \sin 2\pi n t$, the steady state response can be written as

$$x(t) = H(n)F_0 \sin(2\pi n t - \theta) \qquad (D.6)$$

D.2.2 Response to an Arbitrary Load

Let the system described by Eq. (D.2) be subjected to the action of a load equal to the unit impulse function $\delta(t)$ acting at time $t = 0$, that is, to a load defined as follows (Figure D.1):

$$\delta(t) = 0 \qquad\qquad \text{for } t \neq 0 \qquad (D.7)$$

$$\lim_{\Delta t \to 0} \int_0^{\Delta t} \delta(t)dt = 1 \quad \text{for } t = 0. \qquad (D.8)$$

The response of the system to the load $\delta(t)$ depends on time and is denoted by $G(t)$.

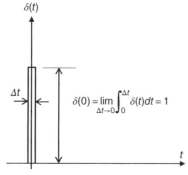

Figure D.1 Unit impulse function.

$\delta(t)$

Δt

$$\delta(0) = \lim_{\Delta t \to 0} \int_0^{\Delta t} \delta(t)dt = 1$$

t

Figure D.2 Load $F(t)$.

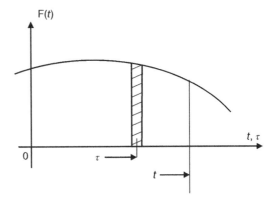

An arbitrary load $F(t)$ (Figure D.2) may be described as a sum of elemental impulses of magnitude $F(\tau')\,d\tau'$ each acting at time τ'. Since the system is linear, the response at time t to each such impulse is $G(t-\tau')F(\tau')\,d\tau'$. The total response is

$$x(t) = \int_{-\infty}^{t} G(t-\tau')F(\tau')d\tau' \tag{D.9}$$

The limits of the integral indicate that all the elemental impulses that have acted before time t have been taken into account. Denoting $\tau = t - \tau'$, Eq. (D.9) becomes

$$x(t) = \int_{0}^{\infty} G(\tau)F(t-\tau)d\tau \tag{D.10}$$

Let $F(t) = F_0 \cos 2\pi nt$. It follows from Eqs. (D.3) and (D.10) that

$$H(n)\cos\theta = \int_{0}^{\infty} G(\tau)\cos 2\pi n\tau d\tau \tag{D.11a}$$

$$H(n)\sin\theta = \int_{0}^{\infty} G(\tau)\sin 2\pi n\tau d\tau \tag{D.11b}$$

Equations (D.11a) and (D.11b) yield Eqs. (D.12a) and (D.12b), whose summation yields Eq. (D.13):

$$H^2(n)\cos^2\theta = \int_{0}^{\infty}\int_{0}^{\infty} G(\tau_1)\cos 2\pi n\tau_1 G(\tau_2)\cos 2\pi n\tau_2 d\tau_1 d\tau_2 \tag{D.12a}$$

$$H^2(n)\sin^2\theta = \int_{0}^{\infty}\int_{0}^{\infty} G(\tau_1)\sin 2\pi n\tau_1 G(\tau_2)\sin 2\pi n\tau_2 d\tau_1 d\tau_2 \tag{D.12b}$$

$$H^2(n) = \int_{0}^{\infty}\int_{0}^{\infty} G(\tau_1)G(\tau_2)\cos 2\pi n\,(\tau_1 - \tau_2)d\tau_1 d\tau_2 \tag{D.13}$$

D.2.3 Response to a Stationary Random Load

Now let the load $F(t)$ be a stationary process with spectral density $S_F(n)$. Using Eqs. (B.20), (B.21), and (D.10), we obtain the spectral density of the system response as follows:

$$S_x(n) = 2\int_{-\infty}^{\infty} R_x(\tau)\cos 2\pi n\tau d\tau$$

$$= 2 \int_{-\infty}^{\infty} \left[\lim_{T \to \infty} \frac{1}{T} \int_{-T/2}^{T/2} x(t) x(t + \tau) dt \right] \cos 2\pi n\tau d\tau$$

$$= 2 \int_{-\infty}^{\infty} \left\{ \lim_{T \to \infty} \frac{1}{T} \int_{-T/2}^{T/2} dt \left[\int_0^{\infty} G(\tau_1) F(t - \tau_1) d\tau_1 \right. \right.$$

$$\left. \times \int_0^{\infty} G(\tau_2) F(t + \tau - \tau_2) d\tau_2 \right] \right\} \cos 2\pi n\tau d\tau$$

$$= 2 \int_0^{\infty} G(\tau_1) \left\{ \int_0^{\infty} G(\tau_2) \left[\int_{-\infty}^{\infty} R_F(\tau + \tau_1 - \tau_2) \cos 2\pi n\tau d\tau \right] d\tau_2 \right\} d\tau_1$$

$$= 2 \int_0^{\infty} \int_0^{\infty} G(\tau_1) G(\tau_2) \cos 2\pi n(\tau_1 - \tau_2) d\tau_1 d\tau_2$$

$$\times \int_{-\infty}^{\infty} R_F(\tau + \tau_1 - \tau_2) \cos 2\pi n(\tau + \tau_1 - \tau_2) d(\tau + \tau_1 - \tau_2)$$

$$+ 2 \int_0^{\infty} \int_0^{\infty} G(\tau_1) G(\tau_2) \sin 2\pi n(\tau_1 - \tau_2) d\tau_1 d\tau_2$$

$$\times \int_{-\infty}^{\infty} R_F(\tau + \tau_1 - \tau_2) \sin 2\pi n(\tau + \tau_1 - \tau_2) d(\tau + \tau_1 - \tau_2) \tag{D.14}$$

where, in the last step, the following identity is used:

$$\cos 2\pi n\tau \equiv \cos 2\pi n[(\tau + \tau_1 - \tau_2) - (\tau_1 - \tau_2)] \tag{D.15}$$

From Eqs. (B.20), (B.23), (D.12a,b), and (D.13), there follows

$$S_x(n) = H^2(n) S_F(n) \tag{D.16}$$

This relation between frequency-domain forcing and response is useful in applications.

D.3 Continuously Distributed Linear Systems

D.3.1 Normal Modes and Frequencies: Generalized Coordinates, Mass and Force

D.3.1.1 Modal Equations of Motion

A linearly elastic structure with continuously distributed mass per unit length $m(z)$ and low damping can be shown to vibrate in resonance with the exciting force if the latter has certain sharply defined frequencies called the structure's *natural frequencies of vibration*. Associated with each natural frequency is a *mode*, or *modal shape*, of the vibrating structure. The first four normal modes $x_i(z)$ ($i = 1, 2, 3, 4$) of a vertical cantilever beam with running coordinate z are shown in Figure 11.3. The natural modes and frequencies are structural properties independent of the loads.

A deflection $x(z,t)$ along a principal axis of a continuous system, due to time-dependent forcing, can in general be written in the form

$$x(z, t) = \sum_i x_i(z) \xi_i(t) \tag{D.17}$$

where the functions $\xi_i(t)$ are called the *generalized coordinates* of the system, and $x_i(z)$ denotes the modal shape in the ith mode of vibration. For a building, similar expressions hold for deflections $y(z,t)$ in the direction of its second principal axis, and for horizontal torsional angles $\varphi(z,t)$. For structures whose centers of mass and elastic centers do not coincide, the x, y, and φ motions are coupled, as is shown in Section D.4, which presents the development of the equations of motion for this general case.

In this section, we limit ourselves to presenting the modal equations of motion corresponding to the particular case of translational motion along a principal axis x:

$$\ddot{\xi}_i(t) + 2\zeta_i(2\pi n_i)\dot{\xi}_i(t) + (2\pi n_i)^2 \xi_i(t) = \frac{Q_i(t)}{M_i} \quad (i = 1, 2, 3, \ldots) \tag{D.18}$$

where ζ_i, n_i, M_i, and Q_i are the ith mode damping ratio, natural frequency, generalized mass, and generalized force, respectively,

$$M_i = \int_0^H [x_i(z)]^2 m(z)dz \tag{D.19}$$

$$Q_i = \int_0^H p(z,t)x_i(z)dz \tag{D.20}$$

where $m(z)$ is the mass of the structure per unit length, $p(z, t)$ is the load acting on the structure per unit length, and H is the structure's height. For a concentrated load acting at $z = z_1$,

$$p(z,t) = F(t)\delta(z - z_1) \tag{D.21}$$

where $\delta(z - z_1)$ is defined, with a change of variable, as in Eq. (D.8),

$$Q_i(t) = \lim_{\Delta z \to 0} \int_{z_1}^{z_1 + \Delta z} p(z,t)x_i(z)dz$$

$$= x_i(z_1)F(t) \tag{D.22}$$

D.3.2   Response to a Concentrated Harmonic Load

If a concentrated load

$$F(t) = F_0 \cos 2\pi nt \tag{D.23}$$

is acting on the structure at a point of coordinate z_1, by virtue of Eq. (D.22) the generalized force in the ith mode is

$$Q_i(t) = F_0 x_i(z_1) \cos 2\pi nt \tag{D.24}$$

and the steady-state solutions of Eq. (D.18) are similar to the solution Eq. (D.3) of a single-degree-of-freedom system under a harmonic load:

$$\xi_i(t) = F_0 x_i(z_1)H_i(n) \cos(2\pi nt - \theta_i) \tag{D.25}$$

where

$$H_i(n) = \frac{1}{4\pi^2 n_i^2 M_i\{[1 - (n/n_i)^2]^2 + 4\zeta_i^2(n/n_i)^2\}^{1/2}} \tag{D.26}$$

$$\theta_i = \tan^{-1}\frac{2\zeta_i(n/n_i)}{1 - (n/n_i)^2} \tag{D.27}$$

The response of the structure at a point of coordinate z is then

$$x(z, t) = F_0 \sum_i x_i(z)x_i(z_1)H_i(n) \cos(2\pi nt - \theta_i) \tag{D.28}$$

It is convenient to write Eq. (D.28) in the form

$$x(z, t) = F_0 H(z, z_1, n) \cos[2\pi nt - \theta(z, z_1, n)] \tag{D.29}$$

where, as follows immediately from Eqs. (B.4a) and (B.4b),

$$H(z, z_1, n) = \left\{ \left[\sum_i x_i(z)x_i(z_1)H_i(n) \cos \theta_i \right]^2 + \left[\sum_i x_i(z)x_i(z_1)H_i(n) \sin \theta_i \right]^2 \right\}^{1/2} \tag{D.30}$$

$$\theta(z, z_1, n) = \tan^{-1} \frac{\sum\limits_i x_i(z)x_i(z_1)H_i(n) \sin \theta_i}{\sum\limits_i x_i(z)x_i(z_1)H_i(n) \cos \theta_i} \tag{D.31}$$

Similarly, the steady state response at a point of coordinate z to a concentrated load

$$F(t) = F_0 \sin 2\pi nt \tag{D.32}$$

acting at a point of coordinate z_1 can be written as

$$x(z, t) = F_0 H(z, z_1, n) \sin[2\pi nt - \theta(z, z_1, n)] \tag{D.33}$$

D.3.3 Response to a Concentrated Stationary Random Load

Let the response at a point of coordinate z to a concentrated unit impulsive load $\delta(t)$ acting at time $t = 0$ at a point of coordinate z_1 be denoted $G(z, z_1, t)$. Following the same reasoning that led to Eq. (D.10), the response $x(z,t)$ to an arbitrary load $F(t)$ acting at a point of coordinate z_1 is

$$x(z, t) = \int_0^\infty G(z, z_1, \tau)F(t - \tau)d\tau \tag{D.34}$$

Note the complete similarity of Eqs. (D.29), (D.33), and (D.34) to Eqs. (D.3), (D.6), and (D.10), respectively. Therefore, the same steps that led to Eq. (D.16) yield the relation between the spectra of the random forcing and the response:

$$S_x(z, z_1, n) = H^2(z, z_1, n)S_F(n) \tag{D.35}$$

D.3.4 Response to Two Concentrated Stationary Random Loads

Let $x(z,t)$ denote the response at a point of coordinate z to two stationary loads $F_1(t)$ and $F_2(t)$ acting at points with coordinates z_1 and z_2, respectively. The autocovariance of $x(z,t)$ is (see Eq. [B.21]):

$$R_x(z, \tau) = \lim_{T \to \infty} \frac{1}{T} \int_{-T/2}^{T/2} x(z, t)x(z, t + \tau)dt$$

$$= \lim_{T \to \infty} \frac{1}{T} \int_{-T/2}^{T/2} \left[\int_0^\infty G(z, z_1, \tau_1) F_1(t - \tau_1) d\tau_1 \right.$$

$$+ \int_0^\infty G(z, z_2, \tau_1) F_2(t - \tau_1) d\tau_1 \bigg]$$

$$\times \left[\int_0^\infty G(z, z_1, \tau_2) F_1(t + \tau - \tau_2) d\tau_2 + \int_0^\infty G(z, z_2, \tau_2) F_2(t + \tau - \tau_2) d\tau_2 \right] dt$$

$$= \int_0^\infty G(z, z_1, \tau_1) \left[\int_0^\infty G(z, z_1, \tau_2) R_{F_1}(\tau + \tau_1 - \tau_2) d\tau_2 \right] d\tau_1$$

$$+ \int_0^\infty G(z, z_2, \tau_1) \left[\int_0^\infty G(z, z_2, \tau_2) R_{F_2}(\tau + \tau_1 - \tau_2) d\tau_2 \right] d\tau_1$$

$$+ \int_0^\infty G(z, z_1, \tau_1) \left[\int_0^\infty G(z, z_2, \tau_2) R_{F_1 F_2}(\tau + \tau_1 - \tau_2) d\tau_2 \right] d\tau_1$$

$$+ \int_0^\infty G(z, z_2, \tau_1) \left[\int_0^\infty G(z, z_1, \tau_2) R_{F_1 F_2}(\tau + \tau_1 - \tau_2) d\tau_2 \right] d\tau_1 \tag{D.36}$$

The spectral density of the displacement $x(z,t)$ is

$$S_x(z, n) = 2 \int_{-\infty}^\infty R_x(z, \tau) \cos 2\pi n\tau d\tau$$

$$= 2 \int_{-\infty}^\infty R_x(z, \tau) \cos 2\pi n[(\tau + \tau_1 - \tau_2) - (\tau_1 - \tau_2)] d(\tau + \tau_1 - \tau_2) \tag{D.37}$$

Substitute the right-hand side of Eq. (D.36) for $R_x(z, \tau)$ in Eq. (D.37). Using the relations

$$H(z, z_i, n) \cos \theta(z, z_i, n) = \int_0^\infty G(z, z_i, \tau) \cos 2\pi n\tau d\tau \tag{D.38a}$$

$$H(z, z_i, n) \sin \theta(z, z_i, n) = \int_0^\infty G(z, z_i, \tau) \sin 2\pi n\tau d\tau \tag{D.38b}$$

(which are similar to Eqs. [D.11a] and [D.11b]), and

$$H(z, z_1, n) H(z, z_2, n) \cos[\theta(z, z_1, n) - \theta(z, z_2, n)]$$

$$= \int_0^\infty \int_0^\infty G(z, z_1, \tau_1) G(z, z_2, \tau_2) \cos 2\pi n(\tau_1 - \tau_2) d\tau_1 d\tau_2 \tag{D.39a}$$

$$H(z, z_1, n) H(z, z_2, n) \sin[\theta(z, z_1, n) - \theta(z, z_2, n)]$$

$$= \int_0^\infty \int_0^\infty G(z, z_1, \tau_1) G(z, z_2, \tau_2) \sin 2\pi n(\tau_1 - \tau_2) d\tau_1 d\tau_2 \tag{D.39b}$$

which are derived from Eqs. (D.38a) and (D.38b), and following the steps that led to Eq. (D.16), there results

$$S_x(z, n) = H^2(z, z_1, n) S_{F_1}(n) + H^2(z, z_2, n) S_{F_2}(n)$$

$$+ 2H(z, z_1, n) \ H(z, z_2, n) \{ S_{F_1 F_2}^C(n) \cos[\theta(z, z_1, n) - \theta(z, z_2, n)]$$

$$+ S_{F_1 F_2}^Q(n) \sin[\theta(z, z_1, n) - \theta(z, z_2, n)] \} \tag{D.40}$$

where $S^C_{F_1F_2}(n)$ and $S^Q_{F_1F_2}(n)$ are the co-spectrum and quadrature spectrum of the forces $F_1(t)$ and $F_2(t)$, defined by Eqs. (B.33) and (B.34), respectively.

D.3.5 Effect of the Correlation of the Loads upon the Magnitude of the Response

Let two stationary random loads $F_1(t) \equiv F_2(t)$ act at points of coordinates z_1 and z_2, respectively. The loads $F_1(t)$ and $F_2(t)$ are *perfectly correlated*. By definition, in this case $S^C_{F_1F_2}(n) = S_{F_1}(n)$, and $S^Q_{F_1F_2}(n) = 0$ (Eqs. [B.21] and [B.29], [B.20] and [B.33]; [B.23] and [B.34]). From Eq. (D.40),

$$S_x(z, n) = \{H^2(z, z_1, n) + H^2(z, z_2, n)$$
$$+ 2H(z, z_1, n)\,H(z, z_2, n)\cos[\theta(z, z_1, n) - \theta(z, z_2, n)]\}S_{F_1}(n) \tag{D.41}$$

If $z_1 = z_2$,

$$S_x(z, n) = 4H^2(z, z_1, n)S_{F_1}(n) \tag{D.42}$$

Consider now two loads $F_1(t)$ and $F_2(t)$ for which the cross-covariance $R_{F_1F_2}(\tau) = 0$. Then, by Eqs. (B.33) and (B.34),

$$S^C_{F_1F_2}(n) = S^Q_{F_1F_2}(n) = 0 \tag{D.43}$$

and, if $S_{F_1}(n) \equiv S_{F_2}(n)$,

$$S_x(z, n) = [H^2(z, z_1, n) + H^2(z, z_2, n)]\,S_{F_1}(n) \tag{D.44}$$

If $z_1 = z_2$,

$$S_x(z, n) = 2H^2(z, z_1, n)\,S_{F_1}(n) \tag{D.45}$$

The spectrum of the response to the action of the two uncorrelated loads is in this case only half as large as in the case of the perfectly correlated loads.

D.3.6 Distributed Stationary Random Loads

The spectral density of the response to a distributed stationary random load can be obtained by generalizing Eq. (D.40) to the case where an infinite number of elemental loads, rather than two concentrated loads, are acting on the structure. Thus, if the load is distributed over an area A, and if it is noted that in the absence of torsion the mechanical admittance functions are independent of the across-wind coordinate y, the spectral density of the along-wind fluctuations may be written as

$$S_x(z, n) = \int_A \int_A H(z, z_1, n)H(z, z_2, n)\{S^C_{p'_1p'_2}(n)\cos[\theta(z, z_1, n) - \theta(z, z_2, n)]$$
$$+ S^Q_{p'_1p'_2}(n)\sin[\theta(z, z_1, n) - \theta(z, z_2, n)]\}dA_1A_2 \tag{D.46}$$

where p'_1 and p'_2 denote pressures acting at points of coordinates (y_1, z_1) and (y_2, z_2), respectively. It can be verified that from Eq. (D.46) there follows[2]

$$
S_x(z, n) = \frac{1}{16\pi^4} \sum_i \sum_j \frac{x_i(z)x_j(z)}{n_i^2 n_j^2 M_i M_j}
$$

$$
\times \frac{1}{\{[1 - (n/n_i)^2]^2 + 4\zeta_i^2(n/n_i)^2\}\{[1 - (n/n_j)^2]^2 + 4\zeta_j^2(n/n_j)^2\}}
$$

$$
\times \left[\left\{\left[1 - \left(\frac{n}{n_i}\right)^2\right]\left[1 - \left(\frac{n}{n_j}\right)^2\right] + 4\zeta_i\zeta_j\frac{n}{n_i}\frac{n}{n_j}\right\}\right.
$$

$$
\times \int_A\int_A x_i(z_1)x_j(z_2)S^C_{p'_1p'_2}(n)dA_1 dA_2
$$

$$
+ \left\{2\zeta_j\frac{n}{n_j}\left[1 - \left(\frac{n}{n_i}\right)^2\right] - 2\zeta_i\frac{n}{n_i}\left[1 - \left(\frac{n}{n_j}\right)^2\right]\right\}
$$

$$
\left. \times \int_A\int_A x_i(z_1)x_j(z_2)S^Q_{p'_1p'_2}(n)dA_1 dA_2 \right] \tag{D.47}
$$

If the damping is small and the resonant peaks are well separated, the cross-terms in Eq. (D.47) become negligible, and

$$
S_x(z, n) = \sum_i \frac{x_i^2(z)\int_A\int_A x_i(z_1)x_i(z_2)S^C_{p'_1p'_2}(n)dA_1 dA_2}{16\pi^4 n_i^4 M_i^2\{[1 - (n/n_i)^2]^2 + 4\zeta_i^2(n/n_i)^2\}} \tag{D.48}
$$

D.4 Example: Along-Wind Response

To illustrate the application of the material presented in Section D.3.6, we consider the along-wind response of tall buildings subjected to pressures per unit area $p(y, z, t) = \bar{p}(z) + p'(y, z, t)$ (Figure D.3).

Mean Response. The along-wind deflection induced by the mean pressures $\bar{p}(z)$ is

$$
\bar{x}(z) = B\sum_i \frac{\int_0^H \bar{p}(z)x_i(z)dz}{4\pi^2 n_i^2 M_i}x_i(z) \tag{D.49}
$$

Consider the case of loading induced by wind with longitudinal speed $U(z,t) = \bar{U}(z) + u(z, t)$ normal to a building face. The sum of the mean pressures $\bar{p}(z)$ acting on the windward and leeward faces of the building is then

$$
\bar{p}(z) = \tfrac{1}{2}\rho(C_w + C_l)B\bar{U}^2(z) \tag{D.50}
$$

---

2 By using Eqs. (D.30) and (D.31), (D.26) and (D.27), and (B.4a,b). For a derivation of Eq. (D.47) in terms of complex variables, see [1].

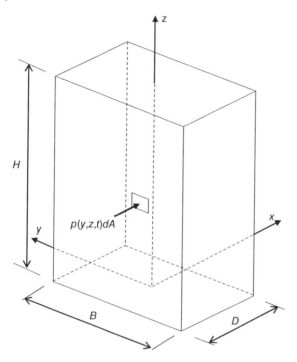

Figure D.3 Schematic view of a building.

where ρ is the air density; C_w and C_l are the values, averaged over the building width B, of the mean positive pressure coefficient on the windward face and of the negative pressure coefficient on the leeward face, respectively; and $\overline{U}(z)$ is the mean wind speed at elevation z in the undisturbed oncoming flow. Equation (D.49) then becomes

$$\overline{x}(z) = \frac{1}{2}\rho(C_w + C_l)B\sum_i \frac{\int_0^H \overline{U}^2(z)x_i(z)dz}{4\pi^2 n_i^2 M_i}x_i(z) \tag{D.51}$$

Fluctuating Response: Deflections and Accelerations. The co-spectrum of the pressures at points M_1, M_2 of coordinates (y_1, z_1), (y_2, z_2), respectively, can be written as

$$S_{p_1'p_2'}^C = S_{p'}^{1/2}(z_1, n)S_{p'}^{1/2}(z_2, n)\mathrm{Coh}(y_1, y_2, z_1, z_2, n)N(n) \tag{D.52}$$

where $S_{p'}^{1/2}(z, n)$ is the spectral density of the fluctuating pressures at point P_i ($i = 1, 2$), $\mathrm{Coh}(y_1, y_2, z_1, z_2, n)$ is the coherence of pressures, both of which are acting on the same building face, and $N(n)$ is the coherence of pressures, one of which is acting on the windward face, while the other is acting on the leeward face of the building. By definition, if both P_1 and P_2 are on the same building face, $N(n) \equiv 1$. Since

$$p(z, t) \approx \frac{1}{2}\rho C[\overline{U}(z) + u(z, t)]^2 \tag{D.53}$$

where C (which is equal to C_w or C_l, depending upon whether the pressure acts on the windward or leeward face) is the average pressure coefficient,

$$S_{p'}(z_i, n) \simeq \rho^2 C^2 \overline{U}^2(z_i)S_u(z_i, n) \tag{D.54}$$

where we used the fact that u^2 is small in relation to $2\overline{U}(z)u(z)$.

Equation (D.48) then becomes

$$S_x(z, n) \approx \frac{\rho^2}{16\pi^4} \sum_i \frac{x_i^2(z)[C_w^2 + 2C_w C_l N(n) + C_l^2]}{n_i^4 M_i^2 \{[1 - (n/n_i)^2]^2 + 4\zeta_i^2(n/n_i)^2\}}$$

$$\times \int_0^B \int_0^B \int_0^H \int_0^H x_i(z_1)x_i(z_2)\overline{U}(z_1)\overline{U}(z_2)$$

$$\times S_u^{1/2}(z_1)S_u^{1/2}(z_2)\text{Coh}(y_1, y_2, z_1, z_2, n)dy_1 dy_2 dz_1 dz_2 \tag{D.55}$$

The coherence $\text{Coh}(y_1, y_2, z_1, z_2, n)$ may be expressed as in Chapter 2. A simple, tentative expression for the function $N(n)$, a measure of the coherence between pressures on the windward and leeward faces, is:

$$N(n) = 1 \quad \text{for } n\overline{U}(z)/D < 0.2 \tag{D.56a}$$

$$N(n) = 0 \quad \text{for } n\overline{U}(z)/D \geq 0.2 \tag{D.56b}$$

where D is the depth of the building (Figure D.3).

The mean square value of the fluctuating along-wind deflection is (Eq. [B.15])

$$\sigma_x^2(z) = \int_0^\infty S_x(z, n)dn \tag{D.57}$$

From Eq. (B.16b) it follows that the mean square value of the along-wind acceleration is

$$\sigma_{\ddot{x}}^2(z) = 16\pi^4 \int_0^\infty n^4 S_x(z, n)dn \tag{D.58}$$

The expected value of the largest peak of the fluctuating along-wind deflection occurring in the time interval T is

$$x_{max} = K_x(z)\sigma_x(z) \tag{D.59}$$

where (see Eqs. [B.52] and [B.47])

$$K_x(z) = [2\ln v_x(z)T]^{1/2} + \frac{0.577}{[2\ln v_x(z)T]^{1/2}} \tag{D.60}$$

$$v_x(z) = \left[\frac{\int_0^\infty n^2 S_x(z, n)dn}{\int_0^\infty S_x(z, n)dn}\right]^{1/2} \tag{D.61}$$

Similarly, the largest peak of the along-wind acceleration is, approximately,

$$\ddot{x}_{max}(z) = K_{\ddot{x}}(z)\sigma_{\ddot{x}}(z) \tag{D.62}$$

$$K_{\ddot{x}}(z) = [2\ln v_{\ddot{x}}(z)T]^{1/2} + \frac{0.577}{[2\ln v_{\ddot{x}}(z)T]^{1/2}} \tag{D.63}$$

$$v_{\ddot{x}}(z) = \left[\frac{\int_0^\infty n^6 S_x(z, n)dn}{\int_0^\infty n^4 S_x(z, n)dn}\right]^{1/2} \tag{D.64}$$

It can be shown that the mean square value of the deflection may be written, approximately, as a sum of two terms: the "background term" that entails no resonant amplification, and is due to the quasi-static effect of the fluctuating pressures, and the "resonant term," which is associated with resonant amplification due to force components with frequencies equal or close to the fundamental natural frequency of the structure, and is inversely proportional to the damping ratio [2, p. 212].

References

1 Robson, J.D. (1964). *An Introduction to Random Vibration*. New York: Elsevier.
2 Simiu, E. and Scanlan, R.H. (1996). *Wind Effects on Structures*, 3rd ed. New York: Wiley.

Appendix E

Structural Reliability

E.1 Introduction

The objective of structural reliability is to develop criteria resulting in acceptably low probabilities that structures will fail to perform adequately under dead, live, and environmental loads. Adequate performance is defined as the non-exceedance of specified limit states.

The following are examples of limit states:

- Demand-to-capacity indexes (DCIs) may not significantly exceed unity (strength limit state).
- Buildings essential from a community resilience point of view (e.g., hospitals, police stations, fire stations, power plants) must not collapse under loads induced by extreme events (collapse limit state).
- Inter-story drift may not exceed a specified limit dependent upon type of cladding and/or partitions (serviceability limit state).
- Accelerations may not exceed a specified peak or r.m.s. value (serviceability limit state).
- The performance of equipment essential to the building functionality must not be affected by the occurrence of an extreme event (serviceability limit state).
- Cladding performance must not result in damage to the structure's contents (serviceability limit state).

Other limit states may be specified, depending upon the building, its contents, and its functions. Associated with the exceedance of any limit state is a minimum allowable mean recurrence interval (MRI). The more severe the consequences of exceeding the limit state, the larger are the minimum allowable MRIs.

Building codes specify strength limit states. For example, the ASCE Standard 7-16 specifies a 700-year MRI of the event that the strength of structural members of typical structures will be exceeded (for critical structures whose failure would cause loss of life the Standard specifies a higher MRIs). The specified MRIs are not based on explicit estimates of failure probabilities, but rather on professional consensus based on experience, intuition, or belief. Limit states not related to life safety and associated MRIs may be established by agreement among the owner, the designer, and the insurer, although some non-structural limit states may require compliance with regulatory requirements.

In the early phases of its development it was believed that structural reliability could assess the performance of any structural system by performing the following steps:

Wind Effects on Structures: Modern Structural Design for Wind, Fourth Edition. Emil Simiu and DongHun Yeo.
© 2019 John Wiley & Sons Ltd. Published 2019 by John Wiley & Sons Ltd.

(i) clear and unambiguous definition of limit states, (ii) specification of design criteria on acceptable probabilities of exceedance of the limit states, and (iii) checking whether, for the structure being designed, those criteria are satisfied.

The clear definition of certain limit states can be a difficult task. For redundant structural systems, as opposed to individual structural members, structural safety assessments via reliability calculations are typically not possible in the present state of the art. In addition, probability distribution tails, which determine failure probabilities, are in many cases unknown. Finally, the specification of the acceptable failure probability for any limit state can be a complex economic or political issue that exceeds the bounds of structural engineering.

In view of apparently insuperable difficulties inherent in the original goals of structural reliability, the discipline has settled for more modest goals. Under the demand inherent in the wind and gravity loads with specified MRIs and/or affected by their respective load factors, each member cross section must experience DCIs lower than or approximately equal to unity. Past experience with wind effects on buildings suggests that the member-by-member approach just described is safe,[1] even though it does not provide any explicit indication of the probability of exceedance of the incipient collapse limit state.

Improved forecasting capabilities, which allow sufficient time for evacuation, have resulted in massively reduced loss of life due to hurricanes, particularly in developed countries. The motivation to perform research into failure limit states has been far stronger for seismic regions than for regions with strong winds, and the ASCE 7 Standard specifies seismic design criteria based on nonlinear analyses, consistent with the requirement that the structure not collapse under a Maximum Considered Earthquake with a 2500-year MRI. The development of similar design criteria and research into nonlinear structural behavior are only beginning to be performed for structures subjected to wind loads. Such development is necessary, among other reasons, because evacuation can be impractical or hampered by traffic problems, hence the need for certain structures to be capable of safely surviving strong winds.

Section E.2 explains why the use of probability distributions of demand and capacity may be problematic in structural engineering practice. Subsequent sections are devoted to Load and Resistance Factor Design (LRFD) and its limitations (Section E.3), structural strength reserve (Section E.4), design MRIs for multi-hazard regions (Section E.5). The calibration of design MRIs for structures experiencing significant dynamic effects or for which errors in the estimation of extreme wind effects are significantly larger than the typical errors accounted for in the ASCE 7-16 Standard is considered in Chapters 7 and 12.

E.2 The Basic Problem of Structural Safety

Assume that the probability distribution of the demand Q and the capacity R

$$P(q, r) = \text{Prob}(Q \leq q, R \leq r) \tag{E.1}$$

is known. The probability that $q < Q < q + dq$ and $r < R < r + dr$ is $f(q, r)dq\,ds$. The probability of failure is the probability that $r < q$ (the shaded area in Figure E.1):

$$P_f = \int_0^\infty dq \int_0^q f(q, r)dr \tag{E.2}$$

---

1 The degree to which this is the case depends upon the structure's strength reserve, see Section E.5.

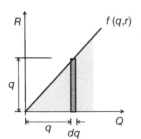

Figure E.1 Domain of integration for calculation of probability of failure.

Since the loads and resistances are independent (i.e., $f(q,r) = f_Q(q)f_R(r)$),

$$P_f = \int_0^\infty f_Q(q) \int_0^q f_R(r)dr\, dq$$

$$= \int_0^\infty f_Q(q)F_R(q)\, dq \qquad \text{(E.3a, b)}$$

where f_Q is the probability density function of the demand (load), and f_R and F_R are the probability density and the cumulative distribution functions of the capacity (resistance), respectively.

The integrand of Eq. (E.3a, b) depends upon the upper and lower tail of the distributions f_Q and F_R, respectively (Figure E.2). Typically, it is not possible to ascertain what those distributions are. For this reason, a fully probabilistic approach to the estimation of structural reliabilities is in most cases not feasible.

E.3 First-Order Second-Moment Approach: Load and Resistance Factors

The first order-second moment (FOSM) approach considered in this section was developed, primarily in the 1970s, following the realization that structural reliability theory based on explicit estimation of failure probabilities is not achievable in practice.

E.3.1 Failure Region, Safe Region, and Failure Boundary

Consider a member subjected to a load Q, and let the load that induces a given limit state (e.g., first yield) be denoted by R. Both Q and R are random variables that define the *load space*. Failure occurs for any pair of values for which

$$R - Q < 0 \qquad \text{(E.4)}$$

The *safe region* is defined by the inequality

$$R - Q > 0 \qquad \text{(E.5)}$$

The *failure boundary* separates the failure and the safe regions, and is defined by the relation

$$R - Q = 0 \qquad \text{(E.6)}$$

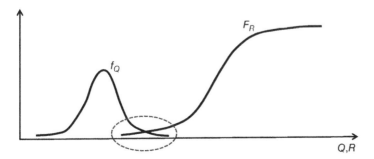

Figure E.2 Probability density function f_Q and cumulative distribution function F_R.

Relations similar to Eqs. (E.4)–(E.6) hold in the *load effect space*, defined by the variables Q_e and R_e, where Q_e is an effect (e.g., the stress) induced in a member by the load Q, and R_e is the corresponding limit state (e.g., the yield stress). The failure boundary is then

$$R_e - Q_e = 0 \tag{E.7}$$

Henceforth we use for simplicity the notations Q, R for both the load space and the load effect space. In general, Q and R are functions of independent random variables $X_1, X_2,..., X_n$ (e.g., terrain roughness, aerodynamic coefficients, wind speeds, natural frequencies, damping ratios, strength) called *basic variables*, that is,

$$Q = Q(X_1, X_2, \ldots, X_m) \tag{E.8}$$
$$R = R(X_{m+1}, X_{m+2}, \ldots, X_n) \tag{E.9}$$

Substitution of Eqs. (E.8) and (E.9) into Eq. (E.6) yields the failure boundary in the space of the basic variables, defined by the equation (Figure E.3):

$$g(X_1, X_2, \ldots, X_n) = 0 \tag{E.10}$$

It can be useful in applications to map the failure region, the safe region, and the failure boundary onto the space of variables Y_1 and Y_2, defined by transformations

$$Y_1 = \ln R \tag{E.11}$$
$$Y_2 = \ln Q \tag{E.12}$$

On the failure boundary, $R = Q$, so in the coordinates Y_1, Y_2 the failure boundary is $Y_1 = Y_2$.

E.3.2 Safety Indexes

Denote by S the failure boundary in the space of the reduced variables $x_{i\,\text{red}} = (X_i - \overline{X_i})/\sigma_{xi}$, where the variables X_i are mutually independent, and $\overline{X_i}$ and σ_{xi} are, respectively, the mean and standard deviation of X_i. (The subscript "*red*" stands for "*reduced.*") The *reliability index*, denoted by β, is defined as the shortest distance in this space between the origin (i.e., the image in the space of the reduced variables of

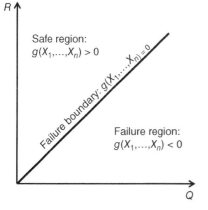

Figure E.3 Safe region, failure region, and failure boundary.

the point with coordinates $\overline{X_i}$) and the failure boundary S. The point on the boundary S closest to the origin, and its image in the space of the original basic variables X_i, are called the *design point*. For any given structural problem, the numerical value of the safety index depends upon the set of variables being considered.

Assume that the load Q and resistance R follow the normal distribution. It is convenient to express the random variables in non-dimensional terms as follows:

$$q_{\text{red}} = \frac{Q - \overline{Q}}{\sigma_Q},$$

$$r_{\text{red}} = \frac{R - \overline{R}}{\sigma_R}. \qquad\qquad (\text{E.13a, b})$$

The failure surface (Eq. [E.6]) has the following expression in the space of the reduced coordinates:

$$\sigma_S s_{\text{red}} - \sigma_Q q_{\text{red}} + \overline{R} - \overline{Q} = 0 \qquad\qquad (\text{E.14})$$

The coordinates of points A, B in Figure E.4 are, respectively:

$$\left(\frac{\overline{R} - \overline{Q}}{\sigma_Q}, 0\right) \text{ and } \left(0, \frac{-(\overline{R} - \overline{Q})}{\sigma_R}\right) \qquad\qquad (\text{E.15})$$

The slope of failure surface line is

$$\alpha = \tan^{-1}(OB/OA) = \tan^{-1}(\sigma_Q/\sigma_R) \qquad\qquad (\text{E.16})$$

The slope of line (L) normal to the failure surface is $-1/\tan^{-1}(\sigma_Q/\sigma_R)$. The design point D is the intersection of the failure surface and line (L). Its coordinates $(q^*_{\text{red}}, r^*_{\text{red}})$ are

$$q^*_{\text{red}} = \beta \sin\alpha = (\overline{R} - \overline{Q})\frac{\sigma_Q}{\sigma_R^2 + \sigma_Q^2}$$

$$r^*_{\text{red}} = -\beta \cos\alpha = -(\overline{R} - \overline{Q})\frac{\sigma_R}{\sigma_R^2 + \sigma_Q^2} \qquad\qquad (\text{E.17a, b})$$

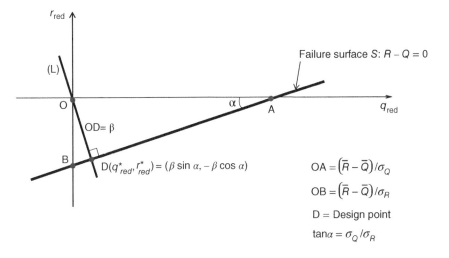

Figure E.4 Index β in the space of the reduced variables q_{red} and r_{red}.

where the distance β between the origin and the failure surface in the space of reduced variables is defined as the *safety index*. From Eqs. (E.17) if follows that

$$\beta = \frac{\overline{R} - \overline{Q}}{(\sigma_R^2 + \sigma_Q^2)^{1/2}} \tag{E.18}$$

Example E.1   Assume that the resistance is deterministic, that is, $R \equiv \overline{R}$. The mapping of the failure boundary

$$Q - \overline{R} = 0$$

onto the space of the reduced variate $q_{red} = (Q - \overline{Q})/\sigma_Q$ is a point q_{red}^* such that $Q = \overline{R}$, that is, $q_{red}^* = (\overline{R} - \overline{Q})/\sigma_Q$ (Figure E.5). The asterisk denotes the design point. The origin in that space is the point for which $q_{red} = 0$, and corresponds to $Q = \overline{Q}$. The distance $\sigma_r = 0$ between the origin and the failure boundary is the safety index $\beta = (\overline{R} - \overline{Q})/\sigma_Q$, since in this case in Eq. (E.18). The case $Q - \overline{R} > 0$ (load larger than resistance) corresponds to failure. In the space of the reduced variable failure occurs for $q_{red} > q_{red}^*$, that is $q_{red} > \beta$.

The larger the ratio $\beta = (\overline{R} - \overline{Q})/\sigma_Q$, the smaller is the probability of failure. The reliability index thus provides an indication on a member's safety. However, this indication is largely qualitative, unless information is available on the probability distribution of the variate Q.

Instead of operating in the load space R, Q, consider the failure boundary in the transformed space defined by Eqs. (E.11) and (E.12). If Q and R are assumed to be mutually independent and lognormally distributed, the distribution of Y_1, Y_2, and $Y_2 - Y_1$ (i.e., $\ln Q$, $\ln R$, and $\ln(R/Q)$, respectively) will be normal. Following the same steps as in the normal distribution case, but applying them to the variables Y_1 and Y_2, the safety index becomes

$$\beta = \frac{\overline{Y}_1 - \overline{Y}_2}{(\sigma_{Y1}^2 + \sigma_{Y2}^2)^{1/2}} \tag{E.19}$$

Expansion in a Taylor series yields the expression

$$Y_1 = \ln \overline{R} + (R - \overline{R})\frac{1}{\overline{R}} - \frac{1}{2}(R - \overline{R})^2 \frac{1}{\overline{R}^2} + \dots \tag{E.20}$$

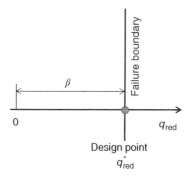

Figure E.5 Index β for member with random load and deterministic resistance.

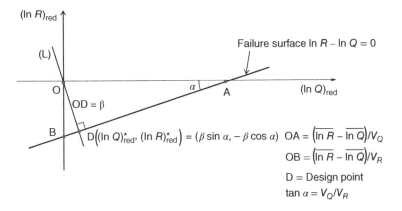

Figure E.6 Index β for member with random load and random resistance in the space of the reduced variables $(\ln R)_{red}$ and $(\ln Q)_{red}$.

and a similar expression for Y_2. Averaging these expressions, neglecting second and higher order terms, and using the notations $\sigma_R/\overline{R} = V_R$, $\sigma_Q/\overline{Q} = V_Q$, the safety index can be expressed as:

$$\beta \approx \frac{\ln \overline{R} - \ln \overline{Q}}{(V_R^2 + V_Q^2)^{1/2}} \tag{E.21}$$

Figure E.6 is the counterpart of Figure E.4 obtained by substituting in Eq. (E.18) $\ln \overline{R}$, $\ln \overline{Q}$, V_R, V_Q for $\overline{R}$, $\overline{Q}$, σ_R, σ_Q, respectively.

Note: The approach wherein only means and standard deviations (or coefficients of variation) are used is called the "first-order second moment" (FOSM) approach.

E.3.3 Reliability Indexes and Failure Probabilities

The probability of failure is

$$P_f = \text{Prob}[(R - Q) \leq 0]$$
$$= \text{Prob}(g \leq 0) \tag{E.22}$$

If the variates R and Q are *normally distributed*, the probability distribution of $R - Q = g$ is also normal. It follows then from Eq. (E.22) that

$$P_f = F_g(0) \tag{E.23}$$

(Figure E.7) where F_g is the Gaussian cumulative distribution of g, or

$$P_f = \text{Pr}(g \leq 0)$$
$$= \Phi\left(\frac{0 - \overline{g}}{\sigma_g}\right)$$
$$= \Phi\left(-\frac{\overline{R} - \overline{Q}}{(\sigma_R^2 + \sigma_Q^2)^{1/2}}\right)$$
$$= \Phi(-\beta)$$
$$= 1 - \Phi(\beta) \tag{E.24a,b,c,d,e}$$

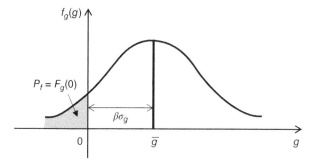

Figure E.7 Probability distribution function f_g (g) of variate $g = R - Q$. The probability of failure is equal to the area under the curve f_g (g) for $g < 0$.

where Φ is the standard normal cumulative distribution function and β is defined by Eq. (E.18).

If the variates R and Q are *lognormally distributed* (meaning that their logarithms are normally distributed), the probability of failure is

$$P_f = \Pr(\ln R - \ln Q \le 0)$$

$$= 1 - \Phi\left(\frac{\ln \overline{R} - \ln \overline{Q}}{(\sigma_{\ln R}^2 + \sigma_{\ln Q}^2)^{1/2}}\right)$$

$$\approx 1 - \Phi(\beta) \qquad\qquad\text{(E.25a,b,c)}$$

where the fraction in Eq. (E.25b) is equal to the safety index defined in Eq. (E.21). The usefulness of Eq. (E.24e) and (E.25c) is limited by the fact that typically neither the load nor the resistance is normally or lognormally distributed.

E.3.4 Partial Safety Factors: Load and Resistance Factor Design

Consider a structure characterized by a set of variables with means $\overline{X}_i$ and standard deviations σ_i, and design points X_i^* $(i = 1, 2, \dots, n)$ in the space of the original variables. By definition

$$X_i^* = \overline{X}_i + \sigma_{X_i} x_{i\,\text{red}}^* \qquad\qquad\text{(E.26)}$$

Equation (E.26) can be written in the form

$$X_i^* = \gamma_{\overline{X}_i} \overline{X}_i \qquad\qquad\text{(E.27)}$$

where

$$\gamma_{\overline{X}_i} = 1 + V_{X_i} x_{i\,\text{red}}^* \qquad\qquad\text{(E.28)}$$

and $V_{X_i} = \sigma_{X_i}/\overline{X}_i$; the asterisk denotes the design point. Let $i = 1, 2$; $X_1 = Q$, $X_2 = R$; and $\gamma_{\overline{X}_1} \equiv \gamma_{\overline{Q}}$, $\gamma_{\overline{X}_2} \equiv \varphi_R$. The quantities $\gamma_{\overline{Q}}$, φ_R are called the load and the resistance factor, respectively.

We consider now the case $Y_1 = \ln R$ and $Y_2 = \ln Q$, on which current design practice is based. The counterpart to Eqs (E.26) is

$$(\ln Q)_{\text{red}}^* = \frac{(\ln Q)^* - \ln \overline{Q}}{V_Q} \qquad\qquad\text{(E.29)}$$

Since $(\ln Q)^* = \beta \sin\alpha$, where $\tan\alpha = V_Q/V_R$ (see Figure E.6) and β is defined by Eq. (E.21), it follows that

$$\ln(Q^*/\overline{Q}) = V_Q\beta\sin\alpha \tag{E.30}$$

(see Figure E.6). Therefore

$$\frac{Q^*}{\overline{Q}} = \exp(V_Q\beta\sin\alpha). \tag{E.31}$$

Since $Q^* = \gamma_{\overline{Q}}\overline{Q}$, the load factor is

$$\gamma_{\overline{Q}} = \exp(V_Q\beta\sin\alpha). \tag{E.32}$$

Similarly, the resistance factor is

$$\varphi_R = \exp(-V_R\beta\cos\alpha). \tag{E.33}$$

In Eqs. (E.29)–(E.33), β is defined by Eq. (E.21).

The following linear approximation to Eq. (E.26) has been developed for use in standards [1]:

$$\gamma_{\overline{Q}} = 1 + 0.55\beta V_Q \tag{E.34}$$

Equation (E.34) can in many instances be a poor approximation to Eq. (E.32).

E.3.5 Calibration of Safety Index β, Wind Directionality, and Mean Recurrence Intervals of Wind Effects

Because the approach to the calculation of the safety index by methods that presuppose the universal validity of the lognormal distribution can be unsatisfactory, load factors specified explicitly or implicitly in the ASCE 7 Standard have been calibrated against past practice using uncertainty estimates and engineering judgment – see Chapters 7 and 12.

If wind directionality is considered by explicitly taking into account the directional distribution of the wind speeds at the building site, rather than by using wind directionality factors as specified by the ASCE 7 Standard, MRIs of the design wind effects are no longer equal to the MRIs of the design wind speeds – see Chapter 13 for details.

E.4 Structural Strength Reserve

The design of structural members by LRFD methods ensures that they do not experience unacceptable behavior as they attain the respective strength limit states. However, it is desirable that, even if those limit states are exceeded, the performance of the structure remains acceptable in some sense. A structure with large strength reserve is one for which this is the case for wind effects with MRIs significantly larger than the MRIs inducing strength limit states.

Strength reserve can be assessed by estimating MRIs of incipient collapse or other appropriate performance measures. Sections E.4.1 and E.4.2 provide such estimates for portal frames for a single wind direction and by considering the effect of all wind directions, respectively, and note a thorough study of post-elastic behavior of tall buildings subjected to wind [2].

E.4.1 Portal Frame Ultimate Capacity Under Wind with Specified Direction

For low-rise industrial steel buildings with gable roofs and portal frames, nonlinear push-over studies have been conducted in which the buildings were subjected to two sets of wind pressures [3]. One set consisted of wind pressures based on aerodynamic information specified for low-rise structures in the ASCE 7 Standard. The second set consisted of simultaneous wind pressures measured and recorded in the wind tunnel at a large number of taps on the building model's surface. The structural design of the frames was based on ASCE 7 Standard loads and the Allowable Stress Design approach. The objectives of the studies were: (i) to compare the strength reserve levels estimated by using (a) the simplified wind loads inherent in the ASCE Standard, and (b) recorded time series of wind tunnel pressures, and (ii) to examine the degree to which the strength reserve can be increased by the adoption of alternative designs. The following alternative features of the lateral bracing and joint stiffening were considered:

1) (a) 2.5 m spacing, and (b) 6 m spacing of lateral bracing of rafter bottom flanges.
2) Knee (a) horizontal and vertical stiffeners, and (b) horizontal, vertical, and diagonal stiffeners.
3) Ridge (a) without and (b) with vertical stiffener at ridge.

Strength analyses were performed for the load combinations involving wind. Calculations were performed of the ratio λ between ultimate and allowable wind load for each load combination being considered, the ultimate wind load corresponding to incipient failure through local or global instability as determined by using a finite element analysis program. Reducing the distance between bracings of the rafter's lower flanges increased the strength reserve more effectively than providing diagonal stiffeners in the knee joint (Figure E.8). Significant differences were found between the values of λ obtained under loading by pressures specified in the ASCE 7 Standard provisions and loading by the more realistic pressures measured in the wind tunnel. For details see [4].

E.4.2 Portal Frame Ultimate Capacity Estimates Based on Multi-Directional Wind Speeds

The following methodology was developed for the estimation of MRIs of ultimate wind effects by accounting for wind directionality [5]:

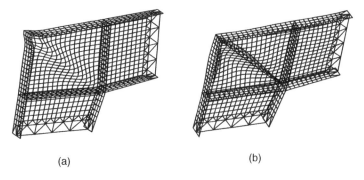

(a)   (b)

Figure E.8 Local buckling in knee (a) with and (b) without diagonal stiffener: industrial building steel portal frame.

1) Using recorded wind tunnel pressure data, obtain the loads that induce peak internal forces (axial forces, bending moments, shear forces) at a number of cross sections deemed to be critical. Obtain the loads corresponding to a unit wind speed at 10 m above ground over open terrain for, say, 16 or 36 wind directions spanning the 360° range. These loads, multiplied by the square of wind speeds U considered in design, are used in step 2.

2) Using nonlinear finite element analyses, determine the wind speed from each direction θ_i that causes the frame to experience incipient failure, defined as the onset of deformations that increase so fast under loads that implicit nonlinear finite element analyses fail to converge to a solution.

3) From available wind climatological data create, by simulation, time series of directional wind speeds with length t_d that exceeds the anticipated MRIs of the failure events [6].

4) Count the number n_f of cases in which directional wind speeds in the time series created in step 3 exceed the directional wind speeds determined in step 2 to produce incipient failure events. The MRI in years of the failure event is estimated as $\overline{N} = t_d/n_f$.

This methodology was applied to an industrial low-building portal frame located in a hurricane-prone region. The frame was strengthened by triangular stiffeners at the column supports and by haunches and horizontal, vertical, and diagonal stiffeners at the knee joints. Owing to such strengthening the estimated failure MRI was in this case quite high (100,000 years, corresponding to a nominal 1/1000 probability that the frame will fail during a 100-year life).

E.4.3 Nonlinear Analysis of Tall Buildings Under Wind Loads

An extensive study of post-elastic behavior of high-rise buildings subjected to wind loads is presented in [2], which incorporates and adapts methods and results obtained for structures that behave nonlinearly under seismic loads. Future research may consider the possibility that, under the strong wind loading inducing nonlinear behavior in structural members, tall buildings might experience aeroelastic (or, to introduce a new but apposite term, aeroplastic) effects.

E.5 Design Criteria for Multi-Hazard Regions

E.5.1 Strong Winds and Earthquakes

Structures in regions subjected to both strong earthquakes and strong winds are currently designed by considering separately loads induced by earthquakes and by winds, and basing the final design on the more demanding of those loads. The rationale for this approach has been that the probability of simultaneous occurrence of both earthquakes and high winds is negligibly small. It is shown in this section that implicit in this approach are probabilities of failure that can be greater by a factor of up to two than their counterparts for structures exposed to wind only or to earthquakes only.

An intuitive illustration of this statement follows. Assume that a motorcycle racer applies for insurance against personal injuries. The insurer will calculate a rate commensurate with the probability that the racer will be injured in a motorcycle accident. Assume now that the motorcycle racer is also a high-wire artist. The insurance rate would then be increased, since the probability that an injury will occur during a specified period of time, either in a motorcycle or high-wire accident, will be greater than the probability associated with risk due to only one of these types of accident. This is true even though the nature of the injuries in the two types of event may differ. This argument is expressed formally as

$$P(s_1 \cup s_2) = P(s_1) + P(s_2) \tag{E.35}$$

where $P(s_1)$ = annual probability of event s_1 (injury due to a motorcycle accident), $P(s_2)$ = annual probability of event s_2 (injury due to a high-wire accident), and $P(s_1 \cup s_2)$ = probability of injury due to a motorcycle *or* a high-wire accident.

Equation (E.35) is applicable to structures as well, particularly to members experiencing large demands under lateral loads (e.g., columns in lower floors). For details and case studies, see [7, 8].

E.5.2 Winds and Storm Surge

Unlike earthquakes and windstorms, winds and storm surge are not independent events. Therefore, for some applications it is necessary to consider their simultaneous effects. This entails the following steps: (i) select a stochastic set of hurricane storm tracks in the region of interest; (ii) use the selected storm tracks to generate time histories of wind speeds and corresponding time histories of storm surge heights at sites affected by those wind speeds; (iii) use those time histories to calculate time series of wind and storm surge effects; and (iv) obtain from those time series estimates of joint effects of wind and storm surge with the mean recurrence intervals of interest [9, 10]. In this approach the calculations are performed in the load effect space.

An important factor in the estimation of storm surge heights is the bathymetry at and near the site of interest. To be realistic, storm surge intensities must be based on current information on local bathymetry, which can change significantly over time.

References

1 Ravindra, M.K.G., Theodore, V., and Cornell, C.A. (1978). Wind and snow load factors for use in LRFD. *Journal of the Structural Division* 104: 1443–1457.

2 Mohammadi, A., "Wind performance-based design of high-rise buildings," Doctoral dissertation, Department of Civil and Environmental Engineering, Florida International University, 2016.

3 Jang, S., Lu, L.W., Sadek, F., and Simiu, E. (2002). Database-assisted wind load capacity estimates for low-rise steel frames. *Journal of Structural Engineering* 128: 1594–1603.

4 Duthinh, D. and Fritz, W.P. (2007). Safety evaluation of low-rise steel structures under wind loads by nonlinear database-assisted technique. *Journal of Structural Engineering* 133: 587–594. (https://www.nist.gov/wind).

5 Duthinh, D., Main, J.A., Wright, A.P., and Simiu, E. (2008). Low-rise steel structures under directional winds: mean recurrence interval of failure. *Journal of Structural Engineering* 134: 1383–1388.

6 Yeo, D. (2014). Generation of large directional wind speed data sets for estimation of wind effects with long return periods. *Journal of Structural Engineering* 140: 04014073. https://www.nist.gov/wind.

7 Duthinh, D. and Simiu, E. (2010). Safety of structures in strong winds and earthquakes: multihazard considerations. *Journal of Structural Engineering* 136: 330–333. https://www.nist.gov/wind).

8 Crosti, C., Duthinh, D., and Simiu, E. (2010). Risk consistency and synergy in multi-hazard design. *Journal of Structural Engineering* 137: 1–6.

9 Phan, L. T., Simiu, E., McInerney, M. A., Taylor, A. A., and Powell, M. D., "Methodology for the Development of Design Criteria for Joint Hurricane Wind Speed and Storm Surge Events: Proof of Concept," NIST Technical Note 1482, National Institute of Standards and Technology, Gaithersburg, MD, 2007. https://www.nist.gov/wind.

10 Phan, L. T., Slinn, D. N., and Kline, S. W., "Introduction of Wave Set-up Effects and Mass Flux to the Sea, Lake, and Overland Surges from Hurricanes (SLOSH) Model," NISTIR 7689, National Institute of Standards and Technology, Gaithersburg, MD, 2010. https://www.nist.gov/wind.

Appendix F

World Trade Center Response to Wind

A Skidmore Owings and Merrill Report

Note. The material that follows reproduces NIST document NCSTAR1–2, Appendix D, dated April 13 2004, (http://wtc.nist.gov/NCSTAR1/NCSTAR1-2index.htm) submitted by Skidmore, Owings and Merrill LLP, Chicago, Illinois (wtc.nist.gov). The documents listed in Sections F.1, F.2, and F.3 are not in the public domain, but are believed to be obtainable under the provisions of the Freedom of Information Act. The material illustrates difficulties encountered by practicing structural engineers in evaluating wind engineering laboratory reports, and contains useful comments on the state of the art in wind engineering at the time of its writing. The text that follows is identical to the text of the Skidmore, Owings and Merrill report, except for numbering of the headings.

F.1   Overview

F.1.1   Project Overview

The objectives for Project 2 of the WTC Investigation include the development of reference structural models and design loads for the WTC Towers. These will be used to establish the baseline performance of each of the towers under design gravity and wind loading conditions. The work includes expert review of databases and baseline structural analysis models developed by others as well as the review and critique of the wind loading criteria developed by NIST.

F.1.2   Report Overview

This report covers work on the development of wind loadings associated with Project 2. This task involves the review of wind loading recommendations developed by NIST for use in structural analysis computer models. The NIST recommendations are derived from wind tunnel testing/wind engineering reports developed by independent wind engineering consultants in support of insurance litigation concerning the WTC towers. The reports were provided voluntarily to NIST by the parties to the insurance litigation.

As the third party outside experts assigned to this Project, SOM's role during this task was to review and critique the NIST-developed wind loading criteria for use in computer analysis models. This critique was based on a review of documents provided by NIST, specifically the wind tunnel/wind engineering reports and associated correspondence from independent wind engineering consultants and the resulting interpretation and recommendations developed by NIST.

Wind Effects on Structures: Modern Structural Design for Wind, Fourth Edition. Emil Simiu and DongHun Yeo.
© 2019 John Wiley & Sons Ltd. Published 2019 by John Wiley & Sons Ltd.

F.2   NIST-Supplied Documents

F.2.1   Rowan Williams Davies Irwin (RWDI) Wind Tunnel Reports

Final Report, Wind-Induced Structural Responses World Trade Center – Tower 1, New York, New York, Project Number: 02-1310A, October 4, 2002; Final Report, Wind-Induced Structural Responses World Trade Center – Tower 2, New York, New York, Project Number:02-1310B, October 4, 2002.

F.2.2   Cermak Peterka Petersen, Inc. (CPP) Wind Tunnel Report

Wind-Tunnel Tests – World Trade Center, New York, NY
  CPP Project 02-2420
  August 2002

F.2.3   Correspondence

Letter dated October 2, 2002 from Peter Irwin/RWDI to Matthys Levy/Weidlinger Associates, Re: Peer Review of Wind Tunnel Tests
World Trade Center, RWDI Reference #02-1310

Weidlinger Associates Memorandum dated March 19, 2003 from Andrew Cheung to Najib Abboud, Re: Errata to WAI Rebuttal Report

Letter dated September 12, 2003 from Najib N. Abboud/Hart-Weidlinger to S. Shyam Sunder and Fahim Sadek/NIST, Re: Responses to NIST's Questions on *"Wind-Induced Structural Responses, World Trade Center*, Project Number 02-1310A and 02-1310B, October 2002, by RWDI, Prepared for Hart-Weidlinger"

Letter dated April 6, 2004
From: Najib N. Abboud /Weidlinger Associates
To: Fahim Sadek and Emil Simiu
Re: Response to NIST's question dated March 30, 2004 regarding "Final Report, Wind- Induced Structural Responses, World Trade Center – Tower 2, RWDI, Oct 4, 2002"

F.2.4   NIST Report, Estimates of Wind Loads on the WTC Towers, Emil Simiu and Fahim Sadek, April 7, 2004

F.3   Discussion and Comments

F.3.1   General

This report covers a review and critique of the NIST recommended wind loads derived from wind load estimates provided by two independent private sector wind engineering groups, RWDI and CPP. These wind engineering groups performed wind tunnel testing and wind engineering calculations for various private sector parties involved in insurance litigation concerning the destroyed WTC Towers in New York. There are substantial disparities (greater than 40%) in the predictions of base shears and base overturning moments between the RWDI and CPP wind reports. NIST has attempted to reconcile these differences and provide wind loads to be used for the baseline structural analysis.

F.3.2   Wind Tunnel Reports and Wind Engineering

The CPP estimated wind base moments far exceed the RWDI estimates. These differences far exceed SOM's experience in wind force estimates for a particular building by independent wind tunnel groups.

In an attempt to understand the basis of the discrepancies, NIST performed a critique of the reports. Because the wind tunnel reports only summarize the wind tunnel test data and wind engineering calculations, precise evaluations are not possible with the provided information. For this reason, NIST was only able to approximately evaluate the differences. NIST was able to numerically estimate some corrections to the CPP report but was only able to make some qualitative assessments of the RWDI report. **It is important to note that wind engineering is an emerging technology and there is no consensus on certain aspects of current practice.** Such aspects include the correlation of wind tunnel tests to full-scale (building) behavior, methods and computational details of treating local statistical (historical) wind data in overall predictions of structural response, and types of suitable aeroelastic models for extremely tall and slender structures. It is unlikely that the two wind engineering groups involved with the WTC assessment would agree with NIST in all aspects of its critique. This presumptive disagreement should not be seen as a negative, but reflects the state of wind tunnel practice. It is to be expected that well-qualified experts will respectfully disagree with each other in a field as complex as wind engineering.

SOM's review of the NIST report and the referenced wind tunnel reports and correspondence has only involved discussions with NIST; it did not involve direct communication with either CPP or RWDI. SOM has called upon its experience with wind tunnel testing on numerous tall building projects in developing the following comments.

F.3.2.1   CPP Wind Tunnel Report

The NIST critique of the CPP report is focused on two issues: a potential overestimation of the wind speed and an underestimation of load resulting from the method used for integrating the wind tunnel data with climatic data. NIST made an independent estimate of the wind speeds for a 720-year return period. These more rare wind events are dominated by hurricanes that are reported by rather broad directional sectors (22.5°). The critical direction for the towers is from the azimuth direction of 205–210°. This wind direction is directly against the nominal "south" face of the towers (the plan north of the site is rotated approximately 30 degrees from the true north) and generates dominant cross-wind excitation from vortex shedding. The nearest sector data are centered on azimuth 202.5° (SSW) and 225° (SW). There is a substantial drop (12%) in the NIST wind velocity from the SSW sector to the SW sector. The change in velocity with direction is less dramatic in the CCP 720-year velocities or in the ARA hurricane wind roses included in the RWDI report. This sensitivity to directionality is a cause for concern in trying to estimate a wind speed for a particular direction. However, it should be noted that the magnitude of the NIST interpolated estimated velocity for the 210 azimuth direction is similar to the ARA wind rose. The reduction of forces has been estimated by NIST based on a square of the velocity, however, a power of 2.3 may be appropriate based on a comparison of the CPP 50-year (nominal) and 720-year base moments and velocities.

The NIST critique of the CPP use of sector by sector approach of integrating wind tunnel and climatic data is fairly compelling. The likelihood of some degree of under-estimation is high but SOM is not able to verify the magnitude of error (15%), which is estimated by NIST. This estimate would need to be verified by future research, as noted by NIST.

F.3.2.2   RWDI Wind Tunnel Report

The NIST critique of RWDI has raised some issues but has not directly estimated the effects. These concerns are related to the wind velocity profiles with height used for hurricanes and the method used for up-crossing.

NIST questioned the profile used for hurricanes and had an exchange of correspon-dence with RWDI. While RWDI's written response is not sufficiently quantified to permit a precise evaluation of NIST's concerns, significant numerical corroboration on this issue may be found in the April 6 letter (Question 2) from N. Abboud (Weidlinger Associates) to F. Sadek and E. Simiu (NIST).

NIST is also concerned about RWDI's up-crossing method used for integrating wind tunnel test data and climatic data. This method is computationally complex and verifica-tion is not possible because sufficient details of the method used to estimate the return period of extreme events are not provided.

F.3.2.3   Building Period used in Wind Tunnel Reports

SOM noted that both wind tunnel reports use fundamental periods of vibrations that exceed those measured in the actual (north tower) buildings. The calculation of building periods are at best approximate and generally underestimate the stiffness of a building thus overestimating the building period. The wind load estimates for the WTC tow-ers are sensitive to the periods of vibration and often increase with increased period as demonstrated by a comparison of the RWDI base moments with and without P-Delta effects. Although SOM generally recommends tall building design and analysis be based on P-Delta effects, in this case even the first order period analysis (without P-Delta) exceeds the actual measurements. It would have been desirable for both RWDI and CPP to have used the measured building periods.

F.3.2.4   NYCBC Wind Speed

SOM recommends that the wind velocity based on a climatic study or ASCE 7-02 wind velocity be used in lieu of the New York City Building Code (NYCBC) wind velocity. The NYCBC wind velocity testing approach does not permit hurricanes to be accommo-dated by wind tunnel testing as intended by earlier ASCE 7 fastest mile versions because it is based on a method that used an importance factor to correct 50-year wind speeds for hurricanes. Because the estimated wind forces are not multiplied by an importance factor, this hurricane correction is incorporated in analytical methods of determining wind forces but is lost in the wind tunnel testing approach of determining wind forces.

F.3.2.5   Incorporating Wind Tunnel Results in Structural Evaluations

It is expected that ASCE 7 load factors will also be used for member forces for evaluating the WTC towers. Unfortunately, the use of ASCE 7 with wind tunnel-produced loadings is not straightforward. Neither wind tunnel report gives guidance on how to use the provided forces with ASCE 7 load factors.

The ASCE 7 load factors are applied to the nominal wind forces and, according to the ASCE 7 commentary, are intended to scale these lower forces up to wind forces associated with long return period wind speeds. The approach of taking 500-year return period wind speeds and dividing the speeds by the square root of 1.5 to create a nominal design wind speed; determining the building forces from these reduced nominal design wind speeds; and then magnifying these forces by a load factor (often 1.6) is, at best, convoluted. For a building that is as aerodynamically active as the WTC, an approach of directly determining the forces at the higher long return period wind speeds would be preferred. The CPP data did provide the building forces for their estimates of both 720-years (a load factor of 1.6) and the reduced nominal design wind speeds. A comparison of the wind forces demonstrates the potential error in using nominal wind speeds in lieu of directly using the underlying long period wind speeds.

It should also be noted that the analytical method of calculating wind forces in ASCE 7 provides an importance factor of 1.15 for buildings such as the WTC in order to provide more conservative designs for buildings with high occupancies. Unfortunately, no similar clear guidance is provided for high occupancy buildings where the wind loads are determined by wind tunnel testing. Utilizing methods provided in the ASCE 7 Commentary would suggest that a return period of 1800 years with wind tunnel-derived loads would be comparable to the ASCE 7 analytical approach to determining wind loads for a high occupancy building.

It would be appropriate for the wind tunnel private sector laboratories or NIST, as future research beyond the scope of this project, to address how to incorporate wind tunnel loadings into an ASCE 7-based design.

F.3.2.6   Summary

The NIST review is critical of both the CPP and RWDI wind tunnel reports. It finds substantive errors in the CPP approach and questions some of the methodology used by RWDI. It should be noted that boundary layer wind tunnel testing and wind engineering is still a developing branch of engineering and there is not industry-wide consensus on all aspects of the practice. For this reason, some level of disagreement is to be expected.

Determining the design wind loads is only a portion of the difficulty. As a topic of future research beyond the scope of this project, NIST or wind tunnel private sector laboratories should investigate how to incorporate these wind tunnel-derived results with the ASCE 7 Load Factors.

F.3.3   NIST Recommended Wind Loads

NIST recommends a wind load that is between the RWDI and CPP estimates. The NIST recommended values are approximately 83% of the CPP estimates and 115% of the RWDI estimates. SOM appreciates the need for NIST to reconcile the disparate wind tunnel results. It is often that engineering estimates must be done with less than the desired level of information. In the absence of a wind tunnel testing and wind engineering done to NIST specifications, NIST has taken a reasonable approach to estimate appropriate values to be used in the WTC study. However, SOM is not able to independently confirm the precise values developed by NIST.

The wind loads are to be used in the evaluation of the WTC structure. It is therefore recommended that NIST provide clear guidelines on what standards are used in the evaluations and how they are to incorporate the provided wind loads.

Index

Wind Effects on Structures: Modern Structural Design for Wind, Fourth Edition. Emil Simiu and DongHun Yeo.
© 2019 John Wiley & Sons Ltd. Published 2019 by John Wiley & Sons Ltd.

Printed and bound by CPI Group (UK) Ltd, Croydon, CR0 4YY

16/04/2025

14658554-0007